T0325451

NON-SELFADJOINT OPERATORS IN QUANTUM PHYSICS

NON-SELFADJOINT OPERATORS IN QUANTUM PHYSICS

Mathematical Aspects

Editors:

FABIO BAGARELLO
Università di Palermo and INFN

JEAN PIERRE GAZEAU
Université Paris Diderot, Sorbonne Paris Cité
Centro Brasileiro de Pesquisas Físicas, Rio de Janeiro

FRANCISZEK HUGON SZAFRANIEC
Uniwersytet Jagielloński

MILOSLAV ZNOJIL
Ústav jaderné fyziky AV ČR, Řež

Published by John Wiley & Sons, Inc., Hoboken, New Jersey
Published simultaneously in Canada

Library of Congress Cataloging-in-Publication Data:

Non-selfadjoint operators in quantum physics : mathematical aspects / editors: Fabio Bagarello, Jean Pierre Gazeau, Franciszek Hugon Szafraniec, Miloslav Znojil.
pages cm
Includes index.
ISBN 978-1-118-85528-7 (cloth)
1. Nonselfadjoint operators. 2. Spectral theory (Mathematics) 3. Quantum theory–Mathematics. 4. Hilbert space. I. Bagarello, Fabio, 1964- editor. II. Gazeau, Jean-Pierre, editor. III. Szafraniec, Franciszek Hugon, editor. IV. Znojil, M. (Miloslav), editor.
QA329.2.N67 2015
530.1201′515724–dc23

2014048325

Typeset in 10/12pt, TimesLtStd by SPi Global, Chennai, India

Printed in the United States of America

10 9 8 7 6 5 4 3 2 1

1 2015

To Charles Hermite
with apologies

CONTRIBUTORS

Sergio Albeverio Institut für Angewandte Mathematik, Universität Bonn, Bonn, Germany CERFIM, Locarno, Switzerland Department of Mathematics and Statistics, King Fahd University of Petroleum and Minerals, Dhahran, Saudi Arabia

Jean-Pierre Antoine Institut de Recherche en Mathématique et Physique, Université catholique de Louvain, Louvain-la-Neuve, Belgium

Fabio Bagarello Università di Palermo and INFN, Torino, Italy

Emanuela Caliceti Dipartimento di Matematica, Università di Bologna, Bologna, Italy

Sandro Graffi Dipartimento di Matematica, Università di Bologna, Bologna, Italy

Sergii Kuzhel AGH University of Science and Technology, Kraków, Poland

David Krejčiřík Nuclear Physics Institute, ASCR, Řež, Czech Republic

Petr Siegl Mathematical Institute, University of Bern, Bern, Switzerland

Franciszek Hugon Szafraniec Instytut Matematyki, Uniwersytet Jagielloński, Kraków, Poland

Camillo Trapani Dipartimento di Matematica e Informatica, Università di Palermo, Palermo, Italy

Miloslav Znojil Nuclear Physics Institute, ASCR, Řež, Czech Republic

CONTENTS IN BRIEF

CONTENTS

PREFACE

Although it is widely accepted common wisdom that the discussions between mathematicians and physicists are enormously rewarding and productive, it is usually much easier to select illustrative examples from the past than to convert such a nice-sounding observation into a concrete and constructive project or into a proposal of a collaboration between a mathematician and a physicist.

For people involved, the reasons are more than obvious: in contrast to physics in which one may always appeal to experiments, the mathematicians feel free to ask (and study) time-independent questions. Consequently, the communication among physicists is usually full of urgency and with emphasis on novelty, while the language used by mathematicians is perceivably different, more explicit and much less hasty. All of the statements in mathematics must be rigorous and made after precise definitions.

In this sense, the persuasive success of the mutual interaction between mathematics and physics appears slightly puzzling, because of the different goals and habits rather than the language itself. Obviously, it requires not only a lot of mutual tolerance and openness but also a change of the language. Fortunately, it is equally obvious that the necessary efforts almost always pay off. This is also one of the main reasons why we decided to collect a few members of the mathematical physics community and to compose an edited book in which an account of one of the most interesting developments in contemporary theoretical physics would be retold using mathematical style.

Our selection of the subject of the applicability and applications of non-self-adjoint operators in Hilbert spaces was dictated, first of all, by the current status of the development of the field in the context of physics and, in particular, of quantum physics. In parallel, many of the ideas involved in these recent developments may be identified as not so new in mathematics. For this reason, we believe that our current book could fill one of the increasingly visible gaps in the existing literature. We believe that the current emergence of multiple new ideas connected with the concepts of non-self-adjointness in physics will certainly profit from a less speedy return to the older knowledge and to the roots of at least some of these ideas in mathematics.

Naturally, the message delivered by our current book is far from being complete or exhaustive. We decided to prefer a selection of a few particular subjects, giving the authors more space for the presentation of their review-like summaries of the existing knowledge as well as of their own personal interpretation of the history of the field as well as of its expected further development in the nearest future.

F. BAGARELLO, J. P. GAZEAU,
F. H. SZAFRANIEC, M. ZNOJIL

Palermo, Paris, Rio, Kraków, Prague
September, 2014

ACRONYMS

CCM	coupled clusters method
$\mathcal{D}$-PB(s)	pseudo-bosons(s) on a dense subspace $\mathcal{D}$
EEP	extreme exceptional point
EP	exceptional point
HO	harmonic oscillator
IB	interacting bosons
KLMN	Kato, Lax, Lions, Milgram, Nelson
LHS	lattice of Hilbert spaces
LBS	lattice of Banach spaces
MHD	magnetohydrodynamics
MP	Mathematical Physics/Mathematical Physicists
MPI	Max Planck Institute
NiTheP	National Institute of Theoretical Physics
PB(s)	pseudo-boson(s)
PF(s)	pseudo-fermion(s)
PIP space	partial inner product space
PHHQP	Pseudo-Hermitian Hamiltonians in Quantum Physics (conference series)
PT	see $\mathcal{PT}$ in Symbols
PT symmetric	invariant under space reflection and complex conjugation
QM	quantum mechanics
RS	Rayleigh-Schrödinger

SIGMA	Symmetry, Integrability and Geometry: Methods and Applications
SSUSY	second-order supersymmetry
SUSY	supersymmetry
THS	three-Hilbert-space
WKB	Wentzel-Kramers-Brillouin

GLOSSARY

Hermitian	denoting or relating to a matrix in which those pairs of elements that are symmetrically placed with respect to the principal diagonal are complex conjugates (from Oxford Dictionaries, `http://www.oxforddictionaries.com/`)
non-Hermitian	logical negation of "Hermitian"
Hermitesch	*so heißt ein linearer Operator R selbstadjungiert oder Hermitesch...* J. v. Neumann 1930, Allgemeine Eigenwerttheorie Hermitescher Funktionaloperatoren, *Mathematische Annalen* **102** 49-131
nicht Hermitesch	logical negation of "self-adjoint"
self-adjoint	*A densely defined operator T on a Hilbert space is called* ***symmetric*** *(or* ***Hermitian****) if $T \subset T^*$ and* ***self-adjoint*** *if $T = T^*$* (from M. Reed and B. Simon, Methods of Mathematical Physics, Vol 1: Functional Analysis Academic Press 1980)
non-Hermitian	logical negation of "symmetric"
non-selfadjoint	logical negation of "self-adjoint"

SYMBOLS

$\mathcal{D}$	dense subspace
$\mathcal{F}_\varphi$	$\mathcal{G}$-quasi base (Chapter 3)
$\mathcal{F}_\Psi$	$\mathcal{G}$-quasi base (Chapter 3)
$\mathcal{G}$	dense subspace (Chapter 3)
$G(t)$	non-Hermitian generator of quantum evolution in the time-dependent dynamical regime of Eqs. 1.5.2 and 1.5.3
$\mathfrak{h}$	for operator H, one of the isospectral self-adjoint partners living in a "prohibited", third Hilbert space $\mathcal{H}^{(P)}$ (Chapter 1)
H	Hamiltonian
$\mathcal{H}$	Hilbert space
$H^\ddagger$	the adjoint of operator H in "the second", physical Hilbert space $\mathcal{H}^{(S)}$ (Eq. 1.4.5)
$\langle u, v\rangle$	inner product **linear** in the **first** argument (Chapter 2)
$\langle u, v\rangle$	inner product **linear** in the **second** argument (Chapter 3)
(u, v)	inner product **linear** in the **second** argument (Chapter 5)
$[u, v]$	indefinite inner product **linear** in the **first** argument (Chapter 6)
(u, v)	inner product **linear** in the **first** argument (Chapters 6, 7)
$\langle u\|v\rangle$	inner product **linear** in the **first** argument (Chapter 7)
Ω	Dyson-map factor of metric $\Theta = \Omega^\dagger\Omega$ (Eq. 1.4.7)
$\|\psi_n\rangle\rangle$	in "the first", unphysical Hilbert space $\mathcal{H}^{(F)}$, any biorthogonal partner of eigenket $\|\psi_n\rangle$ of Hamiltonian $H \neq H^\dagger$ (Chapter 1)
$\mathcal{PT}$	parity times time-reversal
$\rho(A)$	resolvent set of operator A
$\sigma(A)$	spectrum of operator A
$\sigma_p(A)$	point spectrum of operator A

$\sigma_r(A)$	residual spectrum of operator A
$\sigma_c(A)$	continuous spectrum of operator A
$\sigma_{\mathrm{disc}}(A)$	discrete spectrum of operator A
$\sigma_{\mathrm{ess}}(A)$	essential spectrum of operator A
$\Theta(A)$	numerical range of operator A (Chapter 5)
Θ	metric operator
G	metric operator (Chapter 7)

INTRODUCTION

F. BAGARELLO, J.P. GAZEAU, F. SZAFRANIEC AND M. ZNOJIL
Palermo, Paris, Rio, Kraków, Prague

The overall conception of this multipurpose book found one of its sources of inspiration in a comparatively new series of international conferences "Pseudo-Hermitian Hamiltonians in Quantum Physics" (1). This series offered, from its very beginning in 2003, a very specific opportunity of confrontation of the mathematical and phenomenological approaches to the concepts of the non-self-adjointness of operators. At the same time, the recent meetings on this series (the conferences in Paris (2) and Istanbul (3)) seemed, to us at least, to convert this confrontation to a sort of just a polite coexistence.

We (i.e., our team of guest editors of this book) came to the conclusion that it is just time to complement the usual written outcome of these meetings (i.e., typically, the volumes of proceedings or special issues as published, more or less regularly, in certain physics-oriented journals) by a few more mathematically oriented texts, reviews, and/or studies.

The idea of collecting the contributions forming this volume came out from the workshop "Non-Hermitian operators in quantum physics," held in Paris, in August 2012. It was the 11th meeting in the PHHQP series. Keeping track of the contemporary development of Quantum Physics, either monitoring the publications or attending conferences in diverse areas, we have realized that, in order to stimulate properly further progress as well as optimize the scientific efforts undertaken by researchers in the field, a résumé of mathematical methods used so far would surely be beneficial. As Mathematics is unquestionably a basic tool, people working in Quantum Physics should be aware of its applicability, deepening insight and widening perspectives. Therefore, we thought that any update in this direction should be welcome, particularly topics that refer to "non-self-adjoint operators," primarily those involved in $\mathcal{PT}$-symmetric Hamiltonians (4) and in their extensions. We are convinced that this relatively wide subject will attract the attention of many scientists, from mathematics to theoretical and applied physics, from functional analysis to operator algebras.

This mathematically oriented *state of the art* book is a result of these reflections and efforts. It includes a general survey of $\mathcal{PT}$-symmetry, and invited chapters, reviewing, in a self-consistent way, various mathematical aspects of non-Hermitian or non-self-adjoint operators in mushrooming Quantum Physics. It is composed of contributions of several representative authors (or groups of authors) who accepted the challenge and who tried to promote the currently available physics – emphasizing accounts of the current status of the field to a level of more rigorous mathematical standards in the following areas:

- Functional analytic methods for non-self-adjoint operators
- Algebraic methods for non-self-adjoint lattices of Hilbert spaces
- Perturbation theory
- Spectral theory
- Krein space theory
- Metric operators and lattices of Hilbert spaces

The organization of the book follows more or less faithfully the aforementioned list of subjects. Each chapter can be read independently of the others and has its own references at the end.

Chapter 1 is thought as a comprehensive historical description of motivations and developments of those "non-hermitian" explorations and/or transgressions of self-adjointness, a crucial requirement for physical observability and dynamical evolution, lying at the heart of Von Neumann quantum paradigm. Its content reflects the selection of topics that are covered by the more mathematically oriented rest of the book. It intends, through the Hilbertian trilogy $\mathcal{H}^{(P)}$, $\mathcal{H}^{(F)}$, $\mathcal{H}^{(S)}$, to restrict the readership attention to a few moments at which a cross-fertilizing interaction between the phenomenological and formal aspects of the use of non-self-adjoint operators in physics proved particularly motivating and intensive.

Chapter 2 is intended to give those "operators" considered in mathematical physics a form of operators as mathematicians would like to see them. This in turn creates a need of having the commutation relations properly understood. As all this refers to the quantum harmonic oscillator and its relatives, the operators involved are rather nonsymmetric. The class of operators they belong to as well as their spatial properties are described in some detail. As a matter of fact, and besides isometries, there are only two classes of Hilbert space operators that are commonly recognizable in Quantum Mechanics: symmetric (essential self-adjoint, self-adjoint) and generators of different kinds of semigroups. Other important operators, for instance, those appearing in the quantum harmonic oscillator seem to be not categorized, at least unknown to the bystanders. One of the goals of this survey is to expose their role, enhancing the most distinctive features. The main "non-self-adjoint" object is the class of (unbounded) *subnormal* operators. This is compelling, and as such it determines our *modus operandi:* "spatial" approach rather than Lie group/algebra connections. A natural consequence is to refresh the meaning traditionally given to commutation relations.

Chapter 3 shows how a particular class of biorthogonal bases arises out of some deformations of the canonical commutation and anticommutation relations. The deformed raising and lowering operators define extended number operators, which are not self-adjoint but are related by a certain intertwining operator, which can also be used to introduce a new scalar product in the Hilbert space of the theory. The content of this chapter clarifies some of the questions raised by such deformations by making use of a rather general structure, with central ingredient being the so-called $\mathcal{D}$-pseudo-bosons ($\mathcal{D}$-PBs) or their fermionic counterparts, the pseudo-fermion (PFs). This structure is unifying as many examples introduced along the years in the literature on $\mathcal{PT}$-quantum mechanics and its relatives can be rewritten in terms of $\mathcal{D}$-PBs or of PFs.

Chapter 4 is a review presenting some simple criteria, mainly of perturbative nature, entailing the reality or the complexity of the spectrum of various classes of $\mathcal{PT}$-symmetric Schrödinger operators. These criteria deal with one-dimensional operators as well as multidimensional ones. Moreover, mathematical questions such as the diagonalizabilty of the $\mathcal{PT}$-symmetric operators and their similarity with self-adjoint operators are also discussed, also through the technique of the convergent quantum normal form. A major mathematical problem in $\mathcal{PT}$-symmetric quantum mechanics is to determine whether or not the spectrum of any given non-self-adjoint but $\mathcal{PT}$-symmetric Schrödinger operator is real. Clearly, in this connection, an equally important issue is the spontaneous breakdown of the $\mathcal{PT}$-symmetry, which might occur in a $\mathcal{PT}$-symmetric operator family. The spontaneous violation of the $\mathcal{PT}$-symmetry is defined as the transition from real values of the spectrum to complex ones at the variation of the parameter labeling the family. Its occurrence is referred to also as the $\mathcal{PT}$-symmetric phase transition. This chapter is a review of the recent results concerning these two mathematical points, within the standard notions of spectral theory for Hilbert space operators.

Chapter 5 focuses on spectral theory. It is an extremely rich field, which has found applications in many areas of classical as well as modern physics and most notably in quantum mechanics. This chapter gives an overview of powerful spectral-theoretic methods suitable for a rigorous analysis of non-self-adjoint operators. It collects some classical results as well as recent developments in the field in one place, and it illustrates the abstract methods by concrete examples. Among other things, the notions of *quasi-Hermiticity, pseudo-Hermiticity, similarity to normal and self-adjoint operators, Riesz-basicity*, and so on, are recalled and treated in a unified manner. The presentation is accessible for a wide audience, including theoretical physicists interested in $\mathcal{PT}$-symmetric models. It is a useful source of tools for dealing with physical problems involving non-self-adjoint operators.

Chapter 6 presents a variety of Krein-space methods in studying $\mathcal{PT}$ symmetric Hamiltonians and outlines possible developments. It bridges the gap between the growing community of physicists working with $\mathcal{PT}$ symmetry (4) with the community of mathematicians who study self-adjoint operators in Krein spaces for their own sake. The general mathematical properties of $\mathcal{PT}$-symmetric operators are discussed within the Krein spaces framework, focusing on those aspects of the Krein spaces theory that may be more appealing to mathematical physicists. This supports the idea that every $\mathcal{PT}$-symmetric operator corresponding to a quantum observable should be a self-adjoint operator in a suitably chosen Krein space and that a proper investigation of a $\mathcal{PT}$-symmetric Hamiltonian A involves the following stages: interpretation of A as a self-adjoint operator in a Krein space $(\mathcal{H}, [\cdot,\cdot])$; construction of an operator C for A; interpretation of A as a self-adjoint operator in the Hilbert space $(\mathcal{H}, (\cdot,\cdot)_C)$.

Chapter 7 analyzes the possible role and structure of the generalized metric operators G, which are allowed to be unbounded. As early as 1960, Dieudonné already tried to introduce and analyze such a concept. In the context of mathematics of Hilbert spaces he found, to his disappointment, that the properties of the operators A, which he suggested to be called quasi-Hermitian and which had to satisfy the generalized Hermiticity relation of the form $GA = A^*G$, appeared not so attractive. Later, the class of the admissible G's has been narrowed by physicists. They found that once the G's are just bounded and strictly positive self-adjoint operators with bounded inverse, the quasi-Hermitian operators A reacquire virtually all of the properties that are needed in quantum mechanics. Unfortunately, in a number of examples including, in particular, many $\mathcal{PT}$-symmetric models (4), the latter requirements proved too restrictive. Their moderate mathematical generalization appeared necessary. In Chapter 7, therefore, several generalizations of the notion of quasi-Hermiticity are introduced and the questions of the preservation of the spectral properties of operators are examined.

Canonical lattices of Hilbert spaces generated by unbounded metric operators are then considered. Such lattices constitute the simplest case of a partial inner product space (*PIP* space), and this justifies the employment of the technique of *PIP* space operators. Some of the previous results are applied to operators on a particular *PIP* space, namely, the scale of Hilbert spaces generated by a single metric operator. Finally, the notion of pseudo-Hermitian operators is reformulated in the preceding formalism.

As a concluding remark, the material presented in our book will certainly draw the attention of the reader to a well-known occurrence in the mutual irrigation of Mathematics and Physics, namely, the existence of basic, even trivial operations or properties leading to nontrivial developments in both disciplines. In the present case, there are two (very) discrete involutions in inner product complex vector spaces with countable basis $\{e_n\}$, namely, antilinear complex conjugation of vectors $(\xi_n) \mapsto (\overline{\xi_n})$ and linear parity $(\xi_n) \mapsto ((-1)^n \xi_n)$. The next 400 pages are recurrent symphonic variations around that $\mathcal{P}$e$\mathcal{T}$*ite phrase musicale* (5) pervading our lost $\mathcal{P}$roustian $\mathcal{T}$ime.

REFERENCES

1. Available at http://gemma.ujf.cas.cz/%7Eznojil/conf/index.html
2. Available at http://phhqp11.in2p3.fr/Home.html Accessed 2014 Nov 13.
3. Available at http://home.ku.edu.tr/%7Eamostafazadeh/workshop_2012/index.html
4. Bender CM, Boetcher S. Real spectra in non-Hermitian Hamiltonians having PT symmetry. *Phys Rev Lett* 1998;80:5243–55246.
5. Proust M. *Vinteuil Sonate in Un amour de Swann, A la recherche du Temps Perdu*. Grasset and Gallimard, Paris, 1913, p 1871–1922.

1

NON-SELF-ADJOINT OPERATORS IN QUANTUM PHYSICS: IDEAS, PEOPLE, AND TRENDS

MILOSLAV ZNOJIL
Nuclear Physics Institute, ASCR, Řež, Czech Republic

1.1 THE CHALLENGE OF NON-HERMITICITY IN QUANTUM PHYSICS

1.1.1 A Few Quantum Physics' Anniversaries, for Introduction

The year of writing this history-oriented chapter on the appeal of non-Hermiticity was also the year of several minor but interesting anniversaries in quantum physics. So let us start by recalling some of these dates.

1.1.1.1 Hundred Years of the Bohr's Model In 2013, on occasion of the centenary of the Bohr's model of atom (1) (marking, in a way, the birth of quantum theory), one should appreciate the multitude of results of the first hundred years of our study of quantum world. One of typical characteristics of these developments may be seen in an incessant emergence of dramatic innovations and changes in our perception of what is measurable. The process still remains unfinished. Even the Nobel Prize in Physics for year 2012 was not awarded for the fresh, expensive, and long expected discovery of the Higgs boson in particle physics (which had to wait for one more year) but rather for the invention of "ground-breaking" methods of quantum measurements (2).

One must underline that during the century, the fundamental quantum physics remained a vivid discipline and that its experimental side never ceased to be a

Non-Selfadjoint Operators in Quantum Physics: Mathematical Aspects, First Edition.
Edited by Fabio Bagarello, Jean-Pierre Gazeau, Franciszek Hugon Szafraniec and Miloslav Znojil.

topical subject. What should be appreciated, in parallel, is the fact that none of these innovations ever disproved any of the apparently counterintuitive basic principles of the theory. One must admire the robust nature of the basic mathematical ideas.

In particular, it was not necessary to change the theory after Herman Feshbach (3) succeeded in describing the usual processes of quantum scattering and reactions in atomic nuclei (*including* the elastic ones) by means of a *complex* effective potential. On this occasion, the "exotic" non-self-adjoint (*alias*, in the physicist's language, non-Hermitian) operators seem to have entered the scene.

1.1.1.2 Fifty-five Years of the Feshbach's Non-self-adjoint Hamiltonians For stable quantum systems, the evolution in time is usually assumed generated by a physical Hamiltonian $H = H^{(P)}$ which is defined as acting in a suitable representation $\mathcal{H}^{(P)}$ of the physical Hilbert space of states. It is very important that the popular principle of correspondence, albeit vaguely defined, often enables us to choose, in realistic models, constructively tractable versions of spaces $\mathcal{H}^{(P)}$ and Hamiltonians $H^{(P)}$.

The practical feasibility of calculations quickly decreases during transition to more complicated systems. One may recall multiple examples, say, in nuclear physics where the computer-assisted numerical determination of the bound-state energies hardly remains sufficiently routine even in the lightest nuclei. For the heavier nuclei, the growth of complexity of calculations may be perceived as one of the fundamental methodical challenges in nuclear physics.

As we already mentioned, one of the productive tools of an amendment of the algorithms has been proposed by Feshbach (3). In his considerations, he admits that even if one knows Hamiltonians $H = H^{(P)}$, many time-independent Schrödinger equations $H\,|\psi\rangle = E\,|\psi\rangle$ describing bound states (with E = real) or resonant states (with E = complex) prove prohibitively difficult to solve in practice. He recalled that in the majority of applications just the knowledge of the low-lying spectrum of energies is asked for. This led him to the conclusion that a judicious restriction of physical space $\mathcal{H}^{(P)}$ to a suitable subspace $\mathcal{H}^{(R)}$ should be performed in such a way that the reduction of Hamiltonian $H^{(P)} \to H^{(R)}$ remains compatible with the requirement of an at least partial isospectrality of the two operators.

Ambitious as the project might have seemed, its analysis resulted into a recipe which proved enormously popular and successful in practice (4). In fact, its basic idea is fairly elementary. One simply partitions the "big" Hilbert space $\mathcal{H}^{(P)}$ into two subspaces via projectors Q (on an "irrelevant" part of the bigger Hilbert space $\mathcal{H}^{(P)}$) and $R = I - Q$ (in our present notation, the projector on *the* model space $\mathcal{H}^{(R)}$). This yields the partitioned Schrödinger equation

$$(R+Q)\,H\,(R+Q)\,|\psi\rangle = E\,(R+Q)\,|\psi\rangle$$

and formula

$$Q\,|\psi\rangle = Q\,[E\,I - Q\,H\,Q]^{-1}\,Q\,H\,|\phi\rangle\,, \qquad |\phi\rangle = R\,|\psi\rangle$$

for the Q-projection of the exact wave function, which is, by assumption, less relevant. Its elimination provides the ultimate compactified, nonlinear, "effective" Schrödinger eigenvalue problem

$$H_{\rm eff}(E)\,|\phi\rangle = E\,|\phi\rangle \qquad (1.1.1)$$

defined inside the subspace $\mathcal{H}^{(R)}$. The action of the effective Hamiltonian

$$H_{\rm eff}(E) = R\,H\,R + R\,H\,Q\;[E\,I - Q\,H\,Q]^{-1}\;Q\,H\,R$$

is energy dependent but it remains restricted just to the relevant, R-projected subspace $\mathcal{H}^{(R)}$.

One must emphasize that the required strict isospectrality between H and $H_{\rm eff}(E)$ is guaranteed. Unfortunately, owing to the manifest energy dependence of $H_{\rm eff}(E)$, the costs grew high. They became even higher in the original nuclear-reaction context in which the physical values of energies lied in the essential part of the spectrum. One must then generalize the definition of the operator pencil $H_{\rm eff}(z)$ and start working with the complex values of the variable parameter $z \in \mathbb{C}$.

In the energy range of interest, the spectral shift caused by the presence of z or E in denominators is often being ignored as not too relevant in practice. Still, it must be reemphasized that the simplified effective-Hamiltonian operator $H_{\rm eff}(z)$ is *manifestly non-self-adjoint* in general.

With such an observation, the projection-operator studies of quantum systems rarely remain restricted to the mere stable dynamical regime with real spectra and unitary evolution in time. The loss of the self-adjoint nature of the effective Hamiltonians is usually interpreted as implying a *necessary* loss of the reality of the energies. Strictly speaking, such a deduction is not always correct. As a counterexample, one may recall, for example, the so-called $\mathcal{PT}$-symmetric systems and Hamiltonians.

1.1.1.3 Fifteen Years of $\mathcal{PT}$-symmetry alias Pseudo-Hermiticity In the broader context of preceding paragraph, the abstract formalism of quantum theory encountered an unexpected challenge circa 15 years ago, after several parallel innovative proposals of inclusion, in the mainstream formalism, of certain manifestly non-self-adjoint Hamiltonian-like operators $H \neq H^\dagger$ possessing *strictly real*, bound-state-like spectra. Fortunately, during the subsequent years, the acceptance of such a class of models proved fully compatible with the first principles of quantum theory. Moreover, an intensified study of mathematics of manifestly non-self-adjoint candidates for observables became an inseparable part of quantum theory.

An "official" start of studies of the possibility of having non-self-adjoint operators in a unitary theory may be dated back to 1998 when Bender and Boettcher published their letter (5). Its title "Real spectra in non-Hermitian Hamiltonians having $\mathcal{PT}$-symmetry" sounded truly provocative at that time because, according to conventional wisdom, the spectra of non-Hermitian operators can hardly be purely real. The explicit construction of a non-Hermitian quantum Hamiltonian $H \neq H^\dagger$

with real spectrum sounded, therefore, like a joke or paradox rather than like a serious scientific proposal[1].

Later on, the situation and attitudes have changed. A sample of the progress is to be reported in this book. As long as the main emphasis will be laid, in the forthcoming chapters, upon the mathematical aspects of the theory, the collected material will be preceded and, in some sense, interconnected by this chapter offering a compact outline of the field, with particular emphasis upon historical and phenomenological context.

Our considerations will reflect the selection of topics to be covered by the more mathematically oriented rest of the book. In a sketchy and incomplete, selective outline of key ideas, we intend to restrict our attention to a few moments at which a cross-fertilizing interaction between the phenomenological and formal aspects of the use of non-self-adjoint operators in physics proved particularly motivating and intensive.

1.1.2 Dozen Years of Conferences Dedicated to Pseudo-Hermiticity

Letter (5) inspired a lot of research activities. In the literature, a number of emerging paradoxes was spotted, exposed to a thorough scrutiny—and shown to disappear. At present, one can say that from the point of view of the recent history of quantum physics, the date of publication of this remarkable letter may be perceived as one of the most important turning points. Not only within the theory itself (in which, later, the concepts of stability and evolution were thoroughly revisited and clarified under its influence) but also in experiments.

At present, the progress and publications of the related results in prestigious physics Journals may be followed online, via dedicated bookkeeping webpage as maintained by Daniel Hook (8). Through this page, one can trace the recent history of the field. One can download a lot of papers that caused the change (or rather a complete reversal) of attitude of the international scientific community toward the real spectra in non-Hermitian Hamiltonians.

The change of the paradigm was due to the work by multiple active authors, so the results cannot be summarized easily (cf. review papers (9–11) for many details and references). Among these results, one should recollect, first of all, the introduction of the influential concepts of $\mathcal{PT}$-symmetry of a quantum Hamiltonian H (i.e., of the rule $H\mathcal{PT} = \mathcal{PT}H$ where symbols $\mathcal{P}$ and $\mathcal{T}$ denote parity and time reversal, respectively) or, after a slight generalization, of the η-pseudo-Hermiticity (or, briefly, pseudo-Hermiticity) of an observable Λ (i.e., of the rule $\Lambda^\dagger \eta = \eta\, \Lambda$ written in terms of a suitable generalization η of the parity operator).

Although the notion of $\mathcal{PT}$-symmetry may be already found in 1993 paper (12), it only played a marginal role there. Probably, the concept was in current use even earlier (13). Anyhow, its heuristic relevance and productivity remained practically unknown before 1998.

[1] *Pars pro toto*, the imminent voices of criticism may be sampled by the R. F. Streater's comments, propagated via his webpage (6) and book (7).

Several years after 1998, symbol $\mathcal{P}$ entering the $\mathcal{PT}$-symmetry relation was mostly perceived as the mere parity in one-dimensional bound-state Schrödinger equations

$$\left(-\frac{d^2}{dx^2} + V(x)\right) \psi_n(x) = E_n \, \psi_n(x), \qquad \psi_n(x) \in L_2(-\infty, \infty) \tag{1.1.2}$$

while symbol $\mathcal{T}$ was strictly identified with an antilinear operator of time reversal. Even under these constraints, the appeal of the innovative notion grew quickly. Its use led to conjectures of multiple toy models (1.1.2) with quantum Hamiltonians of the "usual" form $H = p^2 + V(x)$ (and real spectra) but with an "unusual" non-self-adjointness property $H \neq H^\dagger$ in the underlying "friendly" Hilbert space $\mathcal{H}^{(F)}$.

In the contemporary context of quantum model-building practice, the majority of innovations sounded strangely. One of their presentations to a larger audience during an international scientific conference took place in Paris in 2002. On this occasion, the invited speaker (naturally, Carl Bender) plus four other authors (cf. the written form of the talks in proceedings (14)) discussed the response and concluded that the subject might deserve a separate series of dedicated conferences.

Supported by several other enthusiasts, the dedicated series really started, a year later, by the meeting of 27 participants from as many as 13 different countries in Prague (cf. (15)). The next, similarly compact international workshop followed within a year. Subsequently, the number of participants jumped up (i.e., close to one hundred) in 2005, after the transfer of the meeting from Villa Lanna in Prague to the Universities in Istanbul (Turkey, June 2005) and Stellenbosch (South Africa, November 2005), etc[2].

The proceedings of the PHHQP series of conferences were mostly published in the form of a dedicated and refereed special issue (cf. (17–19), etc.). These materials may be recalled as offering a compact (i.e., introductory, history-oriented, and time-ordered) sample of a few characteristic results, mainly in the field of quantum physics.

Even such a restricted inspection of the history of acceptance of non-self-adjoint operators in quantum physics reveals a sequence of "ups" (i.e., of the periods of a more or less uninterrupted growth), which were followed by "downs," characterized by a sudden emergence of serious obstacles and crises.

1.2 A PERIODIZATION OF THE RECENT HISTORY OF STUDY OF NON-SELF-ADJOINT OPERATORS IN QUANTUM PHYSICS

1.2.1 The Years of Crises

In retrospective, one of the least expected observations resulting from the recollection of history of $\mathcal{PT}$-symmetry and pseudo-Hermiticity shows an amazing regularity in the occurrence of crises. Several less regular precursors of these crises may even

[2]Interested readers may click for more details on the global webpage (16) of the series Pseudo-Hermitian Hamiltonians in Quantum Physics (PHHQP).

be dated before the above-mentioned year 1998. One may recollect, for example, that on the basis of multiple numerical experiments with the *purely imaginary* cubic interaction

$$V(x) = V^{(BZJ)}(x) = \mathrm{i}x^3, \qquad x \in \mathbb{R} \tag{1.2.1}$$

Daniel Bessis with Zinn-Justin already believed, in 1992 at the latest (20), in the strict reality of the bound-state spectrum. Anyhow (was it a crisis?), up to the present author' knowledge, they never published anything on their observations before the years when Carl Bender did.

1.2.1.1 The Year 2001: the First, Spectral-reality-proof Crisis The first "regular" crisis came after circa 3 years of intensified studies of various $\mathcal{PT}$-symmetric potentials and, in particular, of the Bender's and Milton's (21) (or, if you wish, of the Bender's and Boettcher's (5)) extremely popular model

$$V(x) = V^{(BM)}(x, \delta) = x^2\,(\mathrm{i}x)^\delta\,, \qquad \delta \geq 0\,. \tag{1.2.2}$$

In this potential, the choice of the general power-law x–dependence appeared to support an extension of the above-mentioned $\delta = 1$ hypothesis of the reality of the spectrum. During the years 2000 and 2001, nevertheless, people started feeling more and more aware of the lasting absence of a reliable, rigorous proof.

The first proof applicable to the important family of $\mathcal{PT}$-symmetric models (1.2.2) already appeared in 2001 (22). Just in time. What followed was a quick acceptance of the promising perspective of a noncontradictory existence of the real bound-state energies even when obtained from non-Hermitian Hamiltonians.

The much-required clarification of this point encouraged the community to make the next step and to return, *i.a.*, to the Streater's criticism (6, 7) and to the related doubts about the possible physics behind the new and highly nonstandard models as exemplified by eqs (1.2.1) or (1.2.2). During the first years after the first crisis, the decisive suppression of these doubts has been achieved, basically, via a discovery (or rather rediscovery) of the possibility of using an *ad hoc* inner product and of defining a *different* Hilbert space of states in which a correct probabilistic interpretation of the models can be provided. In this text, we denote such a "second" or "standard" Hilbert space by dedicated symbol $\mathcal{H}^{(S)}$.

1.2.1.2 The Year 2004: the Metric-ambiguity Crisis In the amended Dirac's notation of Ref. (23), the generalized inner products may be defined as overlaps $\langle \phi_1 | \Theta | \psi_2 \rangle$ where operator $\Theta \neq I$ may be called "Hilbert-space metric." During a year or two, the discovery has been slightly reformulated and found equivalent to a resuscitation of the three-Hilbert-space (THS) quantum-system representation as known and used, in nuclear physics, as early as in 1992 (24)[3]. In this manner, an overall, sketchy formulation of the theory was more or less completed.

[3]One could even trace the idea back to the Dyson's 1956 paper (25) on models of ferromagnetism. Thus, the Dyson's forthcoming 90th birthday (viz., on December 15, 2013) would also be eligible as an item for inclusion in the list of relevant anniversaries.

The ultimate moment of acceptance of $\mathcal{PT}$-symmetric Hamiltonians by physicists may be identified with the year 2004 of publication of erratum (26). During this year, *any* nontrivial metric $\Theta \neq I$ (known, in the conventional physical terminology, as "non-Dirac" metric) became perceived as a fundamental ingredient in the description of quantum system, in principle at least.

The nontrivial-metric-dependent THS background of the theory was accepted, redirecting attention to the next open problem, namely, to the immanent ambiguity of the THS recipe. People imagined that the assignment of the desirable Hilbert-space metric Θ to a preselected (and, say, $\mathcal{PT}$-symmetric) Hamiltonian $H \neq H^{\dagger}$ is far from unique. Deep crisis number two followed almost immediately.

At a comparable speed, the crisis was suppressed, mainly due to an overall acceptance of an additional postulate of observability of a new quantity $\mathcal{C}$ called quasi-parity (27) or charge (28). It has been clarified that the requirement of the observability of charge $\mathcal{C}$ makes the metric unique. A new wave of optimism followed.

1.2.1.3 The Year 2007: the Nonlocality Crisis The third crisis emerged 3 years later, in 2007, when Jones noticed several counterintuitive features and obstructions to realization of a quantum-scattering-type experiment in $\mathcal{PT}$−symmetrized quantum mechanics arrangement (29). This discovery forced people to reanalyze the concepts of $\mathcal{PT}$-symmetry and of the $\mathcal{CPT}$-symmetry, with the latter name being, for physicists, just the most common alias to the above-mentioned compatibility of the THS representation of quantum systems (using a special metric $\Theta_0 = \mathcal{PC}$) with the standard textbooks on quantum theory where only too often, just for the sake of simplicity, the metric is being set equal to the identity operator.

The simplest way out of the Jones' trap was proposed by Jones himself. He conjectured that the local complex potentials $V(x)$ should be perceived as "effective," say, in the Feshbach's subspace projection sense. In such an approach, the key scattering-unitarity assumption is declared redundant. Although this implies that, say, the flow of mass need not be conserved, that is, certain *deus ex machina* of sinks and sources of particles is admitted, one simply accepts an overdefensive argument that the evolution processes and, in particular, the scattering in a given quantum system is in fact just partially controlled and described by our effective local potentials $V(x)$. In other words, one admits that our information about the quantum dynamics is incomplete.

During the subsequent growth of theoretical efforts, one of the most consequent resolutions of the Jones' apparent paradoxes was offered in Ref. (30). The main source of misunderstanding has been identified as lying in an inconsistence of our assumptions. In the context of a conventional *unitary* quantum scattering theory, one simply asks for too much when demanding that the scattering potential may be chosen complex *and* local. In a way based on the construction of several explicit models, a transition to nonlocal $\mathcal{PT}$-symmetric complex forces V was recommended (see more details in the following sections).

1.2.1.4 The Year 2010: the Construction-difficulty Crisis The fourth crisis may be localized, roughly, to the year 2010 when it became clear that the emergent necessity of working with nonlocal potentials might lead to an enormous increase in technical difficulties during the applications of the unitary THS recipe. The same danger of encountering obstacles was found connected with the need of working with very complicated metrics $\Theta \neq I$ even for originally not too complicated Hamiltonians.

A better profit has been found provided by a "transfer of technologies" beyond quantum physics. Around the year 2010, the active research in the area of quantum theory dropped perceivably down, therefore. This tendency was clearly reflected by a drastic decrease in the number of foreign participants in the meeting PHHQP IX in 2010 in China (31).

During the crisis, it became clear that the transfer of the concept of $\mathcal{PT}$-symmetry out of quantum theory may in fact prove not only necessary but also unexpectedly rewarding. A real boom followed, for example, in experimenting with simulations of $\mathcal{PT}$-symmetry in various gain-or-loss media in nonquantum settings (*pars pro toto*, let us mention here just the quick growth of popularity of $\mathcal{PT}$-symmetry in nonquantum optics (32)).

Within quantum physics, the boom was paralleled by a reenhancement of interest in the traditional studies of open quantum systems and also of unstable complex systems, with the dynamics controlled by the presence of resonances. Symptomatically, Nimrod Moiseyev's monograph "Non-Hermitian Quantum Mechanics" (33) dealing with these topics appeared published in 2011.

During the same, first-after-the-crisis year, the place of the jubilee tenth meeting PHHQP (34) (viz., MPI in Dresden) was packed by participants up to the roof again. The conference lasted much longer than usual (viz., full two weeks) and marked the onset of a new period of growth. The meeting was truly successful in putting emphasis on the closeness of connections between the quantum and nonquantum worlds. In addition, the not-entirely-expected influx of a lot of specialists from open-system quantum phenomenology contributed, by feedback, to the subsequent new growth of activities in all of the neighboring fields.

For mathematical physicists, attention has been redirected to the formal aspects of energy-dependent (i.e., nonlinear, effective, and subspace projected) non-self-adjoint Hamiltonians. Last but not least one should mention that as many as two separate special issues of Journals were needed to play the role of proceedings of the remarkable, direction-changing jubilee conference in 2011.

1.2.1.5 At Present: the Ill-defined-metrics Mathematical Crisis Toward the end of 2013, we already know that we are just in the middle of another, fifth serious crisis. Its roots may be traced back to several, not always noticed critical comments on the THS representation formalism as made, in the recent past, by mathematicians (35). The essence of the problem is connected with several failures of the physicists to check the validity of certain necessary mathematical assumptions in their models.

The problem is serious—even for the most popular examples, unexpected "no-go" theorems were proved in the second half of the year 2012 (36). Several parallel attempts at circumventing the obstacles followed (37, 38). Still, the essence of the conflict remains unresolved at present. A final outcome of the last crisis is not yet known. We only have to stay optimistic, recollecting that up to now, all of the previous crises were overcome. What followed the crisis was always a new phase of the growth of the research activities.

1.2.2 The Periods of Growth

In our present compact review of history of "$\mathcal{PT}$-symmetry and all that," we may take the above-listed crises as separators between the five independent triennials of incorporation of non-self-adjoint operators in standard quantum theory, accompanied by an increasing sophistication of underlying mathematics.

1.2.2.1 The First Period: Between 1998 and 2001 The years between 1998 and 2001 were the years of heuristic innovation of quantum theory which was partially lacking mathematical grounds. The research of this period was all guided by the surprisingly efficient recipes of tentative replacement of the current self-adjointness of Hamiltonians $H = H^\dagger$ by $\mathcal{PT}$−symmetry or $\mathcal{P}$-pseudo-Hermiticity. As long as the self-adjointness is usually recalled as implying the reality of the spectrum of bound states, it was rather surprising to reveal, purely empirically, that the same desirable features of stability appeared exhibited by many non-Hermitian Hamiltonians.

The update of the theoretical perspective proved particularly useful for a deeper understanding of field theories with ϕ^3 interactions (14). In quantum mechanics of one-dimensional systems, people studied the possibility of letting the coordinate x in eq. (1.1.2) complex. This led to a perceivable extension of the class of exactly solvable models (cf. (39, 40)).

A number of basic questions remained open. First of all, the theory with $H \neq H^\dagger$ remained challenged by the Stone's theorem and by the apparent loss of the manifest unitarity of the evolution in time. Secondly, the possible physical interpretation of the models having complex coordinates remained temporarily unclear. The first answers to both of these questions only started appearing during the year 2001.

1.2.2.2 The Second Period: Between 2001 and 2004 In a way discussed, reviewed and partially summarized by the participants of the first PHHQP workshop in 2003, the compatibility of the $\mathcal{PT}$-symmetric quantum models with the Stone's theorem was achieved via an upgrade and completion of the theory. Although the general pattern was already published and used, in nuclear physics, for more than 10 years (24), the key idea of introduction of a nontrivial Hilbert-space metric Θ and its factorization $\Theta = \Omega^\dagger\Omega$ (i.e., of its reinterpretation as a superposition of mappings Ω and $\Omega^\dagger$) was rediscovered.

The factorization was readapted to the needs of the schematic one-dimensional $\mathcal{PT}$-symmetric quantum models and to their prospective applications, first of all, in quantum field theory (41). Along several parallel lines of research people then developed multiple new models using, typically, the piecewise constant or point interactions in order to simplify technicalities (42).

The new wave of interest in Hilbert spaces with nontrivial metrics followed. It opened a path toward the innovative studies of many-body systems as well as toward the first genuine physical applications of the theory in cosmology (43). A clarification has been achieved of the full internal consistency of quantum physics (say, of pionic atoms), which is based on the zero-spin relativistic Klein–Gordon version of Schrödinger equation (44).

1.2.2.3 The Third Period: Between 2004 and 2007 One of the important news announced during the 2004 crisis was that the physicist's definition of $\mathcal{PT}$-symmetry more or less coincides with the common mathematical concept of Krein-space Hermiticity. This discovery had far-reaching consequences and it accelerated the research. People found motivation for turning attention to the nonunitary formulations of scattering theory (45). The successful implementations of the use of the concept of $\mathcal{PT}$-symmetry were extended to classical physics, ranging from mechanics (46) up to magnetohydrodynamics (MHD, (47)). The first elementary experiments were proposed in classical optics (48). Within classical electrodynamics, the first microwave simulations of certain characteristic consequences of the transition to $\mathcal{PT}$-symmetric quantum Hamiltonians $H \neq H^\dagger$ were performed (49).

1.2.2.4 The Fourth Period: Between 2007 and 2010 One of the key challenges encountered during the year 2007 was the apparently incurable loss of the unitarity property in the $\mathcal{PT}$-symmetry-controlled systems "at large distances," that is, typically, in the most common quantum-scattering experiments (29). During the subsequent 3 years, one of the solutions was found in the transition from local to nonlocal, "smeared" interaction potentials (50).

In parallel, during these three years of analogies, experiments and simulations people kept paying enhanced attention to classical physics. Thus, experimentalists revealed the practical appeal of certain models of propagation without unitarity, say, within classical electrodynamics. The emergence of spectral singularities has been found relevant and related to the theory of lasers (51). In an even broader context of nonlinear integrable field equations, a transition to $\mathcal{PT}$-symmetric deformations proved inspiring (52).

The numerical and computer-assisted analyses were found productive (53). After a return to quantum world, it has been revealed, inter alia, that even the explicit construction of a horizon of observability of a pseudo-Hermitian quantum system need not always remain prohibitively difficult (54).

A methodical help started to be sought in an ample use of finite-matrix models. New physical meaning has been assigned to the traditional mathematics of the Jordan-block canonical structures. The related confluence of eigenvalues and their subsequent complexification were connected directly with the physics of the loss of observability of a quantum system in question. Last but not least, the "brachistochrone" paradox has been resolved (55) and multiple "ghost problems" were clarified. On the level of pure theory, the time-dependent version of the THS formalism has finally been formulated (23).

The era between 2007 and 2010 could have been called the era of return to the concept of the Kato's exceptional points (56). The enhancement of knowledge of the older and new exceptional-point-related phenomena was summarized during dedicated conference "The Physics of Exceptional Points" as organized by Dieter Heiss in November 2010 in Stellenbosch (57).

1.2.2.5 ***The Fifth Period: Between 2010 and 2013*** During many years of study of quantum models possessing elementary metrics Θ, it became clear that whenever a given, manifestly non-Hermitian observable (say, a Hamiltonian such that $H \neq H^\dagger$ in the "first" Hilbert space $\mathcal{H}^{(F)}$) is considered, and once it is assigned a "Hermitizing" Hilbert-space metric Θ, we may say that

- the original Hilbert space $\mathcal{H}^{(F)}$ (endowed, by definition, with the natural Dirac's trivial metric $\Theta^{(F)} = I$) must be declared "false";
- the role of the physical Hilbert space is transferred to the "second" Hilbert space $\mathcal{H}^{(S)}$, which is, by definition, endowed with the nontrivial Hermitizing metric $\Theta = \Theta^{(S)}(H) \neq I$;
- strictly speaking, our observables cannot be called non-self-adjoint as they should be all perceived as living in "sophisticated" $\mathcal{H}^{(S)}$ rather than in "friendly but false" Hilbert space $\mathcal{H}^{(F)}$.

We believe that in place of calling the observable H non-Hermitian (plus, if applicable, $\mathcal{PT}$-symmetric or pseudo-Hermitian), it makes much better sense to call it "Hermitian in $\mathcal{H}^{(S)}$" or, shortly, in a way promoted by Smilga (58), "crypto-Hermitian".

In this context, mathematicians revealed that the metrics may be often found ill-defined (59). They concluded that our community had to turn attention to the specific features of unbounded operators in infinite-dimensional Hilbert spaces. The decisive relevance of mathematically rigorous definitions and proofs was underlined. The concept of the Riesz basis has been shown to play the particularly important role in the analysis.

All of these considerations are characteristic for the fifth stage of development of the THS theory which lasts up to now. In the context of physics, this phase became strongly interdisciplinary. Behind the above-mentioned return to the study of open quantum systems, one encounters numerous surprises: for example, nonlinear phenomena are studied, typically, in connection with the phenomena of Bose–Einstein condensation (60). New and new theoretical consequences of the presence of exceptional points are being revealed, the traditional WKB analyses acquire various more advanced forms, and so on. Systems with multiple, degenerate exceptional points are considered as forming quantum parallels to the classical instabilities as classified by the Thom's theory of catastrophes (61, 62).

In experimental physics, anything but a sketchy review (see the following text) is beyond the scope of this text. At present, one is witnessing intensive laser-beam (and other) simulations of non-Hermiticity-related phenomena and the development of multiple other nonquantum systems such as (perhaps, chaotic) billiards, innovative forms of the classical electromagnetic waveguides as well as their formally simplified versions living on the thin wires and forming the so-called quantum graphs (63), and so on.

In theoretical physics, in parallel, the formal problems with metrics (which may not exist) were addressed in a way closely related to the pragmatic analyses of instabilities. A turn of attention to pseudospectra and/or to certain new forms of our understanding of instabilities may be noticed to emerge: *Pars pro toto*, Trefethen's and Embree's book (53) may be recommended as a rich source of information on these subjects.

Before we delve in the literature for more details let us emphasize that now, in the middle of the last crisis, the time seems ripe for a unification of forces of mathematicians and physicists. The point is that the running crisis might prove more serious than ever. From this perspective, the material presented in the subsequent chapters of this book may be perceived as truly topical, indeed.

1.3 MAIN MESSAGE: NEW CLASSES OF QUANTUM BOUND STATES

1.3.1 Real Energies via Non-Hermitian Hamiltonians

1.3.1.1 Successful Heuristics: $\mathcal{PT}$-Symmetric Toy Models Nowadays, whenever one returns to the past and whenever one recalls the first years of peaceful coexistence of the unusual reality of spectra with the anomalous, non-self-adjoint Hamiltonians exhibiting $\mathcal{PT}$-symmetry, one most often recollects the Bender's and Boettcher's letter (5). In this work, the *real* bound-state energies were obtained from *complex* potentials using WKB approximation complemented by a routine, brute-force numerical solution of the standard one-dimensional (i.e., ordinary differential) Schrödinger equations (1.1.2) and (1.2.2). The δ-dependence of the numerically evaluated energies $E_n = E_n(\delta)$ which were numbered by $n = 0, 1, \ldots$ was sampled, in Ref. (5), in graphical form. One might suspect that the picture of the $\delta-$dependence of the spectrum became one of the most often cited, recalled, represented, and reprinted figures of all of the recent history of quantum mechanics.

1.3.1.2 Search for Terminology Among the less pleasant features of the new direction of research, people found multiple terminological misunderstandings, often attributable to a sharp contrast between the enthusiasm of physicists and a definite scepticism of mathematicians. With time, the gap broadened. People started feeling a need of return to a deeper analysis of mathematical aspects of the field (26, 28, 64).

Fortunately, the birth of terminological conflict was quickly suppressed. First, Mostafazadeh (65) pointed out that in the majority of papers on $\mathcal{PT}$-symmetry the mathematical meaning of $\mathcal{T}$ was equivalent to Hermitian conjugation yielding a simpler rule $H^\dagger \mathcal{P} = \mathcal{P} H$ of a pseudo-Hermiticity rather than of a symmetry. Langer and Tretter (66) added that the $\mathcal{PT}$-symmetry *alias* pseudo-Hermiticity of H may conveniently be reread as the property of H being self-adjoint in Krein space, provided only that one endows the latter space with pseudometric that coincides with operator $\mathcal{P}$.

Ultimately (cf., e.g., Refs. (67, 68) for further references), all of these developments climaxed in a translation of the physics-oriented concepts into the more traditional mathematical language. A new field of research seemed to be firmly established. Some of its important roots have subsequently been traced to exist already in the past[4].

1.3.1.3 Search for Methods: Perturbation Theory The reasons for the choice of the particular Bender's and Milton's *alias* Bender's and Boettcher's potential $V^{(\mathrm{BM})}(x, \delta)$ appear entirely formal, in the retrospective at least. Almost certainly, the decision was based on an exceptional amenability of such a toy model to perturbation expansions in a small parameter δ. The method was called, by the authors, delta expansions (cf. Ref. (21) for older references).

The original paper (21) itself remained firmly rooted in physics and, in particular, in quantum field theory. Only its title "Nonperturbative calculation ..." might seem to be in contradiction with what we just wrote about the method. An explanation of the paradox is fairly easy and purely terminological since by "perturbative calculation" people almost invariably mean an application of the most common and traditional Rayleigh-Schrödinger (RS) perturbation-expansion recipe.

In the latter RS recipe, the numerical quantities of interest (say, the bound-state energies of eq. (1.1.2)) are not interpreted as functions of a dynamics-determining parameter (like the above-mentioned exponent δ) but rather as complex functions of an *auxiliary* complex variable $\lambda \in \mathbb{C}$. This RS variable is defined via a scaling generalization $V \rightarrow \lambda V$ of the interaction potential in eq. (1.1.2). Then, one makes use of the simplification of the problem at $\lambda = 0$. Next, after an explicit construction of a suitable Taylor or asymptotic series, say, for the generalized bound-state energies $E_n = E_n(\lambda)$, one finally sets the auxiliary RS variable λ equal to its original, "physical" value of $\lambda = 1$ (69).

The emphasis put, in Ref. (21), on the distinction between the status of being "perturbative" or "nonperturbative" has historical roots. For many years, the specific RS expansions played a key constructive role in relativistic quantum field theory in $D = 3$ dimensions. Naturally, as long as the simplified case of $D = 1$ coincides with quantum mechanics (70), it is not too surprising that one finds, in the related literature, multiple methodically motivated perturbative RS studies of eq. (1.1.2) with various specific potentials (71).

It is particularly interesting to notice that among the latter references there exist several older, mathematically motivated RS-type studies of individual models which are, by our present understanding, $\mathcal{PT}$-symmetric. In a way pointed out by Alvárez (72) who studied, in 1995, potentials $V(x) = V^{(BM)}(x, \delta)$ at $\delta = 1$, one should certainly return to the 1980 paper (73) by Caliceti et al. These authors proved, rigorously and long before the boom of $\mathcal{PT}$-symmetry studies, the reality of a set of bound-state energies generated by the asymptotically imaginary cubic anharmonic oscillator well $V(x) = x^2 + V^{(BM)}(x, \delta)$ with specific $\delta = 1$.

The latter proof was based on an RS resummation trick which was rediscovered in 1997 (cf. paper (21)). By Buslaev and Grecchi (12), another $\mathcal{PT}$-symmetric model

[4]For more detail, the Krein-space interpretation of $\mathcal{PT}$-symmetry is explained in Chapter 6.

with $V(x) = x^2 + V^{(BM)}(x, \delta)$ and with a larger exponent $\delta = 2$ was considered and given a correct physical interpretation as early as in 1993. These authors listed several other, more than 10 years older references on the $\delta = 2$ model using the RS-expansion techniques. Even these older papers already reported several numerically supported unexpected discoveries of the reality of the spectrum. Alas, the phenomenon had been considered incidental by these authors.

For the localization of the boundaries of the domains of parameters (or multi-parameters) $\delta \in \mathcal{D}$ yielding real spectra, the choice of the RS-resummations and WKB- or δ−expansions was fortunate. In comparison, the applications of alternative perturbation recipes, for example, of the strong-coupling expansion techniques of Ref. (74) proved by far less suitable for the purpose.

1.3.1.4 Search for the Simplest Models with Real Spectra: Numerical Experiments Owing to the apparent simplicity of quartic potential (1.2.2) with exponent $\delta = 2$, the model attracted attention not only in Ref. (12) from 1993 but also, independently, in 2006 (75). The authors of these papers emphasized that one should study all of the members of the $\delta = 2$ family of potentials $V^{(BM)}(x, 2) + \mathfrak{o}(x^4)$ which behave, asymptotically, like a quartic anharmonic oscillator with wrong sign. This means that all of these counterintuitive, repulsive-looking interactions require, in contrast to their predecessors with $\delta < 2$, a particularly careful interpretation and delicate mathematical treatment.

With the growth of the exponent δ in opposite direction, beyond $\delta = 2$, one notices that many simple-minded recipes including most of the above-listed perturbation methods cease to be applicable. One of the main reasons is that the choice of $\delta = 2$ represents a natural boundary of applicability of Schrödinger equation in its real-axis form (1.1.2). Beyond this boundary, one must start treating Schrödinger differential equation as defined at complex “coordinates” x (cf. (76) for basic information).

At any nonnegative exponent δ, one can calculate the eigenvalues $E_n = E_n^{(BM)}(\delta)$ as certain smooth real functions of exponent in the whole interval of $\delta \in (0, \infty)$. It is merely necessary to define the wave functions $\psi_n(x)$, via Schrödinger differential equation, as square-integrable along an *ad hoc* complex curve of coordinates $x = \xi^{(\delta)}(t) \in \mathbb{C}$, with a real parameter $t \in (-\infty, \infty)$. According to Refs. (75, 77), these complex curves are deformable but, for the sake of simplicity, their preferred shape may be chosen elementary, say, in the form of a left–right symmetric hyperbola with downward-running asympotics in complex plane.

With the growth of δ, the latter recommended asymptotics must both get closer and closer to the negative imaginary half-axis. In the purely numerical study (77), it has been shown that whenever these “complex-coordinate” asymptotics are *not* turned sufficiently down at a fixed value of δ, a *different* spectrum is obtained.

For each $\delta < 2$, the wave functions $\psi_n(x)$ could have been constructed, from eq. (1.1.2), as solutions which remain square integrable along the real axis of x. For larger exponents $\delta > 2$, on the contrary, each choice of an element $\xi(t)$ of the appropriate class of the deformable nontrivial contours forces us to replace the original Hilbert space $L_2(-\infty, \infty)$ by another, more suitable Hilbert space of functions over the contour. Then, in order to avoid confusion, it makes sense to change the notation conventions and denote the new Hilbert space by a dedicated symbol, say, $\mathcal{H}^{(F)}$. As mentioned earlier, the superscript $^{(F)}$ may be read as marking "friendly" space.

At any positive $\delta > 0$ in $V^{(BM)}(x, \delta)$, the contour-dependence property of the alternative $\mathcal{PT}$-symmetric models may be spotted up to the very square-well-shaped limit of $\delta \to \infty$. In this limit, the bound states of $V^{(BM)}(x, \infty)$ with real energies are simply found living on a thin and down-turned U-shaped complex contour of $x \in \xi^{(\infty)}(t)$ (78). In a slightly more general setting, one could contemplate toboggan-shaped, multilooped generalizations of the contours $\xi(t)$ living, admissibly, on topologically nontrivial, multisheeted Riemann surfaces of x (79).

The contour-dependence phenomenon does not seem to be restricted to the original choice of potentials $V^{(BM)}(x, \delta)$ in eq. (1.1.2). Very similar, numerically supported observations were made in Ref. (80) where we explored an alternative family of the asymptotically exponentially quickly growing trigonometric $\mathcal{PT}$-symmetric potentials $V^{(FGRZ)}(x, \alpha) = -(i \sinh x)^{\alpha}$. The main merit of the choice was found in the possibility of return to the standard real line of $x \in (-\infty, \infty)$. Thus, the reality of the coordinate rendered the particle position observable, in principle at least.

1.3.2 Analytic and Algebraic Constructions

Besides the purely numerical nature of the determination of the spectra of bound states in the δ-dependent and $\mathcal{PT}$-symmetric potentials $V^{(BM)}(x, \delta)$, another shortcoming of the model consisted, for a long time, in the absence of any persuasive nonnumerical demonstration that the whole spectrum is strictly real. The analysis leading to the rigorous proof took several years and the proof itself appeared fairly complicated (22). Let us, therefore, break the tradition and let us recall several simpler parameter-dependent models, which prove solvable in closed form while carrying many of the basic features of a generic $\mathcal{PT}$-symmetric scenario.

1.3.2.1 Exactly Solvable Example: Harmonic-Oscillator One of the most natural, harmonic-oscillator-resembling candidates for the simplest possible non-Hermitian model with real bound-state spectrum has the following, manifestly $\mathcal{PT}$-symmetric form as proposed in 1999 (81),

$$V^{(c)}(x, \delta) = (x - ic)^2 + \frac{\delta}{(x - ic)^2}, \quad c > 0, \quad \delta \geq 0, \quad x \in (-\infty, \infty). \qquad (1.3.1)$$

The key merit of such a choice lies in the obvious closed-form solvability of the corresponding Schrödinger's eigenvalue problem (1.1.2). At any positive $c > 0$ and noninteger square root $\alpha = \alpha(\delta) = \sqrt{\delta + 1/4} \notin \mathbb{Z}$, one obtains the complete spectrum of the quantum bound-state energies in c-independent form,

$$E = E_{n,q}^{(\text{HO})}(\delta) = 4n + 2 - 2q\,\alpha(\delta). \qquad (1.3.2)$$

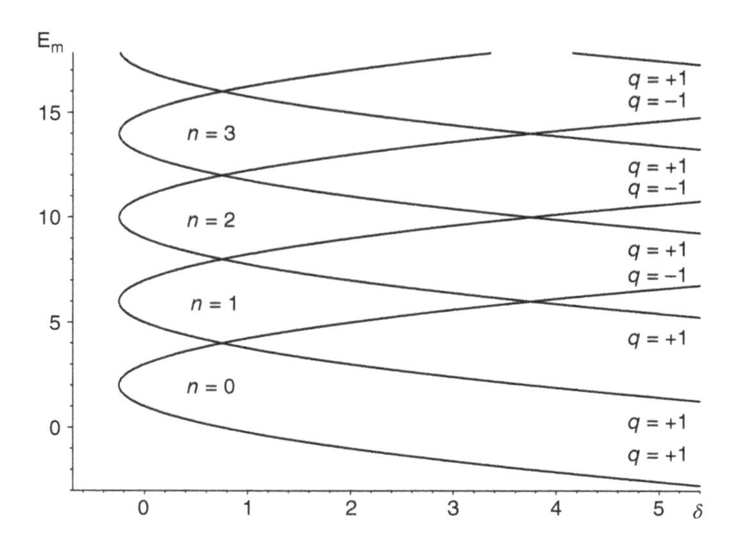

Figure 1.1 The δ-dependence (1.3.2) of the low-lying bound states in the exactly solvable $\mathcal{PT}$-symmetric potential (1.3.1).

These levels (cf. Fig. 1.1) are numbered by a multiindex $m = \{n, q\}$, which is composed of an integer $n = 0, 1, 2, \ldots$ and of the so-called quasi-parity $q = \pm 1$. The related normalizable wave functions are also known in closed form, proportional to Laguerre polynomials (81).

The inspection of the picture reveals that at the exceptional-point (EP) values of $\delta_k^{(EP)} = k^2 - 1/4$ one encounters the unavoided level crossings at $k = 0, 1, \ldots$. At these EP coupling strengths, the algebraic and geometric multiplicities of the eigenvalues become different (56). In the language of physics, the EP versions of the model become manifestly unphysical because our generalized harmonic oscillator Hamiltonian $H_\delta^{(GHO)} = -\partial_x^2 + V^{(c)}(x, \delta)$ ceases to be diagonalizable at all of the exceptional couplings $\delta = \delta_k^{(EP)}$.

The mathematical reason lies in the instantaneous EP confluence (i.e., linear dependence, parallelization) of the respective Laguerre-polynomial wave functions of each pair of eigenstates which cross (cf. (81)). In other words, the number of independent eigenstates of each EP Hamiltonian $H_k^{(EP)} = H_{k^2-1/4}^{(GHO)}$ drops to one half. Their set ceases to form a basis in the infinite-dimensional topological vector spaces $\mathcal{H}^{(F)}$ or $\mathcal{H}^{(S)}$. The canonical matrix form of each exceptional $H_k^{(EP)}$ becomes composed of infinitely many two-by-two Jordan blocs. For this reason, one can only speak about *independent* HO models $H_{\delta,k}^{(GHO)}$ which are numbered by $k = 0, 1, \ldots$ and which may be parametrized just by a coupling δ out of a finite, k-dependent open interval of $\delta + 1/4 \in (k^2, (k+1)^2)$.

To the left from the leftmost, $k = 0$ EP value of $\delta_0^{(EP)} = -1/4$, *none* of the energies remains real. Owing to eq. (1.3.2), *all* of them suddenly form the complex conjugate pairs with quasi-parities $q = \pm 1$.

In a historical remark, let us add that the concept of quasi-parity was introduced in Ref. (27). Such a choice of name originated from the coincidence of q with the usual parity of wave functions in the self-adjoint limit of the vanishing coupling $\delta \to 0$. In the context of quantum physics where one prefers speaking about eigenvalues of operators of observables (24), the value of q may be reinterpreted as an eigenvalue of a charge C for each $H_{\delta,k}^{(\mathrm{GHO})}$ (cf. Ref. (28)).

1.3.2.2 Algebraic Solvability: Truncated Anharmonic Oscillators For the next-to-trivial harmonic-oscillator potential (1.3.1) the above-mentioned exact, closed-form analytic solvability of $\mathcal{PT}$-symmetric Schrödinger equation (1.1.2) is not too surprising because this differential equation can be given the Gauss hypergeometric form. For phenomenological purposes, the model does not seem too exciting, either. It merely offers two mutually shifted copies of the usual equidistant harmonic-oscillator spectrum.

In applications, one usually asks for more. Typically, an enhanced flexibility of a weakly anharmonic model is often welcome. In a suitable basis, one then deforms the equidistant spectrum using a suitable perturbation,

$$H^{(AHO)}(g) = H^{(AHO)}(0) + \mathcal{O}(g)\,,$$

$$H^{(AHO)}(0) = H^{(HO)} = \begin{bmatrix} 1 & 0 & 0 & \dots \\ 0 & 3 & 0 & \dots \\ 0 & 0 & 5 & \ddots \\ \vdots & \vdots & \ddots & \ddots \end{bmatrix}.$$

One may recall, for example, the real and antisymmetric bidiagonal (i.e., nearest-neighbor) anharmonicity $V(g) = H^{(AHO)}(g) - H^{(AHO)}(0)$ in the form of truncated and $\mathcal{PT}$−symmetrized N by N matrices as introduced in Ref. (54). Once we omit, for the sake of brevity, the discussion of the cases with odd matrix dimensions $N = 2J + 1$, we arrive at the series of tilded real Hamiltonian matrices with $N - 1 = 2J - 1$ free parameters,

$$\tilde{H}^{(2)}_{(a)} = \begin{bmatrix} 1 & a \\ -a & 3 \end{bmatrix}, \quad \tilde{H}^{(4)}_{(a,b,c)} = \begin{bmatrix} 1 & b & 0 & 0 \\ -b & 3 & a & 0 \\ 0 & -a & 5 & c \\ 0 & 0 & -c & 7 \end{bmatrix}, \dots .$$

For illustration purposes, let us further shift the energy scale and impose an additional symmetry upon perturbations. The resulting untilded operators $H^{(2J)}$ will then depend on J real parameters,

$$H^{(2)} = \begin{bmatrix} -1 & a \\ -a & 1 \end{bmatrix}, \quad H^{(4)} = \begin{bmatrix} -3 & b & 0 & 0 \\ -b & -1 & a & 0 \\ 0 & -a & 1 & b \\ 0 & 0 & -b & 3 \end{bmatrix}, \dots . \tag{1.3.3}$$

At any positive integer J, the key advantage of the model is that after an ample use of MAPLE-assisted symbolic-manipulation algebra (cf. (82) for technical details), one can qualitatively determine the shape of the boundaries $\partial\mathcal{D}^{(J)}$ of the compact J−dimensional real domains $\mathcal{D}^{(J)}$ of parameters $a, b, \ldots$ for which the spectrum of energies remains real and nondegenerate.

This feature makes the model truly exceptional. The computer-assisted form of its solvability makes the model particularly appealing for applications as it enables us to realize multiple alternative scenarios of the loss of the reality of the spectrum (i.e., of an onset of instability) via the whole family of nonnumerical benchmark explicit realizations (cf. (62) for more details and for a more extensive discussion).

1.3.3 Qualitative Innovations of Phenomenological Quantum Models

1.3.3.1 Exceptional Points as Points of a Quantum Horizon One of the most remarkable mathematical aspects of the sequence (1.3.3) of the real-matrix N by N toy-model Hamiltonians may be seen in the feasibility of a partially nonnumerical description of their spectra, especially near their exceptional-point degeneracies (54, 82). In this context, a typical result of symbolic-manipulation analysis is an "optimal," λ-parametric, and dimension-dependent reparametrization $a \to A$, $b \to B$, … yielding the same sequence of matrices in new form, namely,

$$H^{(2)}_{[A]}(\lambda) = \left[\begin{array}{cc} -1 & \sqrt{1-A\,\lambda} \\ -\sqrt{1-A\,\lambda} & 1 \end{array}\right], \quad H^{(4)}_{[A,B]}(\lambda) =$$

$$= \left[\begin{array}{cccc} -3 & \sqrt{3}\sqrt{\lambda'-B\lambda^2} & 0 & 0 \\ -\sqrt{3}\sqrt{\lambda'-B\lambda^2} & -1 & 2\sqrt{\lambda'-A\lambda^2} & 0 \\ 0 & -2\sqrt{\lambda'-A\lambda^2} & 1 & \sqrt{3}\sqrt{\lambda'-B\lambda^2} \\ 0 & 0 & -\sqrt{3}\sqrt{\lambda'-B\lambda^2} & 3 \end{array}\right]$$

(where we abbreviated $\lambda' = \lambda'(\lambda) = 1 - \lambda$), etc.

In the weakly non-Hermitian dynamical regime with small off-diagonal elements $a, b, \ldots$, the spectra of eigen-energies remain safely real. In contrast, the spectra of the more strongly non-Hermitian matrices $H^{(2J)}_{[A,B,\ldots,Z]}(\lambda)$ may be expected to complexify at some critical couplings—take just $J = 1$ with energies $E_\pm = \pm\sqrt{1-a^2} = \pm\sqrt{A\lambda}$ for illustration.

By analogy, one may expect (and, if asked for, prove (82)) that at any dimension $N = 2J$ the complexifications are always encountered during a decrease in the apparently redundant auxiliary real parameter $\lambda > 0$. In this manner, one can determine the $(J-1)$-dimensional exceptional-point boundary $\partial\mathcal{D}^{(2J)}$ of the J-dimensional "physical" domain $\mathcal{D}^{(2J)}$ of parameters $A, B, \ldots, Z$ for which the spectrum stays real (i.e., in which the time evolution of the quantum model in question remains unitary).

After one exempts the $J = 1$ model as oversimplified, one may conclude that in order to determine the points of the horizon of quantum stability (i.e., of the hypersurface $\partial\mathcal{D}^{(2J)}$) at arbitrary dimension $N = 2J$ (*mutatis mutandis*, analogous conclusions apply to odd $N = 2J + 1$ (82)), one only has to evaluate the Kato's exceptional points $\lambda^{(\mathrm{EP})} = \lambda^{(\mathrm{EP})}(A, B, \ldots, Z)$ as functions of the J new variable couplings $A, B, \ldots, Z$.

The main purpose and benefit of the above-outlined sophisticated though still fully nonnumerical reparametrization of the model is connected with the heavily computer-assisted guarantee (54) that a complete confluence (plus subsequent complexification) of *all* of the $2J$ real eigenvalues E takes place at the vanishing uppercase couplings $A = B = \ldots = Z = 0$ *plus* at the vanishing "extreme" exceptional point $\lambda^{(EEP)} = 0$.

In the language of quantum phenomenology, the latter extreme exceptional point may be perceived as representing a quantum catastrophe or collapse, resembling the quantum versions of phenomena such as Big Bang or Big Crunch in quantum gravity (see Ref. (62) for more details). After a return to mathematics and to the questions of geometry of the space of original parameters $a, b, \ldots, z$, one may conclude that at any dimension $N = 2J$, the quantum horizon has the shape of the hypersurface of a hypercube with protruded vertices.

An analysis of the geometry of these hypersurfaces near their full-confluence EEP vertices is feasible nonnumerically. At any dimension $N = 2J$ and at the sufficiently small λ, there exist $\lambda \to 0$ limits of certain two smooth real functions $\mu_N^2(\lambda)$ and $\nu_N^2(\lambda)$ such that the *complete* reality of the spectrum is guaranteed by the *single* inequality sampled by the following first few relations

$$-\mu_4^2(0) \le 2A/2 - B \le +\nu_4^2(0)\,, \quad N = 4,$$

$$-\mu_6^2(0) \le 6A/2 - 4B + C \le +\nu_6^2(0)\,, \quad N = 6,$$

$$-\mu_8^2(0) \le 20A/2 - 15B + 6C - D \le +\nu_8^2(0)\,, \quad N = 8,$$

and so on, with $\mu_4^2(0) = 1/4$ and $\nu_4^2(0) = 4/9$, and so on (cf. (82) for more details).

1.3.3.2 The Feshbach's Effective Hamiltonians Revisited In one of the areas in which the old and new ideas merged, the applicability of non-self-adjoint operators has been sought and found in an immediate transfer of the concept of evolution in time from the usual physical Hilbert space $\mathcal{H}^{(P)}$ in which the evolution is unitary to its alternative $\mathcal{H}^{(F)}$ in which the unitarity would be violated. It should be pointed out that one of the most natural visualizations of such a transfer is provided by the above-mentioned Feshbach's reduction of $\mathcal{H}^{(P)}$ to a subspace $\mathcal{H}^{(R)}$ playing the role of the false space $\mathcal{H}^{(F)}$.

For one of the most immediate applications of the idea of the loss of Hermiticity of Hamiltonians during space-reduction $\mathcal{H}^{(P)} \to \mathcal{H}^{(R)}$, we may return to letter (55). Its authors explained the impact of the reduction of spaces on the measurable properties of the so-called quantum brachistochrone, the name of which derives from the Greek "brachistos" (= the shortest) and "chronos" (= time). The core of the explanation lied in the fact that the full Hilbert space $\mathcal{H}^{(P)}$ may be perceived as carrying additional degrees of freedom, a dynamical coupling to which may be made responsible for the loss of Hermiticity (as well as for its multiple consequences) inside subspace $\mathcal{H}^{(R)}$.

In another, more traditional illustration of the space-reduction scenario, one may speak about the so-called open quantum systems in which, typically, the products of a decay of an unstable system simply move out of the reduced Hilbert space. Naturally, the instability gets immediately removed via a return to the full space.

Whenever one decides to stay inside the reduced, false space $\mathcal{H}^{(F)} = \mathcal{H}^{(R)}$, the reference to the above-cited Feshbach's projector-operator isospectrality may remain just vague and implicit during the concrete approximative calculations. In applications to atomic and molecular systems, the underlying reduction-by-projection simplifications may be found in the background of multiple approximate calculations (cf. their detailed account in (33)). As a conceptual guide to qualitative analyses of resonance and various other phenomena emerging in unstable atomic nuclei, precisely the same idea may also be used (cf., e.g., review paper (83) giving further references).

As already mentioned, the popularity of the latter type of correspondence climaxed in 2011 when, for all of the above-listed reasons, the specialists in the study of open systems and the models with effective Hamiltonians $H = H_{eff}(E)$ happened to form the majority of participants of the tenth non-self-adjoint-operator conference PHHQP 2011 (34) in Dresden. The extension of the phenomenological scope to complex-energy quantum systems proved clearly beneficial because it reattracted the attention of theoreticians to the analysis of the behavior of quantum systems near the horizons of their stability (84).

In the subsequent years, the last crisis emerged but it did not hit the models using effective Hamiltonians $H = H_{eff}(E)$ because their phenomenological interpretation does not require any construction of the metric Θ at all. As a consequence, many people started paying attention to the deeper analysis of the phase transitions connected with the spontaneous breakdown of $\mathcal{PT}$-symmetry (cf. the following sections in this respect).

For illustration of the possible (although, up to now, not too much explored) survival of the overlap between quantum model-building strategies based on the traditional (i.e., effective, trivial-metric-requiring) and on the $\mathcal{PT}$-symmetric (i.e., non-self-adjoint, nontrivial-metric-requiring)), Hamiltonians H let us now recall Ref. (85). The puzzling energy dependence of the effective Hamiltonian $H_{eff}(E)$ has been reanalyzed there via its possible rearrangement and rigorous transformation to the two alternatives of the effective Hamiltonian denoted as mappings $H_{eff}(E) \leftrightarrow K \neq K^{\dagger}$ and $H_{eff}(E) \leftrightarrow L \neq L^{\dagger}$.

The main conclusion was that one can make both of the new isospectral partner Hamiltonians non-self-adjoint but, by construction, energy-independent, $K \neq K(E)$ and $L \neq L(E)$. The nonlinear eigenvalue problem (1.1.1) has been, in this manner, linearized. What is important is that the change did not destroy the immanent, phenomenologically appealing possibility of the loss of the reality of the bound-state energies E at the boundary $\delta \in \mathcal{D}$ of the domain of the stability-supporting variable parameters.

1.3.3.3 Systems Exhibiting Nonlinear Supersymmetry The elementary and exactly solvable non-self-adjoint harmonic oscillator (1.3.1) has been found to play an interesting role in the framework of the so-called supersymmetric (SUSY) quantum mechanics (86). Initially, people found it impossible to parallel, strictly, the existing self-adjoint constructions and to implement any non-self-adjoint operator into the usual SUSY-representation pattern. Nevertheless, a partial success has been achieved after a transition to the algebras of the so-called nonlinear SUSY *alias* second-order SUSY (SSUSY, (87, 88)). In the latter, broadened framework, it appeared feasible to generalize the well-known correspondence between the exact solvability of Schrödinger equation (1.1.2) and a SUSY-related property of shape invariance of the corresponding potential $V(x)$. The results of these efforts may be found summarized, for example, in paper (39).

In such a context, potential (1.3.1) reacquired the status of one of the simplest special cases of the whole shape-invariant and/or SSUSY-algebra-related family of solvable interactions. For this reason, it makes sense to return to this example, in the new SSUSY context, in more detail. Skipping the majority of technicalities and physics interpretations that may be, after all, found explained in Refs (88) and (89), let us merely display the most common SUSY generators formed by Hamiltonian $\mathcal{G}$ and by the two usual SUSY charges,

$$\mathcal{G} = \begin{bmatrix} H_{(\text{Left})} & 0 \\ 0 & H_{(\text{Right})} \end{bmatrix}, \quad \mathcal{Q} = \begin{bmatrix} 0 & 0 \\ A & 0 \end{bmatrix}, \quad \tilde{\mathcal{Q}} = \begin{bmatrix} 0 & B \\ 0 & 0 \end{bmatrix} \tag{1.3.4}$$

(cf., e.g., eqs Nr. (15)–(17) in Ref. (86), respectively). Once we further postulate that $H_{(\text{Left})} = B \cdot A$ and $H_{(\text{Right})} = A \cdot B$, the triplet of operators (1.3.4) will generate, by construction, a representation of SUSY algebra sl(1/1) based on the following linear anticommutator and commutator relations,

$$\{\mathcal{Q}, \tilde{\mathcal{Q}}\} = \mathcal{H}, \qquad \{\mathcal{Q}, \mathcal{Q}\} = \{\tilde{\mathcal{Q}}, \tilde{\mathcal{Q}}\} = 0, \qquad [\mathcal{H}, \mathcal{Q}] = [\mathcal{H}, \tilde{\mathcal{Q}}] = 0.$$

This representation may be admitted non-self-adjoint. For this purpose, we have to return to the above-introduced complex quasi-coordinates $q \in \mathbb{C}$ and to define, formally, the first-order linear differential operators $A = \partial_q + W$ and $B = -\partial_q + W$ in terms of a one-parametric complex function called superpotential,

$$W = W^{(\gamma)}(q) = q - \frac{\gamma + 1/2}{q}, \qquad q = q(x) = x - \mathrm{i}c\,, \qquad \gamma, x, c \in \mathbb{R}\,. \tag{1.3.5}$$

Once we denote

$$H^{(\xi)} = -\frac{d^2}{dq^2} + \frac{\xi^2 - 1/4}{q^2} + q^2$$

we may verify that

$$H^{(\gamma)}_{(\text{Left})} = H^{(\alpha)} - 2\gamma - 2, \qquad H^{(\gamma)}_{(\text{Right})} = H^{(\beta)} - 2\gamma, \tag{1.3.6}$$

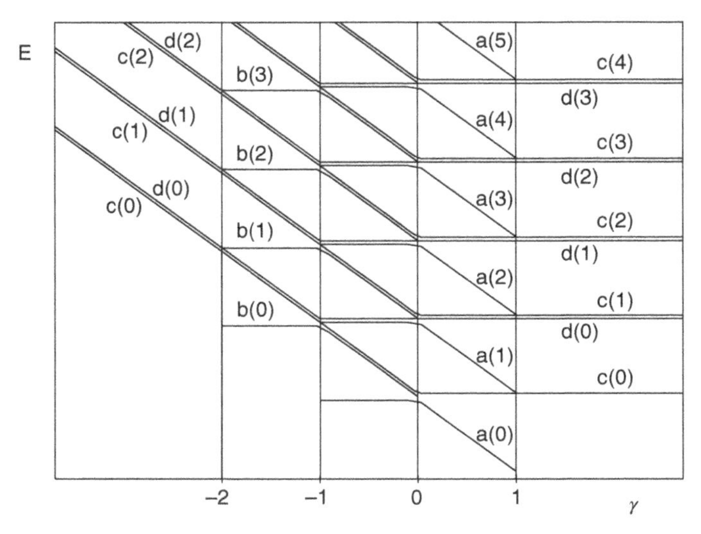

Figure 1.2 The spectrum of the non-self-adjoint harmonic-oscillator SUSY Hamiltonian $\mathcal{G}$ of eqs (1.3.4) and (1.3.6).

provided only that we define quantities $\alpha = \alpha(\gamma) = |\gamma|$ and $\beta = \beta(\gamma) = |\gamma + 1|$ as nonnegative functions of the real input parameter γ.

As long as our benchmark SUSY Hamiltonian $\mathcal{G}$ of eq. (1.3.4) is a direct sum of two non-self-adjoint harmonic oscillators (1.3.6), the determination of its spectrum (cf. Fig. 1.2) is straightforward. The discussion of its properties leads to interesting and unusual observations. For example, we may always treat this spectrum as an infinite sequence of quadruplets of energies that share the main index $n = 0, 1, \ldots$ and that may differ just by their superscripts α or β and by their quasi-parities,

$$a(n) \le b(n) \le c(n) \le d(n)\,. \tag{1.3.7}$$

Naturally, the usual and well-known self-adjoint special case reemerges here at the unique value of $\gamma = -1/2$ for which the numbers $\alpha = \beta = 1/2$ become equal. In review (86) (see, in particular, Section 2.1 in *loc. cit.*), the corresponding self-adjoint harmonic-oscillator SUSY model with $\alpha = \beta = 1/2$ is called "linear SUSY."

At any other $\gamma \neq -1/2$, the model ceases to be self-adjoint. Thus, at every main quantum number n, we may reinterpret the four corresponding eigen-energies as an ordered quadruplet (1.3.7). We may recall the discussion in Ref. (89) and conclude that sufficiently far from the self-adjoint special case with $\gamma = -1/2$, that is, more precisely, out of the interval of $\gamma \in (-1, 0)$, some of the energy levels of eq. (1.3.7) cease to exist.

The pattern of such a disappearance of states is made visible in Fig. 1.2. Your count of the levels is guided there by a small artificial shift of one of the levels in the cases of the well- known SUSY-related degeneracies[5]. After the thorough inspection of the picture, we may conclude that in contrast to the dominant levels $c(N)$ and $d(N)$ that correspond to the normalizable bound states at any real $\gamma \in \mathbb{R}$, the existence of the remaining two levels $b(N)$ and $a(N)$ is restricted to the mere finite intervals of $\gamma \in (-2, 0)$ and $\gamma \in (-1, 1)$, respectively.

The phenomenon of the disappearance of levels itself is caused by the loss of normalizability of wave functions at some integer values of γ. The critical phenomenon of the disappearance always involves infinitely many states, but such a "phase transition" only occurs at one of the four privileged forbidden values of $\gamma = -2$, $\gamma = -1$, $\gamma = 0$, and $\gamma = 1$.

1.4 PROBABILISTIC INTERPRETATION OF THE NEW MODELS

1.4.1 Variational Constructions

One of the typical features of the differential operators A and B of eq. (1.3.4) is that they are unbounded. In principle, this would entail the necessity of a very careful specification of the domains on which they are defined. For this reason, it has been proposed, in the very first review (24) of the THS formalism by Scholtz et al., that whenever possible, our attention should remain restricted just to the bounded-operator Hamiltonians $H \in \mathcal{B}(\mathcal{H})$ or even to the mere truncated, N-dimensional matrix representations of all of the operators of quantum observables.

1.4.1.1 Nuclear Physics Example Under the above-mentioned constraints, many mathematical challenges would either disappear or become inessential. In particular, whenever the finite matrix-truncation dimensions $N < \infty$ are used, the sophisticated formalism of functional analysis may be replaced by the mere linear algebra. This also seems to explain why the first (viz., variational) applications of the THS strategy to heavy nuclei took place perceivably earlier that in field theory.

In the $\mathcal{PT}$-symmetry studying community, a return of attention to the bounded-operator or even linear-algebraic mathematical origins came rather late. One of the important by-products of this simplification of technicalities may be seen in the subsequent opening of communication channels between different areas of physics and, in particular, in a rediscovery of the natural factorizability of the metrics, $\Theta = \Omega^\dagger\Omega$.

The introduction of the factor operator Ω (which must be, in all of the nontrivial cases of our present interest, nonunitary) is usually attributed to Freeman Dyson (see one of the footnotes earlier). In the context of nuclear physics, Janssen et al. (90) were probably the first who imagined that the nonunitarity assumption $\Omega^\dagger \neq \Omega^{-1}$ opens a truly unexplored space for possible simplification of the computing.

Let us recollect that the Dyson-inspired nuclear physics concrete choice of Ω was in fact just an intuition-guided mapping between the most common Hilbert space

[5]The illustration was prepared as a part of Ref. (89). Unfortunately, the Figure itself was not included in the printed version of *loc. cit.*, due to an inadvertent technical omission by the Editors.

$\mathcal{H}^{(P)}$ of protons and neutrons (forming a heavy atomic nucleus—the superscript $^{(P)}$ hints that the space is "physical" as well as "primary") and another, very different (and, incidentally, much friendlier!) bosonic representation space $\mathcal{H}^{(F)}$.

In the latter space, one works with preselected and, presumably, more relevant effective degrees of freedom carried by quasi-particles called "interacting bosons" (IB). In contrast to the naive and prohibitively complicated form of the initial, strictly microscopic and realistic Hamiltonian (for the sake of clarity let us denote it by the "dedicated" lower-case symbol $\mathfrak{h}$), its isospectral IB partner $H = \Omega^{-1}\mathfrak{h}\Omega$ constructed via an educated guess of Ω proved, in many a respect, simpler after truncation. One should not be too surprised because the very nature of the dynamical input information about the nucleon–nucleon forces is such that it patches the fermionic nucleons to form certain correlated pairs. *A priori* these pairs may be expected to exhibit bosonic features.

The practical numerical tests of the Dyson-inspired IB models of atomic nuclei confirmed the expectations and were successful. The non-unitary-mapping-mediated transition to a definitely friendlier bosonic Hilbert space rendered the calculation of certain spectra of heavy nuclei feasible, easier and usually much more quickly convergent in the realistic limit of matrix dimension $N \to \infty$.

1.4.1.2 Quantum Theory Using Bounded Observables One of the paradoxes which resulted from the lack of a sufficiently intensive communication between mathematicians and physicists manifested itself in a deplorable mutual incompatibility of the respective vocabularies. Besides the above-mentioned parallelism and duplicity between the concepts of $\mathcal{PT}$-symmetry in quantum physics and Krein-space Hermiticity in operator-theory mathematics, another hidden source of possible misunderstandings may be found in the use of term "quasi-Hermiticity." It carries, traditionally, an entirely different meaning in mathematics and physics. Thus, in the light of the definition as introduced, in 1960, by Dieudonné (91), the quasi-Hermiticity of H need not imply the quasi-Hermiticity of its adjoint $H^\dagger$ (or, as mathematicians would write it, H^*). In the context of physics, on the contrary, such an implication is *always* required and guaranteed. Owing to the purpose-adapted and better-designed definition as promoted by Scholtz et al. (24), nuclear physicists assume that the quasi-Hermitian quantum observable H is a *bounded* operator.

Using this basic assumption, review paper (24) presented an overall theoretical framework for an update of the unitary quantum theory guaranteeing the consistent use of non-self-adjoint representations of observable quantities. We already emphasized above that in the subsequent preparatory step of the THS representation theory, once a *given* Hamiltonian H (or possibly another quantum observable represented by a non-self-adjoint operator $Q \neq Q^\dagger$ with real spectrum (92)) appears non-Hermitian in a *given* Hilbert space $\mathcal{H}^{(F)}$, the latter a space must be declared ill-chosen and unphysical, manifestly "false."

Let us now add that in practice one preserves the topological vector space $\mathcal{H}^{(F)}$ as the *exclusive* working space. In other words, the "correct" Hilbert space $\mathcal{H}^{(S)}$ is solely present *via* its representation in $\mathcal{H}^{(F)}$. Thus, one works with the two metrics

$\Theta^{(F)} = I$ and $\Theta^{(S)} \neq I$ in parallel and solely in $\mathcal{H}^{(F)}$. In such a setting, it suffices to introduce just a new double-ket symbol or abbreviation $|\psi\rangle\rangle$ for the representation of the updated dual vectors $\Theta|\psi\rangle$, which are identified with the elements of the dual vector space $\left(\mathcal{H}^{(S)}\right)'$.

Unless the metric operator itself appears unbounded (which would require a more sophisticated construction (38)), one can return to popular abbreviations. Thus, the change $\mathcal{H}^{(F)} \to \mathcal{H}^{(S)}$ is then most concisely characterized as an *ad hoc* upgrade of the inner product (11),

$$\langle\psi|\chi\rangle^{(F)} \to \langle\psi|\chi\rangle^{(S)} \equiv \langle\psi|\Theta|\chi\rangle^{(F)} . \tag{1.4.1}$$

The necessary assumptions are that the symbol $\Theta = \Theta^{(S)}$ denotes a nontrivial and self-adjoint, bounded and strictly positive operator possessing a bounded inverse in $\mathcal{H}^{(F)}$.

In a pragmatic perspective of postulates of unitary quantum theory, the consistent formulation of the idea is concise. One merely stops making the measurable predictions directly in the auxiliary Hilbert space $\mathcal{H}^{(F)}$ and replaces them by the predictions containing the metric. Then, the same operator of any quantum observable (be it a Hamiltonian or not) is made acceptable and self-adjoint (in the properly amended physical representation space $\mathcal{H}^{(S)}$) and its mean values are simply always calculated between brabras and kets.

In most of the older applications of the idea, the above-mentioned mathematical condition of boundedness was not always sufficiently carefully considered. This did not necessarily imply that the construction was mathematically incorrect. For example, let us return to the "wrong-sign" quartic anharmonic oscillator Hamiltonian $H \neq H^\dagger$ of Buslaev and Grecchi (12) for which these authors found the manifestly self-adjoint "correct-sign" quartic-oscillator double-well isospectral partners $\mathfrak{h} = \mathfrak{h}^\dagger$ without any recourse to the metric.

We are now prepared to summarize the message: The abstract quantum theory using *bounded* non-Hermitian representations of its operators of observables may be perceived as a well-defined and mathematically consistent foundation of the standard quantum picture of reality. The trick is that the specific three-Hilbert-space recipe based on the use of a nontrivial metric $\Theta \neq I$ enables us to treat the "first" Hilbert space $\mathcal{H}^{(F)}$, simultaneously, as most friendly as well as manifestly unphysical. And while the advantage is simply kept unchanged, the shortcoming is easily suppressed by the metric-mediated redefinition $\mathcal{H}^{(F)} \to \mathcal{H}^{(S)}$ yielding the metric-equipped second space $\mathcal{H}^{(S)}$, which is, by construction, unitarily equivalent to the primary $\mathcal{H}^{(P)}$.

As long as we very often choose space $\mathcal{H}^{(F)}$ in the most common form $L^2(\mathbb{R}^d)$ of the space of the quadratically integrable wave functions in d dimensions (69), a truly unpleasant increase in the complexity of the necessary constructive mathematics may be expected to follow from the transition to $\mathcal{H}^{(S)}$ (i.e., to the complicated metrics Θ) in general.

Last but not least, also the development of reliable and efficient techniques of proving the reality (i.e., the observability) of the spectrum of preselected Hamiltonians H (which used to be, for any self-adjoint $H = H^\dagger$ in $L^2(\mathbb{R}^d)$, trivial) becomes a serious technical obstacle (77). In order to support such a necessity of search for updates of mathematical methods, once more, the nice older study of the imaginary cubic anharmonicities by Caliceti et al. (73) may be recalled as an illustrative example: The reality of the energy levels has been shown there to follow from a clever modification of the usual perturbation-expansion techniques[6].

1.4.2 Non-Dirac Hilbert-Space Metrics $\Theta \neq I$

1.4.2.1 Hilbert Space $\mathcal{H}^{(S)}$ Many non-Hermitian Hamiltonians $H \neq H^\dagger$ as sampled above are, strictly speaking, merely non-Hermitian in $\mathcal{H}^{(F)}$ but not in $\mathcal{H}^{(S)}$. In any case, these operators must be required diagonalizable (11). In the amended Dirac's notation of Ref. (23), this means that one has to solve two rather than one time-independent Schrödinger equation,

$$H\,|\psi_n\rangle = E_n\,|\psi_n\rangle\,, \qquad H^\dagger\,|\psi_m\rangle\!\rangle = E_m^*\,|\psi_m\rangle\!\rangle\,, \quad m,n = 0,1,\ldots\,. \tag{1.4.2}$$

For finite, $(N+1)$-dimensional Hilbert spaces $\mathcal{H}^{(F)}$ and for the Hamiltonian matrices H with real, nondegenerate and discrete spectrum[7], this implies that the respective vectors $|\psi_n\rangle$ and $|\psi_m\rangle\!\rangle$ are mutually orthogonal in $\mathcal{H}^{(F)}$ for $m \neq n$. After a suitable rescaling these eigenvectors of the two respective operators in (1.4.2) will form a biorthonormal basis in $\mathcal{H}^{(F)}$. The following two spectral formulae become available,

$$I = \sum_{n=0}^{N} |\psi_n\rangle\langle\!\langle\psi_n|\,, \qquad H = \sum_{n=0}^{N} |\psi_n\rangle E_n\,\langle\!\langle\psi_n|\,. \tag{1.4.3}$$

Under the same assumptions, we introduce an auxiliary $(N+1)$ by $(N+1)$ matrix, which is $(N+1)$–parametric, invertible, positive-definite and Hermitian,

$$\Theta = \sum_{n=0}^{N} |\psi_n\rangle\!\rangle \kappa_n^2 \langle\!\langle\psi_n| = \Theta^\dagger > 0\,. \tag{1.4.4}$$

Once we interpret this matrix as a metric operator, we obtain a new, "standard" Hilbert space $\mathcal{H}^{(S)}$. In this $(N+1)$-dimensional space, the Hermitian conjugation of any operator Λ becomes defined in terms of the given metric as follows,

$$\mathcal{T}^{(S)}_{(oper.)} : \Lambda \to \Lambda^\ddagger := \Theta^{-1}\,\Lambda^\dagger\,\Theta\,. \tag{1.4.5}$$

Thus, in the amended space $\mathcal{H}^{(S)}$, the habitual Hermitian conjugation

$$\mathcal{T}^{(F)}_{(oper.)} : \Lambda \to \Lambda^\dagger \tag{1.4.6}$$

as defined in $\mathcal{H}^{(F)}$ must be replaced by the amended definition (1.4.5).

[6]Chapter 4 of the present book should be read as a topical review of the freshmost updates of these efforts.
[7]For more general cases, a few relevant comments may be found in (93).

The main consequence of the introduction of metric (1.4.4) is that Hamiltonian H becomes, by construction, self-adjoint in $\mathcal{H}^{(S)}$. The new Hilbert space $\mathcal{H}^{(S)}$ may be declared physical. We have $H = H^{\ddagger}$ so that in the new space the time evolution generated by H becomes unitary. In parallel, the original, "friendly" Hilbert space $\mathcal{H}^{(F)}$ in which the time evolution generated by H remains manifestly nonunitary must be declared unphysical and "false."

On this linear-algebraic methodical background, one may decide to move to the more usual separable, infinite-dimensional Hilbert spaces. In some cases, it proves sufficient to set simply $N = \infty$ in the above-listed formulae. Unfortunately, there exist only too many models in which much more mathematical care is needed and necessary. Perhaps, this is one of the key reasons why the present book had to be written.

1.4.2.2 The Reasons for Factorization of the Metrics Let us recall the family of quantum models exhibiting $\mathcal{PT}$-symmetry *alias* parity-pseudo-Hermiticity $H^{\dagger}\mathcal{P} = \mathcal{P}H$. For them, we immediately see the link between the empirical observation of the reality of their spectra and a general mathematical framework offered by the THS formalism. In one direction, the reality of the spectrum gets explained as a consequence of the Hermiticity of H in some not yet known Hilbert space $\mathcal{H}^{(S)}$. In opposite direction, the knowledge of the Dyson-type mapping between upper-case operator H and lower-case operator $\mathfrak{h}$ leads immediately to the identification of the metric Θ with the superposition of the Dyson's mapping with its conjugate,

$$\Theta = \Omega^{\dagger}\Omega\,. \tag{1.4.7}$$

In the unitary-mapping extreme, one would have $\Omega^{\dagger} = \Omega^{-1}$ so that the whole THS-represented quantum theory would degenerate to its textbook special case in which one never needs to work with any auxiliary unphysical Hilbert space $\mathcal{H}^{(F)}$ and with any non-self-adjoint representation H of the Hamiltonian.

The overall belief in applicability of nontrivial metrics Θ proved amazingly productive in a number of phenomenological and heuristic considerations. The approach helped to clarify, first of all, the consistent probabilistic tractability of the relativistic quantum-mechanical Klein–Gordon systems with spin zero (44). Secondly, formal analogs of Klein–Gordon systems were found in the gravity-quantizing Wheeler-De-Witt equations (43). Finally, very similar conclusions were achieved for the less known Proca systems with spin one (94), and so on.

For a number of $\mathcal{PT}$-symmetric systems, a key to the progress in applications was found in a simplification of the factorization of the metric in which it has been assumed that $\Theta = \Omega_s^{\dagger}\Omega_s$ where $\Omega_s = \Omega_s^{\dagger}$. This assumption may be reread as formula $\Omega_s = \sqrt{\Theta}$ that defines a special Dyson map Ω_s and removes a part of the ambiguity from the correspondence between Hilbert spaces $\mathcal{H}^{(F)}$ and $\mathcal{H}^{(P)}$ for a given Hamiltonian H.

1.5 INNOVATIONS IN MATHEMATICAL PHYSICS

1.5.1 Simplified Schrödinger Equations

From the knowledge of the triplet of operators H, Θ, and Ω, one may reconstruct the primitive physical Hamiltonian $\mathfrak{h} = \Omega^{-1} H \Omega$, which is assumed prohibitively complicated (otherwise, the whole THS machinery would not be needed at all). The latter operator is still, by construction, kept self-adjoint in its own Hilbert space, that is, $\mathfrak{h} = \mathfrak{h}^\dagger$ in $\mathcal{H}^{(P)}$. As long as the inner product is postulated trivial in $\mathcal{H}^{(P)}$, the mathematically and computationally much more complicated representation $\mathfrak{h}$ of the Hamiltonian often offers, paradoxically, the most natural background for the physical interpretation of the results of calculations. The vital point of the whole THS recipe is that two Hilbert spaces $\mathcal{H}^{(S)}$ and $\mathcal{H}^{(P)}$ are unitarily equivalent. This makes them indistinguishable from the point of view of physical predictions so that one can work in more friendly $\mathcal{H}^{(S)}$.

1.5.1.1 Starting from an Overcomplicated, Hermitian Schrödinger Equation As we already mentioned, the general abstract pattern of quantum theory as it emerges from such a THS arrangement found its important application in nuclear physics (24). The implementation of the ideas started from an introductory step in which one defines effective non-self-adjoint matrix $H \neq H^\dagger$ (of any dimension $N + 1 \leq \infty$) as a simplified isospectral partner of a realistic input (i.e., known but prohibitively complicated) Hamiltonian $\mathfrak{h}$. As long as the details of the construction may be found in Ref. (24), let us merely summarize the pattern as the following THS flowchart,

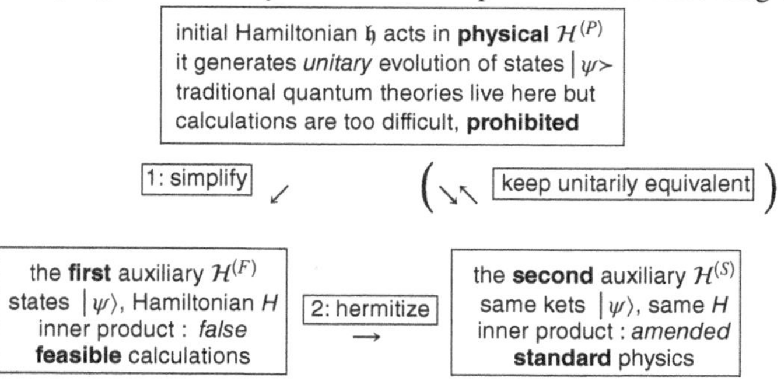

It is necessary to add that while $\mathcal{H}^{(F)}$ and $\mathcal{H}^{(S)}$ were, in the nuclear physics implementation, the perceivably simpler Hilbert spaces of bosons (called "interacting bosons" in the related extensive literature), the "initial" space $\mathcal{H}^{(P)}$ was the most common Hilbert space of fermionic nucleons. For this reason, the definition of the "prohibitively complicated" Hamiltonian $\mathfrak{h}$ was more or less routine (i.e., based on the principle of correspondence). In contrast, the choice of the innovative, *nonunitary* form of the mapping Ω between spaces (defined via formula $|\psi\succ = \Omega\,|\psi\rangle$) required a truly ingenious insight in the underlying physics.

1.5.1.2 Starting Directly from a Non-Hermitian Schrödinger Equation In a broader methodical perspective, the success of simplification $\mathfrak{h} \to H = \Omega\,\mathfrak{h}\,\Omega^{-1}$ may be rephrased as an acceptability of non-Hermiticity during practical model building. As long as the main purpose of the mapping lies in the feasibility of the determination of the spectrum, one can also proceed in opposite direction, trying to deal with a given Hamiltonian operator $H \neq H^\dagger$. This operator living in an auxiliary Hilbert space $\mathcal{H}^{(F)}$ should only be assumed maximally elementary, at the expense of being allowed manifestly non-self-adjoint.

Such an operator H may then be related, via a suitable *family of the operators* $\Omega = \Omega(H)$, to a *multiplet* of its eligible isospectral partners $\mathfrak{h} = \mathfrak{h}^\dagger$. The individual members of the latter family live in the respective metric-dependent (i.e., different) "third" Hilbert spaces $\mathcal{H}^{(T)}$. The overall THS flowchart should be rearranged and redirected as follows:

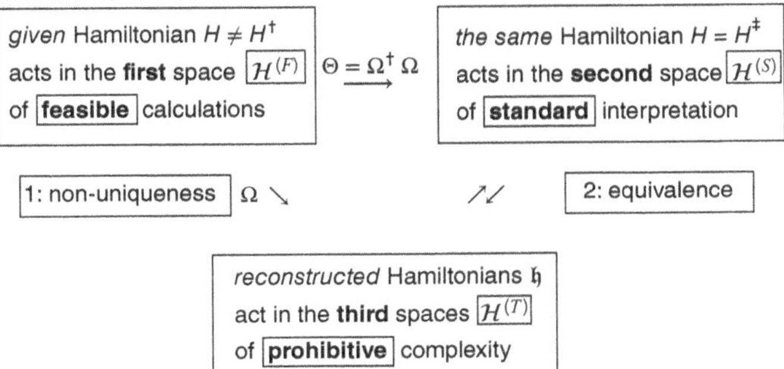

Just one of the reconstructed "third" scenarios should be selected as representing the expected observable physical phenomena, $\mathcal{H}^{(T)} \to \mathcal{H}_0^{(T)} \equiv \mathcal{H}^{(P)}$. The ambiguity of the mapping Ω to the third space implies the nonequivalence of the alternative physical interpretations compatible with the initial information about the quantum system in question. Formally, all of the options are equally acceptable while admitting very broad variability of their phenomenological and observable consequences. More information about the system must necessarily be provided (see the extensive discussion of this point in Ref. (24)).

1.5.1.3 The Concept of Charge C An introduction of a *privileged and unique* physical operator $\mathfrak{h}_0$ assigned to a given toy-model Hamiltonian H belongs to the most impressive merits of the very specific $\mathcal{PT}$-symmetric quantum theory as summarized by Bender (10). This formalism makes *two additional* assumptions by asking for the existence of a charge C (yielding the unique special metric $\Theta_0 = \mathcal{P}C$) and for the unique self-adjoint specification $\Omega_{s,0} = \sqrt{\Theta_0}$ of the preferred Dyson-type mapping. These two physics-determining postulates imply a complete suppression of the ambiguity of the reconstruction of the isospectral self-adjoint representation $\mathfrak{h}_0$ of the Hamiltonian for the conservative systems and in the finite-dimensional Hilbert spaces at least[8].

The essence of the uniqueness of the reconstruction $H \to \mathfrak{h}_0$ in $\mathcal{PT}$-symmetric quantum mechanics is closely connected to another, implicit assumption that a generic $\mathcal{PT}$-symmetric quantum system with Hamiltonian $H = p^2 + V(x)$ lives, in general, on a *complex* curve of coordinates x. In other words, the values of x cannot

[8]cf. following paragraph.

certainly be treated as eigenvalues of an operator, Hermitian or crypto-Hermitian, of an observable, anymore. Thus, the concept of locality is lost. For compensation, in a way inspired by quantum field theory, one introduces the $\mathcal{CPT}$-symmetry of H *plus* the observability status of the operator C called charge. Review paper (10) may be recommended as one of the most exhaustive physics-oriented comprehensive accounts of the related technical and interpretation details.

1.5.2 Nonconservative Systems and Time-Dependent Dyson Mappings

In the THS notation of review (23), one marks each element ϕ of the primary physical Hilbert space by the "spiked" ket symbol, $|\phi\succ \in \mathcal{H}^{(P)}$. In parallel, the representants of the same state in both the first and second auxiliary Hilbert spaces may remain marked by the same, usual ket symbol, $|\phi\rangle \in \mathcal{H}^{(F,S)}$. The conjugate or bra-vector symbols retain their traditional forms just in the two spaces, with $\prec\phi| \in \left(\mathcal{H}^{(P)}\right)'$ and with $\langle\phi| \in \left(\mathcal{H}^{(F)}\right)'$. Thus, the notation convention only remains unclear for the dual *alias* Hermitian-conjugate elements within the second physical Hilbert space $\mathcal{H}^{(S)}$.

In the current THS-related literature, the apparent puzzle is easily resolved by keeping the latter Hilbert space "absent," that is, *exclusively* present *just* via its representation in $\mathcal{H}^{(F)}$. For the purpose, one merely changes, whenever necessary, the inner product. In other words, the transitions $\mathcal{H}^{(F)} \leftrightarrow \mathcal{H}^{(S)}$ are mediated by the mere replacements of the metric operators, $\Theta^{(F)}$ $(= I = \Theta^{(Dirac)}) \leftrightarrow \Theta^{(S)}$ $(= \Theta \neq I)$. Thus (93), the linear functionals *alias* bra-vector elements of the third dual vector space $\left(\mathcal{H}^{(S)}\right)'$ may always be chosen and defined, inside the friendly representation space $\mathcal{H}^{(F)}$ and in terms of the *ad hoc* metric Θ, as products $\langle\phi|\Theta \in \left(\mathcal{H}^{(S)}\right)'$. In fact, while staying in $\mathcal{H}^{(F)}$, we are permitted to abbreviate the related (i.e., $F-$space conjugate) ket-vector products $\Theta|\phi\rangle$ by a dedicated double-ket symbol, $\Theta|\phi\rangle := |\phi\rangle\rangle$, with the $F-$space conjugate $\langle\langle\phi| = \langle\phi|\Theta$ as long as $\Theta = \Theta^\dagger$ by construction.

1.5.2.1 Paradox of Two Hamiltonians In the above-mentioned notation, it is very easy to extend the THS scheme to cover the cases where the Hamiltonian is allowed to be time dependent in the primary physical Hilbert space $\mathcal{H}^{(P)}$. Naturally, the prohibitively complicated time-dependent Schrödinger equation

$$\mathrm{i}\,\partial_t|\varphi(t)\succ \;=\; \mathfrak{h}(t)\,|\varphi(t)\succ \tag{1.5.1}$$

is still correct and it determines the unitary evolution of the wave functions of the system. In contrast, one encounters a thorough weakening of the physical relevance of the eigenvalues of $\mathfrak{h}(t)$ because the very concept of observable energy only retains its full physical sense and meaning in adiabatic approximation. Secondly, also the role of the general time-dependent isospectral "Hamiltonian" operator $H(t) = \Omega^{-1}(t)\mathfrak{h}(t)\Omega(t)$ becomes influenced by the emergence of time dependence in the Dyson mapping, $\Omega = \Omega(t)$.

According to Ref. (23), the combined effect of time dependence in $\mathfrak{h} = \mathfrak{h}(t)$ and in $\Omega = \Omega(t)$ only manifests itself via the non-Hermitian operator

$$G(t) = H(t) - \mathrm{i}\Omega^{-1}(t)\dot{\Omega}(t),$$

where the over-dot represents the derivative with respect to time t. Precisely this operator plays the role of the generator of the time evolution of wave functions, that is, of the ket vector and its Hermitian conjugate brabra dual vector in the "standard" physical Hilbert space $\mathcal{H}^{(S)}$. For this reason, we have to replace eq. (1.5.1) by the doublet of the time-evolution equations

$$i\partial_t|\Phi(t)\rangle = G(t)\,|\Phi(t)\rangle\,, \tag{1.5.2}$$

$$i\partial_t|\Phi(t)\rangle\!\rangle = G^\dagger(t)\,|\Phi(t)\rangle\!\rangle \tag{1.5.3}$$

represented and to be solved in the friendly Hilbert space $\mathcal{H}^{(F)}$ as usual.

1.5.2.2 Heisenberg-like Frames Naturally, also in the manifestly time-dependent scenario, one may try to proceed in opposite direction, from G to $\mathfrak{h}$. In such a case, the usual self-adjoint Hamiltonian $\mathfrak{h}(t)$ acting in a primary Hilbert space $\mathcal{H}^{(P)}$ (where, typically, the principle of correspondence may form the background of the preparation of measurements) is not known in advance. In principle, it must be reconstructed from a given and, necessarily, *particularly elementary* non-self-adjoint generator $G(t)$ defined as acting in an auxiliary, preselected but manifestly unphysical Hilbert space $\mathcal{H}^{(F)}$ via eqs (1.5.2) and (1.5.3).

The general scenario has only been discussed in unpublished preprint (95) (cf. also (96)). The simplest choice with time-independent $G(t) = G(0)$ was recently introduced, in (97), as an immediate generalization of the usual fundamental and universal Heisenberg representation of quantum systems. In such a special case, the explicit respective solutions

$$|\phi(t)\rangle = \exp(-iG(0)\,t)\,|\phi(0)\rangle\,, \qquad |\phi(t)\rangle\!\rangle = \exp(-iG^\dagger(0)\,t)\,|\phi(0)\rangle\!\rangle \tag{1.5.4}$$

of non-Hermitian Schrödinger equations (1.5.2) and (1.5.3) degenerate immediately to the common (69) Heisenberg representation rule at the trivial choice of $G^{(Heis.)}(0) = 0$.

1.6 SCYLLA OF NONLOCALITY OR CHARYBDIS OF NONUNITARITY?

The message of traditional textbooks on quantum mechanics may be read as a declaration of necessity of avoiding the nonunitarity of the evolution in time by all means including even the acceptance of a nonlocality of the interactions. This immediately reminds us about the Homer's story about the six-headed monster Scylla which, together with the all-destroying whirlpool of Charybdis controlled the pass through the Strait of Messina. No surprise that in the mythology of old Greeks the fabulous Odysseus followed the advise by Kirke and that, in order to avoid the deadly Charybdis, he rather sacrificed six of his men to Scylla while saving the rest. In this sense, the "acceptable evil" of Scylla finds its traditional quantum analog in nonlocality. The "Charybdis" of nonunitarity was perceived much more deadly a danger.

Certainly so before the year 1998. For this reason, the Bender's and Boettcher's proposal (5) of working with non-Hermitian generators H of evolution in quantum physics looked, initially, so courageous and counter-Kirke-ish. Still, even their proposal finds a support and parallels in the same old myth. In fact, the message sent by old Greeks is not as straightforward as it might have seemed since during his second trip, Odysseus changed his mind and, having kept safely out of the reach of the Scylla, he still managed to navigate out of Charybdis without any losses at all.

In the present analogy with the story of non-Hermitian Hamiltonians in quantum physics, the second message is certainly encouraging: It allows us to reclassify the Bender's and Boettcher's apparent acceptance of the Charybdis-resembling evils of nonunitarity in 1998 as a promising and well-working strategy.

Of course, it is true that a highly nonstandard care is needed to get through, especially when the most traditional local-potential interactions are used as a dynamical input. These days (i.e., the additional 15 years later, in the year 2013) an updated conventional wisdom still reconfirms that the local models need not necessarily be the most easily tractable ones.

The first signs of a return of pendulum and the first words of warning against too stubborn an insistence on the locality of interactions already appeared around the second, metric-ambiguity crisis in 2004. The threat of the Scylla-resembling nonlocality of the interaction was circumvented and appeared tractable as well. In the influential letter (28), the status of the main energy-complementing physical observable has been transferred from the coordinate x to another quantity C carrying all of the characteristics of a *global, delocalized* charge.

On a higher level of analysis, the dilemma of the choice between the Scylla of nonlocality and the Charybdis of nonunitarity has been reopened. It also reappeared at the very core of the third crisis in 2007 and the fifth crisis in 2013.

1.6.1 Scattering Theory

In the years 1998–2007, we witnessed a virtually uninterrupted success of physics-oriented studies of $\mathcal{PT}$-symmetric quantum systems defined by non-self-adjoint toy-model quantum Hamiltonians of the traditional form $H = p^2 + V(x)$. During the dedicated series of international conferences "Pseudo-Hermitian Hamiltonians in Quantum Physics" (PHHQP, (16)), people reported the reality of bound-state spectra for a surprising number of various local complex potentials $V(x) \in \mathbb{C}$ defined along suitable complex curves of $x \in \mathbb{C}$. An exhaustive list of these results would have ranged from the original field-theory-inspired models with forces $V(x) = x^2(\mathrm{i}x)^\delta$ using an arbitrary nonnegative exponent δ (5, 10) up to an amazingly extensive family of their simplified alternatives (39, 42) including even supersymmetric generalizations (98), many-body systems (99), and deformed-symmetry dynamics (52).

It is rather surprising that during the first years of the study of such a quantum-theory-enriching class of toy-model interactions $V \neq V^\dagger$, practically nobody[9] paid

[9]i.e., up to a few marginal exceptions (100, 101)

attention to the parallel problem of scattering. Naturally, in a longer perspective, such an omission was untenable.

1.6.1.1 ***Cross-Shaped-Matrix Metrics and the Loss of Causality*** Keeping in mind the enormous heuristic success of the concept of $\mathcal{PT}$-symmetric bound states, it was necessary to turn attention to the phenomenological as well as theoretical aspects of possible further extensions of the theory beyond bound states. Not quite expectedly, a number of new problems emerged.

A definite failure of the theory took place commencing with the remarkable lecture by Jones during PHHQP VI in London in 2007 ((102); cf. also his two subsequent publications (29)). He pointed out that the description of a *unitary* scattering process caused by a *local* $\mathcal{PT}$-symmetric interaction potential, for example, a complex square well $V(x)$ presents quantum theory with a dilemma. Either one changes the experimental setting *nonlocally* (i.e., in fact, everywhere), or one must accept (like, typically, the authors of Ref. (45)) that the probability is not conserved.

For the needs of theoretical physicists, the Jones' argument has been based on the Mostafazadeh's choice and study of one of the simplest possible toy-model interactions, namely, of its delta-function form $V(x) \sim \delta(x - x_0)$ (11, 103). This rendered possible an approximate spectral-method construction of the related metric (or rather metrics) Θ. The result appeared enormously important because the construction led to the metric operators, which were *all* strongly nondiagonal in the most common coordinate representation.

The use of such a representation was, naturally, vital for the Jones' considerations about scattering in which such a diagonal plus antidiagonal (i.e., "cross-shaped") matrix form of the metric implied that in the most elementary one-dimensional arrangement of the scattering experiment the two separate asymptotic regions (viz., with $x \ll -1$ and $x \gg +1$, respectively) appeared mutually coupled by the above-introduced matrices Θ (of the metric) and/or Ω (of the Dyson-type mapping), which enter the physical mean values and the measurable scattering cross sections.

Consequently, the causality of the scattering process appeared strongly violated. Such a conclusion requalified all metrics *or* all local potentials $V(x)$ as unacceptable in the scattering models exhibiting $\mathcal{PT}$-symmetry. In other words, it was necessary to reclassify all of the models based on a local $V(x)$ as deeply nonunitary. For experimental physicists, this simply meant that these models can only be used as "effective" in the sense that the information about the dynamics remains, in general, incomplete.

Up to now, a fully satisfactory way out of such a quandary has not been found. One of the most promising keys to the resolution of the puzzle has been found using a discretization of the coordinates in papers (30). It has been demonstrated there that the theory may remain unitary (i.e., the probability may remain conserved) provided only that one allows for a suitable *restricted, short-ranged* form of nonlocality in the underlying $\mathcal{PT}$-symmetric interaction potential.

1.6.2 Giving up the Locality of Interaction

The emergence of mathematical difficulties with $\mathcal{PT}$-symmetric scattering proves intimately related to the necessity of an overall upgrade of the specification of what we mean by locality and unitarity in scattering experiments. One has to keep in mind, first of all, the mutual connection between the traditional time-independent formulation of the scattering theory and its proper time-dependent upgrades (cf. (23, 104) for an outline of some most relevant technicalities in the latter case).

In the nearest future, the very concept of quantum scattering must be, after transition to generalized, non-self-adjoint forms of the Hamiltonians, duly and thoroughly rethought and modified, therefore. Up to now, the latter problem has only been tackled either from the "extremely sceptical" point of view (sampled by Ahmed (45) and giving up unitarity) or from the nondifferential, discrete-coordinate, "drastically simplified" point of view of Ref. (30).

In the current literature, the majority of papers seems to prefer the preservation of the locality of $V(x)$ to the unitarity of the process. Certainly, for a consistent build-up of the unitary theories, it will be necessary to pay *simultaneous* attention to the judicious specification of a minimally nonlocal interaction as well as to the construction and selection of such a nontrivial metric operators, which would allow us to return to the properly generalized position and momentum observables. Their operators should certainly retain at least a part of their role of observable quantities, in the remote, asymptotic spatial domain at least[10].

1.6.2.1 The First Trick: Equidistant Discretization of the Real Line of x In a methodical *Gedankenexperiment* aimed at a recovery of unitarity, one would have to consider a one-dimensional quantum scattering mediated by a real and slightly nonlocal, integral-operator scatterer U. The underlying integro-differential Hamiltonian $H = -\triangle + \int U(x, \cdot)$ will de admitted non-Hermitian, say, in the conventional Hilbert space $L^2(\mathbb{R}) := \mathcal{H}^{(F)}, H \neq H^\dagger$. In the light of Ref. (30), these operators might probably be studied as limits of their discrete approximations. We would expect that one can still keep the scattering unitary via a suitably self-consistent delocalization or "smearing trick" leading to an integral operator $U \neq U^\dagger$. The smearing itself should be short ranged, that is, achieved, in general, via an *ad hoc* replacement $V \to U$ of the conventional local-interaction operator $V(x, x') = \delta(x - x')V(x)$ by its nonlocal generalization $U(x, x')$ such that, roughly speaking, the value of the latter function of two real variables becomes small whenever the distance $|x - x'|$ becomes large.

Up to now, the problem remains open. From the isolated source (30), we merely know that the related constructive considerations (based, say, on a systematic refinement of the grid) will not be easy in general. In *loc. cit.*, the work was decisively simplified via the replacement of the real axis of coordinates by keeping the distance h between "coordinate grid points" x_n of the *infinite* and equidistant discrete lattice fixed. In the next step, one should allow this distance to shrink to zero, $h \to 0$, while guaranteeing still the unitarity of the scattering.

1.6.2.2 The Second Trick: the Use of Two-center Bidiagonal Interactions In order to see the whole setting in a more concrete form, let us now add a few details on the construction with fixed h. Firstly, let the kinetic energy operator be represented

[10]The authors of Chapters 2 and 3 will say more in this context.

by the conventional negative discrete Laplacian, that is, by the tridiagonal, doubly infinite matrix T with just three nonvanishing constant diagonals,

$$T_{nn} = 2\,, \quad T_{n,n+1} = T_{n+1,n} = -1\,, \qquad n = \ldots, -2, -1, 0, 1, 2, \ldots\,. \tag{1.6.1}$$

In such a discretization approximation, the fundamental requirement of the unitarity of the quantum evolution has been decisive during the successful search for several scattering-unitarity-supporting crypto-Hermitian Hamiltonians $H = T + U = H^{\ddagger}$.

Recalling the results of our paper (105) for illustration, the aim has been reached by studying a class of tridiagonal matrix interactions U, which varied with a single coupling g and which mimicked the existence of the two interaction centers that were separated by a positive distance $\sim 2M$. Naturally, a guarantee of the necessary mutual compatibility between the two nonlocal operators $H \neq H^{\dagger}$ and $\Theta = \Theta^{\dagger} > 0$ then became a true technical challenge. Fortunately, after an ample use of computer-assisted symbolic-manipulation experiments, we were able to satisfy all the requirements using the following family of interaction matrices of the (suitably partitioned) form

$$U = \left(\begin{array}{cc|cc|cc|cc|cc}
 & \ddots & & & & & & & & \\
\ddots & & 0 & & & & & & & \\
\hline
 & 0 & & -g & & & & & & \\
 & & g & & 0 & & & & & \\
\hline
 & & & 0 & & \ddots & & & & \\
 & & & & \ddots & & 0 & & & \\
\hline
 & & & & & 0 & & g & & \\
 & & & & & & -g & & 0 & \\
 & & & & \multicolumn{2}{c|}{\underbrace{\qquad\qquad}} & & & & \\
\hline
 & & & & \multicolumn{2}{c|}{2M-3} & & 0 & & \ddots \\
 & & & & \multicolumn{2}{c|}{\text{columns}} & & & \ddots &
\end{array} \right). \tag{1.6.2}$$

Let us just summarize the main steps of the construction. First of all, Hamiltonian and metric were required to satisfy a doubly infinite set of linear algebraic relations $H^{\dagger}\Theta = \Theta H$. In parallel, according to Jones (29), we had to eliminate the obstinate causality-violating antidiagonal components out of the metric. Finally, what was truly nontrivial was the implementation of the asymptotic causality constraint requiring that the nonnegligible matrix elements of the doubly infinite matrix Θ_{ij} of the metric only occurred in a band of indices such that $|i-j| < K$ at some finite positive integer K and at all of the sufficiently large absolute values of subscripts, namely, at $|i| \gg 1$ and $|j| \gg 1$.

Owing to the lack of space, let us skip the abstract recurrence relations analysis and, instead, let us jump directly to the supportive "proof by example" as provided by Ref. (105). Its essence lies in the choice of an elementary although still sufficiently flexible doubly infinite ansatz for the metric, which would make our Hamiltonian crypto-Hermitian, $H = H^{\ddagger}$. For this purpose, we first picked up $M = 1$ and obtained the following almost trivial $K = 1$ solution $\Theta = \Theta_1$, which only differed from the unit matrix in its single, "central" matrix element

$$\Theta_{00} = \frac{(1+g)}{(1-g)}\,. \tag{1.6.3}$$

Naturally, this matrix remains positive and bounded, with bounded inverse, if and only if $|g| < 1$. In its turn, this inequality guarantees that the scattering is unitary, in $\mathcal{H}^{(S)}$, for $K = 1$.

The main pedagogical weakness of the $K = 1$ example is that the metric may still be perceived as a mere rescaled discrete version of the Dirac's delta-function. This weakness has been removed by the next choice of $K = 2$ for which the algebraic solution of the Hermiticity condition leads us to the first nontrivial and fully satisfactory tridiagonal candidate for the metric,

$$\Theta = \Theta_2 = \left[\begin{array}{ccccccccc}
 & \ddots & & & & & & & \\
\ddots & \ddots & 1 & & & & & & \\
 & 1 & 0 & 1 & & & & & \\
 & & 1 & 0 & 1+g & & & & \\
 & & & 1+g & 0 & 1+g & & & \\
 & & & & 1+g & 0 & 1 & & \\
 & & & & & 1 & 0 & 1 & \\
 & & & & & & 1 & \ddots & \ddots \\
 & & & & & & & \ddots &
\end{array}\right].$$

In a suitable superposition with diagonal Θ_1, this matrix yielded a positive definite and bounded metric with bounded inverse and, hence, the unitarity of $S-$matrix in the corresponding physical Hilbert space $\mathcal{H}^{(S)}$.

In order to strengthen the illustration of the nontriviality of the construction, let us further display the pentadiagonal $K = 3$ component of the metric,

$$\Theta_3 = \left[\begin{array}{ccccccccc}
\ddots & \ddots & \ddots & & & & & & \\
\ddots & 1 & 0 & 1 & & & & & \\
\ddots & 0 & 1 & 0 & 1+g & & & & \\
 & 1 & 0 & 1-g^2 & 0 & 1-g^2 & & & \\
 & & 1+g & 0 & \frac{(1+g)(1-2g^2)}{1-g} & 0 & 1+g & & \\
 & & & 1-g^2 & 0 & 1-g^2 & 0 & 1 & \\
 & & & & 1+g & 0 & 1 & 0 & \ddots \\
 & & & & & 1 & 0 & 1 & \ddots \\
 & & & & & & \ddots & \ddots & \ddots
\end{array}\right].$$

In Ref. (105), interested readers may find closed formulae for all the finite integers K. In this manner, it is demonstrated that the causal and unitary scattering may also be caused by the non-Hermitian Hamiltonians of the usual form $H = p^2 + U$ containing just a short-range nonlocal interaction term U.

1.6.3 The Threat of the Loss of Unitarity

Mathematically slightly different difficulties emerge in connection with *bound states* generated by differential-operator Hamiltonians $H = T + V(x) \neq H^\dagger$. The obvious phenomenological relevance of the rigorous proofs of reality of the energy spectra of these operators has already been emphasized earlier. The same care must be devoted to the analysis of spectra of candidates Q for other observables (e.g., coordinates). We need not feel bothered by the non-self-adjoint nature of Qs in the false Hilbert space $\mathcal{H}^{(F)}$ where $Q \neq Q^\dagger$ in general. Nevertheless, during the building of a consistent quantum model living in the standard Hilbert space $\mathcal{H}^{(S)}$ with metric Θ, all of these candidates Q must pass the test of reacquired Hermiticity, $Q = Q^\ddagger := \Theta^{-1}Q^\dagger\Theta$.

In paper (24) which may be read as the basic complement of the current textbooks on quantum theory, it has been emphasized that by far the most visible challenge is found in the necessity of the construction of the metric Θ compatible with the Hermitization of *all* of the given set of operators of observables Q. Thus, one cannot perceive the construction of the necessary (and, in practice, rarely easy) *concrete* representations of the underlying standard physical observables as just a technical exercise.

By physics community, it is widely being accepted that the successful use of a diagonalizable non-self-adjoint operator H with real spectrum in quantum theory requires just a successful transition from the false, unphysical Hilbert space $\mathcal{H}^{(F)}$ to its physical amendment $\mathcal{H}^{(S)}$. Although such a step requires a technically complicated assignment of a Hermitizing metric $\Theta^{(S)}$ to the given, possibly even non-self-adjoint observable $H \neq H^\dagger$, physicists generally believe in the feasibility and applicability of such an approach.

The word of strong warning is due. First of all, one should get reminded of the key role of the well-known principle of the classical-quantum correspondence. This is connected with the necessity of an interpretation of measurements. On mathematical side, the loss of Hermiticity of H and all Qs in $\mathcal{H}^{(F)}$ may immediately lead to the necessity of a certain weakening of our requirements of the direct and sharp observability of the coordinate and/or the momentum. *Vice versa*, whenever we insist on the observability status of the coordinate, specific mathematical difficulties may arise[11].

1.6.3.1 Metrics May Be Unbounded The possibility of a generic occurrence of metric-related class of conceptual difficulties has been first mentioned, in their unpublished note from 2001, by Kretschmer and Szymanowski (106). They noticed the existence of an inconsistency in the model-building strategy based on the brute-force redefinition (1.4.1) of the inner products. They emphasized that in only too many concrete toy models the formal inner-product-modification and metric-assignment $H \to \Theta^{(S)}$ leads to unbounded operators $\Theta^{(S)}$. They pointed out that such a choice is in sharp contradiction with the requirements of quantum theory and functional analysis.

[11] Chapter 3 will throw more light on the problem from the algebraic point of view, with emphasis on the role of the canonical commutation (or anticommutation) relations.

These authors recommended that at least some of the serious technical problems of this type could have been circumvented via the above-mentioned formal factorization of the metrics $\Theta = \Omega^\dagger\Omega$. In our present notation, such a recommendation would imply the necessity of the replacement of H, given as acting in $\mathcal{H}^{(S)}$, by $\mathfrak{h}$, which is to be constructed as acting in $\mathcal{H}^{(P)}$. Naturally, the merits of working in "standard space" $\mathcal{H}^{(S)}$ or even "friendly space" $\mathcal{H}^{(F)}$ (i.e., with simpler initial H) would be lost. At the same time, this loss would only concern the toy models H for which the physical Hilbert-space representation $\mathcal{H}^{(S)}$, strictly speaking, does not exist at all.

For an explicit illustration of the latter idea, one could recall examples (although, usually, just the ones working with the finite-dimensional matrices H), which may even prove tractable in closed form. There even exists an infinite-dimensional illustration that connects H and $\mathcal{H}^{(S)}$ with $\mathfrak{h}$ and $\mathcal{H}^{(P)}$ in constructive manner. It has been described in the older study (12) where the authors studied the quartic anharmonic oscillator possessing the wrong sign in the asymptotic $|x| \gg 1$ domain, with $V(x) = -x^4 + \mathcal{O}(x^3)$. Owing to the simplicity of the model, the authors were able to choose operators Ω in a next-to-elementary Fourier plus change-of-variables form. The self-adjoint partner Hamiltonian $\mathfrak{h} = \mathfrak{h}^\dagger$ in $\mathcal{H}^{(P)}$ has been then obtained in a closed form of another *local* quartic oscillator. Finally, the correct physical content of the model was fairly easy to specify (cf. a rediscovery of the methodical merits of this model in Ref. (75)).

1.6.3.2 The Metric's Inverse May Be Unbounded Excluding the models without bounded metric as incomplete, the formal, purely mathematical sources of the possible failure of the whole three-Hilbert-space recipe remain threefold. Firstly, in purely technical sense, even the most popular (i.e., typically, the most common linear differential second-order) input-physics operators $H \neq H^\dagger$ may fail to admit the *feasible* construction of *any* sufficiently tractable S-space metric. For illustration see, for example, the constructions presented in Ref. (107) where even for the most elementary toy-model choices of H, the representations of $\Theta = \Theta(H)$ remain overcomplicated and formal, not sufficiently clearly defined.

Secondly, in phenomenological context, even a truly extreme simplicity of the toy-model H enabling one to find some (or even all—cf. (108)) eligible metrics in a tractable form encounters the challenging ambiguity of the assignment $H \to \Theta(H)$. This point (i.e., a true physical background of the model in question) is usually not discussed with due care. In principle, an additional information about the dynamics is *always* required (24).

Thirdly, even if one deals with a theory in which the *pair* of operators H (= observable) and Θ (= physical Hilbert-space metric) is given as a complete dynamical information input in advance, one has to verify the compatibility of these operators with several *mathematical* postulates as imposed by functional analysis. As we saw earlier, it is not even so easy to prove the necessary reality and semi-bounded nature of the spectrum of H itself[12]. In parallel, a double trouble occurs when one must add a

[12]Cf., in this respect, refer Chapters 4 and 5.

more subtle guarantee of the positivity and the boundedness of the metric operator Θ and its inverse.

During the eleventh PHHQP meeting in Paris in 2012 (109), the attention of attendees happened particularly intensively concentrated on the latter set of difficulties. Consequently, after circa 15 years of an apparently peaceful coexistence of the PT-symmetry-studying mathematicians and physicists, the idyll definitely ended. The community of specialists became split in the group of "traditional optimists" and "deep sceptics." During 2012, the latter group took the initiative.

The rigorous treatment of the above-mentioned questions resulted into a new and particularly deep crisis in the field. In particular, Siegl and Krejčiřík (36) discovered a very subtle although utterly decisive gap in the habitual description of the imaginary cubic oscillator, that is, of one of the most popular and appealing benchmark illustrations of the theory. Using the rigorous tools of operator theory, these authors revealed and demonstrated that their critique applies to many other popular physical examples, based on the use of the *local* $\mathcal{PT}$-symmetric interactions when considered from the semi-classical-analysis point of view. In the language of mathematics, the main source of such a subtle criticism of multiple apparently crypto-Hermitian Hamiltonians lies in the innocent-looking violation of the condition of the bounded nature of the inverse-metric operator Θ^{-1}. At present, the true consequences of such an observation are still to be explored[13].

1.7 TRENDS

Immediately after the PHHQP 2012 meeting in Paris, it became clear that it would be truly necessary to initiate an intensification of communication between mathematicians and physicists. A few months later, our present book has been designed, therefore, as a project which could sufficiently efficiently map the theoretical terrain and which could reunify the community while emphasizing the rigorous interpretation and resolution of the existing problems.

At present, a new round of attempted amendments of the formulation of the theory started (cf., e.g., (37) or (38)). Currently, we are witnessing a new wave of quick developments in the area[14].

1.7.1 Giving Up Metrics

One of the almost amusing paradoxes which paralleled the emergence of the above-listed mathematical difficulties with ill-defined metrics may be seen in a few parallel declarations of their almost negligible relevance and impact on practical applications of quantum theory. For illustration, let us recall a few samples of such "heresies," starting from many-body quantum physics again.

[13] An updated account of the current situation may be found described in Chapters 6 and 7.

[14] In this sense, Chapter 7 may be read as sampling a few new horizons for future research.

1.7.1.1 Coupled-Cluster Variational Method Naturally, the IB Dyson-inspired mappings Ω are amenable to various generalizations. Even in the comparatively narrow domain of nuclear physics, the IB model-building pattern based on the use of non-self-adjoint Hamiltonians finds a nonequivalent analog in the so-called coupled-cluster method (CCM, cf. Ref. (110) for a truly thorough review and for an extensive list of older references).

Besides a partial parallelism in the assumptions (the CCM approach also starts from the knowledge of $\mathfrak{h}$ defined in $\mathcal{H}^{(P)}$), the main differences lie in the purpose. The CCM calculations seem less ambitious—their first aim is a maximally precise determination of the ground state of a given many-body quantum system. In such a sense, the CCM calculations are less universal. They offer an algorithm that merely requires an elementary input guess of the ground-state eigenket of the prospective non-Hermitian operator H (i.e., of $|0\rangle \neq |0\rangle\rangle$). The rest of the CCM construction proceeds via a systematic reconstruction of an optimal Dyson-resembling map Ω. As long as the latter operator is sought via a specific exponential ansatz $\Omega = \exp S$, the method is often called "exp S method" in the literature.

1.7.1.2 Non-Hermitian Proto-Metrics $\tilde{\Theta} \neq \Omega^{\dagger}\Omega$ The high and well-tested practical numerical efficiency of the CCM approach seems more than compensated by several of its theoretical shortcomings. One of them may be seen in the fact that the method puts too heavy an emphasis on the optimal Hamiltonian-adapted reconstruction of the operator $\Omega^{(CCM)} = \exp S$. This is achieved at a cost of just very rough and approximate reconstruction of a far-from optimal operator $\tilde{S}$ playing, in effect, the role of the metric, $\tilde{S} = \Theta^{(CCM)} \neq I$.

From the abstract and consequent THS point of view, the latter metric is not sufficiently well defined. Still, the numerical efficiency of the CCM recipe leads us to the belief that there must exist a way out of the difficulty. For this purpose, it would be necessary to circumvent, first of all, the obstacles represented by the manifest non-Hermiticity of the CCM "proto-metric" $\tilde{S} \neq \tilde{S}^{\dagger}$. A return to Hermiticity of this operator may be perceived as one of the most important challenges in the further development of the CCM formalism (111).

1.7.2 Giving Up Unitarity

In 2010, the weakening of emphasis on the explicit constructions of metrics Θ led, in parallel, to a certain loss of concentration on the problem of the stable quantum systems and to a perceivable growth of activities in the neighboring areas of physics and mathematics. In the latter, mathematical context, more attention has been paid to the subtleties of differences between the role and behavior of bases and operators in finite- and infinite-dimensional Hilbert spaces $\mathcal{H}$. Some of the emerging new ideas may be found, for example, in recent reviews (53) or (112).

1.7.2.1 Open Quantum Systems In spite of a strict concentration of the forthcoming chapters of this book on the rigorous mathematical study of problems *inside* quantum theory of stable systems, the potential future impact of these results will involve, paradoxically, also several other branches of physics *outside* such a restricted area. Such an effect is well known in all of the science and it certainly will apply also in the present, narrower "hidden Hermiticity" context. In this chapter, unfortunately, we were only able to devote a very limited space to a few comments and references concerning the open quantum systems possessing, typically, complex spectra (cf. Section 1.3).

1.7.2.2 Resonances This being said, it still makes sense to add that already as early as during the third PHHQP meeting in 2005 (113) the interdisciplinary appeal of the concepts of $\mathcal{PT}$-symmetry and/or pseudo-Hermiticity attracted attention of nonspecialists. The first consequence was that the traditional participants of this series of meetings were complemented by mathematicians and/or physicists who were mainly interested in the apparently remote fields of nonreal spectra and/or the unstable, resonant quantum states, respectively.

The progress in the research certainly profited from the resulting cooperations and from the extension of the field[15].

1.7.2.3 Phase Transitions Once the attention of the community has been reattracted to the intriguing possibility of simulation of phase transitions caused by the loss of the reality of spectrum during a *smooth* variation of couplings or some other parameters that control the dynamics, a new field of research has been opened. In the language of mathematics, many Hamiltonians H and many other operators Q of quantum observables may lose the reality of their eigenvalues. By definition, this occurs at the Kato's (56) exceptional-point value of the parameter. In the context of quantum physics, the very new and very important distinctive feature of the use of manifestly non-self-adjoint operators of observables $Q = Q(\lambda) \neq Q^\dagger$ is that such a phase-transition-mediating exceptional point value of parameter may prove *real*, that is, in principle, accessible to manipulations and to an *ad hoc* variation in experimental setups.

1.7.3 Giving Up Quantization

While the (real or complex) variable parameter λ crosses an exceptional-point value $\lambda^{(EP)}$ for a realistic quantum Hamiltonian $H = H(\lambda)$, we may speak about the transmutation of bound states into resonances. As long as such a step returns us back to the huge area of traditional quantum phenomenology, it is not too surprising that in quantum physics, the specialists did not see much difference between the real and complex values of $\lambda^{(EP)}$. Thus, many people admitted that the complexification of the spectrum of energies may keep the underlying quantum system phenomenologically interesting in spite of the obvious fact that the evolution of the system in time ceased to be unitary.

[15]cf. the proceedings of the third PHHQP conference which were published as a special issue of the Czech Journal of Physics in September 2005 (18).

Unexpectedly, the phenomenon of the system's crossing the EP boundary attracted the attention of experimentalists who found a new and fairly exciting field for *simulations* of the phenomenon in *classical*, nonquantum systems.

1.7.3.1 Interdisciplinary Aspects of Exceptional Points In the motivating and experiment-related context of physics and phenomenology, people always paid attention to the detailed analyses of various quantum and nonquantum systems in a close vicinity of their exceptional-point singularities. Still, in a way inspired by an unprecedented increase in popularity of quantum systems exhibiting $\mathcal{PT}$-symmetry during the past 15 years, many older studies of the physics near exceptional points acquired, suddenly, an entirely new relevance (cf., e.g., the talks for a dedicated Workshop, in NiTheP in Stellenbosch, in 2010 (57)). Mutual connections were found between the behavior of the atoms and molecules in external fields as produced, say, by lasers. In addition, the states of the classical electromagnetic field in a microwave billiard were found to share the same mathematical structures (114). The control of the schematic classical models of magnetohydrodynamics (115) was found similar to the quantum models of phase transitions, and so on.

The theoretical study of exceptional points acquired an undeniably interdisciplinary character. At random, an introductory review of a randomly selected sample of references may start, say, in the domain of classical optics. A strict formal analog of the usual quantum Schrödinger equations has been found there, shortly after the PHHQP 2007 meeting in London (116), in the so-called paraxial version of the classical equation for the diffraction, say, in two dimensions (117) (one should also refer to the older paper (101) in this respect). In the role of a $\mathcal{PT}$-symmetric quantum potential $V(x)$, one finds here, typically, a position-dependent complex refractive index $n(x)$, which remains constant along the $z-$axis (this axis mimics the time in Schrödinger equation) while exhibiting a suitable elementary gain-or-loss spatial structure along the transverse, "spatial" $x-$axis.

In this comparison, the classical electric-field envelope of the laser beam (which is, clearly, observable) plays the role of the quantum wave function (which was, naturally, NOT observable). It is not too surprising, therefore, that one immediately arrives at the entirely new observable features of the beam including the double refraction patterns and their power oscillations as well as nonreciprocal diffractions.

What followed the first papers and simulations in the field led not only to an experimental heyday (for a randomly selected sample of references cf. (118)) but also to the inspiration and feedback for the quantum theory of nuclei (119) and for quantum electronics (120), and so on. In addition, the search for a reunification of vocabularies of mathematics and physics continued. People moved, spontaneously, beyond the rather restricted area of study of non-self-adjoint operators with real spectra. After the complexity of the energies was admitted, one of the most natural phenomenological targets for extension of the projects has been found in an extensive search for nonquantum analogs of the traditional unstable quantum systems. The reasons for staying restricted to the field of quantum physics faded away. A number of applications of new project-setting ideas has been published, during the past 3 or 4 years, in the subdomains of physics as remote as quantum condensed-matter physics or classical optics.

Once we return back to mathematics, the quick growth of the overlap between the pragmatically and theoretically oriented studies implied an improvement of our understanding of the properties of the self-adjoint operators in the Krein or Pontriagin spaces. The contemporary forms of the applied functional analysis and spectral theory proved useful for the purpose. Thus, nowadays one finds a true challenge in presenting the new physics in the fully rigorous language of standard mathematics.

1.7.3.2 Outlook The quantum interpretation of the Kato's exceptional points as the values of parameters at which the unitarity of the evolution gets lost initiated multiple new theoretical studies beyond quantum physics. In most of these cases, such a loss of stability is still being interpreted, in the newly invented terminology, as a manifestation of the spontaneous breakdown of PT symmetry.

It is not surprising that under the new name, the old phenomenon immediately attracted new attention. Thus, experimentalists are currently trying to simulate such a form of phase transitions in multiple traditional contexts. At present, the subject of these studies ranges from optical absorbers (121) to Bose–Hubbard dimers (122) to spin chain models (123) or, if you wish, from quantum lattice models (124) to higher derivative Hamiltonians (125), and so on. In the literature, one even finds applications using the concepts of pomeron (126) or of the coupling of reaction channels (127), and so on.

The study of related mathematical methods received an equally strong impulse for the development of new ideas about coherent-state quantization techniques (128) and about their path-integral realizations (129). The rotation-operator analogs of parity (130) were studied, and the updated view of the role of the transfer matrices was presented in Ref. (131). Other methodical proposals involved also the density matrices in thermodynamics (132) or the random matrix ensembles in stochastic systems (133), and so on.

New perspectives were opened and the related upgrade of the research in mathematics blossomed. Besides the innovative new applications of the traditional Krein-space theory (134), people made a successful use of the formalism of Moyal products (135) and the ideas of the canonical classical-quantum correspondence (136). Needless to add that the applicability-oriented support of the latter efforts remained strong and broadening. New demand of rigorous background is now coming from classical electrodynamics and from the theory as well as increase in sophistication of lasers (cf. (137)), and so on. Once one admits, in mathematical formulation, the admissibility of the loss of the robust reality of the spectrum of an operator, the notion of a quantum bound state must be replaced, in quantum-theory setting, by phenomenologically more appropriate concepts that need not originate just from the needs of quantum physics.

At present, unfortunately, the global scientific community remains often separated into multiple narrow-minded subcommunities. This may be noticed to lead to a slow loss of a stable unifying communication ground and common language. Typically, having separated physicists into groups of "theoretical physicists" and "mathematical physicists," Roman Kotecký, in his essay (138), tried to analyze the points of contact between similar subcommunities and he emphasized the need of their collaboration. We share this view. Besides the collaboration between mathematicians and physicists during conferences (16), there exists a hope that these two subcommunities will often break the traditional isolacionistic bad habits and that they will keep communicating, in the future, also via some shared publication efforts sampled by this book.

REFERENCES

1. Bohr N. On the constitution of atoms and molecules. *Philos Mag* 1913;26(1):476.
2. Available at http://www.nobelprize.org/nobel_prizes/physics/laureates/2012/ Accessed 2015 Jan 11.
3. Feshbach H. Unified theory of nuclear reactions. *Ann Phys (NY)* 1958;5:357–390.
4. (a) Feshbach H. A unified theory of nuclear reactions II. *Ann Phys (NY)* 1962;19:287–313; (b) Löwdin P-O. Studies in perturbation theory. IV. Solution of eigenvalue problem by projection operator formalism. *J Math Phys* 1962;3:969–982.
5. Bender CM, Boettcher S. Real spectra in non-Hermitian Hamiltonians having PT symmetry. *Phys Rev Lett* 1998;80:5243–5246.
6. Available at http://www.mth.kcl.ac.uk/%7Estreater/lostcauses.html/#XIII Accessed 2015 Jan 11.
7. Streater RF. *Lost Causes in and Beyond Physics*. Berlin: Springer; 2007.
8. Available at http://ptsymmetry.net Accessed 2014 Nov 11.
9. Dorey P, Dunning C, Tateo R. The ODE/IM correspondence. *J Phys A Math Theor* 2007;40:R205.
10. Bender CM. Making sense of non-Hermitian Hamiltonians. *Rep Prog Phys* 2007;70:947–1018.
11. Mostafazadeh A. Pseudo-Hermitian representation of quantum mechanics. *Int J Geom Methods Mod Phys* 2010;7:1191–1306.
12. Buslaev V, Grechi V. Equivalence of unstable anharmonic oscillators and double wells. *J Phys A Math Gen* 1993;26:5541–5549.
13. (a) Andrianov AA. The large-N expansion as a local perturbation theory. *Ann Phys (NY)* 1982;140:82–100; (b) Hatano N, Nelson DR. Localization transitions in non-Hermitian quantum mechanics. *Phys Rev Lett* 1996;77:570–573.
14. Gazeau J-P, Kerner R, Antoine J-P, Métens S, Thibon J-Y, editors. *GROUP 24: Physical and Mathematical Aspects of Symmetries*. London: Taylor & Francis; 2003.
15. Available at http://gemma.ujf.cas.cz/~znojil/conf/index01.html Accessed 2014 Nov 11.
16. Available at http://gemma.ujf.cas.cz/%7Eznojil/conf/index.html Accessed 2015 Jan 11.
17. (a) Znojil M, editor. *Czech J Phys* (special issue) 2004;54(1):1–156; (b) Znojil M, editor. *Czech J Phys* (special issue) 2004;54(10):1005–1148.
18. Znojil M, editor. *Czech J Phys* (special issue) 2005;55(9):1049–1192.
19. Znojil M, editor. *Czech J Phys* (special issue) 2006;56(9):885–1064.
20. D. Bessis private communication to MZ in summer, 1992. Similar communication is cited in [5].

21. Bender CM, Milton KA. Nonperturbative calculation of symmetry breaking in quantum field theory. *Phys Rev* 1997;D 55:3255–3259.
22. Dorey P, Dunning C, Tateo R. Spectral equivalences, Bethe Ansatz equations and reality properties in PT-symmetric quantum mechanics. *J Phys A Math Gen* 2001;34: 5679–5704.
23. Znojil M. Three-Hilbert-space formulation of Quantum Mechanics. *Symm Integr Geom Methods Appl (SIGMA)* 2009;5 001:19p (arXiv overlay: 0901.0700).
24. Scholtz FG, Geyer HB, Hahne FJW. Quasi-Hermitian operators in quantum mechanics and the variational principle. *Ann Phys (NY)* 1992;213:74–101.
25. Dyson FJ. General theory of spin-wave interactions. *Phys Rev* 1956;102:1217–1230.
26. Bender CM, Brody DC, Jones HF. Complex extension of quantum mechanics. *Phys Rev Lett* 2002;89:270401; Erratum: *Phys Rev Lett* 2004;92:119902.
27. Znojil M. Conservation of pseudo-norm in PT symmetric quantum mechanics. *Rendic Circ Mat Palermo (Ser II Suppl)* 2004;72:211–218 (arXiv:math-ph/0104012).
28. Bender CM, Brody DC, Jones HF. Complex extension of quantum mechanics. *Phys Rev Lett* 2002;89:270401 (arXiv:quant-ph/0208076).
29. (a) Jones HF. Scattering from localized non-Hermitian potentials. *Phys Rev* 2007;D 76:125003; (b) Jones HF. Interface between Hermitian and non-Hermitian Hamiltonians in a model calculation. *Phys Rev* 2008;D 78:065032.
30. (a) Znojil M. Scattering theory with localized non-Hermiticities. *Phys Rev* 2008;D 78:025026; (b) Znojil M. Discrete PT-symmetric models of scattering. *J Phys A Math Theor* 2008;41:292002.
31. Available at http://gemma.ujf.cas.cz/%E7znojil/conf/2010.htm Accessed 2014 Nov 11.
32. (a) Rüter CE, Makris R, El-Ganainy KG, Christodoulides DN, Segev M, Kip D. Observation of parity-time symmetry in optics. *Nat Phys* 2010;6:192; (b) Longhi S. Optical realization of relativistic non-Hermitian Quantum Mechanics. *Phys Rev Lett* 2010;105:013903; (c) Chong YD, Ge L, Douglas Stone A. PT-symmetry breaking and laser-absorber modes in optical scattering systems. *Phys Rev Lett* 2011;106:093902.
33. Moiseyev N. *Non-Hermitian Quantum Mechanics*. Cambridge: Cambridge University Press; 2011.
34. Available at http://www.pks.mpg.de/%E7phhqpx11 Accessed 2014 Nov 11.
35. (a) Aslanyan A, Davies E-B. Spectral instability for some Schrödinger operators. *Numer Math* 2000;85:525–552; (b) Davies B. Wild spectral behaviour of anharmonic oscillators. *Bull London Math Soc* 2000;32:432–438; (c) Davies B, Kuijlaars A. Spectral asymptotics of thenon-selfadjoint harmonic oscillator. *J London Math Soc* 2004;70:420–426.
36. Siegl P, Krejcirik D. On the metric operator for the imaginary cubic oscillator. *Phys Rev* 2012;D 86:121702(R).
37. Bagarello F, Znojil M. Non linear pseudo-bosons versus hidden Hermiticity. II: the case of unbounded operators. *J Phys A Math Theor* 2012;45:115311.
38. Mostafazadeh A. Pseudo-Hermitian quantum mechanics with unbounded metric operators. *Philos Trans R Soc London Ser A* 2013;371:20120050.

39. Lévai G, Znojil M. Systematic search for PT symmetric potentials with real energy spectra. *J Phys A Math Gen* 2000;33:7165–7180.
40. (a) Znojil M. Shape invariant potentials with PT symmetry. *J Phys A Math Gen* 2000;33:L61–L62; (b) Znojil M, Tater M. Complex Calogero model with real energies. *J Phys A Math Gen* 2001;34:1793–1803.
41. Bender CM, Brody DC, Jones HF. Extension of PT-symmetric quantum mechanics to quantum field theory with cubic interaction. *Phys Rev* 2004;D 70:025001; Erratum-ibid. 2005;D 71:049901.
42. (a) Znojil M. PT symmetric square well. *Phys Lett* 2001;A 285:7–10; (b) Albeverio S, Fei S-M, Kurasov P. Point interactions, PT-Hermiticity and reality of the spectrum. *Lett Math Phys* 2002;59:227–242; (c) Znojil M, Jakubský V. Solvability and PT-symmetry in a double-well model with point interactions. *J Phys A Math Gen* 2005;38:5041–5056.
43. (a) Mostafazadeh A. Quantum mechanics of Klein-Gordon-type fields and quantum cosmology. *Ann Phys (NY)* 2004;309:1–48; (b) Andrianov AA, Cannata F, Kamenshchik AY. Phantom universe from CPT symmetric QFT. *Int J Mod Phys* 2006;D 15:1299.
44. (a) Mostafazadeh A. Hilbert space structures on the solution space of Klein-Gordon type evolution equations. *Class Quantum Grav* 2003;20:155–171; (b) Znojil M. Relativistic supersymmetric quantum mechanics based on Klein-Gordon equation. *J Phys A Math Gen* 2004;37:9557–9571; (c) Znojil M. Solvable relativistic quantum dots with vibrational spectra. *Czech J Phys* 2005;55:1187–1192; (d) Mostafazadeh A, Zamani F. Quantum mechanics of Klein-Gordon fields I: Hilbert space, localized states, and chiral symmetry. *Ann Phys (NY)* 2006;321:2183–2209.
45. (a) Ahmed Z. Handedness of complex PT-symmetric potential barriers. *Phys Lett* 2004;A 324:152–158; (b) Cannata F, Dedonder J-P, Ventura A. Scattering in PT-symmetric quantum mechanics. *Ann Phys (NY)* 2007;322:397–437; (c) Mehri-Dehnavi H, Mostafazadeh A, Batal A. Application of pseudo-Hermitian quantum mechanics to a complex scattering potential with point interactions. *J Phys A Math Theor* 2010;43:145301; (d) Lin Z, Ramezani H, Eichelkraut T, Kottos T, Cao H, Christodoulides DN. Unidirectional invisibility induced by PT-symmetric periodic structures. *Phys Rev Lett* 2011;106:213901; (e) Albeverio S, Kuzhel S. On elements of the Lax-Phillips scattering scheme for PT-symmetric operators. *J Phys A Math Theor* 2012;45:444001.
46. (a) Stehmann T, Heiss WD, Scholtz FG. Observation of exceptional points in electronic circuits. *J Phys A Math Gen* 2004;37:7813–7820; (b) Heiss WD. Exceptional points - their universal occurrence and their physical significance. *Czech J Phys* 2004;54:1091–1100.
47. Günther U, Stefani F, Znojil M. MHD α^2–dynamo, Squire equation and PT-symmetric interpolation between square well and harmonic oscillator. *J Math Phys* 2005;46:063504.
48. Berry MV. Physics of nonhermitian degeneracies. *Czech J Phys* 2004;54(10):1039-1047.
49. Dembowski C, Dietz B, Gräf H-D, Harney HL, Heine A, Heiss WD, Richter A. Observation of a chiral state in a microwave cavity. *Phys Rev Lett* 2003;90:034101.
50. Znojil M. Scattering theory using smeared non-Hermitian potentials. *Phys Rev* 2009;D 80(4):045009.

51. Mostafazadeh A. Spectral singularities of complex scattering potentials and infinite reflection and transmission coefficients at real energies. *Phys Rev Lett* 2009;102:220402.
52. Fring A. Particles versus fields in PT-symmetrically deformed integrable systems. *Pramana J Phys* 2009;73(2):363–374.
53. Trefethen LN, Embree M. *Spectra and Pseudospectra. The Behavior of Nonnormal Matrices and Operators*. Princeton (NJ): Princeton University Press; 2005.
54. Znojil M. Maximal couplings in PT-symmetric chain-models with the real spectrum of energies. *J Phys A Math Theor* 2007;40:4863–4875.
55. Günther U, Samsonov B. The Naimark dilated PT-symmetric brachistochrone. *Phys Rev Lett* 2008;101:230404.
56. Kato T. *Perturbation Theory for Linear Operators*. Berlin: Springer-Verlag; 1966.
57. Available at http://www.nithep.ac.za/2g6.htm Accessed 2014 Nov 11.
58. Smilga AV. Cryptogauge symmetry and cryptoghosts for crypto-Hermitian Hamiltonians. *J Phys A Math Theor* 2008;41:244026.
59. (a) Bagarello F. Algebras of unbounded operators and physical applications: a survey. *Rev Math Phys* 2007;19(3):231–272; (b) Henry R. Spectral instability of some non-selfadjoint anharmonic oscillators. *C R Acad Sci Paris, Ser I* 2012;350:1043–1046; (c) Bagarello F, Inoue A, Trapani C. Weak commutation relations of unbounded operators: nonlinear extensions. *J Math Phys* 2012;53:123510; (d) Bagarello F, Fring A. A non selfadjoint model on a two dimensional noncommutative space with unbound metric. *Phys Rev* 2013;A 88:042119; (e) Bagarello F. From selfadjoint to non-selfadjoint harmonic oscillators: physical consequences and mathematical pitfalls. *Phys Rev* 2013;A 88:032120; (f) Mityagin B, Siegl P. Root system of singular perturbations of the harmonic oscillator type operators (arXiv:1307.6245); (g) Mityagin B, Siegl P, Viola J. Differential operators admitting various rates of spectral projection growth (arXiv:1309.3751).
60. (a) Graefe EM. Stationary states of a PT symmetric two-mode Bose-Einstein condensate. *J Phys A Math Theor* 2012;45:444015; (b) Cartarius H, et al. Stationary and dynamical solutions of the Gross-Pitaevskii equation for a Bose-Einstein condensate in a PT symmetric double well. *Acta Polytech* 2013;53(3):259–267.
61. (a) Zeeman EC. *Catastrophe Theory - Selected Papers 1972–1977*. Reading (MA): Addison-Wesley; 1977; (b) Arnold VI. *Catastrophe Theory*. Berlin: Springer; 1984.
62. Znojil M. Quantum catastrophes: a case study. *J Phys A Math Theor* 2012;45:444036.
63. (a) Exner P, Keating JP, Kuchment P, Teplyaev A, editors. *Analysis on Graphs and Its Applications*. Providence (RI): American Mathematical Society; 2008; (b) Znojil M. Fundamental length in quantum theories with PT-symmetric Hamiltonians. II. The case of quantum graphs. *Phys Rev* 2009;D 80:105004; (c) Znojil M. Cryptohermitian Hamiltonians on graphs. *Int J Theor Phys* 2011;50:1052–1059.
64. (a) Bagchi B, Quesne C, Znojil M. Generalized continuity equation and modified normalization in PT-symmetric quantum mechanics. *Mod Phys Lett* 2001;A 16:2047–2057; (b) Bagchi B, Mallik S, Quesne C. PT-symmetric square well and the associated SUSY hierarchies. *Mod Phys Lett* 2002;A 17:1651–1664.
65. Mostafazadeh A. Pseudo-Hermiticity versus PT symmetry. *J Math Phys* 2002;43:205.

66. Langer H, Tretter Ch. A Krein space approach to PT symmetry. *Czech J Phys* 2004;54:1113–1120.

67. Gohberg IC, Krein MG. *Introduction to the Theory of Linear Non-Selfadjoint Operators*. Providence (RI): American Mathematical Society; 1969.

68. (a) Blasi A, Scolarici G, Solombrino L. Pseudo-Hermitian Hamiltonians, indefinite inner product spaces and their symmetries. *J Phys A Math Gen* 2004;37:4335; (b) Tanaka T. General aspects of PT-symmetric and P-selfadjoint quantum theory in a Krein space. *J Phys A Math Gen* 2006;39:14175; (c) Azizov TYa, Trunk C. PT symmetric, Hermitian and P-self-adjoint operators related to potentials in PT quantum mechanics. *J Math Phys* 2012;53:012109; (d) Zelezny J. The Krein-space theory for non-Hermitian PT-symmetric operators [MSc thesis]. Prague: FNSPE CTU; 2011; (e) Krejcirik D, Siegl P, Zelezny J. On the similarity of Sturm-Liouville operators with non-Hermitian boundary conditions to selfadjoint and normal operators. *Complex Anal Oper Theory* 2014;8:255–281.

69. Messiah A. *Quantum Mechanics I*. Amsterdam: North-Holland; 1961.

70. Brown LS. *Quantum Field Theory*. Cambridge: Cambridge University Press; 1992.

71. Simon B. Large orders and summability of eigenvalue perturbation theory: a mathematical overview. *Int J Quant Chem* 1982;21(1):3–26.

72. Alvarez G. Bender-Wu branch points in the cubic oscillator. *J Phys A Math Gen* 1995;27:4589–4598.

73. Caliceti E, Graffi S, Maioli M. Perturbation theory of odd anharmonic oscillators. *Commun Math Phys* 1980;75:51–66.

74. Fernández FM, Guardiola R, Ros J, Znojil M. Strong-coupling expansions for the PT-symmetric oscillators $V(r) = aix + b(ix)^2 + c(ix)^3$. *J Phys A Math Gen* 1998;31:10105–10112.

75. Jones HF, Mateo J. Equivalent Hermitian Hamiltonian for the non-Hermitian $-x^4$ potential. *Phys Rev* 2006;D 73:085002.

76. Sibuya Y. *Global Theory of a Second Order Linear Ordinary Differential Equation with a Polynomial Coefficient*. Amsterdam: North-Holland; 1975.

77. Bender CM, Boettcher S, Meisinger PN. PT-symmetric quantum mechanics. *J Math Phys* 1999;40:2201–2229.

78. Bender CM, Boettcher S, Jones HF, Savage VM. Complex square well - a new exactly solvable quantum-mechanical model. *J Phys A Math Gen* 1999;32:6771–6781.

79. (a) Znojil M. Quantum toboggans with two branch points. *Phys Lett* 2007;A 372(5):584–590; (b) Znojil M. Quantum toboggans: models exhibiting a multisheeted PT symmetry. *J Phys Conf Ser* 2008;128:012046; (c) Znojil M. Topology-controlled spectra of imaginary cubic oscillators in the large-L approach. *Phys Lett* 2010;A 374:807–812.

80. Fernández FM, Guardiola R, Ros J, Znojil M. A family of complex potentials with real spectrum. *J Phys A Math Gen* 1999;32:3105–3116.

81. Znojil M. PT symmetric harmonic oscillators. *Phys Lett* 1999;A 259:220–223.

82. (a) Znojil M. Tridiagonal PT-symmetric N by N Hamiltonians and a fine-tuning of their observability domains in the strongly non-Hermitian regime. *J Phys A Math Theor* 2007;40:13131–13148; (b) Znojil M. PT-symmetric quantum chain models. *Acta Polytech* 2007;47:9–14.

83. Rotter I. A non-Hermitian Hamilton operator and the physics of open quantum systems. *J Phys A Math Theor* 2009;42:153001.
84. (a) Znojil M. Conditional observability. *Phys Lett* 2007;B 650:440–446; (b) Znojil M. Horizons of stability. *J Phys A Math Theor* 2008;41:244027.
85. Znojil M. Linear representation of energy-dependent Hamiltonians. *Phys Lett* 2004;A 326:70–76.
86. Cooper F, Khare A, Sukhatme U. Supersymmetry and quantum mechanics. *Phys Rep* 1995;251(5/6):267–388.
87. Znojil M, Cannata F, Bagchi B, Roychoudhury R. Supersymmetry without hermiticity within PT symmetric quantum mechanics. *Phys Lett* 2000;B 483:284–289.
88. Znojil M. Non-Hermitian SUSY and singular PT-symmetrized oscillators. *J Phys A Math Gen* 2002;35:2341–2352.
89. Znojil M. Re-establishing supersymmetry between harmonic oscillators in $D \neq 1$ dimensions. *Rendic Circ Mat Palermo (Ser II Suppl)* 2003;71:199–207.
90. Janssen D, Dönau F, Frauendorf S, Jolos RV. Boson description of collective states: (I) Derivation of the boson transformation for even fermion systems. *Nucl Phys* 1971;A 172:145–165.
91. Dieudonne J. *Quasi-Hermitian operators*. In: *Proceedings of International Symposium on Linear Spaces*. Oxford: Pergamon; 1961. p 115–122.
92. Znojil M. The cryptohermitian smeared-coordinate representation of wave functions. *Phys Lett* 2011;A 375:3176–3183.
93. Znojil M. On the role of the normalization factors κ_n and of the pseudo-metric P in crypto-Hermitian quantum models. *Symm Integr Geom Methods Appl (SIGMA)* 2008;4:001 (arXiv overlay: 0710.4432v3).
94. (a) Jakubský V, Smejkal J. A positive-definite scalar product for free Proca particle. *Czech J Phys* 2006;56:985; (b) Zamani F, Mostafazadeh A. Quantum mechanics of Proca fields. *J Math Phys* 2009;50:052302.
95. (a) Bíla H. Adiabatic time-dependent metrics in PT-symmetric quantum theories, e-print (arXiv:0902.0474); (b) Gong J-B, Wang Q-H. Time-dependent PT-symmetric quantum mechanics. *J Phys A Math Theor* 2013;46:485302.
96. Gong J-B, Wang Q-H. Geometric phase in PT-symmetric quantum mechanics. *Phys Rev* 2010;A 82:012103.
97. Znojil M. Crypto-unitary forms of quantum evolution operators. *Int J Theor Phys* 2013;52:2038–2045.
98. (a) Andrianov AA, Ioffe MV, Cannata F, Dedonder J-P. SUSY quantum mechanics with complex superpotentials and real energy spectra. *Int J Mod Phys* 1999;A 14:2675–2688 (arXiv:quant-ph/9806019); (b) Znojil M. PT symmetrized SUSY quantum mechanics. *Czech J Phys* 2001;51:420–428; (c) Lévai G, Znojil M. The interplay of supersymmetry and PT symmetry in quantum mechanics: a case study for the Scarf II potential. *J Phys A Math Gen* 2002;35:8793–8804; (d) Sinha A, Roychoudhury R. Isospectral partners of a complex PT-invariant potential. *Phys Lett* 2002;A 301:163; (e) Bagchi B, Banerjee A, Caliceti E, Cannata F, Geyer HB, Quesne C, Znojil M. CPT-conserving Hamiltonians and their nonlinear supersymmetrization using differential charge-operators C. *Int*

J Mod Phys 2005;A 20:7107–7128; (f) Quesne C, Bagchi B, Mallik S, Bila H, Jakubsky V, Znojil M. PT-supersymmetric partner of a short-range square well. *Czech J Phys* 2005;55:1161–1166; (g) Sinha A, Roy P. Pseudo supersymmetric partners for the generalized Swanson model. *J Phys A Math Theor* 2008;41:335306; (h) Znojil M, Jakubsky V. Supersymmetric quantum mechanics living on topologically nontrivial Riemann surfaces. *Pramana J Phys* 2009;73:397–404; (i) Shamshutdinova VV. Construction of the metric and equivalent Hermitian Hamiltonian via SUSY transformation operators. *Phys Atom Nucl* 2012;75:1294; (j) Ghosh PK. Supersymmetric many-particle quantum systems with inverse-square interactions. *J Phys A Math Theor* 2012;45:183001; (k) Miri M-A, Heinrich M, Christodoulides DN. Supersymmetry-generated complex optical potentials with real spectra. *Phys Rev* 2013;A 87:043819.

99. (a) Fring A, Znojil M. PT-symmetric deformations of Calogero models. *J Phys A Math Theor* 2008;41:194010; (b) Assis PEG, Fring A. From real fields to complex Calogero particles. *J Phys A Math Theor* 2009;42:425206; (c) Fring A, Smith M. Non-Hermitian multi-particle systems from complex root spaces. *J Phys A Math Theor* 2012;45:085203.
100. Znojil M. Spiked potentials and quantum toboggans. *J Phys A Math Gen* 2006;39:13325–13336.
101. Ruschhaupt A, Delgado F, Muga JG. Physical realization of PT-symmetric potential scattering in a planar slab waveguide. *J Phys A Math Gen* 2005;38:L171.
102. Available at http://www.staff.city.ac.uk/%E7fring/PT/Hugh-Jones.pdf Accessed 2014 Nov 11.
103. Mostafazadeh A. Metric operator in pseudo-Hermitian quantum mechanics and the imaginary cubic potential. *J Phys A Math Gen* 2006;39:10171–10188.
104. Znojil M. Time-dependent version of cryptohermitian quantum theory. *Phys Rev* 2008;D 78:085003.
105. Znojil M. Cryptohermitian picture of scattering using quasilocal metric operators. *Symm Integr Geom Methods Appl (SIGMA)* 2009;5:085 (arXiv overlay: 0908.4045).
106. (a) Kretschmer R, Szymanowski L. The interpretation of quantum-mechanical models with non-Hermitian Hamiltonians and real spectra (arXiv:quantph/0105054); (b) Kretschmer R, Szymanowski L. Quasi-Hermiticity in infinite-dimensional Hilbert spaces. *Phys Lett* 2004;A 325:112–115; (c) Kretschmer R, Szymanowski L. The Hilbert-space structure of non-Hermitian theories with real spectra. *Czech J Phys* 2004;54:71–75.
107. Ghatak A, Mandal BP. Comparison of different approaches of finding the f metric in pseudo-Hermitian theories. *Commun Theor Phys* 2013;59:533.
108. (a) Znojil M. Fundamental length in quantum theories with PT-symmetric Hamiltonians. *Phys Rev* 2009;D 80:045022; (b) Znojil M. Complete set of inner products for a discrete PT-symmetric square-well Hamiltonian. *J Math Phys* 2009;50:122105.
109. Available at http://phhqp11.in2p3.fr/Home.html Accessed 2014 Nov 11.
110. (a) Bishop RF. An overview of coupled cluster theory and its applications in physics. *Theor Chim Acta* 1991;80(2/3):95–148; (b) Bishop RF, Li PHY. Coupled-cluster method: a lattice-path-based subsystem approximation scheme for quantum lattice models. *Phys Rev* 2011;A 83:042111.

111. (a) Arponen J. Constrained Hamiltonian approach to the phase space of the coupled-cluster method. *Phys Rev* 1997;A 55:2686–2700; (b) Bishop RF, Znojil M. The coupled-cluster approach to quantum many-body problem in a three-Hilbert-space reinterpretation. *Acta Polytech* 2014;54:85–92.

112. Davies EB. Non-self-adjoint differential operators. *Bull Lond Math Soc* 2002;34:513–532.

113. Available at http://home.ku.edu.tr/%7Eamostafazadeh/workshop_2005/post_workshop.htm Accessed 2015 Jan 11.

114. Bittner S, Dietz B, Günther U, Harney HL, Miski-Oglu M, Richter A, Schaefer F. PT symmetry and spontaneous symmetry breaking in a microwave billiard. *Phys Rev Lett* 2012;108:024101.

115. Günther U, Kirillov ON. A Krein space related perturbation theory for MHD α^2–dynamos and resonant unfolding of diabolical points. *J Phys A Math Gen* 2006;39:10057.

116. Available at http://www.staff.city.ac.uk/%7Efring/PT Accessed 2015 Jan 11.

117. Makris KG, El-Ganainy R, Christodoulides DN, Musslimani ZH. Beam dynamics in PT symmetric optical lattices. *Phys Rev Lett* 2008;100:103904.

118. (a) West CT, Kottos T, Prosen T. PT-symmetric wave chaos. *Phys Rev Lett* 2010;104:054102; (b) Schomerus H. Quantum noise and self-sustained radiation of PT-symmetric systems. *Phys Rev Lett* 2010;104:233601; (c) Longhi S, Della Valle G. Photonic realization of PT-symmetric quantum field theories. *Phys Rev* 2012;A 85:012112; (d) Ge L, Chong YD, Stone AD. Conservation relations and anisotropic transmission resonances in one-dimensional PT-symmetric photonic heterostructures. *Phys Rev* 2012;A 85:023802; (e) Suchkov SV, Dmitriev SV, Malomed BA, Kivshar YS. Wave scattering on a domain wall in a chain of PT-symmetric couplers. *Phys Rev* 2012;A 85:033825; (f) Lin Z, Schindler J, Ellis FM, Kottos T. Experimental observation of the dual behavior of PT-symmetric scattering. *Phys Rev* 2012;A 85:050101.

119. Rotter I. Environmentally induced effects and dynamical phase transitions in quantum systems. *J Opt* 2010;12:065701.

120. (a) Chtchelkatchev NM, Golubov AA, Baturina TI, Vinokur VM. Stimulation of the fluctuation superconductivity by PT symmetry. *Phys Rev Lett* 2012;109:150405; (b) Schindler J, Lin Z, Lee JM, Ramezani H, Ellis FM, Kottos T. PT-symmetric electronics. *J Phys A Math Theor* 2012;45:444029.

121. Longhi S. PT-symmetric laser absorber. *Phys Rev* 2010;A 82:031801.

122. Graefe E-M, Korsch HJ, Niederle AE. Quantum-classical correspondence for a non-Hermitian Bose-Hubbard dimer. *Phys Rev* 2010;A 82:013629.

123. Castro-Alvaredo OA, Fring A. A spin chain model with non-Hermitian interaction: the Ising quantum spin chain in an imaginary field. *J Phys A Math Theor* 2009;42:465211.

124. (a) Korff C, Weston RA. PT symmetry on the lattice: the quantum group invariant XXZ spin-chain. *J Phys A Math Theor* 2007;40:8845–8872; (b) Joglekar YN, Scott D, Babbey M, Saxena A. Robust and fragile PT-symmetric phases in a tight-binding chain. *Phys Rev* 2010;A 82:030103; (c) Znojil M. Gegenbauer-solvable quantum chain model. *Phys*

Rev 2010;A 82:052113; (d) Jin L, Song Z. Wave-packet dynamics in a non-Hermitian PT symmetric tight-binding chain. *Commun Theor Phys* 2010;54:73; (e) Jin L, Song Z. Physics counterpart of the PT non-Hermitian tight-binding chain. *Phys Rev* 2010;A 81:032109.

125. Bender CM, Mannheim PD. No-ghost theorem for the fourth-order derivative Pais-Uhlenbeck oscillator model. *Phys Rev Lett* 2008;100:110402.
126. Braun MA, Vacca GP. PT symmetry and Hermitian Hamiltonian in the local supercritical pomeron model. *Eur Phys J* 2008;C 59:795.
127. Znojil M. Coupled-channel version of PT-symmetric square well. *J Phys A Math Gen* 2006;39:441–455.
128. (a) Roy B, Roy P. Coherent states of non-Hermitian quantum systems. *Phys Lett* 2006;A 359:110; (b) Siegl P. *Non-Hermitian quantum models, indecomposable representations and coherent states quantization* [PhD thesis]. Prague: University Paris Diderot & FNSPE CTU; 2011.
129. Rivers RJ. Path integrals for quasi-Hermitian Hamiltonians. *Int J Mod Phys* 2011;D 20:919.
130. Zhang XZ, Song Z. Non-Hermitian anisotropic XY model with intrinsic rotation-time-reversal symmetry. *Phys Rev* 2013;A 87:012114.
131. Monzón JJ, Barriuso AG, Montesinos-Amilibia JM, Sánchez-Soto LL. Geometrical aspects of PT-invariant transfer matrices. *Phys Rev* 2013;A 87:012111.
132. (a) Jakubský V. Thermodynamics of pseudo-Hermitian systems in equilibrium. *Mod Phys Lett A* 2007;22:1075–1084; (b) Brody DC, Graefe E-M. Mixed-state evolution in the presence of gain and loss. *Phys Rev Lett* 2012;109:230405.
133. Jarosz A, Nowak MA. Random Hermitian versus random non-Hermitian operators - unexpected links. *J Phys A Math Gen* 2006;39:10107–10122.
134. Günther U, Langer H, Tretter Ch. On the spectrum of the magnetohydrodynamic mean-field α^2–dynamo. *SIAM J Math Anal* 2010;42:1413–1447.
135. (a) Scholtz FG, Geyer HB. Moyal products - a new perspective on quasi-Hermitian quantum mechanics. *J Phys A Math Gen* 2006;39:10189–10206; (b) Figueira de Morrison Faria C, Fring A. Isospectral Hamiltonians from Moyal products. *Czech J Phys* 2006;56:899–908.
136. (a) Graefe E-M, Hoening M, Korsch HJ. Classical limit of non-Hermitian quantum dynamics—a generalized canonical structure. *J Phys A Math Theor* 2010;43:075306; (b) Mostafazadeh A. Real description of classical Hamiltonian dynamics generated by a complex potential. *Phys Lett* 2006;A 357:177.
137. Mostafazadeh A, Loran F. Propagation of electromagnetic waves in linear media and pseudo-Hermiticity. *Europhys Lett* 2008;81:10007.
138. Kůrka P, Matoušek A, Velický B, editors. *Spor o matematizaci světa (Dispute on the mathematization of world)*. Prague: Pavel Mervart; 2011 (in Czech).

2

OPERATORS OF THE QUANTUM HARMONIC OSCILLATOR AND ITS RELATIVES

FRANCISZEK HUGON SZAFRANIEC
Instytut Matematyki, Uniwersytet Jagielloński, Kraków, Poland

> ... quanto maior é a diferença, maior será igualdade, e quanto maior é a igualdade, maior a diferença será ...
>
> —José Saramago (1, p.96)

In point of fact, there are only two classes of Hilbert space operators that are commonly recognizable in Quantum Mechanics: symmetric (essential self-adjoint, and self-adjoint) and generators of different kind of semigroups. Other important operators, for instance those appearing in the quantum harmonic oscillator, seem to be not classified, at least unknown to the bystanders. One of our goals of this survey is to expose their role enhancing the most distinctive features. Our main "non-self-adjoint" object is the class of (unbounded) **subnormal** operators. This is compelling, and as such, it determines our *modus operandi:* **spatial** approach rather than Lie group/algebra connections. As a natural consequence, we have to refurnish the meaning that commutation relations are traditionally given to.

[†]the author acknowledges support of the MNSzW Grant No. NN201 546438

Non-Selfadjoint Operators in Quantum Physics: Mathematical Aspects, First Edition.
Edited by Fabio Bagarello, Jean-Pierre Gazeau, Franciszek Hugon Szafraniec and Miloslav Znojil.

2.1 INTRODUCING TO UNBOUNDED HILBERT SPACE OPERATORS

2.1.1 How to Understand an Unbounded Operator

An unbounded Hilbert space operator, as it appears usually in both Mathematics and Physics[1] is (or rather has to be if one refers to the last mentioned community) composed of two items

$$\begin{array}{c}\text{formula according to which the operator, say } A\text{, acts;}\\ \text{its domain, denoted here by } \mathcal{D}(A).\end{array} \tag{2.1.1}$$

These two _together_ legitimate thinking and speaking of a Hilbert space operator[2]. Unfortunately, in most of MP literature, the second of (2.1.1) is even not mentioned; "it is clear" is the typical authors' claim, if they are asked for what a domain of an operator they consider is—this kind of opportunism or rather nonchalance may have sometimes unpredicted consequences.

One of the ways to choose a domain of an operator is to collect all vectors to which the formula is applicable. This in a sense maximal domain is usually convenient from the theoretical point of view however in practice may create problems: it is too big and because of that too difficult to be described and, consequently, to be handled with. The compromise is as follows: choose as a domain a set (usually a linear subspace) of vectors on which the formula is conveniently executable and get from somewhere the information the would-be operator is closable (this as well as the other notion, that of core, is explained in Section 2.1.2.5 which follows). However "to be executable" may lead to different choices for domains, hence determine different operators (cf. Section 1.3 of (2)). The question whether they still concern the same operator can be tested by means of being a core. In conclusion, any core of a closable operator has equal rights to serve as its domain. Therefore, two domains attached to the formula (we refer here directly to (2.1.1)) determine the same operator if the closures of their corresponding graphs coincide.

The article of faith

Never consider unbounded operators to be closed unless this is an emergency case. Instead _always_ get ensured the operator in question is closable and take the advantage of choosing the most convenient domain among all its possible cores.

2.1.2 Very Basic Notions and Facts

All the spaces from now are Hilbert ones and their inner products are linear in the first variable[3]; the inner product is denoted[4] by $\langle \cdot , - \rangle$; sometimes, a subscript indicating

[1] The acronym MP will be used as gently as possible to make an appeal to either the discipline or the community of its followers.

[2] In mathematical papers, it is a kind of rule, more orthodox authors even reverse the order in (2.1.1).

[3] This is my mathematical habit, I do not expect MPs to have any problem at this stage.

[4] Dirac's "bra-ket" notations is deliberately avoided here.

the space in question is added. The operators are supposed to be linear and

densely defined

"Subspaces" are used for linear (not necessarily closed) subsets, if closed ones are considered we always make it explicit.

If an operator is everywhere defined we say it is *on*, otherwise it is *in*.

An operator is very often identified with its graph as a subspace in $\mathcal{H} \times \mathcal{H}$, $\mathcal{H}$ is the underlying Hilbert space; therefore, the set theoretical symbolism is sometimes used here without any warning. A kind of sample is the notion $A \subset B$ which can be read as $\mathcal{D}(A) \subset \mathcal{D}(B)$ and $Af = Bf$ for $f \in \mathcal{D}(A)$. This approach justifies the term *A-graph topology* considered in $\mathcal{D}(A)$ and expressed in the language of inner product as

$$\langle f, g \rangle_A \overset{\text{def}}{=} \langle f, g \rangle_{\mathcal{H}} + \langle Af, Ag \rangle_{\mathcal{H}}, \quad f, g \in \mathcal{D}(A). \tag{2.1.2}$$

2.1.2.1 Closures and Adjoints Besides $\mathcal{D}(A)$ already defined we attach to A two other basic subspaces: the *range* (sometimes called *image*) denoted by $\mathcal{R}(A)$ and the *null space* $\mathcal{N}(A)$ (sometimes called a kernel of A, confusing!).

$\bar{A}$ stands for the *closure*[5] of A meant in the graph sense (the easiest possible and the fastest way to describe). An operator A is said to be *closed* if $A = \bar{A}$; this amount to $\mathcal{D}(A)$ to be complete in the A-graph topology. A is said to be *closable* if there is a closed operator B such that $A \subset B$; the smallest B in the inclusion sense is just $\bar{A}$. As the notion of closability of an operator is of great importance[6] let us notice that A is closable if and only if for any sequence $(x_n)_n \subset \mathcal{D}(A)$ convergent to 0 and such that $(Ax_n)_n$ is a Cauchy sequence, $(Ax_n)_n$ must converge to 0 as well.

Given a densely defined operator A, consider the set

$$\{g \in \mathcal{H} : \text{ there is } h \in \mathcal{H} \text{ such that } \langle Af, g \rangle = \langle f, h \rangle \text{ for all } f \in \mathcal{D}(A)\} \tag{2.1.3}$$

Notice that because A is densely defined, the vector h in (2.1.3) is uniquely determined for each $g \in \mathcal{D}(A)$. Therefore, we can define an adjoint operator, denoted customarily by A^* with domain equal to the space (2.1.3). The relationship between A and A^* is usually expressed as

$$\langle Af, g \rangle = \langle f, A^* g \rangle, \quad f \in \mathcal{D}(A),\ g \in \mathcal{D}(A^*).$$

The domain (2.1.3) of $\mathcal{D}(A^*)$ can be characterized as the space of those g's for which the functional $f \to \langle Af, g \rangle$ is continuous or, still equivalently, of those g's for which there is c_g such that $|\langle Af, g \rangle| \leqslant c_g \|f\|$ for all $f \in \mathcal{D}(A)$.

[5]The dash $^-$ is going to stand for closures while the asterisk * will be used to denote conjugates of any kind (like those of scalars or Hilbert space operators).

[6]It looks like MPs never take care of this as they (MPs) believe that the supreme authorities do this for them.

REMARK **1** In MP's tradition, the role of A^* is usually played by $A^\dagger$, the latter called "Hermitian adjoint" of the previous. It is hardly ever defined rigorously also because the domain of A is commonly not indicated. This make some sense as long as A and $A^\dagger$ are formal algebraic objects rather than Hilbert space operators; a serious warning comes from (3) and (4). Let us mention that "Hermitian adjoint" is settled in (finite) matrix theory; in modern operator theory, it is used rather for operators which are not densely defined, cf. (5, p. 72). ♣

The adjoint A^* is always a closed operator, it is densely defined if and only if A is closable. If A is closable, then $(\bar{A})^* = A^*$ and $\bar{A} = A^{**} \stackrel{\text{def}}{=} (A^*)^*$. Moreover, with convention

$$\mathcal{D}(A+B) \stackrel{\text{def}}{=} \mathcal{D}(A) \cap \mathcal{D}(B), \tag{2.1.4}$$

$$\mathcal{D}(AB) \stackrel{\text{def}}{=} \{f \in \mathcal{D}(B) : \; Bf \in \mathcal{D}(A)\}, \tag{2.1.5}$$

we have

(a) $A^* + B^* \subset (A+B)^*$ provided $\mathcal{D}(A+B)$ is dense, with equality when one of the operators is bounded and densely defined;
(b) if $\mathcal{D}(AB)$ and $\mathcal{D}(A)$ are dense then $B^*A^* \subset (AB)^*$, with equality if A is bounded and everywhere defined.

2.1.2.2 C^∞-Vectors

For an operator A set

$$\mathcal{D}^\infty(A) \stackrel{\text{def}}{=} \bigcap_{n=0}^{\infty} \mathcal{D}(A^n),$$

$$\mathcal{D}^\infty(A, A^*) \stackrel{\text{def}}{=} \bigcap_{\substack{A_1,\dots A_n \in \{A,A^*\} \\ \text{any finite choice}}} \mathcal{D}(A_1 \cdots A_n).$$

It is customary to refer to vectors in any of these two classes as to C^∞-ones.

One has to notify that

$$\mathcal{D}^\infty(A^*, A) = \mathcal{D}^\infty(A, A^*) \subset \mathcal{D}^\infty(A^*) \cap \mathcal{D}^\infty(A).$$

If $f \in \mathcal{D}^\infty(A)$, then $p(A)f \in \mathcal{D}^\infty(A)$ for any polynomial $p \in \mathbb{C}[Z]$ as well, if $f \in \mathcal{D}^\infty(A^*, A)$ then $p(A^*, A)f \in \mathcal{D}^\infty(A^*, A)$ for any polynomial $p \in \mathbb{C}[Z, \bar{Z}]$; the latter regardless any commutativity property between A and A^*.

A vector $f \in \mathcal{D}^\infty(A)$ may belong to one of the following classes: $\mathcal{B}(A)$ (*bounded*), $\mathcal{A}(A)$ (*analytic*), $\mathcal{E}(A)$ (*entire*) or $\mathcal{Q}(A)$ (*quasianalytic*). While the second and the fourth are rather pretty well known the other two are much less frequent.

Let us recall all the definitions. Customarily, $f \in \mathcal{D}^\infty(A) \overset{\text{def}}{=} \bigcap_{n=0}^\infty \mathcal{D}(A^n)$ is

a *bounded* vector if there are $a > 0$ and $b > 0$ such that

$$\|A^n f\| \leqslant ab^n \text{ for } n = 0, 1, \ldots ;$$

an *analytic* vector of A if there is $t > 0$ such that

$$\sum_{n=0}^\infty \frac{t^n}{n!}\|A^n f\| < +\infty; \tag{2.1.6}$$

an *entire* vector if (2.1.6) holds for all $t > 0$;

a *quasianalytic* vector of A if $\sum_{n=0}^\infty \|A^n f\|^{-1/n} = +\infty$.

Let us introduce the self-evident notation $\mathcal{B}(A)$, $\mathcal{A}(A)$, $\mathcal{E}(A)$, and $\mathcal{Q}(A)$ for the consecutive classes. The first three are always linear while the last may not be. Nevertheless, the following inclusions are transparent

$$\mathcal{B}(A) \subset \mathcal{E}(A) \subset \mathcal{A}(A) \subset \mathcal{Q}(A). \tag{2.1.7}$$

It may happen that even $\mathcal{Q}(A)$ is a zero space. Nevertheless, density of $\operatorname{lin}\mathcal{Q}(A)$[7] implies that of any other in the chain of inclusions (2.1.7) and it helps a lot.

2.1.2.3 *Invariant and Reducing Subspaces* A <u>closed</u> subspace $\mathcal{L}$ of $\mathcal{H}$ is *invariant* for A if $A(\mathcal{L} \cap \mathcal{D}(A)) \subset \mathcal{L}$; then the *restriction* $A{\restriction}_{\mathcal{L}} \overset{\text{def}}{=} A|_{\mathcal{L}\cap\mathcal{D}(A)}$ as an operator in $\mathcal{L}$ becomes clear. If A is a closable (closed) operator in $\mathcal{H}$ then so is $A{\restriction}_{\mathcal{L}}$; this happens because the notions are topological, with topology in the graph space. The closed subspace $\mathcal{L}$ is invariant for A if and only if

$$PAP = AP \tag{2.1.8}$$

where P is the orthogonal projection of $\mathcal{H}$ onto $\mathcal{L}$.

On the other hand, a <u>linear</u> subspace $\mathcal{D} \subset \mathcal{D}(A)$ is said to be *invariant* for an operator A in $\mathcal{H}$ if $A\mathcal{D} \subset \mathcal{D}$; this is a standard set theoretical notion. If this occurs and $\mathcal{D}$ is not dense in $\mathcal{H}$, we consider the *restriction* $A|_{\mathcal{D}}$ as a densely defined operator in $\overline{\mathcal{D}}$. However, if $\mathcal{D}$ is a dense in $\mathcal{H}$, then $A|_{\mathcal{D}}$ is still a densely defined operator in $\mathcal{H}$. If density happens, the only difference between this and the previous case is in the latter $\mathcal{R}(A) \subset \mathcal{D}$.

[7]lin stands for the linear span while clolin does for the closed linear one.

The concepts of invariance and restriction look much alike although in general they are not. For bounded, everywhere defined operators they coincide. If a linear subspace $\mathcal{D}$ is invariant for A, then $A|_{\mathcal{D}} \subset A\restriction_{\overline{\mathcal{D}}}$, whereas $\overline{A|_{\mathcal{D}}} = \bar{A}\restriction_{\overline{\mathcal{D}}}$ provided A is closable[8]. This makes the difference more transparent.

A step further, a closed subspace $\mathcal{L}$ *reduces* an operator A if both $\mathcal{L}$ and $\mathcal{L}^{\perp}$ are invariant for A as well as $P\mathcal{D}(A) \subset \mathcal{D}(A)$; all this is the same as to require

$$PA \subset AP. \tag{2.1.9}$$

The restriction $A\restriction_{\mathcal{L}}$ is called a *part* of A in $\mathcal{L}$.

The above-mentioned concepts have a matrix interpretations, which correspond to (2.1.8) and (2.1.9) accordingly

$$\begin{pmatrix} A\restriction_{\mathcal{L}} & 0 \\ * & A\restriction_{\mathcal{L}^{\perp}} \end{pmatrix} \text{ INVARIANCE, } \quad \begin{pmatrix} A\restriction_{\mathcal{L}} & 0 \\ 0 & A\restriction_{\mathcal{L}^{\perp}} \end{pmatrix} \text{ REDUCIBILITY.} \tag{2.1.10}$$

The second matrix representation allows as to write the orthogonal decomposition

$$A = A\restriction_{\mathcal{L}} \bigoplus A\restriction_{\mathcal{L}^{\perp}}.$$

A is called *irreducible* if there is no nontrivial closed subspace $\mathcal{L}$ which reduces A, it is synonymous to impossibility of the decomposition (2.1.9) to hold.

2.1.2.4 Extensions and Dilations In this section, we collect some facts that are scattered in the literature, sometimes written in a slightly old-fashioned way. For some, it would be nothing but an occasion to refresh their memory, for others a moment of getting new thoughts.

This paragraph is rather scratchy, its only purpose is to evoke some forgotten or overlooked associations rather than provide precise statements; they will follow. Consider two Hilbert spaces $\mathcal{H}$ and $\mathcal{K}$ with $\mathcal{H} \subset \mathcal{K}$ isometrically. Suppose $\boldsymbol{A}$ and $\boldsymbol{B}$ are families of operators in $\mathcal{H}$ and $\mathcal{K}$, respectively. Moreover, suppose both $\boldsymbol{A}$ and $\boldsymbol{B}$ are composed of pairwise commuting operators, which is meant pointwise where ever executable[9]. The two notions can be defined along with: for any finite collection $A_1, \dots, A_n$ in $\boldsymbol{A}$ there correspond $B_1, \dots, B_n$ in $\boldsymbol{B}$ such that

$$A_1 \cdots A_n \subset \begin{cases} PB_1 \cdots B_n & \text{DILATION} \\ \text{or} & \\ B_1 \cdots B_n & \text{EXTENSION} \end{cases}. \tag{2.1.11}$$

[8] Identifying operators with their graphs we can write $A|_{\mathcal{D}} = A \cap (\mathcal{D} \times \mathcal{D})$ and $A\restriction_{\mathcal{D}} = A \cap (\overline{\mathcal{D}} \times \overline{\mathcal{D}})$. Hence, the equality follows.

[9] If they are not supposed to commute than keeping the order is important.

where P stands for the orthogonal projection of $\mathcal{K}$ onto $\mathcal{H}$. An operator $B \in \boldsymbol{B}$ which is a dilation[10] or an extension of $A \in \boldsymbol{A}$ has a matrix decomposition with respect to $\mathcal{K} = \mathcal{H} \bigoplus \mathcal{H}^{\perp}$

$$\begin{pmatrix} A & * \\ \& & \# \end{pmatrix} \quad \text{DILATION}, \qquad \begin{pmatrix} A & * \\ 0 & \# \end{pmatrix} \quad \text{EXTENSION}, \tag{2.1.12}$$

respectively.

It is clear that extensions are dilations; it is also clear extensions of singletons are in a sense enough to be considered in (2.1.11). The recipe (2.1.11) can be decoded as follows:

$$\mathcal{D}(A_1 \cdots A_n) \subset \mathcal{D}(B_1 \cdots B_n) \text{ and for } f \in \mathcal{D}(A_1 \cdots A_n),\ g \in \mathcal{H}$$

$$\langle A_1 \cdots A_n f, g \rangle = \langle B_1 \cdots B_n f, g \rangle \quad \text{DILATION}$$

or

$$A_1 \cdots A_n f = B_1 \cdots B_n f \quad \text{EXTENSION}$$

respectively. Thus, dilations can be thought of as weak extensions.

While extensions seem to be easier to accept dilations at first glance may create some resistance. The following is in order to help in bringing closer the idea of dilation to those who are not accustomed to it.

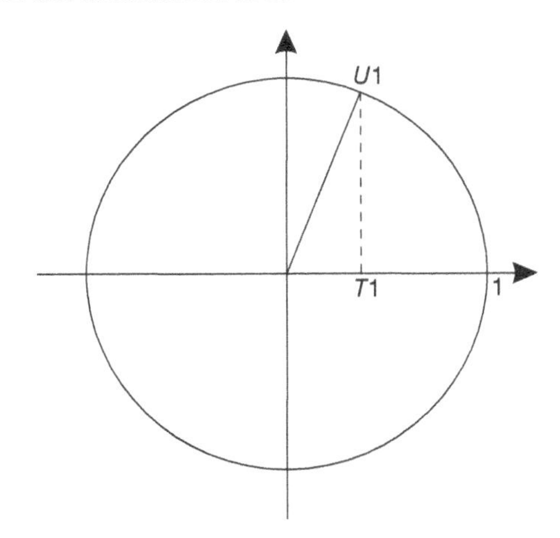

The space $\mathcal{H}$ is represented here by the x-axis, and the space $\mathcal{K}$ is the two-dimensional real plane. To obtain $T1$, one has to project orthogonally the action of the unitary (rotation) U on the vector 1 onto the space $\mathcal{H} = x$-axis. In other words, $T1 = PU1$.

[10]The term "dilation" refers to operators and is completely different from what is used elsewhere, for instance in frame theory; the latter the author remembers from his school years as "homothety"—progress in terminology.

The dilation pictured above was done in the case of an arbitrary Hilbert space by Julia (6, pp. 339–348) . It, as a rule, doubles the dimension of the input space, that is $\dim \mathcal{K} = 2 \dim \mathcal{H}$ whatever it means. As a disadvantage, it does not dilate T and T^2 simultaneously to U and U^2.

A look at the matrix representations (2.1.10) as compared with (2.1.12) makes it clear that the current contents is a kind of "the other way around" of that of Section 2.1.2.3.

The main reason for considering dilations and extensions is to deduct some properties of operators in the input space, say $\mathcal{H}$, from those acting in the output space, like $\mathcal{K}$ here. Therefore, the problem consists in

given $\boldsymbol{A}$, determine a class $\boldsymbol{B}$ with properties well-suited to those of the class $\boldsymbol{A}$ we are particularly interested in.

A simple though instructive example of such deduction is the case when operators inherit closedness or closability from that of their extensions.

REMARK **2** *If B is an extension of A which is closable then so is A. If B is a dilation of A which is closed then so is A.*

Extendibility of an operator forces dilatability of its adjoint.

Fact 2.1.1 *If B is an extension of A from $\mathcal{H}$ to $\mathcal{K}$ then $A^* = PB^*|_{\mathcal{H}}$, where P is the orthogonal projection of $\mathcal{K}$ onto $\mathcal{H}$.*

Representation (2.1.12) shows that dilatability is $*$-invariant while extendibility is not unless the operators in $\boldsymbol{A}$ are symmetric. They also suggest the rough proof of the following statement, which is a sort of converse to Fact 2.1.1.

Fact 2.1.2 *B is an extension of A if and only if B^* is a dilation of A^* and B^*B is a dilation of A^*A, and, equivalently, if and only if $B^{*j}B^i$ is a dilation of $A^{*j}A^i$ for $i,j = 0, 1, \ldots$*

This if made precise can be proved in much the way as in (7, p. 20) for bounded operators (more on dilations and extension, see (8)).

Fact 2.1.2 makes it clear why the difference between dilations and extensions in the case of symmetric operators becomes invisible.

In general, we do not exclude the case $\mathcal{H} = \mathcal{K}$. Then, the notion of dilation becomes irrelevant though that of extension is flourishing. On the other hand, for the extension questions, everywhere defined operators do not bring in any novelty; the unbounded operators are the only players left—as also this volume shows very rewarding players.

However, there is one important and beautiful instance when dilations and extensions (or, that is to say: bounded and unbounded) go together—that is Naĭmark's.

2.1.2.5 Core Now is the right time to introduce the notion which has already appeared here, i.e. that of core.

A subspace $\mathcal{D} \subset \mathcal{D}(A)$ is said to be a *core* of a closable operator A if

$$\overline{A|_{\mathcal{D}}} = \bar{A}.$$

It is plain that $\mathcal{D}(A)$ itself is a core of A. However, the essence of this notion is in fostering a chance to change the *a priori* given domain if a need of some more appropriate one appears, without demolishing the operator at large. This may happen for instance (which is the case here) if one wants to have a domain to be invariant for A. The maximal domain, which usually makes the operator closed, must not be used for that purpose because; owing to some results (cf. (9)), in most cases, it would force the operator to be bounded.

$\mathcal{D}$ is a core of A if it is dense in $\mathcal{D}(\bar{A})$ in the $\bar{A}$-graph topology. Referring more explicitly to (2.1.2), we have $\mathcal{D}$ to be a core of A if and only if

$$f \in \mathcal{D}(\bar{A}) \ \& \ \langle f, g\rangle_{\mathcal{H}} + \langle \bar{A}f, Ag\rangle_{\mathcal{H}} = 0 \text{ for all } g \in \mathcal{D} \text{ implies } f = 0. \tag{2.1.13}$$

2.1.2.6 Spectra and Spectral Measures To think of spectra of unbounded operators, the emergency case mentioned in "The article of faith," p. 60, has to be taken into account in a sense, at least one may define diverse parts of the spectrum of the closure of the operator in question; this is to avoid trivialities. While the definition of the spectrum causes no doubt, to keep away from any confusion, we want to state explicitly how we understand different parts of the spectrum of an operator A in $\mathcal{H}$, cf. (10, p. 104). Denote by $\mathrm{sp_p}(A)$ the *point spectrum* of A, that is, the set of all $\lambda \in \mathbb{C}$ for which there is a nonzero $f \in \mathcal{D}(A)$ such that $Af = \lambda f$; then λ is called an *eigenvalue* of A with the corresponding *eigenvector* f. The *residual spectrum* $\mathrm{sp_r}(A)$ of A is composed of these λ's for which $\mathcal{N}(\lambda - A) = \{0\}$ and $\mathcal{R}(\lambda - A)$ is not dense in $\mathcal{H}$; if λ is such that $\mathcal{N}(\lambda - A) = \{0\}$, $\mathcal{R}(\lambda - A)$ is dense in $\mathcal{H}$ and the inverse $(\lambda - A)^{-1}$ is not continuous, then λ belongs to the *continuous spectrum* $\mathrm{sp_c}(A)$ of A; the *approximate point spectrum* $\mathrm{sp_{ap}}(A)$ of A is composed of these λ's for which there is a sequence $(f)_{n=0}^{\infty} \subset \mathcal{D}(A)$ of unit vectors such that $(\lambda - A)f_n \longrightarrow 0$ as $n \longrightarrow \infty$. There is another description of $\mathrm{sp_{ap}}(A)$, namely $\lambda \notin \mathrm{sp_{ap}}(A)$ if and only if there is $c > 0$ such that $\|(\lambda - A)f\| \geqslant c\|f\|$ for $f \in \mathcal{D}(A)$. *Resolvent set* $\rho(A)$ of A is composed of all λ's for which $\mathcal{N}(\lambda - A) = \{0\}$, $\mathcal{R}(\lambda - A)$ is dense in $\mathcal{H}$ and $(\lambda - A)^{-1}$ is bounded; in case, A is closed this means precisely $\lambda - A$ is boundedly invertible. The *spectrum* of A is the set $\mathrm{sp}(A) \overset{\text{def}}{=} \mathbb{C} \setminus \rho(A)$.

Notify two decompositions of $\mathrm{sp}(A)$

$$\mathrm{sp}(A) = \mathrm{sp_p}(A) \cup \mathrm{sp_c}(A) \cup \mathrm{sp_r}(A), \quad \mathrm{sp}(A) = \mathrm{sp_{ap}}(A) \cup \mathrm{sp_r}(A), \tag{2.1.14}$$

where the summands in the first decomposition are pairwise disjoint. It is clear that unitary equivalent (or even similar) operators have the same spectra.

It is evident that

$$\mathrm{sp_p}(A) \subset \mathrm{sp_{ap}}(A).$$

This and the second of (2.1.14) affect spectral properties of those operators which appear here. Let us mention also that for the topological boundary of sp(A) we have $\partial\mathrm{sp}(A) \subset \mathrm{sp}_{\mathrm{ap}}(A)$, see the beginning of the proof of Theorem 2 in (11).

Let us consider for simplicity a σ-algebra $\mathfrak{B}(X)$ of Borel sets of a given topological space X. By a *semispectral measure* (=POVM) on $\mathfrak{B}(X)$), we understand a bounded positive operator valued function F on $\mathfrak{B}$, which is σ-additive in the weak (or, equivalently, strong) operator topology and $F(X) = I$. If a semispectral measure is orthogonal projection valued then we call it a *spectral measure*; this is equivalent to $F(\rho \cap \sigma) = F(\rho)F(\sigma)$, $\rho, \sigma \in \mathfrak{B}$. The latter shows that the values of a spectral measure commute.

The celebrated Naĭmark dilation theorem (12) relates the two kinds of operator valued measures each to the other, see (8) and (13).

Theorem 2.1.3 *F is a semispectral measure in $\mathcal{H}$ if and only if there exists a Hilbert space $\mathcal{H} \subset \mathcal{K}$ (isometric embedding) and a spectral measure E in $\mathcal{K}$ such that*

$$\langle F(\sigma)f, g\rangle_{\mathcal{H}} = \langle E(\sigma)f, g\rangle_{\mathcal{K}}, \quad f, g \in \mathcal{H}. \tag{2.1.15}$$

(2.1.15) can be read as $F(\sigma) = PE(\sigma)|_{\mathcal{H}}$, where P is the orthogonal projection of $\mathcal{K}$ onto $\mathcal{H}$, which refers directly to how dilations look like. Dilations of semispectral measures when they satisfy some minimality condition are determined uniquely up to unitary isomorphism.

REMARK **3** It is important to notice that Naĭmark theorem dilates not necessarily commuting operators (the values of semispectral measures) to commuting ones (such are the values of spectral measures). This, in the context of Cartesian decomposition, is hidden behind subnormality.

On the other hand, extensions share multiplicativity, hence commutativity with the operators they extend. ♣

2.1.2.7 From Symmetric and Self-adjoint Operators … An operator A is called

$$\begin{aligned} &\textit{symmetric} \text{ if } \mathcal{D}(A) \subset \mathcal{D}(A^*) \text{ and } Af = A^*f \text{ for } f \in \mathcal{D}(A), \\ &\textit{self-adjoint} \text{ if it is symmetric and } \mathcal{D}(A) = \mathcal{D}(A^*). \end{aligned} \tag{2.1.16}$$

In other words, symmetricity reads as $A \subset A^*$ while self-adjointness as $A = A^*$. If $\bar{A} = A^*$, A is called *essentially self-adjoint*; this gives a freedom of choice between possible domains. A symmetric operator is essentially self-adjoint if and only if both its deficiency indices are equal to 0; density of any class of C^∞ vectors form a nothing but a *sufficient* condition for essential self-adjointness.

There are two standard procedures which lead to self-adjoint extensions of symmetric operators: von Neumann's (work in the same space if deficiency indices are equal) and Naĭmark's[11] (go beyond the space even if deficiency indices are equal).

[11] Do not confuse with Naĭmark dilations.

For an arbitrary operator we have the *Cartesian decomposition*

$$Af = \mathfrak{Re}\,Af + \mathrm{i}\,\mathfrak{Im}\,Af, \quad A^*f = \mathfrak{Re}\,Af - \mathrm{i}\,\mathfrak{Im}\,Af, \quad f \in \mathcal{D}(A) \cap \mathcal{D}(A^*) \tag{2.1.17}$$

after defining $\mathfrak{Re}\,A \overset{\text{def}}{=} \frac{1}{2}(A + A^*)$ and $\mathfrak{Im}\,A \overset{\text{def}}{=} \frac{1}{2\mathrm{i}}(A - A^*)$ with (2.1.4) in mind. If $\mathcal{D}(A) \cap \mathcal{D}(B)$ is dense, both $\mathfrak{Re}\,A$ and $\mathfrak{Im}\,A$ are symmetric.

2.1.2.8 ... to Formally Normal and Normal Ones An operator N is called

$$\begin{array}{c} \textit{formally normal} \text{ if } \mathcal{D}(N) \subset \mathcal{D}(N^*) \text{ and } \|Nf\| = \|N^*f\| \text{ for } f \in \mathcal{D}(N), \\ \textit{normal} \text{ if it is formally normal and } \mathcal{D}(A) = \mathcal{D}(A^*). \end{array} \tag{2.1.18}$$

In other words, with (2.1.5) in mind for N to be formally normal is equivalent to $N^*N \subset NN^*$ while to be normal is $NN^* = N^*N$ as well as N closed, cf. (14). N is said to be *essentially normal* if $\overline{N}$ is normal.

REMARK **4** Because N^* is always closed, formally normal operator N must necessarily be closable. ♣

The interrelations between them can be visualized by the diagram

$$\begin{array}{ccc} \text{self-adjoint} & \Longrightarrow & \text{symmetric} \\ \Downarrow & & \Downarrow \\ \text{normal} & \Longrightarrow & \text{formally normal} \end{array} \tag{2.1.19}$$

There is a clear analogy between (2.1.16) and (2.1.18) which makes the notions parallel; even more, they look much alike. However, there is a dramatic difference between these two notions. Let us demonstrate it by answering two questions.

- Suppose $A \subset B$ and B is self-adjoint. Should A necessarily be symmetric? Answer: **yes**, of course.
  Suppose $M \subset N$ and N is normal. Should M necessarily be formally normal? Answer: **no**. Why? Because the closure of $\mathcal{D}(M)$ may not reduce N.
- Suppose B is symmetric. Does it always have a self-adjoint extension? Answer: **yes**, always[12], according to the classics (von Neumann and Naĭmark).
  Suppose M is formally normal. Does it always have a normal extension? Answer: it may **not**. Counterexample: (15), see also (16).

Decomposing a formally normal operator as a Cartesian product (2.1.17) results in a kind of Pythagorean theorem, $\|Nf\|^2 = \|\,\mathfrak{Re}\,Nf\|^2 + \|\,\mathfrak{Im}\,Nf\|^2$, and vice versa. This in turn is equivalent to $\mathfrak{Re}\,N$ and $\mathfrak{Im}\,N$ are weakly commuting, i.e. $\langle \mathfrak{Re}\,Nf, \mathfrak{Im}\,Ng \rangle = \langle \mathfrak{Im}\,Ng, \mathfrak{Re}\,Nf \rangle$ for f, g belonging to the common domain of these two. On the other hand for normality of N, it is required both $\mathfrak{Re}\,N$ and $\mathfrak{Im}\,N$

[12]Here and below we think of extension *either in the same or a larger space.*

to *commute spectrally* (means their spectral measure have to commute). Therefore, the famous Nelson counterexample (17) makes another, serious obstacle which has to be considered when thinking of existence of normal extensions.

Because formal normality does not guarantee existence of normal extensions, the primary question is

which operators in general do have normal extensions.

This question, of the upmost importance, is going to be challenged in Section 2.1.3. For what refers to formally normal operators, a complete (geometrical in nature) characterization of those which do have normal extensions consult (18, 19).

Notice that for bounded, everywhere defined operators there is no distinction between symmetric and self-adjoint operators, there is no distinction between formally normal and normal operators either.

2.1.2.9 Spectral Representation of Normal Operators First of all notice that for a normal operator N, one has

$$\mathrm{sp}(N) = \mathrm{sp}_{\mathrm{ap}}(N). \tag{2.1.20}$$

Theorem 2.1.4 (Spectral theorem, the extras included) *An operator N is normal if and only if it is a spectral integral of the identity function on $\mathbb{C}$ with respect to a spectral measure E on $\mathbb{C}$, that is*

1° $\langle Nf, g\rangle = \int_{\mathbb{C}} z\langle E(\mathrm{d}z)f, g\rangle$ *for all $f \in \mathcal{D}(N)$ and $g \in \mathcal{H}$.*

Moreover, if this happens then

2° $\mathcal{D}(N) = \{f \in \mathcal{H} : \ \int_{\mathbb{C}} |z|^2\langle E(\mathrm{d}z)f,f\rangle < +\infty\}$;

3° *for every Borel measurable nonnegative function ϕ on $\mathbb{C}$ and $f \in \mathcal{D}(N)$*

$$\int_{\mathbb{C}} \phi(x)\langle E(\mathrm{d}z)Nf, Nf\rangle = \int_{\mathbb{C}} \phi(x)\,|z|^2\langle E(\mathrm{d}z)f,f\rangle,$$

in particular,

$$\|Nf\|^2 = \int_X |z|^2\langle E(\mathrm{d}z)f,f\rangle, \quad f \in \mathcal{D}(N)$$

and

$$E(\sigma)N \subset NE(\sigma) \text{ for all Borel sets } \sigma;$$

4° *the spectral measure E is uniquely determined and its closed support coincides with the spectrum of N;*

5° *the space*

$$\mathcal{B}(N) = \mathrm{lin}\{E(\sigma)f : \ f \in \mathcal{H},\ \sigma \in \mathfrak{B} \text{ compact}\}$$

is an invariant core of N.

The converse goes as follows: given a spectral measure E such that the linear space

$$\mathcal{D}_E \stackrel{\mathsf{def}}{=} \mathcal{D}(N) = \{f \in \mathcal{H} : \int_{\mathbb{C}} |z|^2 \langle E(\mathrm{d}z)f, f\rangle < +\infty\}$$

appearing in 2° is dense in $\mathcal{H}$, then operator N defined in 1° with $\mathcal{D}(N) = \mathcal{D}_E$ is normal and all other consequences follow (see (2, p. 103) for some details as well as (20)).

As a consequence of 5°, a normal operator decomposes as an orthogonal sum of a sequence of bounded normal operators.

In addition to (2.1.20), we have

$$\mathrm{sp}(N) = \mathrm{sp}_{\mathrm{ap}}(N) = \mathrm{supp}\, E.$$

Suppose now μ is a (positive = nonnegative in this case) measure on the complex plane[13] $\mathbb{C}$, with finite moments. Suppose moreover that the set

$$\{f \in \mathcal{L}^2(\mu) : \int_{\mathbb{C}} |zf|^2 \mu(\mathrm{d}z)\} < +\infty \tag{2.1.21}$$

is dense in $\mathcal{L}^2(\mu)$. Then, the operator multiplication N_Z defined by $(N_Z f)(z) \stackrel{\mathsf{def}}{=} zf(z)$, $z \in \mathbb{C}$, with domain given by (2.1.21) is normal. The spectral theorem says that N_Z is generic for some class of normal operators.

Call a normal operator $*$-*cyclic* if there is a vector $e \in \mathcal{D}^\infty(N, N^*)$ such that

$$\{p(N, N^*)e : \ p \in \mathbb{C}[Z, Z^*]\} \tag{2.1.22}$$

is a core of N.

If N is normal then the closure of the space (2.1.22) reduces N to a normal operator; therefore, N decomposes as an orthogonal sum of $*$-cyclic normal operators, cf. Section 2.1.2.3.

Theorem 2.1.5 (Spectral representation of $*$-cyclic normal operators) *Suppose that N is a $*$-cyclic normal operator with the cyclic vector e and E its spectral measure. Then, there is a unique unitary operator $U : \mathcal{H} \to \mathcal{L}^2(\mu)$ with $\mu \stackrel{\mathsf{def}}{=} \langle E(\cdot)e, e\rangle$ acting as*

$$Up(N, N^*)e = p, \quad p \in \mathbb{C}[Z, Z^*].$$

Consequently, $UNf = N_Z Uf$ for $f \in \mathcal{D}(N)$. Moreover, for $k = 0, 1, \ldots,$ one has $UN^k f = (N_Z)^k Uf$, $f \in \mathcal{D}^\infty(N, N^*)$ at least.

2.1.2.10 Weighted Shifts Consider a *separable* Hilbert space $\mathcal{H}$ and its orthonormal basis[14] $(e_n)_{n=0}^\infty$. Given a sequence $(\sigma_n)_{n=0}^\infty$ of positive numbers (called *weights*),

[13]The support of μ, which is closed by definition, needs not be the whole of $\mathbb{C}$

[14]That is an orthonormal and *complete* sequence.

an operators S is called a *(forward) weighted shift* (with respect to the given data) if, with $\mathcal{D} \stackrel{\text{def}}{=} \operatorname{lin}(e_n)_{n=0}^{\infty}$

$$\mathcal{D}(S) \stackrel{\text{def}}{=} \mathcal{D}, \quad Se_n \stackrel{\text{def}}{=} \sigma_n e_{n+1}, \quad n = 0, 1, \dots$$

The *(backward) weighted shift* T is defined as

$$\mathcal{D}(T) \stackrel{\text{def}}{=} \mathcal{D}, \quad Te_0 \stackrel{\text{def}}{=} 0, \quad Te_n \stackrel{\text{def}}{=} \sigma_{n-1} e_{n-1}, \; n = 1, \dots .$$

Sometimes, an emphasis is put on them to be called *unilateral*, or better *onesided*, the others, *bilateral* or *twosided* come into side when the data are indexed from $-\infty$ to $+\infty$.

Fact 2.1.6 *Let S be a forward weighted shift and let $C(S)$ stand for one of the classes in* (2.1.7). *Then*[15]

1° *if e_0 does not belong to $C(S)$ then $C(\bar{S}) = \{0\}$;*
2° *if $e_0 \in C(S)$ then* lin $C(S)$ *is dense in $\mathcal{H}$.*

In contrast to the aforementioned always

Fact 2.1.7 *$\mathcal{D}(T) \subset \mathcal{B}(T)$ for any backward shift.*

Weighted shifts are always irreducible, cf. Section 2.1.2.3. This is a basic fact. Moreover, we have the following description:

$$\mathcal{D}(\bar{S}) = \sum_{n=0}^{\infty} |\xi_n|^2 |\sigma_n|^2 < +\infty, \; \bar{S}f = \sum_{n=0}^{\infty} \xi_n \sigma_n e_{n+1}, \; f = \sum_{n=0}^{\infty} \xi_n e_n \in \mathcal{D}(\bar{S});$$
$$\mathcal{D}(\bar{T}) = \sum_{n=0}^{\infty} |\xi_n|^2 |\sigma_{n-1}|^2 < +\infty, \; \bar{T}f = \sum_{n=0}^{\infty} \xi_n \sigma_{n-1} e_{n-1}, \; f = \sum_{n=0}^{\infty} \xi_n e_n \in \mathcal{D}(\bar{T}).$$

There are some immediate consequences of the above, see (21, Propositions 3 and 4). The most important is

$$S = T^*|_{\mathcal{D}} \text{ and } T = S^*|_{\mathcal{D}}. \tag{2.1.23}$$

The others are as follows:

the operator S, and hence T, is a bounded operator if and only if $\sup_n \sigma_n < +\infty$; $\mathcal{D}(\bar{S}) = \mathcal{D}(S^*)$ if and only if $c^{-1}|\sigma_{n+1}|^2 \leqslant 1 + |\sigma_n|^2 \leqslant c(2 + |\sigma_{n+1}|^2)$ for some $c > 0$.

[15] Cf. (21, Proposition 5)

REMARK **5** It is worthy of our mention that the recipe to have

$$Te_0 = 0 \tag{2.1.24}$$

can be achieved in two ways:

- either to define a weigted shift S and then use the second of (2.1.23) for this,
- or, *à rebours* to define S and T independently which results in getting (2.1.24) from the very definition of T.

This, rather trivially looking observation, has some *methodological* meaning when one thinks of the way "vacuum" or "ground state" e_0 may appear, which happens more or less explicitly here too. ♣

2.1.3 Subnormal Operators

2.1.3.1 Subnormal Operators: the General Case A densely defined operator S is called *subnormal* if it has a normal extension, cf. (11, 22–26). In the graph notation, it can be written as $S \subset N$. The latter shows evidently that a subnormal operator must necessarily be closable because its normal extension is such.

The only characterization of subnormality which does not impose any constrain on behavior of domains of the operator is that via semispectral measures (27) or (28).

Theorem 2.1.8 *An operator S is subnormal if and only if there is a semispectral measure F on Borel sets of $\mathbb{C}$ such that*[16]

$$\langle S^m f, S^n g\rangle = \int_{\mathbb{C}} z^m \bar{z}^n \langle F(\mathrm{d}z)f, g\rangle, \quad m, n = 0, 1, \quad f, g \in \mathcal{D}(S). \tag{2.1.25}$$

Notice that semispectral measures related to a particular subnormal operator may not be uniquely determined, see (29) for an explicit example. As spectral measures of normal extensions come via dilating semispectral measures, according Naĭmark's dilation theorem, cf. (12), we may have quite a number of them as well. This foretells somehow the problem with uniqueness (and minimality) we are going to expose a little bit later.

Minimality in the unbounded case becomes a very sensitive issue. Let us start with a definition: call N *minimal extension of spectral type* if $\mathcal{H} \subset \mathcal{K}_1 \subset \mathcal{K}$ and $N\restriction_{\mathcal{K}_1}$ turns out to be normal then either $\mathcal{K}_1 = \mathcal{H}$ or $\mathcal{K}_1 = \mathcal{K}$. Owing to Proposition 1 in (11), it is equivalent to

$$\mathcal{K} = \operatorname{clolin}\{E(\sigma)f : \ \sigma \text{ Borel subset of } \mathbb{C}, f \in \mathcal{H}\},$$

E being the spectral measure of N.

[16] If (2.1.25) holds only for $m = 1$ and $n = 1$ ($m = n = 0$ is a triviality), then S has a normal dilation exclusively and *vice versa*. In this case, the fourth condition encoded in (2.1.25) downgrades to the inequality $\langle Sf, Sg\rangle \leqslant \int_{\mathbb{C}} |z|^2 \langle F(\mathrm{d}z)f, g\rangle, f, g \in \mathcal{D}(S)$.

The sad news is that minimal normal extensions of spectral type may not be $\mathcal{H}$-equivalent[17], see (29) or Example 1 in (11) much further developed in (30); therefore, no uniqueness can be expected at this stage. The good news is the welcomed spectral inclusion

$$\mathrm{sp}(N) \subset \mathrm{sp}(S) \tag{2.1.26}$$

is always preserved; as a consequence of (2.1.26), notify $\mathrm{sp}(S) \neq \emptyset$. A list of further spectral properties is in Theorem 1 of (11).

Old friends in the new environment The following draft shows how all the notions described so far in this or another way interplay; each of the implications may be *irreversible*[18].

$$\begin{array}{ccccc} \textit{self-adjoint} & \Longrightarrow & \textit{symmetric} & \Longrightarrow & \textit{formally normal} \\ \Downarrow & & \Downarrow & & \\ \text{normal} & \Longrightarrow & \textbf{subnormal} & & \\ \Downarrow & & & & \\ \textit{formally normal} & & & & \end{array}$$

This graph is complementary to (2.1.19). Notice that the formally normal operators are positioned somehow apart and this corresponds to their role.

2.1.3.2 Subnormal Operators with Invariant Domain From now onward, we declare

$$\boxed{S\mathcal{D}(S) \subset \mathcal{D}(S)}$$

Now formula (2.1.25) can be stated as

$$\langle S^m f, S^n g\rangle_{\mathcal{H}} = \int_{\mathbb{C}} z^m \bar{z}^n \langle F(\mathrm{d}z)f, g\rangle_{\mathcal{H}} = \int_{\mathbb{C}} z^m \bar{z}^n \langle E(\mathrm{d}z)f, g\rangle_{\mathcal{K}},$$

$$m, n = 0, 1, \dots, \quad f, g \in \mathcal{D}(S).$$

This gives a good reason for saying that

subnormal operators inherit from normal ones the *holomorphic* portion of their functional calculus. (2.1.27)

More precisely, among all possible cores of the operator S, we choose one which is invariant and appoint it as the domain. This does not allow the operator to be closed,

[17] It is right time to give the definition: two extensions B_1 and B_2 in spaces $\mathcal{K}_1$ and $\mathcal{K}_2$ of A in $\mathcal{H}$ are called *$\mathcal{H}$-equivalent* if there is a unitary operator $U : \mathcal{K}_1 \to \mathcal{K}_2$ such that $U\mathcal{K}_1 = \mathcal{K}_2 U$ and $U{\restriction}_{\mathcal{H}} = I_{\mathcal{H}}$.

[18] In the finite dimensional case subnormals, formally normals and normals coincide.

in most cases. An important consequence of the aforementioned is that we can still consider, like done in (31) for bounded operators, the following positive definiteness condition

$$\sum_{i,j} \langle S^i f_j, S^j f_i \rangle \geqslant 0 \text{ for any finite sequence } (f_i)_i \text{ in } \mathcal{D}(S). \tag{H}$$

Theorem 2.1.9 *An operator S in $\mathcal{H}$ has a formally normal extension N in $\mathcal{K}$, say such that*[19]

$$N\mathcal{D}(N) \subset \mathcal{D}(N) \text{ and } N^*\mathcal{D}(N) \subset \mathcal{D}(N)$$

if and only if it satisfies (H).

If this happens, N can be chosen to satisfy

$$\mathcal{D}(N) = \text{lin}\{N^{*k}f : \ k = 0, 1, \dots, f \in \mathcal{D}(S)\}. \tag{2.1.28}$$

The converse is not true.

Corollary 2.1.10 *N in* Theorem 2.1.9 *is essentially normal if and only if*

$$x \in \mathcal{D}(N^*) \ \& \ \langle x, y\rangle + \langle x, N^*Ny\rangle = 0 \ \forall \ y \in \mathcal{D}(N) \implies x = 0.$$

Remark 6 As we already know from Remark 4 N is closable and, by Remark 2, so is S satifying the assumptions of Theorem 2.1.9—everything goes in accordance with "The article of faith," p. 60. ♣

Theorem 2.1.9 is in (24) and it may give an intermediate step toward subnormality of S. A typical result based on existence of a rich set of C^∞-vectors looks like this.

Theorem 2.1.11 *Suppose S satisfies* (H). *If any of the classes in* (2.1.7) *is dense (in the case of $\mathcal{Q}(S)$ its linear span is dense), then S is subnormal.*

This is a two-step construction. On the first step, one can prove that the C^∞-vectors can be lifted in Theorem 2.1.9 to the formally normal extension N. The second level uses available techniques ensuring normality of the closure of N.

Theorem 2.1.11 by no means gives necessary condition for subnormality.

It follows from Fact 2.1.2 that N^* is a dilation of S^*. If, moreover, $\mathcal{D}(S)$ is invariant for S^*, then the result is stronger. If N is a normal extension of S, then, cf. (Fact D)(32)

$$\mathcal{D}(N^{*i}) = \mathcal{D}(N^{i*}), P\mathcal{D}(N^{i*}) \in \mathcal{D}(S^{*i}) \text{ and } S^{*i}Px = PN^{*i}x.$$

[19] If S is a bounded operator then N is bounded as well, hence normal, cf. (33) and (34).

2.1.3.3 Cyclic Subnormal Operators An operator A is said to be *cyclic* if there is a vector $e \in \mathcal{D}^\infty(A)$ such that $\{p(A)e: \ p \in \mathbb{C}[Z]\}$ is a core of A. Notice that $\mathcal{D}^\infty(A)$ must necessarily be a core too.

We say that $(c_{m,n})_{m,n=0}^\infty$ is a *complex moment* (bi)sequence if

$$c_{m,n} = \int_{\mathbb{C}} z^m z^{*n} \mu(\mathrm{d}z), \quad m, n = 0, 1, \ldots$$

with some positive measure μ on $\mathbb{C}$.

Theorem 2.1.12 (cf. (24)) *A cyclic operator S with a cyclic vector e is subnormal if and only if $(\langle S^m e, S^n e\rangle)_{m,n=0}^\infty$ is a complex moment (bi)sequence.*

With the remark in the following in mind compare Theorem 2.1.12 with Theorem 2.1.8. Anyway, Theorem 2.1.12 brings us closer to construct an analytic functional models for cyclic subnormal operators mentioned in (2.1.27).

REMARK **7** Suppose S and N are as in Theorem 2.1.9. If S is cyclic with a cyclic vector e, then N is $*$-cyclic with respect to the same vector e and (2.1.28),

$$\mathcal{D}(N) = \mathrm{lin}\{N^{*k} N^l e: \ k, l = 0, 1, \ldots\}.$$

Conversely, if N is $*$-cyclic with the cyclic vector e and $e \in \mathcal{D}(S)$, the S is cyclic with vector e. ♣

As already mentioned, Theorem 2.1.9 is nothing but an intermediate step toward subnormality. Because N is just formally normal the uncertainty is still ahead. The only known result which is an unbounded version of that of Bram (33) is as follows, cf. (24)[20].

Theorem 2.1.13 *Suppose S is a weighted shift. Then S is subnormal if and only if it satisfies the positive definiteness condition* (H).

Theorem 2.1.12 allows us to display an important result from [95] characterizing subnormality of operators with invariant domain. In the cyclic case, everything can be describe easier just on the integer lattice points of the plane because a necessary condition for $(c_{m,n})_{m,n}^\infty$ to be a complex moment sequence is a kind of positive definiteness

$$\sum_{m,n,p,q} c_{m+q,n+p} \lambda_{m,n} \bar{\lambda}_{p,q} \geqslant 0, \quad \text{for any finite sequence } (\lambda_{m,n}) \text{ in } \mathbb{C}. \qquad (2.1.29)$$

Instead of stating it formally we explain the idea behind the result by a sequence of three drawings, they refer to the cyclic case when (H) can be thought of as (2.1.29) (Figures 2.1–2.3). Only Figure 2.2 needs some comment. It refers to the situation of positive definiteness like (2.1.29) defined on the lattice points $\mathbb{Z} \times \mathbb{Z}$ (notice: the dotted point are those in use). In this case, we do get a solution and the extra conclusion that the measure involved does not have 0 in its support.

[20]Confront it with Theorem 2.1.11 and Fact 2.1.7.

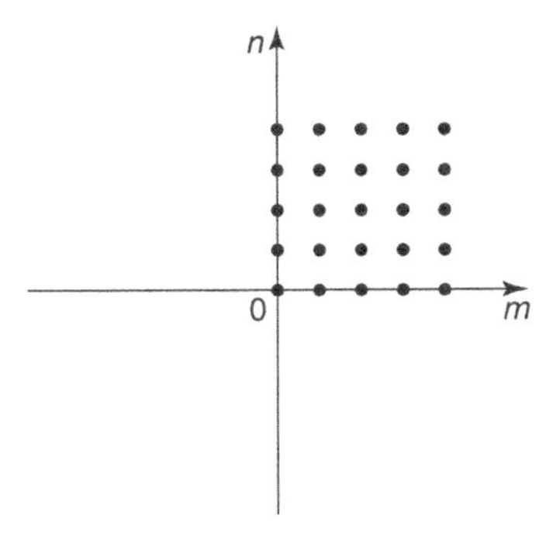

Figure 2.1 Positive definiteness (2.1.29): *too Little.*

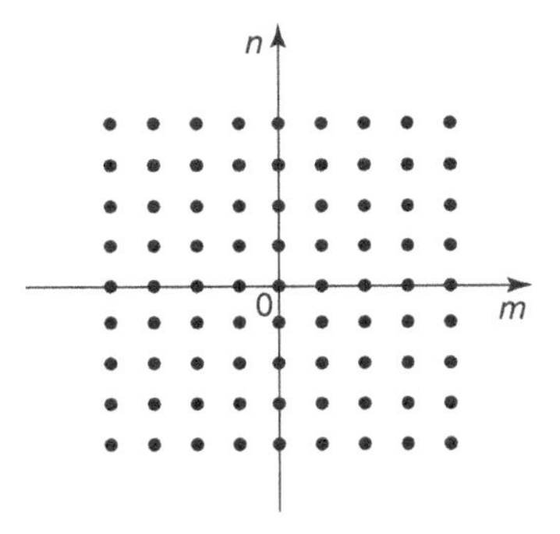

Figure 2.2 Very extended positive definiteness: *too much.*

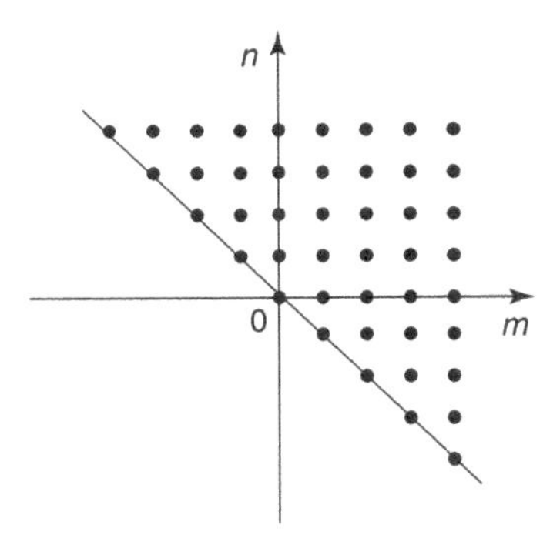

Figure 2.3 "Half-plane" extended positive definiteness: *that's it!*

It is worthy to mention a matrix construction of normal extensions given in (25), which in turn follows that for bounded operators invented in (35).

2.1.3.4 Minimality and Uniqueness Having already defined minimality of spectral type, p. 73, we can consider another, which is in close relation with (2.1.28). Call N to be a *minimal normal extension of cyclic type* if

$$\operatorname{lin}\{N^{*k}f: \ k = 0, 1, \dots, f \in \mathcal{D}(S)\}$$

is a core of N.

First of all, minimal normal extensions may not exist; if they do then all minimal normal extensions (these of spectral type and those of cyclic type) coincide, which means that they are $\mathcal{H}$-equivalent. If this happens, one can think of *uniqueness* of normal extension. In other words, existence of minimal normal extensions of cyclic type is responsible for uniqueness.

REMARK 8 Among the results already considered, which guarantee subnormality, the only ensuring uniqueness is Theorem 2.1.11. In addition to the latter, a complete characterization of subnormal operators enjoying the uniqueness extension property is in (Theorems 4 and 5) (36). ♣

2.1.4 Operators in the Reproducing Kernel Hilbert Space

2.1.4.1 Reproducing Kernel and Its Hilbert Space. The Basics Let X be a set. Given two objects: a function $K : X \times X \to \mathbb{C}$ (called kernel) and a Hilbert space $\mathcal{H}$ of complex valued functions on X merging by means of the so-called *reproducing kernel formula*. It articulates the relation

$$K_x \in \mathcal{H} \text{ and } f(x) = \langle K_x, f\rangle \text{ for all } x \in X \text{ and } f \in \mathcal{H}, \tag{2.1.30}$$

where $K_x \stackrel{\text{def}}{=} K(\cdot, x)$, sometimes called a *kernel function*[21]. This point of view has been disseminated by the present author on several occasions like in (35–39, 94, 95) and (40). Anyway, we call $(K, \mathcal{H})$ the *reproducing kernel couple*.

An immediate consequence of (2.1.30) is that the linear space

$$\mathcal{D}_K \stackrel{\text{def}}{=} \operatorname{lin}\{K_x: \ x \in X\}$$

is dense in $\mathcal{H}$. The following two consequences are pretty often mixed up and in fact are benchmarks of the whole theory of RKHS:

(α) the linear functional $\mathcal{H}_K \ni f \mapsto f(x) \in \mathbb{C}$ is continuous for any $x \in X$;

(β) the kernel K is positive definite, that is

$$\sum_{i,j} K(x_i, x_j)\xi_i\xi_j^* \geqslant 0 \text{ for any finite collection of } \xi_i\text{s in } \mathbb{C} \text{ and } x_i\text{s in } X. \tag{PD}$$

[21] It ought to be distinguished from the kernel itself which traditionally in the mathematical literature stands for a function of <u>two</u> variables.

As (α) refers exclusively to $\mathcal{H}$ and (β) does to K, each of them marks independent way to determine the other member of the couple. All this is described in detail in Refs (37–39, 94 and 95) and (93, 95) as well as in the master source (41) of the topic.

The following turns out to be useful to determine the kernel, it should to be called the *Zaremba decomposition*, cf. (42).

Fact 2.1.14 *If* $(\phi_\alpha)_{\alpha\in A}$ *is an orthonormal basis*[22] *in* $\mathcal{H}$*, then*

$$K(x,y) = \sum_{\alpha\in A} \phi_\alpha(x)\phi_\alpha(y)^*, \quad x,y \in X. \tag{2.1.31}$$

Conversely, if $(\phi_\alpha)_{\alpha\in A}$ *is any family of functions in* $\mathcal{H}$ *satisfying* (2.1.31)*, then it is an orthonormal basis of* $\mathcal{H}$ *provided*

$$(\xi_\alpha)_{\alpha\in A} \in \ell^2(A) \textit{ and } \sum_{\alpha\in A} \xi_\alpha\phi_\alpha(x) = 0 \textit{ for all } x \textit{ imply all } \xi_\alpha\textit{'s are 0.} \tag{2.1.32}$$

Remark 9 It may turn out to be productive to know that convergence of a sequence of functions in $\mathcal{H}_K$ forces its pointwise convergence which becomes uniform on each set on which $x \to K(x,x)$ is bounded.

Permuting the basis $(\phi_\alpha)_{\alpha\in A}$ does not impact the left-hand side in (2.1.31) ensures us that the (pointwise) convergence of the series therein is unconditional. This comment makes sense when the index set A is countable, if it cardinality is greater that the kind of convergence is included in its definition. ♣

From the reproducing kernel property, we get that the Fourier coefficients of K_x are $\langle K_x, \phi_\alpha\rangle = \phi_\alpha(x)^*$ arriving in this way at a version of Zaremba decomposition.

Fact 2.1.15 $K_x = \sum_\alpha \phi_\alpha(x)^*\phi_\alpha$, $x \in X$, *with convergence being that*[23] *in* $\mathcal{H}_K$*. Consequently, the reproducing kernel property* (2.1.30) *can be restated as*

$$f(x) = \sum_\alpha \phi_\alpha(x)\langle f, \phi_\alpha\rangle, \quad f \in \mathcal{H}_K$$

with convergence as indicated in Remark 9.

Remark 10 $\mathcal{L}^2$ spaces rather do not share the reproducing kernel property except those of ℓ^2 type; the kernel is the Kronecker function[24] then. ♣

[22] Cardinality of the index set A is that of the Hilbert space dimension of $\mathcal{H}$.

[23] Occasionally, we decorate $\mathcal{H}$ with the subscript as $\mathcal{H}_K$ if any danger of confusion may appear.

[24] Do not confuse with Dirac's delta; a common sin.

2.1.4.2 The Basic Operators Consider two natural couples of operators in $\mathcal{H}_K$; one is a multiplication operator, the other a kind of composition one.

1°. Suppose a function $\varphi : X \to \mathbb{C}$ is given. Define a linear operator

$$M_\varphi : f \to \varphi f,$$

with the maximal domain that is

$$\mathcal{D}(M_\varphi) \stackrel{\text{def}}{=} \{f \in \mathcal{H}_K : \ \varphi(\cdot)f(\cdot) \in \mathcal{H}_K\}. \tag{2.1.33}$$

Supposing f and φf are both in $\mathcal{H}$ and using the reproducing property, we get via (2.1.35)

$$\langle M_\varphi f, K_z\rangle = \langle \varphi f, K_z\rangle = \varphi(z)\langle f, K_z\rangle = \langle f, \varphi(z)^* K_z\rangle, \quad z \in X.$$

Proposition 2.1.16 *M_φ is always a closed operator. M_φ is densely defined if and only if the formula*

$$K_z \to \varphi(z)^* K_z \quad z \in X$$

defines an operator, which, in turn, is an adjoint of M_φ.

The kernel functions K_z are eigenvectors of M_φ^* corresponding to the eigenvalues $\varphi(z)^*$.

2°. For a semigroup $\mathfrak{S}$ of actions on X, a complex function f defined on X set $f_{[\mathfrak{s}]}(x) \stackrel{\text{def}}{=} f(\mathfrak{s}x)$, getting in this way the function $f_{[\mathfrak{s}]} : X \ni x \mapsto f_{[\mathfrak{s}x]} \in \mathbb{C}$, $s \in \mathfrak{S}$.

The operator

$$S_\mathfrak{s} : f \to f_{[\mathfrak{s}]}, \quad \mathfrak{s} \in \mathfrak{S} \tag{2.1.34}$$

is a well-defined linear operator although it may not be densely defined. The other operator of the second couple, $T_\mathfrak{s}$ say, also depending on $\mathfrak{s} \in \mathfrak{S}$, which we would like to have acting as

$$T_\mathfrak{s} : K_x \to K_{\mathfrak{s}x} \quad x \in X \tag{2.1.35}$$

may not extend linearly the formula (2.1.35) although its prospective domain $\mathcal{D}_K$ is clearly linear and dense. These doubts can be put in one.

Proposition 2.1.17 *$S_\mathfrak{s}$ is always a closed operator. $S_\mathfrak{s}$ is densely defined if and only if the operator $T_\mathfrak{s}$ is well defined via* (2.1.35). *If this happens $T_\mathfrak{s} = S_{u\mathfrak{s}}^*$.*

Proof: All this comes out directly from (here both f and $f_{[\mathfrak{s}]}$ are supposed to be in $\mathcal{H}_K$)

$$\langle f_{[\mathfrak{s}]}, K_x\rangle = \langle f, K_{\mathfrak{s}x}\rangle, \quad x \in X,$$

which in turn is an immediate consequence of the reproducing formula (2.1.30). ∎

The difference between these two couples of operators will become more clear in Section 2.1.4.3, which follows. In principle, they appear in dilation and extension theory put in a fairly general framework, cf. (42) and also (36, 44).

2.1.4.3 Assorted Kernels

The $\ell^2(\mathbb{N})$ The simplest though useful RKHS is when $X = \mathbb{N}$. The kernel then is $K(m, n) = \delta_{m,n}$, $m, n \in \mathbb{N}$, and its space is just $\ell^2 = \ell^2(\mathbb{N})$.

If $(\varepsilon_k)_{k=0}^{\infty}$ stands for the canonical, zero-one basis in ℓ^2, that is $\varepsilon_n(k) = \delta_{n,k}$ then the Zaremba decomposition (2.1.31) reads as

$$K(m, n) = \sum_{k=0}^{\infty} \delta_{m,k}\delta_{n,k}, \quad m, n \in \mathbb{N}.$$

The operator M_φ is usually unbounded unless φ is bounded[25]. On the other hand, with $\mathfrak{S} = X = \mathbb{N}$, the operator S_1 defined by (2.1.34) (warning: now the action of $\mathfrak{S}$ is written additively) is a forward shift operator, which is an isometry[26].

The Zaremba decomposition is clearly not unique, any orthonormal basis can serve for it. Here, we point out another basis which is going to be of service later on.

Let $\alpha > 0$ be a parameter. Define in two steps[27] the *Charlier sequences* $c_n^{(\alpha)}$, $n = 0, 1, \ldots$ as functions in discrete variable

$$\tilde{c}_n^{(\alpha)}(x) = \alpha^{-\frac{n}{2}}(n!)^{-\frac{1}{2}} C_n^{(\alpha)}(x) \mathrm{e}^{-\frac{a}{2}} \alpha^{\frac{x}{2}} \begin{cases} (x!)^{-\frac{1}{2}}, & \text{for } x \geqslant 0 \\ 1 & \text{for } x < 0 \end{cases}; \tag{2.1.36}$$
$$c_n^{(\alpha)} = \tilde{c}_n^{(\alpha)}|_{\mathbb{N}}, \quad n = 0, 1, \ldots$$

where the Charlier polynomials are determined by, cf. (45),

$$\mathrm{e}^{-\alpha z}(1 + z)^x = \sum_{n=0}^{\infty} C_n^{(\alpha)}(x)\frac{z^n}{n!}. \tag{2.1.37}$$

They are orthogonal with respect to a nonnegative integer supported measure according to

$$\sum_{x=0}^{\infty} C_m^{(\alpha)}(x) C_n^{(\alpha)}(x) \frac{\mathrm{e}^{-\alpha}\alpha^x}{x!} = \delta_{m,n}\alpha^n n!, \quad m, n = 0, 1, \ldots$$

which makes the sequence $(c_n^{(\alpha)})_{n=0}^{\infty}$ orthonormal. As the monic Charlier polynomials satisfy (cf. (45, p. 171)) the appropriate duality relation, we get immediately for Charlier sequences

$$(-1)^x c_x^{(\alpha)}(n) = (-1)^n c_n^{(\alpha)}(x), \quad n, x = 0, 1, \ldots. \tag{2.1.38}$$

[25] This is a kind of general behavior, cf. (39).

[26] As such, it has a unitary extension and this is a prototype of the "subnormal-to-normal" relation as introduced in (31).

[27] This looks a little bit artificial right now but turns out to be a necessary trick, cf. footnote [42].

REMARK **11** Owing to (2.1.38), the orthonormal sequence $(c_n^{(\alpha)})_{n=0}^{\infty}$ becomes a (self-dual) basis in ℓ^2, cf. (46), and another Zaremba decomposition emerges again. More precisely,

$$\delta_{x,y} = \sum_{n=0}^{\infty} c_n^{(\alpha)}(x)c_n^{(\alpha)}(y)^* = \sum_{n=0}^{\infty} c_x^{(\alpha)}(n)c_y^{(\alpha)}(n)^*.$$

The second sum in the aforementioned is nothing but an integral of the function

$$n \mapsto c_x^{(\alpha)}(n)c_y^{(\alpha)}(n)^*$$

with respect to the counting measure on $\mathbb{N}$ and the equality says it has to be equal to $\delta_{x,y}$. If this happens for all $x, y \in \mathbb{N}$ then, according to Fact 2.1.14, $(c_n^{(\alpha)})_{n=0}^{\infty}$ must necessarily be an orthonormal basis in ℓ^2. This is true for any self-dual sequence $(e_n)_{n=0}^{\infty}$ (that is such that $\alpha_x e_x(n) = \alpha_n e_n(x)$ with unimodular α_x's). ♣

The generating formula (2.1.37), after some simple manipulations, takes the form

$$\mathrm{e}^{-\frac{\alpha}{2}-\sqrt{\alpha}z}(\sqrt{\alpha}+z)^x(x!)^{-\frac{1}{2}} = \sum_{n=0}^{\infty} c_n^{(\alpha)}(x)\frac{z^n}{\sqrt{n!}}, \quad x = 0, 1, \dots, \; z \in \mathbb{C}. \tag{2.1.39}$$

Inserting $\alpha = 0$ in the left-hand side of (2.1.39) convinces us that we can extend the range of the parameter α from $\alpha > 0$ to $\alpha \geqslant 0$ by setting

$$c_n^{(0)} \overset{\mathsf{def}}{=} \varepsilon_n, \quad n = 0, 1, \dots . \tag{2.1.40}$$

Define for the extended values of the parameter $\alpha \geqslant 0$

$$A^{(\alpha)}(x, z) \overset{\mathsf{def}}{=} \mathrm{e}^{-\frac{\alpha}{2}-\sqrt{\alpha}z}(\sqrt{\alpha}+z)^x(x!)^{-\frac{1}{2}}, \quad x = 0, 1, \dots, \quad z \in \mathbb{C}.$$

Consequently, by (2.1.39), the operator U_α given by

$$(U_\alpha f)(z) = F(z) = \sum_{x=0}^{\infty} A^{(\alpha)}(x, z)f(x), \quad f \in \ell^2$$

maps ℓ^2 into $\mathcal{L}^2_{\mathrm{hol}}(\pi^{-1}\exp(-|z|^2)\,\mathrm{d}z)$. Even more, because

$$U_\alpha : c_n^{(\alpha)} \mapsto \frac{z^n}{\sqrt{n!}}, \quad n = 0, 1, \dots$$

it is in fact unitary; this is an ℓ^2 analog of the Segal–Bargmann transform defined in (2.1.44), cf. (47, Section 5).

Diagonal kernels Suppose X is a symmetric (with respect to the complex conjugation) subset[28] of the complex plane $\mathbb{C}$. The most popular kernels come from polynomials and they determine spaces of holomorphic functions. Let us describe some of them passing from the simplest to more advanced.

The diagonal kernels are of the form

$$K(z,w) \overset{\text{def}}{=} \sum_{n=0}^{d} \alpha_n z^n w^{*n}, \quad z,w \in X \tag{2.1.41}$$

with α_ns being nonnegative and $d \in \mathbb{N} \cup \{+\infty\}$. This is apparently a positive definite kernel so the reproducing kernel Hilbert space $\mathcal{H}_K$ comes out at once. Taking advantage of Fact 2.1.14, we can infer that

$$(\sqrt{\alpha_n} Z^n)_{n=0}^{d} \tag{2.1.42}$$

is an orthonormal basis provided that the interior of X is not empty. The norm $\|Z^N\| = \frac{1}{\sqrt{\alpha_n}}$.

The two couples of the previous Section can be illustrated by means of this case.

For 1°, consider the operator M_Z of multiplication by the independent variable Z. Because

$$M_Z \sqrt{\alpha_n} Z^n = \sqrt{\frac{\alpha_n}{\alpha_{n+1}}} \sqrt{\alpha_{n+1}} Z^{n+1},$$

the basic vectors (2.1.42) belong to the domain of M_Z and the operator $(M_Z)|_{\mathbb{C}[Z]}$ is a weighted shift according to our definition. Using (2.1.13), one may check that $\mathbb{C}[Z]$ is a core of M_Z. Therefore the two domains of M_Z, this of (2.1.33) and that given on p. 72 must necessarily coincide. Notice that neither of these two domains can be described analytically in a natural way; therefore, the information the polynomials form a core of M_Z is of great advantage. Concerning the adjoint operator, we cannot expect to have any analytic description of $(M_Z)^*$ in the general case. The only thing we know is that the kernel functions K_z, $z \in X$ are its eigenfunctions and that $(M_Z)^*|_{\mathbb{C}[Z]}$ is a backward weighted shift.

For 2°, take $\mathfrak{G}$ to be either $\mathbb{C}$ or $\mathbb{D}$ and X is such that it allows $\mathfrak{G}$ to act. Then, because of the invariance of the kernel K, that is

$$K(uz,w) = K(z,u^*w), \quad z,u,w \in X,$$

one has $(K_z)_{[u]} = K_{u^*z}$ which means that the kernel functions belong to the domains of both S_u and T_u and $S_u K_z = K_{\bar{u}z}$.

Kernels from polynomials orthogonal on $\mathbb{R}$ Suppose $(p_n)_{n=0}^{\infty}$ is a sequence of polynomials <u>*orthonormal*</u> with respect to some μ on $\mathbb{R}$. If the measure μ is

[28] This corresponds to the fact that the quantum physics behind is that of a single particle.

indeterminate (in the sense of moment problems), then, the well-known fact,

$$\sum_{n=0}^{\infty} |p_n(z)|^2 < +\infty, \quad \text{for all } z \in \mathbb{C}.$$

The kernel

$$K(z, w) \stackrel{\text{def}}{=} \sum_{n=0}^{\infty} p_n(z) p_n(w)^* \quad z, w \in \mathbb{C} \tag{2.1.43}$$

is positive definite; hence, the corresponding Hilbert space is composed of entire functions.

Kernels from holomorphic polynomials Here is an example of what may happen within this category, it comes from (48).

The Hermite polynomials H_n, which do not fall in any of the cases mentioned earlier, are now considered as those in a complex variable. Let $0 < A < 1$. Then[29]

$$\int_{\mathbb{R}^2} H_m(x + \mathrm{i}\, y) H_n(x - \mathrm{i}\, y) \exp\left[-(1-A)x^2 - \left(\frac{1}{A} - 1\right)y^2\right] \mathrm{d}x\mathrm{d}y = b_n(A)\delta_{m,n},$$

where

$$b_n(A) = \frac{\pi\sqrt{A}}{1-A}\left(2\frac{1+A}{1-A}\right)^n n!\,.$$

Introducing the Hilbert space $\mathcal{X}_A$ of entire functions f such that

$$\int_{\mathbb{R}^2} |f(x + \mathrm{i}\, y)|^2 \exp\left[Ax^2 - \frac{1}{A}y^2\right] \mathrm{d}x\mathrm{d}y < \infty$$

and defining

$$h_n^{(A)}(z) = b_n(A)^{-1/2}\, \mathrm{e}^{-z^2/2}\, H_n(z), \quad z \in \mathbb{C},$$

it was shown in (48) that $\{h_n^{(A)}\}_{n=0}^{\infty}$ is an orthonormal basis in $\mathcal{X}_A$.

The explicit form of the reproducing kernel is

$$K(z, w) = \frac{1-A^2}{2\pi A} \exp\left[-\frac{1+A^2}{4A}(z^2 + \bar{w}^2) + \frac{1-A^2}{2A} z\bar{w}\right], \quad z, w \in \mathbb{C}.$$

It is clear that

$$0 < A < B < 1 \text{ implies } \mathcal{X}_B \subset \mathcal{X}_A.$$

On the other hand, as shown in (48),

$$\text{any } \mathcal{X}_A \subset \mathcal{L}^2(\mathbb{R})$$

[29] Both $\mathrm{d}x\,\mathrm{d}y$ and $\mathrm{d}z$ have the same meaning, and they stand for two-dimensional Lebesque measure.

if one thinks of this inclusion as taking restrictions of functions in $\mathcal{X}_A$ to the real axis. All these inclusions are continuous.

This case has been exploited first in (37) and then in (49) and (50).

2.1.4.4 Segal–Bargmann Space and Transform This reproduce ng kernel Hilbert space (51) deserves special attention as it carries the standard model of the quantum oscillator off, the fact known for long time. It is composed of all entire functions in[30] $\mathcal{L}^2(\mathbb{C}, \pi^{-1}\exp(-|z|^2\,\mathrm{d}x\mathrm{d}y)$ and is in fact a reproducing kernel Hilbert space with the kernel $(z, w) \mapsto \exp(zw^*)$; keep the notation $\mathcal{L}^2_{\text{hol}}(\pi^{-1}\exp(-|z|^2\,\mathrm{d}x\mathrm{d}y)$ for it[31].

REMARK **12** It was Fock's observation (52) pointing out usefulness of the Leibnitz differentiation rule in the complex domain and sowing it successfully on the quantum mechanics ground (see also footnote[36]). In another words, quoting after (51, p.187, see also footnote 1 there), "... Fock introduced the operator solution $\xi_k = \partial/\partial\eta_k$ of the commutation rule $[\xi_k, \eta_k] = 1$, in analogy to Schrödinger's solution $p_k = -i\partial/\partial q_k$ of the relation $[q_k, p_k] = i$ and applied it to quantum field theory." This brings us closer to the problem of whose name the space has to bear. Pretty often, the space is named after Fock which is confusing because the space in question does not appear in either in Ref. (52) or in Ref. (53). Nowadays when the English translation of Fock's works is available (54), authors can easier confront their vision with reality[32]. ♣

The standard orthonormal basis in $\mathcal{L}^2_{\text{hol}}(\pi^{-1}\exp(-|z|^2)\,\mathrm{d}x\mathrm{d}y)$ is $\{e_n\}_{n=0}^{\infty}$ defined as

$$e_n = \frac{z^n}{\sqrt{n!}}, \quad z \in \mathbb{C}, \quad n = 0, 1, \ldots .$$

What is related to this space is a transform

$$T:\ \mathcal{L}^2(\mathbb{R}) \to \mathcal{L}^2_{\text{hol}}(\pi^{-1}\exp(-|z|^2)\,\mathrm{d}x\mathrm{d}y),$$

which can be implemented as integral transform (by some authors it is called after Bargmann, by others after Segal–Bargmann[33]). More precisely,

$$(Tf)(z) \overset{\text{def}}{=} \pi^{-\frac{1}{4}} \int_{\mathbb{R}^2} \mathrm{e}^{(-z^2-2\sqrt{2}xz-x^2)} f(x)\,\mathrm{d}x, \quad z \in \mathbb{C},\ f \in \mathcal{L}^2(\mathbb{R}). \tag{2.1.44}$$

It maps the nth Hermite function on the nth monomial of (2.1.42).

[30] Another notation used for $\mathrm{d}x\,\mathrm{d}y$ is $\mathrm{d}z$ with $z = x + \mathrm{i}\,y$ behind.

[31] Despite its importance, there is no commonly acceptable notation for it.

[32] The only Hilbert space which appears in (52) is that around formula (10); formula (11) suggests that it to be considered as a kind of ℓ^2 space, which is understandable because it was time for plays with matrix elements.

[33] Look at Section 6.4 "Historical Remarks" in (55) for in-depth comment and choose one you are in favor of.

Just to remind, the nth *Hermite function* h_n is defined as

$$h_n \stackrel{\text{def}}{=} 2^{-n/2}(n!)^{-1/2}\pi^{-1/4}\, \mathrm{e}^{-x^2/2}\, H_n$$

with H_n, the n-Hermite polynomial, coming from the Rodrigez formula

$$H_n(x) = (-1)^n\, \mathrm{e}^{x^2}\, \frac{\mathrm{d}^n}{\mathrm{d}x^n}\, \mathrm{e}^{-x^2}, \quad x \in \mathbb{R}.$$

2.1.4.5 Analytic Models of Cyclic Operators Suppose that A is a cyclic operator in an infinite dimensional Hilbert space $\mathcal{H}$ with a cyclic vector e. Then, the sequence $\{A^n e\}_{n=0}^{\infty}$ is composed of linearly independent vectors and $\dim \mathcal{H} = \aleph_0$. Let $(e_n)_{n=0}^{\infty}$ be an orthonormal basis of $\mathcal{H}$ and $(r_n)_{n=0}^{\infty}$ be a sequence of polynomials from $\mathbb{C}[Z]$ such that

$$\begin{aligned} e_n &= r_n(A)e, \quad n \geqslant 0, \\ \operatorname{lin}(e_n)_{n=0}^{\infty} &= \operatorname{lin}(A^n e)_{n=0}^{\infty}. \end{aligned} \tag{2.1.45}$$

Such a sequence $(r_n)_{n=0}^{\infty}$ can be obtained by applying the Gram–Schmidt orthonormalization procedure to the sequence $\{A^n e\}_{n=0}^{\infty}$. If $\|e\| = 1$, then the polynomials r_n can be given explicitly by $r_0(z) = 1$ and

$$r_n(z) = \frac{1}{\sqrt{G_n G_{n-1}}} \det \begin{bmatrix} \langle e, e\rangle & \langle e, Ae\rangle & \cdots & \langle e, A^{n-1}e\rangle & 1 \\ \langle Ae, e\rangle & \langle Ae, Ae\rangle & \cdots & \langle Ae, A^{n-1}e\rangle & z \\ \vdots & \vdots & \vdots & \vdots & \vdots \\ \langle A^n e, e\rangle & \langle A^n e, Ae\rangle & \cdots & \langle A^n e, A^{n-1}e\rangle & z^n \end{bmatrix}$$

for $z \in \mathbb{C}$ and $n \geqslant 1$, where $G_n \stackrel{\text{def}}{=} \det[\langle A^i e, A^j e\rangle]_{i,j=0}^{n}$ for $n \geqslant 0$.

Proposition 2.1.18 (cf. (11)) *Suppose that A is a cyclic operator in an infinite dimensional Hilbert space $\mathcal{H}$, e is a cyclic vector of A and $\{e_n\}_{n=0}^{\infty}$ is an orthonormal basis of $\mathcal{H}$ satisfying the conditions* (2.1.45) *for some sequence $\{r_n\}_{n=0}^{\infty} \subset \mathbb{C}[Z]$. Then, the following conditions are equivalent*

1° $\lambda^* \in \operatorname{sp_p}(A^*)$,

2° *there exists a real number $c_\lambda > 0$ such that $|p(\lambda)| \leqslant c_\lambda \|p(A)e\|$ for all $p \in \mathbb{C}[Z]$,*

3° *there exists a (unique) vector $h_\lambda \in \mathcal{H}$ such that $p(\lambda) = \langle p(A)e, h_\lambda\rangle$ for all $p \in \mathbb{C}[Z]$,*

4° $\sum_{n=0}^{\infty} |r_n(\lambda)|^2 < \infty$.

If $\lambda^* \in \operatorname{sp_p}(A^*)$, then

5° $h_\lambda = \sum_{n=0}^{\infty} r_n(\lambda)^* e_n$, $\|h_\lambda\|^2 = \sum_{n=0}^{\infty} |r_n(\lambda)|^2$ *and* $A^* h_\lambda = \bar{\lambda} h_\lambda$.

The aforementioned opens the way to build up a model in an RKHS composed of polynomials; the success depends on whether $\operatorname{sp_p}(A^*)$ is nonempty or not.

Suppose $X \stackrel{\text{def}}{=} \mathrm{sp_p}(S^*)^* \neq \emptyset$. Then, owing to Proposition 2.1.18, 4°, the Zaremba decomposition formula (2.1.31) defines a kernel K on X and consequently its Hilbert space $\mathcal{H}_K$. Defining $(Wf)(x) \stackrel{\text{def}}{=} \langle f, h_x \rangle$ for $f \in \mathcal{H}$, we check by 5° that $Wh_\lambda = K_\lambda$. Therefore, W is a unitary map of $\mathrm{clolin}\{h_\lambda : \lambda \in \mathrm{sp_p}(A^*)^*\}$ onto $\mathcal{H}_K$. Finally, $\mathcal{H} = \mathrm{clolin}\{h_\lambda : \lambda \in \mathrm{sp_p}(A^*)^*\}$ is equivalent to show that (2.1.32) holds. If this happens, we can think of an RKHS model of a cyclic operator. Then,

$$WA = M_Z W \text{ and } \mathrm{sp_p}(A^*) = \mathrm{sp_p}(M_Z^*).$$

Another problem is if $\mathcal{H}_K$ is composed of holomorphic function. This is delicate question and topological properties of $\mathrm{sp_p}(A^*)$ play a role in resolving it. Interested reader may consult (11) for all the details in the matter.

2.1.4.6 Integrability of Reproducing Kernel Couples Call a reproducing kernel couple $(\mathcal{H}, K)$ on $X \subset \mathbb{C}$ *integrable* if there is a measure μ such that $\mathcal{H} \subset \mathcal{L}^2(\mu)$ with the inclusion being an isometry.

This is a rough definition, an idea rather, because the meaning of "inclusion" is a bit vague (remember "functions" have to be identified with μ-equivalent classes of functions). It can be made precise for instance when the polynomials are dense in $\mathcal{H}$ and the measure μ is required to be regular. In this case, the Gram–Schmidt orthonormalization applied to monomials leads to the basis consisting of polynomials of degree corresponding to their index and to a kernel of type (2.1.43); it does not mean that the kernel must come from indeterminate orthogonality, that is from polynomials defined in (2.1.43).

Just to start observe that if $(K, \mathcal{H})$ is integrable, that is if the inclusion preserving monomials is isometric, then a necessary condition for this is the sequence $(\|S^n e_0\|)$, provided e_0 is a cyclic vector just for simplicity, is logarithmically convex, that is

$$\|S^{2n} e_0\|^2 \leqslant \|S^{n-1} e_0\| \|S^{n+1} e_0\|, \quad n = 0, 1, \ldots .$$

This may back some guesses of subnormality and can serve for inspecting negative examples for integrability as well; there is a plenty of either possibility in the ℓ^2 case.

Notice that the monomials are never orthogonal over the real line so the diagonal kernels do not fit in the family considered in the definition (2.1.43). However, choosing the α_n's accidentally, one may get either integrable kernels or not. Referring to (2.1.41), it is worthy to mention an example with $\|Z^n\| = \sqrt{n+1}$, which corresponds to Dirichlet space; this is not integrable in our sense.

Proposition 22 in (56) characterizes integrability of diagonal kernels.

Proposition 2.1.19 *A measure μ on $\mathbb{C}$ such that*

$$\mathcal{H} \subset \mathcal{L}^2(\mu) \textit{ isometrically} \tag{2.1.46}$$

exists if and only if there is a Stieltjes moment sequence $(a_n)_{n=0}^{+\infty}$ such that

$$a_{2n} = \alpha_n^{-1}, \quad n = 0, 1, \ldots . \tag{2.1.47}$$

If the circumstances of Proposition 2.1.19 happen, then a measure μ can be chosen to be rotationally invariant[34], that is such that $\mu(\mathrm{e}^{\mathrm{i}t}\sigma) = \mu(\sigma)$ for all t's and σ's. This can be done as follows:

> if $(a_n)_{n=0}^{+\infty}$ is any Stieltjes moment sequence with a representing measure m and satisfying (2.1.47), then the rotationally invariant measure
>
> $$\mu(\sigma) \stackrel{\text{def}}{=} (2\pi)^{-1} \int_0^{2\pi} \int_0^{+\infty} \chi_\sigma(r\,\mathrm{e}^{\mathrm{i}t}) m(\mathrm{d}r)\,\mathrm{d}t, \quad \sigma \text{ Borel subset of } \mathbb{C}$$
>
> makes the imbedding (2.1.46) happen.

On the other hand, if the moment sequence is Stieljes indeterminate, then one can always find measure which is <u>*not*</u> rotationally invariant, such a measure was determined in (57) for the first time and then in (56) where a method of finding a large class of them was presented.

Integrability of reproducing kernel couples is important if one wants to perform Toeplitz–Berezin- $\cdots$ quantization. If a measure is unique—no problem, if it is not or it does not exist the problem of quantization persists.

2.2 COMMUTATION RELATIONS

The Leibnitz rule for differentiation

$$D(fg) = (Df)g + fDg$$

with f being the operator of multiplication by the independent variable acting on g, the running function, can be viewed as a germ of functional models (or realizations, say) of two basic commutation relations of quantum mechanics:

- in the real variable—canonical commutation relation, known as CCR (the $\mathcal{L}^2(\mathbb{R})$ model) ;
- in the complex variable—the commutation of the quantum harmonic oscillator (the Segal–Bargmann space).

Although the Leibnitz rule is a common pattern for both relations, or rather their particular models, it was Fock's formalism[35] from 1928 [36] Fock (52), which made them twin brothers (this brotherhood is not going too far; however, if Hilbert space operators are in action!).

[34] Or *radial* as some authors say.

[35] See (58, p. 293) for a short mathematical resume and also (59, p. 49).

[36] Going back to the beginning of recent century, it was Fischer (60) who considered this instance already in 1917 just from purely mathematical point of view. In any case to place in this context, the quantum harmonic oscillator requires to consider a Hilbert space and this is what Bargmann did in (51).

Let us say a couple of words on the first case and then concentrate on the second one which is the punch line of this survey.

2.2.0.7 CCR—a Brief Opening The CCR can be written *formally* as

$$PQ - QP = I. \tag{2.2.1}$$

The name of Heisenberg is often attached to this.

There are two standard pairs of the operators P and Q, which "satisfy" (2.2.1)

- Schrödinger's, that is

$$(Pg)(x) = -\mathrm{i}\,\frac{\mathrm{d}g(x)}{\mathrm{d}x}, \quad (Qg)(x) = xg(x).$$

  considered in $\mathcal{L}^2(\mathbb{R})$;
- Heisenberg's, that is P and Q, are infinite Jacobi matrices with zeros on the diagonal, not to be given here explicitly. The latter has to be considered in ℓ^2.

These two pairs are unitary equivalent[37]. To see this take for the first pair, the linear span of Hermite functions as a domain, for the second the sequences which are zoroes but a finite number of coordinates. They both are cores of the corresponding pairs. A unitary mapping is on its way and the argument goes through formulae for Hermite polynomials (details have to be implemented).

The operators P and Q satisfying (2.2.1) must not be simultaneously bounded as showed in (61) and (62); this has to be confronted with Paragraph, p 90.

There is also another formulation of (2.2.1) in folk-theoretic language described as "equivalent"

$$U(s)V(t) = \mathrm{e}^{\mathrm{i}\,st}\, V(t)U(s), \quad s, t \in \mathbb{R} \tag{2.2.2}$$

it is named after Weyl. The would-be "equivalence" depends on what kind of conditions on the operators involved are going to be supposed[38]; in general, it is not—confront with Paragraph, p. 90.

Heisenberg's CCR Here, we have the following uniqueness result proved in (65) and improved in (66).

Theorem 2.2.1 (Rellich–Dixmier) *Suppose that P and Q are symmetric and irreducible. The operators P and Q are unitarily equivalent to the Heissenberg (or, equivalently Schrödinger) pair if and only if*

$$\begin{array}{l}\textit{there is a dense subspace } \mathcal{D} \subset \mathcal{D}(PQ) \cap \mathcal{D}(QP) \\ \textit{such that } (P^2 + Q^2)|_{\mathcal{D}} \textit{ is essentially self-adjoint.}\end{array} \tag{2.2.3}$$

[37] It does not mean that Heisenberg's Mechanics is equivalent to that of Schrödinger as this statement is not a mathematical one.

[38] The equivalence "CCR $\Longleftrightarrow$ Weyl" is the matter of (64).

One more assumption is in (65) requiring $\mathcal{D} \subset \mathcal{B}((P^2 + Q^2)|_{\mathcal{D}})$, the bounded vectors of $(P^2 + Q^2)|_{\mathcal{D}}$.

REMARK **13** In a sense, Theorem 2.2.1 goes along the same lines as Rellich's. He, unlike Dixmier, does not suppose the operators to be closed which is in accordance with the principle exposed on p. 60. In this way, some assumptions become stronger than Dixmier's, others weaker—the conclusion remains the same, and what is important, both sets of assumptions (Dixmier's and ours) are necessary conditions for the representation to happen. This may cause some discomfort for hard-line mathematicians, but from the practical side, it is a kind of inevitability: domains of closures of *concrete* operators (and of their adjoints) are mostly not conveniently determined, even if they are so, they may not be handy enough to make go. ♣

Weyl's CCR The classical result of (67, 68) follows.

Theorem 2.2.2 (Stone-von Neumann) *Suppose that* $(U(s))_{s\in\mathbb{R}}$ *and* $(V(t))_{t\in\mathbb{R}}$ *are irreducible groups of unitaries. Their generators P and Q form a Schrödinger couple if and only if the commutation relation* (2.2.2) *is satisfied.*

This is a starting point to consider C^*-algebras of CCR, which is out of our scope here.

An odd example Take $\mathcal{H} = \mathcal{L}^2[0, 1]$ and the Schrödinger pair P, Q now considered in so defined $|hhc$. Then, Q is a bounded operator and P, with its maximal domain, a closed symmetric operator with deficiency indices $(1, 1)$. The commutation relation (2.2.1) is satisfied on $\mathcal{D}(P)$ but the couple is far from being Schrödinger one (cf. (69) and (58, Example 14.5)).

2.2.1 The Commutation Relation of the Quantum Harmonic Oscillator

Our goal in this section is to work out a quantum harmonic oscillator analogue of the theorem of Rellich and Dixmier, in particular to find out a counterpart of the assumption (2.2.3).

2.2.1.1 The Relation For the quantum harmonic oscillator to be dealt, the proper environment is that of Hilbert space. Let us write then the relation using a bit unconventional notation

$$S^*S - SS^* = I. \tag{2.2.4}$$

The relation (2.2.4) has nothing but a symbolic meaning unless one accepts the convention (2.1.4), then the way to read (2.2.4) is rather as

$$S^*S = SS^* + I. \tag{qho}$$

What most people agree to accept as the principal meaning of (qho) is

* Let $\mathcal{D}$ be a linear subspace of $\mathcal{H}$, which is invariant under S. Then, the first meaning of (2.2.4) is

$$\mathcal{D} \subset \mathcal{D}(S^*S) \cap \mathcal{D}(SS^*), \quad S^*Sf - SS^*f = f, \quad f \in \mathcal{D}. \tag{qho$_\mathcal{D}$}$$

This is equivalent to

$$\langle Sf, Sg\rangle - \langle S^*f, S^*g\rangle = \langle f, g\rangle, f, g \in \mathcal{D}(SS^*). \cap \mathcal{D}(S^*S)$$

The commutation relation and positive definiteness The following formalism can be reckoned straightforwardly.

Fact 2.2.3 *If S satisfies* (qho$_\mathcal{D}$) *with* $\mathcal{D}$ *being invariant for both S and* S^*, *then*

$$S^{*i}S^jf = \sum_{k=0}^{\infty} k!\binom{i}{k}\binom{j}{k}S^{j-k}S^{*(i-k)}f, \quad f \in \mathcal{D},\ i,j = 0,1,\dots, \tag{2.2.5}$$

$$\sum_{i,j=0}^{p}\langle S^if_j, S^jf_i\rangle = \sum_{k=0}^{\infty} k!\,\|\sum_{i=0}^{p}\binom{i}{k}S^{*(i-k)}f_i\|^2, \quad f_0,\dots f_p \in \mathcal{D}. \tag{2.2.6}$$

All this under convention $S^l = (S^*)^l = 0$ *for* $l < 0$ *and* $\binom{i}{j} = 0$ *for* $j > i$.

Equation (2.2.6) comes immediately from (2.2.5) and implies positive definiteness of S as required by (H).

An important conclusion, coming from Remark 6, is that S satisfying (qho$_\mathcal{D}$) must necessarily be closable. Good news!

Weak commutation relation Besides (qho$_\mathcal{D}$) we propose another, extreme in a sense, way of looking at the relation (2.2.4), it appear explicitly in (70).

* Thus, the other is

$$\langle Sf, Sg\rangle - \langle S^*f, S^*g\rangle = \langle f, g\rangle, f, g \in \mathcal{D}(S) \cap \mathcal{D}(S^*) \tag{qho$_\mathrm{w}$}$$

and because this is equivalent to

$$\|Sf\|^2 - \|S^*f\|^2 = \|f\|^2, \quad f \in \mathcal{D}(S) \cap \mathcal{D}(S^*). \tag{qho$'_\mathrm{w}$}$$

S must be closable (cf. (2.2.9)), it implies (qho$_\mathrm{w}$) (as well as (qho$'_\mathrm{w}$)) is equivalent in turn to

$$\langle \bar{S}f, \bar{S}g\rangle - \langle S^*f, S^*g\rangle = \langle f, g\rangle, \quad f, g \in \mathcal{D}(\bar{S}) \cap \mathcal{D}(S^*). \tag{qho$''_\mathrm{w}$}$$

Relationships between the different meanings of (qho) The occurring interdependence, which follows, let us play variation on the theme of (2.2.4).

Firstly,

$$(\mathrm{qho}_{\mathcal{D}}) \text{ with } \mathcal{D} \text{ being a core of } S \implies (\mathrm{qho}_{\mathrm{w}}) \text{ and } \mathcal{D}(\overline{S}) \subset \mathcal{D}(S^*). \quad (2.2.7)$$

Indeed, because S is closable as we already know, for $f \in \mathcal{D}(\overline{S})$, there is a sequence $(f_n)_{n=0}^{\infty} \subset \mathcal{D}$ such that $f_n \to f$ and $Sf_n \to \overline{S}f$. Because S^* is closed, we get from $(\mathrm{qho}_{\mathcal{D}})$ $S^*f_n \to S^*f$ and consequently $f \in \mathcal{D}(S^*)$ and (2.2.7) follows.

Secondly,

$$(\mathrm{qho}_{\mathcal{D}}) \text{ with } \mathcal{D} \text{ being a core of } S^* \implies (\mathrm{qho}_{\mathrm{w}}) \text{ and } \mathcal{D}(S^*) \subset \mathcal{D}(\overline{S}). \quad (2.2.8)$$

This uses the same argument as that for (2.2.7).

Thirdly,

$$(\mathrm{qho}_{\mathrm{w}}) \implies (\mathrm{qho}_{\mathcal{D}}) \text{ with } \mathcal{D} = \mathcal{D}(S^*S) \cap \mathcal{D}(SS^*) \implies S \text{ closable}. \quad (2.2.9)$$

This is because $\mathcal{D}(S^*\overline{S}) \cap \mathcal{D}(\overline{S}S^*) \subset \mathcal{D}(\overline{S}) \cap \mathcal{D}(S^*)$.

Fourthly,

$$(\mathrm{qho}_{\mathrm{w}}) \text{ with } \mathcal{D}(\overline{S}) \cap \mathcal{D}(S^*) \text{ a core of } S \text{ and } S^* \implies \mathcal{D}(S^*\overline{S}) = \mathcal{D}(\overline{S}S^*).$$

Take $f \in \mathcal{D}(S^*\overline{S})$. This means $f \in \mathcal{D}(\overline{S})$ and $\overline{S}f \in \mathcal{D}(S^*)$. Because of this, picking $(f_n)_{n=0}^{\infty} \in \mathcal{D}(\overline{S}) \cap \mathcal{D}(S^*)$, we get from $(\mathrm{qho}_{\mathrm{w}})$ in the limit

$$\langle S^*Sf, g\rangle - \langle S^*f, S^*g\rangle = \langle f, g\rangle \quad (2.2.10)$$

for $g \in \mathcal{D}(\overline{S}) \cap \mathcal{D}(S^*)$. Because $g \in \mathcal{D}(\overline{S}) \cap \mathcal{D}(S^*)$ is a core of S^*, we get (2.2.10) to hold for $g \in \mathcal{D}(S^*)$. Finally, $S^*f \in \mathcal{D}(\overline{S})$. The converse uses the same kind of argument.

The aforementioned results in

$$(\mathrm{qho}_{\mathcal{D}}) \text{ with } \mathcal{D}(\overline{S}) = \mathcal{D}(S^*) \iff \overline{S} \text{ satisfies } (\mathrm{qho}_{\mathcal{D}}). \quad (2.2.11)$$

2.2.1.2 ***Creation and Annihilation Operator*** In order to be free of any model as long as there is no need, we define the two basic operators in the very abstract setting. Consider a Hilbert space $\mathcal{H}$ and an orthogonal basis $(e_n)_{n=0}^{\infty}$ in it, and define the *creation* operator a_+ and the *annihilation* operator a_- as follows (cf. Section 2.1.2.10). With $\mathcal{D} \stackrel{\mathrm{def}}{=} \mathrm{lin}(e_n)_{n=0}^{\infty}$

$$\mathcal{D}(a_+) \stackrel{\mathrm{def}}{=} \mathcal{D}, \quad a_+e_n \stackrel{\mathrm{def}}{=} \sqrt{n+1}\, e_{n+1}, \quad n = 0, 1, \ldots$$
$$\mathcal{D}(a_-) \stackrel{\mathrm{def}}{=} \mathcal{D}, \quad a_-e_0 \stackrel{\mathrm{def}}{=} 0, \quad a_-e_n \stackrel{\mathrm{def}}{=} \sqrt{n}\, e_{n-1}, \; n = 1, \ldots.$$

Moreover,

$$a_+ = (a_-)^*|_{\mathcal{D}} \text{ and } a_- = (a_+)^*|_{\mathcal{D}}. \quad (2.2.12)$$

and, due to the characterization given on p. 83,

$$\mathcal{D}(\overline{a_+}) = \mathcal{D}((a_+)^*). \tag{2.2.13}$$

Furthermore, cf. (11, p. 136)

$$\mathrm{sp}(\overline{a_+}) = \mathrm{sp}_\mathrm{r}(\overline{a_+}) = \mathbb{C}, \quad \mathrm{sp}(\overline{a_-}) = \mathrm{sp}_\mathrm{p}(\overline{a_-}) = \mathbb{C}.$$

Fact 2.2.4 *Always $S = a_+$ satisfies (qho$_\mathcal{D}$) with $\mathcal{D} = \mathcal{D}(a_+)$.*

Specifying more the Hilbert space and its orthonormal basis, we can detail some representations (called alternatively *models* here) of (qho).

The $\mathcal{L}^2(\mathbb{R})$ model This is the most classical and the best know[39] realizations of (qho).

Take $\mathcal{D} = \mathrm{lin}(h_n)_{n=0}^\infty$, where the Hermite functions h_n are on p. 86. For the operators, we have

$$a_+ = \frac{1}{\sqrt{2}}\Big(x - \frac{\mathrm{d}}{\mathrm{d}x}\Big), \quad a_- = \frac{1}{\sqrt{2}}\Big(x + \frac{\mathrm{d}}{\mathrm{d}x}\Big), \tag{2.2.14}$$

they are weighted shifts with respect to the basis of Hermite functions.

The Segal–Bargmann model Here $\mathcal{D} \overset{\mathrm{def}}{=} \{\text{analytic polynomials}\}$. The operators[40] can be written as

$$a_+ = M_Z, \quad a_- = \frac{\mathrm{d}}{\mathrm{d}z} \tag{2.2.15}$$

and things go on in the familiar way.

REMARK **14** We know already that in the general, abstract case the creation operator is subnormal, cf. Theorem (2.2.7). Luckily enough the Segal–Bargmann model has the advantage over others to provide the normal extension and its space explicitly (analytically). The extension space of the Hilbert space $\mathcal{L}^2_{\mathrm{hol}}(\pi^{-1}\exp(-|z|^2)\,\mathrm{d}x\mathrm{d}y)$ is $\mathcal{L}^2(\pi^{-1}\exp(-|z|^2)\,\mathrm{d}x\mathrm{d}y)$ and the normal extension N (which is minimal of cyclic type—uniqueness happens) of a_+ is the multiplication by the independent variable Z in $\mathcal{L}^2(\pi^{-1}\exp(-|z|^2)\,\mathrm{d}x\mathrm{d}y)$. Because the latter is *-cyclic, the Spectral Theorems 2.1.4 and 2.1.5 can apply to define functions of N. According to the rough rule

[39]Commonly recognized by MPs as that in the *configuration space*.

[40]Rather the (formal) operators can be called after Fock than the Hilbert space itself. The space $\mathcal{L}^2_{\mathrm{hol}}(\pi^{-1}\exp(-|z|^2)\,\mathrm{d}x\mathrm{d}y)$ was built up analytically in (51) in order to make the operators (2.2.15) Hilbert space ones; cf. Remark 12.

(2.1.27), the exponential function can be defined at least for the kernel functions $K_w : z \to \exp(zw^*)$. More precisely,

$$\langle e^{\alpha a_+} f, g^* \rangle = \int_{\mathbb{C}} e^{\alpha z} f(z) g(z)^* \exp(-|z|^2)\, dxdy, \quad f, g \in \operatorname{lin}(K_w)_{w\in\mathbb{C}}$$

defines the exponential function $e^{\alpha a_+}$ with $\mathcal{D}(e^{\alpha a_+}) \overset{\text{def}}{=} \operatorname{lin}(K_w)_{w\in\mathbb{C}}$. Then, $e^{\alpha a_-}$ can be defined as the adjoint of $e^{\alpha a_+}$. The space $\operatorname{lin}(K_w)_{w\in\mathbb{C}}$ is invariant for both $e^{\alpha a_+}$ and $e^{\alpha a_-}$ and they are acting as groups in α on $\operatorname{lin}(K_w)_{w\in\mathbb{C}}$.

Notice that the two exponential functions can be got alternatively via Section 2.1.4.2 point 1°; cf. also (71). ♣

An evoking illustration: the ℓ^2 models Now, referring to definitions (2.1.36) and (2.1.40), we can exclusively work with the extended range of parameters $\alpha \geqslant 0$. While the case $\alpha = 0$ may serve as a test piece the other model presented in (47) brings in the topic some sprightliness.

For a parameter $\alpha > 0$, we have defined Charlier sequences $c_n^{(\alpha)}$ as members of ℓ^2. We already know that the sequence $\left(c_n^{(\alpha)}\right)_{n=0}^{\infty}$ (of sequences) is an orthonormal basis in ℓ^2. Let

$$\mathcal{D}^{(\alpha)} \overset{\text{def}}{=} \operatorname{lin}(c_n^{(\alpha)})_{n=0}^{\infty}. \tag{2.2.16}$$

The "operational formulae"[41]

$$g(x) = \sqrt{x} f(x-1) - \sqrt{\alpha} f(x), \quad x = 1, 2, \ldots \text{ and } g(0) = -\sqrt{\alpha} f(0); \tag{2.2.17}$$

$$h(x) = \sqrt{x+1} f(x+1) - \sqrt{\alpha} f(x), \quad x = 0, 1, \ldots \tag{2.2.18}$$

do not lead out of the space ℓ^2 as long as f belongs to $\mathcal{D}_{\max}$ defined as

$$\mathcal{D}_{\max} \overset{\text{def}}{=} \{f \in \ell^2 : \sum_{i=0}^{\infty} |\langle f, \varepsilon_n \rangle|^2 (n+1) < +\infty\}. \tag{2.2.19}$$

The right-hand side of formula (2.2.17) sends $c_n^{(\alpha)}$ to $\sqrt{n+1}\, c_{n+1}^{(\alpha)}$. This is in accordance with (72, formula (9.14.8) p. 248)) if one multiplies the nth Charlier polynomial there by $(-\alpha)^{-n}$ to get ours. In the same way, the formula (2.2.18) is related to (9.14.6) in (72).

Formulae (2.2.17) and (2.2.18) look like finite difference analogs of those in (2.2.14). The operators they define for $f = c_n^{(\alpha)}$ turn out to be[42] precisely the creation operators $a_+^{(\alpha)}$ (formula (2.2.17)) and annihilations ones $a_-^{(\alpha)}$ (formula (2.2.18)) with domains chosen according to the rules of Section 2.2.1.

[41] It is safer to say this way than to call them "operators" as the domains are not determined yet.

[42] Here splitting the definition (2.1.36) into two steps plays an essential role, cf. the footnote [27].

REMARK **15** Notice that for the whole range $\alpha \geqslant 0$ we have got, cf. (47), the following spectacular proclamations[43]

$$\boxed{\begin{array}{c} \mathcal{D}(a_+^{(\alpha)}) \cap \mathcal{D}(a_+^{(\beta)}) = \{0\}, \quad \mathcal{D}(a_-^{(\alpha)}) \cap = \mathcal{D}(a_-^{(\beta)}) = \{0\} \text{ if } \alpha \neq \beta \\ \mathcal{D}(a_+^{(\alpha)^-}) = \mathcal{D}(a_-^{(\alpha)^-}) = \mathcal{D}_{\max} \text{ is independent of } \alpha \geqslant 0 \end{array}}$$

To prove the second equality in the second line of the boxed area, one has to compare the definition (2.2.19) with the description of the domain of the closure of a weighted shift given on p. 72.

In other words, we have a one parameter ($\alpha \geqslant 0$) family of oscillators with domains having only 0 as a common vector such that all their closures enjoy the same, independent of a, domain. This is one of the strongest arguments supporting the necessity of clarifying domains, always! ♣

REMARK **16** Introducing the unitary operator $V_{(\alpha)} : \ell^2 \to \ell^2$ defined as

$$V_{(\alpha)} c_n^{(\alpha)} = \varepsilon_n, \quad n = 0, 1, \dots,$$

we get an amazing formula

$$\boxed{V_{(\alpha)}{}^{-1} a_+^{(\alpha)^-} V_{(\alpha)} = a_+^{(\alpha)^-} - \sqrt{\alpha}\, I|_{\mathcal{D}^{(\alpha)}}, \quad \alpha \geqslant 0} \tag{2.2.20}$$

Indeed, from (2.2.17), we get $(a_+^{(\alpha)}\varepsilon_n)(x) = \sqrt{n+1}\,\varepsilon_{n+1}(x) - \sqrt{a}\,\varepsilon_n(x)$.

Passing to the adjoints in (2.2.20), we get the companion formula for the annihilation operator.

Formula (2.2.20) can be rephrased as: *the creation operator is unitarily equivalent to itself "plus" a multiply of the identity operator*. Something like this would never happen to bounded operators (the spectral mapping theorem!). ♣

All in this pParagraph and more is in (47).

A nonrotationally invariant model Here, we consider the space $\mathcal{X}_A$ described in Paragraph 2.1.4.3. With $\mathcal{D} = \operatorname{lin}(h_n)_{n=0}^{\infty}$ the operators defined as

$$(a_+ f)(z) = \sqrt{\frac{1-A}{2(1+A)}}\,[zf(z) - f'(z)],$$

$$(a_- f)(z) = \sqrt{\frac{1+A}{2(1-A)}}\,[zf(z) + f'(z)], \quad z \in \mathbb{C}, \quad f \in \mathcal{D}$$

are the creation and the annihilation operators in $\mathcal{X}_A$.

[43]The little bit scratchy notation in the second line of the frame stands for the closures of $a_+^{(\alpha)}$ and $a_-^{(\alpha)}$ respectively

2.2.1.3 Two Preliminary Lemmas Let us extract from Theorem, p. 328 in (69), some points and collect them in two separate lemmas.

Lemma 2.2.5 *Consider the following conditions.*

(i) *There is an orthonormal basis in $\mathcal{H}$ of the form $(e_n)_{n=0}^{\infty}$ contained in $\mathcal{D}(\bar{S})$ and such that*

$$\bar{S} e_n = \sqrt{n+1}\, e_{n+1}, \quad n = 0, 1, \dots ;$$

(ii) *S is irreducible and it satisfies ($\mathrm{qho}_{\mathcal{D}}$) with some $\mathcal{D}$, which is an invariant core for S and S^*, and such that $\mathcal{D} \subset \mathcal{B}(S^*)$;*

(iii) *S is irreducible and it satisfies ($\mathrm{qho}_{\mathcal{D}}$) with some $\mathcal{D}$, which is an invariant core for S and S^*, and such that $\mathcal{D} \subset \mathcal{A}(S)$.*

Then (i) $\Longrightarrow$ (ii) $\Longrightarrow$ (iii).

Proof: (i) $\Longrightarrow$ (ii). It is a direct consequence of Facts 2.2.4 and 2.1.7. (ii) $\Longrightarrow$ (iii). Take $\mathcal{D} = \mathcal{B}(S^*)$. Then, by (2.2.5), for $f \in \mathcal{D}$ and $i, = 0, 1, \dots$, we have

$$\|S^i f\|^2 = \sum_{k=0}^{\infty} k! \binom{i}{k}^2 \|S^{*(i-k)} f\|^2 \leq a^2 b^{2i} i! \sum_{k=0}^{\infty} \frac{i!}{k!((i-k)!)^2} b^{-2k}$$

$$\leq a^2 b^{2i} i! \sum_{k=0}^{\infty} \frac{1}{k!} b^{-2k} \leq a^2 b^{2i} i!\, \mathrm{e}^{-2b} .$$

This gives us at once f satisfies with $t = (b+1)^{-1}$ the condition for being analytic. ∎

Lemma 2.2.6 *Consider the following conditions.*

(v) *S satisfies ($\mathrm{qho}_{\mathcal{D}}$) with some $\mathcal{D}$ being a core of both S and S^*;*

(vi) *S satisfies ($\mathrm{qho}_{\mathrm{w}}$) and $\mathcal{D}(\bar{S}) = \mathcal{D}(S^*)$;*

(vii) *S satisfies ($\mathrm{qho}_{\mathrm{w}}$) with $\mathcal{D}(\bar{S}) \cap \mathcal{D}(S^*)$ being dense in $\mathcal{H}$, $\mathcal{N}((S-\lambda)^*) \neq \{0\}$ for <u>all</u> $\lambda \in \mathbb{C}$ and $\bar{S}^n(\mathcal{N}((S-\lambda)^*)) \subset \mathcal{D}(\bar{S}) \cap \mathcal{D}(S^*)$ for $n = 0, 1, \dots$;*

(viii) *S satisfies ($\mathrm{qho}_{\mathrm{w}}$) with $\mathcal{D}(\bar{S}) \cap \mathcal{D}(S^*)$ being dense in $\mathcal{H}$, $\mathcal{N}((S-\lambda)^*) \neq \{0\}$ for <u>some</u> $\lambda \in \mathbb{C}$ and $\bar{S}^n(\mathcal{N}((S-\lambda)^*)) \subset \mathcal{D}(\bar{S}) \cap \mathcal{D}(S^*)$ for $n = 0, 1, \dots$.*

Then (v) $\Longrightarrow$ (vi) $\Longrightarrow$ (vii) $\Longrightarrow$ (viii).

Proof: (v) $\Longrightarrow$ (vi). This follows from (2.2.7) and (2.2.8).

(vi) $\Longrightarrow$ (vii). For $\lambda \in \mathbb{C}$ set $S_\lambda \stackrel{\text{def}}{=} S - \lambda$ with $\mathcal{D}(S_\lambda) \stackrel{\text{def}}{=} \mathcal{D}(S)$. Then, because $\mathcal{D}(\overline{S_\lambda}) = \mathcal{D}(\bar{S})$ and $\mathcal{D}(S_\lambda^*) = \mathcal{D}(S^*)$, S_λ satisfies (vi) as well. Fix λ; the polar decomposition for S_λ^* is $S_\lambda^* = V|S_\lambda^*|$ where V is a partial isometry with the initial space $\mathcal{R}(|S_\lambda^*|)$ and the final space $\mathcal{R}(S_\lambda S_\lambda^*)$. Suppose $\mathcal{N}(S_\lambda^*) = \{0\}$. Then, because

$\mathcal{N}(V) = \mathcal{R}(S_\lambda^*|)^\perp = \mathcal{N}(|S_\lambda^*|) = \mathcal{N}(\overline{S_\lambda}S_\lambda^*) = \mathcal{N}(S_\lambda^*)$, V is unitary. As $\overline{S_\lambda} = |S_\lambda^*|V^*$, from (2.2.11), we get $V|S_\lambda^*|^2V^* = |S_\lambda^*|^2 + I$. Consequently,

$$\mathrm{sp}(()|S_\lambda^*|) \subset \mathrm{sp}(()|S_\lambda^*|) + 1 \subset [0, +\infty)$$

which is an absurd. Thus, $\mathcal{N}(S_\lambda^*) \neq \{0\}$.

We show by induction that for $n = 0, 1, \ldots$

$$f \in \mathcal{N}(S_\lambda^*) \Longrightarrow \overline{S_\lambda}^n f \in \mathcal{D}(\overline{S}_\lambda),\ \ \overline{S}_\lambda^{(n-1)} f \in \mathcal{D}(S_\lambda^*)$$

$$\&\ \ S_\lambda^*\overline{S_\lambda}^{n-1} f = (n-1)\overline{S_\lambda}^{n-2} f. \quad (2.2.21)$$

Of course, $\mathcal{N}(S_\lambda^*) \subset \mathcal{D}(\overline{S}_\lambda) = \mathcal{D}(S_\lambda^*)$, which establishes (2.2.21) for $n = 0$. Suppose $\mathcal{N}(S_\lambda^*) \subset \mathcal{D}(\overline{S}_\lambda^n)$ and $S_\lambda^*\overline{S}_\lambda^{n-1} f = (n-1)\overline{S}_\lambda^{n-2} f$. Then, for $g \in \mathcal{D}(\overline{S}_\lambda) = \mathcal{D}(S_\lambda^*)$,

$$\langle S_\lambda^*\overline{S}_\lambda^{n-1} f, S_\lambda^* g\rangle = (n-1)\langle \overline{S}_\lambda^{n-2} f, \overline{S}_\lambda^* g\rangle.$$

Because already $\overline{S}_\lambda^{(n-2)} f \in \mathcal{D}(\overline{S}_\lambda) = \mathcal{D}(S_\lambda^{**})$, we have

$$|\langle S_\lambda^*\overline{S}_\lambda^{n-1} f, S_\lambda^* g\rangle| \leq C\|g\|. \quad (2.2.22)$$

Moreover, because $\overline{S}_\lambda^{(n-1)} f \in \mathcal{D}(\overline{S}_\lambda) = \mathcal{D}(S_\lambda^*)$, we can use (ϙho$_{\mathrm{w}}$) so as to get

$$\langle \overline{S}_\lambda^n f, \overline{S}_\lambda g\rangle = \langle \overline{S}_\lambda\overline{S}_\lambda^{(n-1)} f, \overline{S}_\lambda g\rangle = \langle S_\lambda^*\overline{S}_\lambda^{(n-1)}, S_\lambda^*\rangle + \langle \overline{S}_\lambda^{(n-1)} f, g\rangle.$$

This, by (2.2.22), implies $\overline{S}_\lambda^n f \in \mathcal{D}(S_\lambda^*) = \mathcal{D}(\overline{S}_\lambda)$ and, consequently, by (2.2.21), gives us $S_\lambda^*\overline{S}_\lambda^n f = n\overline{S}_\lambda^{n-1} f$, which completes the induction argument.

A straightforward application of (2.2.21) gives $\overline{S}^n(\mathcal{N}((S-\lambda)^*)) \subset \mathcal{D}(\overline{S}) \cap \mathcal{D}(S^*)$ for $n = 0, 1, \ldots$.

(vii) $\Longrightarrow$ (viii). Any word is needless. ∎

Notice that the labeling in the above is discontinuous at the moment, the item (iv) is missing. Because of its special role, it has been single out and will appear in the following Section.

2.2.1.4 The Key Point The formula (2.2.13), when stated in the general framework, takes the form of (vi). Because of its simplicity, it is tempting it to be deemed the key to the full success. Theoretically yes but from the practical point of view, the merits of Remark 13 appear here the most.

Persisting in labeling the items of Lemmas 2.2.5 and 2.2.6, we add now the missing item (iv), which bridges the gap.

(iv) S is irreducible and it satisfies ($\mathtt{qho}_{\mathcal{D}}$) with some $\mathcal{D}$ being invariant for S and S^* which is a core of S and S^*, S is a **subnormal** operator having a minimal normal extension of cyclic type.

Theorem 2.2.7 *Finally, one has*

$$\textbf{(i)} \Longrightarrow \text{(ii)} \Longrightarrow \text{(iii)} \Longrightarrow \textbf{(iv)} \Longrightarrow \text{(v)} \Longrightarrow \text{(vi)} \Longrightarrow \text{(vii)} \Longrightarrow \text{(viii)} \Longrightarrow \textbf{(i)}$$

provided in conditions (v), (vi), (vii) *and* (viii) *irreducibility of S is added.*

Proof: The only implications that require some proofs are (iii) $\Longrightarrow$ (iv) $\Longrightarrow$ (v) and $\Longrightarrow$ (viii) $\Longrightarrow$ (i). The others have been already included in Lemmas 2.2.5 and 2.2.6.

(iii) $\Longrightarrow$ (iv) is the matter of Theorem 2.1.11 enriched with Remark 13.

(iv) $\Longrightarrow$ (v). Pick $f \in \mathcal{D}(S^*)$, we have to prove that there is $(f_n)_{n=0}^{\infty} \subset \mathcal{D}$ such that

$$\|f_n - f\| + \|S^*(f_n - f)\| \to 0. \tag{2.2.23}$$

As $f \in \mathcal{D}(N^*)$ and N is a minimal normal extension of cyclic type there is $(x_n)_{n=0}^{\infty} \subset \mathcal{D}_N$, where $\mathcal{D}_N$ is defined as in (2.1.28), such that

$$\|x_n - f\| + \|N^*(x_n - f)\| \to 0.$$

Because $\|f - Px_n\| = \|P(f - x_n)\| \leq \|f - x_n\|$ and $\|S^*(f_n - Px_n)\| = \|PN^*(f - x_n)\| \leq \|N^*(f - x_n)\|$, $f_n \overset{\text{def}}{=} Px_n$ makes (2.2.23) to be true.

(v) $\Longrightarrow$ (vi) follows from (2.2.7) and (2.2.8).

(viii) $\Longrightarrow$ (i). Pick a vector f_0 in $\mathcal{N}(S_\lambda^*)$ and set $f_n \overset{\text{def}}{=} S_\lambda^n f_0$. We show that for $n = 0, 1, \dots,$

$$f_m \perp f_n \quad \text{if} \quad m \neq n. \tag{2.2.24}$$

Indeed, because, by (2.2.21), $\operatorname{lin}\{f_n : \ n = 0, 1, \dots\}$ is invariant for both S_λ and S_λ^*, we can use (2.2.5) as follows

$$\begin{aligned}\langle f_m, f_n \rangle &= \langle S_\lambda^m f_0, S_\lambda^n f_0 \rangle = \langle S_\lambda^{n*} S_\lambda^m f_0, f_0 \rangle \\ &= \sum_{k=0}^{\min\{m,n\}} k! \binom{m}{k} \binom{n}{k} \langle S_\lambda^{(n-k)} S_\lambda^{*(m-k)} f_0, f_0 \rangle\end{aligned}$$

and this is equal to 0 if $m > n$, which proves (2.2.24). Because, what follows from ($\mathtt{qho}_{\mathrm{w}}$), S_λ is injective, all f_ns are different from zero. Thus, if $e_n^{(\lambda)} \overset{\text{def}}{=} \|f_n\|^{-1} f^n$ [44], we get $\{e_n^{(\lambda)} : \ n = 0, 1, \dots\}$ to be a set of orthonormal vectors.

[44] Notice that f_n's depend on λ via S_λ.

Set $\tilde{\mathcal{H}} \stackrel{\text{def}}{=} \text{clolin}\{e_n^{(\lambda)} : \ n = 0, 1, \ldots\}$. We show the subspace $\tilde{\mathcal{H}}$ reduces S_λ. Indeed, taking $f \in \mathcal{D}(\overline{S}_\lambda)$, by (2.2.26), we have

$$\begin{aligned} P\overline{S}_\lambda f &= \sum_n \langle \overline{S}_\lambda f, e_n^{(\lambda)} \rangle e_n^{(\lambda)} = \sum_{n=0}^{\infty} \langle f, S_\lambda^* e_n^{(\lambda)} \rangle e_n^{(\lambda)} \\ &= \sum_{n=}^{\infty} \langle f, \frac{n}{\sigma_{n-1}} e_n^{(\lambda)} \rangle = \sum_n \langle f, e_n^{(\lambda)} \rangle \frac{n}{\sigma_{n-1}} e_n^{(\lambda)} = \sum_{n=0}^{\infty} \langle f, e_{n-1}^{(\lambda)} \rangle \overline{S}_\lambda e_{n-1}^{(\lambda)} \\ &= \overline{S}_\lambda P f. \end{aligned}$$

Because $\mathcal{N}((S_a|_{\tilde{\mathcal{H}}})^*) \oplus \mathcal{N}((S_a|_{\tilde{\mathcal{H}}^\perp})^*) = \mathcal{N}(S_\lambda^*)$ and $\mathcal{N}((S_\lambda|_{\tilde{\mathcal{H}}})^*) \subset \mathcal{N}(S_\lambda^*)$, one has $\mathcal{N}((S_\lambda|_{\tilde{\mathcal{H}}^\perp})^*) = \{0\}$. Irreducibility of S forces $\tilde{\mathcal{H}} = \mathcal{H}$, which results in completeness of the set $\{e_n^{(\lambda)} : \ n = 0, 1, \ldots\}$ of orthonormal vectors making it a basis in $\mathcal{H}$.

Because the chain of implications in Lemma 2.2.6 goes logically in the favorable direction, we can make use of (2.2.21) at this moment. Therefore, we have at our disposal the following

$$f \in \mathcal{N}(S_\lambda^*) \Longrightarrow \overline{S}_\lambda^n f \in \mathcal{D}(\overline{S}_\lambda), \ \ \overline{S}_\lambda^{n-1} f \in \mathcal{D}(S_\lambda^*) \ \ \& \ \ S_\lambda^* \overline{S}_\lambda^{n-1} f = (n-1)\overline{S}_\lambda^{n-2} f.$$

Therefore, we get from this

$$S_\lambda{}^* \overline{S}_\lambda^n e_0^{(\lambda)} = n \overline{S}_\lambda^{n-1} e_0^{(\lambda)}, \quad n = 1, 2, \ldots. \tag{2.2.25}$$

Because $f_{n+1} = \overline{S}_\lambda^{n+1} f_0 = \overline{S}_\lambda \overline{S}_\lambda^n f_0 = \overline{S}_\lambda f_0$, we have $\overline{S}_\lambda f_n = \sigma_n f_{n+1}$ with σ_ns being positive (in fact $\sigma_n = \|f_{n+1}\| \|f_n\|^{-1}$).

On the other hand, if $n > 0$, by (2.2.25), $S^* f_n = (\sigma_{n-1} \cdots \sigma_0)^{-1} S^* \overline{S}^n f_0 = n(\sigma_{n-1} \cdots \sigma_0)^{-1} \overline{S}^{(n-1)} f_0 = n \frac{\sigma_{n-1} \cdots \sigma_0}{\sigma_{n-2} \cdots \sigma_0} f_{n-1} = \frac{n}{\sigma_{n-1}} f_{n-1}$. Furthermore,

$$\begin{aligned} \sigma_n = \sigma_n \langle f_{n+1}, f_{n+1} \rangle &= \langle \sigma_n f_{n+1}, f_{n+1} \rangle = \langle \overline{S} f_n, f_{n+1} \rangle \\ &= \langle f_n, S^* f_{n+1} \rangle = \langle f_n, \frac{n+1}{\sigma_n} f_n \rangle = \frac{n+1}{\sigma_n}, \end{aligned}$$

which implies $\sigma_n = \sqrt{n+1}$. Summing up

$$\overline{S}^n e_n^{(\lambda)} = \sqrt{n+1}\, e_{n+1}^{(\lambda)} \text{ and } S^* e_n^{(\lambda)} = \sqrt{n}\, e_{n-1}^{(\lambda)}, \quad n > 0, \ S^* e_n^{(\lambda)} = 0. \tag{2.2.26}$$

What we have got so far is $S_\lambda f_n = (S-\lambda)e_n^{(\lambda)} = \sqrt{n+1}\,e_{n+1}^{(\lambda)}$, but this is still not what we claim in (i); condition (i) concerns the operator S, not[45] S_λ. Let U be the unitary operator from $\mathcal{H}$ onto ℓ^2 which maps $e_n^{(\lambda)}$ to ε_n, where $\{\varepsilon_n\}_{n=0}^\infty \stackrel{\text{def}}{=} \{\{\delta_{k,n}\}_{k=0}^\infty\}$ is nothing but the canonical zero-one basis in ℓ^2. Our first use of U is in writing

$$US_\lambda e_n^{(\lambda)} = U(S-\lambda)e_n^{(\lambda)} = \sqrt{n+1}\,Ue_{n+1}^{(\lambda)}. \tag{2.2.27}$$

What can be found in (47) is just a construction, based on Charlier polynomials[46], of a basis $(c_n^{(\lambda)})_{n=0}^\infty$ in ℓ^2 such that the weighted shift operator $S_{(\lambda)}$ in ℓ^2 which shifts the canonical zero-one basic vectors with weights $\{\sqrt{n+1}\}_{n=0}^\infty$, for example, $S_\lambda\varepsilon_n = \sqrt{n+1}\,\varepsilon_{n+1}$, behaves with respect to the basis $(c_n^{(\lambda)})_{n=0}^\infty$ as $\overline{S}_{(\lambda)}c_n^{(\lambda)} = \sqrt{n+1}\,c_{n+1}^{(\lambda)} - \sqrt{\lambda}\,c_n^{(\lambda)}$. If we set $e_n \stackrel{\text{def}}{=} U^{-1}c_n^{(\lambda)}$, then, following the pattern of (2.2.27), we get

$$\begin{aligned} U(\overline{S}-\sqrt{\lambda})e_n = U\overline{S}_\lambda e_n = U\overline{S}_\lambda U^{-1}c_n^{(\lambda)} &= \sqrt{n+1}\,c_{n+1}^{(\lambda)} - \sqrt{\lambda}\,c_n^{(\lambda)} \\ &= \sqrt{n+1}\,Ue_{n+1} - \sqrt{\lambda}\,Ue_n. \end{aligned}$$

This gives us at once the required $\overline{S}e_n = \sqrt{n+1}\,e_{n+1}$. ∎

Subnormality of the creation operator was acknowledged already in (74, Chapter III, Section 10).

REMARK **17** Conditions (vii) and (viii) can be stated with $\lambda = 0$ like Lemma 2.3.4. We have decided to go this way because it becomes closer to Theorem 2.2.9 and Corollorary 2.2.10; an intermediate step in [82]. ♣

Removing irreducibility of S opens the way to reproduce Theorem 2.2.7 in a suitable number of (orthogonal) copies.

• Our leitmotif – **subnormality** – is flourishing now •

As an excuse, we would like to explain the reason of making so long and detailed presentation of our way of solving the commutation relation (qho). It is twofold: 1° its importance, of course, and 2° an occasion to rectify (some of) the meanders spoiling the basic paper (69)—no one is free of this kind of creation!

[45] One may ask why we insert λ in the game making things looking more complicated. This accusation may be partially justified; however, the only complication that appears is the need to incorporate the last paragraph of the proof. The good point in this is a strong relation of our setting to what appears in Section 2.2.2 and in (73)

[46] If referring to (47) one has to replace the parameter a here by $\sqrt{\lambda}$.

Rellich–Dixmier condition (2.2.3) when stated for the Schrödinger couple is apparently a *second*-order differential, one considered in the real variable. Our condition (iv) as considered in complex variable, that is in the Segal–Bargmann space, becomes a differential condition of order *one*. Condition (iv) is undoubtedly the proper choice.

REMARK **18** There is one more condition that can be wedged between (iv) and (v). This is

(iv′) S satisfies (qho$_{\mathcal{D}}$) with some $\mathcal{D}$ being invariant for S and S^*, which is a core of S and the formally normal extension N of S, which existence stems, due to Fact 2.2.3, from Theorem 2.1.9, satisfies $\mathcal{D}(S^*) = \mathcal{D}(N) \cap \mathcal{H}$.

It has to be very little to change in the proof of (iv) $\Longrightarrow$ (v) so as to split it into (iv) $\Longrightarrow$ (iv') and (iv') $\Longrightarrow$ (v). ♣

Notice that $\mathcal{D}(S^*) = \mathcal{D}(N) \cap \mathcal{H}$ may not hold for arbitrary subnormal operators, see (75) for a counterexample.

2.2.2 Duality

General principle. Amalgam of a self-dual basis and a Hilbert space one Call an orthonormal sequence $(\eta_n)_{n=0}^{\infty} \subset \ell^2$ *self-dual* if for some[47] $\alpha \in \mathbb{C}$ with $|\alpha| = 1$

$$\alpha^x \eta_x(n) = \alpha^n \eta_n(x), \quad n, x = 0, 1, \dots , \tag{2.2.28}$$

for $\eta_n(x)$ to be the xth coordinate of η_n. One can get immediately from Theorem 1 of (46) that a self-dual sequence is complete; hence, it is a basis.

The observation that follows (cf. (73)) is crucial in our presentation; it provides us with a recipe for constructing new bases from a given one relying on a choice of a self-dual basis in ℓ^2, a reciprocal formula is also included.

Proposition 2.2.8 *Suppose $\mathcal{H}$ is a separable Hilbert space and $(e_n)_{n=0}^{\infty}$ is a basis in it. If $(\eta_n)_{n=0}^{\infty}$ is a self-dual basis in ℓ^2 then*

$$f_n \overset{\text{def}}{=} \sum_{k=0}^{\infty} \eta_k(n) e_k, \quad n = 0, 1, \dots , \tag{2.2.29}$$

defines a basis $\{f_n\}_{n=0}^{\infty}$ in $\mathcal{H}$. Moreover, $\{e_n\}_{n=0}^{\infty}$ can be recaptured from $\{f_n\}_{n=0}^{\infty}$ by the formula

$$e_n = \sum_{k=0}^{\infty} \eta_n(k) f_k, \quad n = 0, 1, \dots .$$

[47] Although we can always change the basis so as to have $\alpha = 1$ in some circumstances, it may be convenient to leave it as it is.

Notice that, due to (2.2.28)

$$\sum_{n=0}^{\infty} |\eta_n(x)|^2 = \sum_{n=0}^{\infty} |\eta_x(n)|^2 = 1$$

and because of this the right-hand side of (2.2.29) converges in $\mathcal{H}$ so f_ns are well defined. The proof of Proposition 2.2.8 is in (73, p. 164)

Translational Invariance of the Quantum Harmonic Oscillator Let us begin with passing comment on (qho): if S satisfies it then any (at least formally) $S - \lambda$ does so for any λ, regardless what kind of solution we have in mind.This and the converse hold true if one specifies more the solution in question.

Take η_n to be a Charlier function $c_n^{(a)}$, $n = 0, 1, \ldots$. The matters of Paragraph. p. 94 provide us with necessary information on those functions; in particular it is a self-dual basis in ℓ^2. This allows us to draw from Proposition 2.2.8 a new basis $(e_n^{(a)})_{n=0}^{\infty}$ given as

$$e_n^{(a)} \stackrel{\text{def}}{=} \sum_{k=0}^{\infty} c_k^{(a)}(n) e_k; \; n = 0, 1, \ldots . \tag{2.2.30}$$

Theorem 2.2.9 *Let S be an operator in a separable Hilbert space $\mathcal{H}$.*

(a) *If S is a creation operator with respect to $(e_n)_{n=0}^{\infty}$, then for any $a > 0$, the operator $\bar{S} + \sqrt{a}I$ is a creation operator with respect to $(e_n^{(a)})_{n=0}^{\infty}$ defined by* (2.2.30).

Conversely,

(b) *if S is weighted shift with respect to a basis $(e_n)_{n=0}^{\infty}$ and for some $a > 0$, the operator $\bar{S} + \sqrt{a}I$ is a weighted shift with respect to some other basis $(f_n^{(a)})_{n=0}^{\infty}$, then S is a creation operator with respect to $(e_n)_{n=0}^{\infty}$ and $f_n = e_n$ as well as $f_n^{(a)} = e_n^{(a)}$, $n = 0, 1, \ldots$, where the $e_n^{(a)}$s are given as in* (2.2.30). *Accordingly, $S + \sqrt{a}I$ is a creation operator with respect to*

$$(e_n)_{n=0}^{\infty}$$

for any $a > 0$.

Let us write down a refinement of the above-mentioned Theorem, which is closer to Theorem 2.2.7.

Corollary 2.2.10 *Given a basis $(e_n)_{n=0}^{\infty}$ in a separable complex Hilbert space $\mathcal{H}$, for $a > 0$, the basis $(e_n^{(a)})_{n=0}^{\infty}$ is defined as in* (2.2.30). *For a closed operator S in $\mathcal{H}$, the following conditions are equivalent*

(a) *S is creation operator with respect to $(e_n)_{n=0}^{\infty}$;*
(b) *S is creation operator with respect to $(e_n)_{n=0}^{\infty}$ and for some (or, equivalently, any) $\lambda > 0$, the operator $S + \lambda I$ is a creation operator with respect to $(e_n^{\lambda^2})_{n=0}^{\infty}$;*

(c) *S is a weighted shift with respect to* $(e_n)_{n=0}^{\infty}$ *and for some (or, equivalently, any)* $\lambda > 0$, *the operator* $S + \lambda I$ *is a weighted shift with respect to some basis* $(f_n)_{n=0}^{\infty}$;

(d) *S is a weighted shift with respect to* $(e_n)_{n=0}^{\infty}$ *with weights* $\{\sigma_n\}_{n=0}^{\infty}$ *such that all* σ_n*s are positive and* $\sigma_0 = 1$ *and for some (or, equivalently, any)* $\lambda \in \mathbb{C} \setminus \{0\}$, $S + \lambda I$ *is a weighted shift with respect to a basis* $(f_n^{(\lambda^2)})_{n=0}^{\infty}$.

If the aforementioned happens then $f_n^{\lambda} = e_n^{|\lambda|^2}$ *for* $n = 0, 1, \ldots$

The results presented earlier can be stated for backward shifts as well.

Coherent states This topic fits perfectly into the present environment. Here, we point out the most basic features in short, for much more referring to (76) (and also to (77)).

Here, we try to make known a fairly general, cautious, and economical, by the by, approach to the coherent state (CS in short) formalism, which is done in (78) and in (79); here, we go even further in examining the anatomy of CS than in those papers.

Suppose we are given

(A) two main objects: a set X and a separable Hilbert space $\mathcal{H}$

and

(B) two "subobjects": a reproducing kernel couple $(K, \mathcal{H}_K)$ on X with a fixed family $(\phi_n)_{n=0}^{\infty}$ of functions in $\mathcal{H}_K$ entering the Zaremba decomposition (2.1.31) and an orthonormal basis $(e_n)_{n=0}^{\infty}$ in $\mathcal{H}$.

Under (A) and (B), the coherent state c_x (at $x \in X$) is defined as

$$c_x \overset{\text{def}}{=} \sum_{n=0}^{\infty} \phi_n(x) e_n, \quad x \in X. \tag{2.2.31}$$

Define, moreover, the mapping $B : \mathcal{H} \to \mathcal{H}_K$ by

$$Bh \overset{\text{def}}{=} \sum_{n=0}^{\infty} \langle h, e_n \rangle \phi_n, \quad h \in \mathcal{H}. \tag{2.2.32}$$

For B, we have

$$Be_n = \phi_n, \ n = 0, 1, \ldots, \quad Bc_x = K_x, \ x \in X. \tag{2.2.33}$$

From (2.2.31), we get immediately

$$\boxed{\langle Bg, Bh \rangle_{\mathcal{H}_K} = \langle g, h \rangle_{\mathcal{H}}, \quad g, h \in \mathcal{H}} \tag{2.2.34}$$

making the mapping B an isometry, which due to the first of (2.2.33), is onto. Hence, B becomes unitary[48].

The boxed formula (2.2.34) is nothing else but the so-called resolution of identity[49] for CS as is argued for in (78, p. 5); it is good reason to call the unitary operator B the (generalized) Segal–Bargmann transform[50]. Notice that any of the conditions (2.2.33) can serve as an alternative to definition (2.2.32) of B.

Now assume only (A) to hold. Given a family $\{c_x\}_{x\in X}$ of vectors of $\mathcal{H}$, can we always consider it as a family of coherent states? Yes, this can be done in two ways:

* choosing $(e_n)_{n=0}^\infty$ we get the Fourier coefficients $\phi_n(x) \stackrel{\text{def}}{=} \langle c_x, e_n\rangle$, $n = 0, 1, \ldots$ such that the kernel defined by (2.1.31) legitimize $\{c_x\}_{x\in X}$ to be coherent states;
* having a family of functions $(\phi_n)_{n=0}^\infty$ such that $\sum_{n=0}^\infty |\phi_x|^2 < +\infty$ for all $x \in X$ and (2.1.45) holds, leads to the kernel K and then the second of (2.2.33) determines B, which *a posteriori* becomes the Segal–Bargmann transform, while the first of (2.2.33) does the vectors e_n, $n = 0, 1, \ldots$. This rounds the story out.

Denote by $S_\mathcal{H}$ the weighted shift in $\mathcal{H}$ with respect to the basis $(e_n)_{n=0}^\infty$ and by S_K that in $\mathcal{H}_K$ with respect to the basis $(\phi_n)_{n=0}^\infty$, both with the same weights $(\sigma_n)_{n=0}^\infty$. It is clear that the unitary operator U intertwines between $S_\mathcal{H}$ and S_K and between $S_\mathcal{H}^*$ and S_K^*. Because $\phi_n \in \mathcal{D}(S_K) \cap \mathcal{D}(S_K^*)$, using the criterion presented on p. 72, we can check that $c_x \in \mathcal{D}(\bar{S}_\mathcal{H}) \cap \mathcal{D}(S_\mathcal{H}^*)$. This allows us to prove

$$\boxed{\bar{S}_\mathcal{H} c_x = \sum_{k=1}^\infty (S_K^*\phi_n)(x) e_k, \quad S_\mathcal{H}^* c_x = \sum_{k=0}^\infty (S_K \phi_n)(x) e_k, \quad x \in X.} \tag{2.2.35}$$

Recall the *canonical coherent states* (called Schrödinger coherent states, see (76, p.18)) are usually defined[51] as a function of two variables, which is nothing else than the kernel of the Segal–Bargmann transform (2.1.44); fortunately, some authors go further, they develop the kernel (read: coherent states) in terms of Hermite functions in x, which is in a flavor of our definition (2.2.31), especially when one decides to go still further and separate[52] the role played by each of the variables as done in (2.2.31).

[48]One of the properties required while generalizing coherent states is the so-called *completeness* of their family. No integral representation, even resolution of the identity of any shape, is needed for that. Owing to the definition (2.2.31), completeness of $(c_x)_{x\in X}$ is equivalent to the condition (2.1.45), which is basic for RKHS. It can also be viewed as injectivity of the Segal–Bargmann transform B, which in turn comes from the fact that B is an isometry. That's it.

[49]A reader is kindly asked to get aware that no $\mathcal{L}^2$ space is needed at this stage.

[50]The role played by this transform has been noticed in (80)

[51]The "normalization factor" $\sqrt{K(x,x)}$ often present in the definition of canonical coherent states is deliberately dropped in (2.2.31).

[52]Like in Glauber–Klauder–Sudarshan definition, cf. (76, p. 20).

The second of (2.2.35) generalizes the well-known property required for coherent states saying roughly that z is an eigenvalue of the annihilation operator with eigenvector being the canonical coherent state c_z. Indeed, referring to Paragraph, p. 83, if ϕ_n are monomials like in (2.1.42), then

$$S_{\mathcal{H}}^* c_z = \sum_{k=0}^{\infty} (S_K \phi_n)(z) e_k = \sum_{k=0}^{\infty} (M_Z \phi_n)(z) e_k = z \sum_{k=0}^{\infty} (\phi_n)(z) e_k, \quad z \in \mathbb{C}$$

regardless the weights are; this certainly covers the case of the canonical coherent states. Notice the coherent states are neither in $\mathcal{D}(a_+)$ nor in $\mathcal{D}(a_-)$; one has to consider the domains of closures and adjoints.

It is a good sense in calling (2.2.35) the *duality* formulae for the coherent states (2.2.31).

2.2.2.1 Three Reasons for an Operator to Creation The very *first* reason is the merits of the whole of Theorem 2.2.7, especially condition (iv) which makes subnormality explicit.

The *second* reason makes use of translational invariance encoded in Theorem 2.2.9.

The *third* reason can be explained as follows, see (81). In (25), an attempt of bringing the matrix construction of Andô (35) over to unbounded case was made. This turned out to be partially successful, more on the difficulties arising while dealing with matrices of unbounded operator can be found in (82) (a warning here: taking entrywise closures and adjoints of matrices composed of unbounded operators is far from automatic!). Nevertheless, the case we are going to consider below is fortunate from that point of view. A formal application of the commutation relation (qho) to the aforementioned construction gives us the matrix

$$\begin{pmatrix} S & I & 0 & 0 & \\ 0 & S & \sqrt{2}I & 0 & \ddots \\ 0 & 0 & S & \sqrt{3}I & \ddots \\ & \ddots & \ddots & \ddots & \ddots \end{pmatrix}. \tag{2.2.36}$$

Let N be an operator determined by the matrix (2.2.36) with $\mathcal{D}(N)$ composed of all sequences in $\ell^2(\mathcal{H})$, which are 0 but a finite numbers of elements belonging to $\mathcal{D}(S)$. The operator N so constructed is *formally normal*, cf. (25). The fortunate circumstance we have mentioned is that $\bar{N}$ as well as N^* can be get from (2.2.36) replacing S by $\bar{S}$ or by S^*, respectively; this follows from the fact that each row and each column contains at most one unbounded operator, cf. (25) and (82) for the role played by row and column operators in making entrywise matrix operations on operator matrices possible. Therefore,

$$\mathcal{D}(\bar{S}) = \mathcal{D}(S^*)$$

if and only if N is essentially normal. On the other hand, it is clear that S satisfies (qho$_w$) with $\mathcal{D}$ being its core if and only if N is essentially normal. Now putting things together and using item (vi) of Theorem 2.2.7, which is explicit in Lemma 2.2.6, we arrive at the following

Corollary 2.2.11 *S is the creation operator if and only if S is irreducible and the operator N defined by* (2.2.36) *is essentially normal.*

The essence of Corollary 2.2.11 is in checking essential normality of a concrete operator, the matrix operator (2.2.36) makes up the rest.

A look at the matrix (2.2.36)) brings up the resemblance of the diagonal above the main one to the creation operator. This becomes more transparent if we revoke the fact that $\ell^2(\mathcal{H})$ is isomorphic to $\ell^2 \bigotimes \mathcal{H}$ in a suitable way. This isomorphism moves the operator N to

$$I_{\ell^2} \otimes S + S_0 \otimes I_{\mathcal{H}} \tag{2.2.37}$$

with the appropriate domain; S_0 is the creation operator in ℓ^2 with respect to the standard zero-one basis.

Corollary 2.2.12 *An irreducible operator S is the creation if and only if the operator* (2.2.37) *is essentially normal.*

This version is for those who may like it.

2.3 THE q–OSCILLATORS

The q-commutation relation for a Hilbert space operator S is as follows:

$$S^*S - qSS^* = I. \qquad (q\text{-o})$$

Of course, q must be perforce *real* then; this is what we assume in the paper.

Sample Theorem 1 *If S is a weighted shift with respect to the basis $\{e_n\}_{n=0}^{\infty}$ and*

$$S^*Sf - qSS^*f = f, \quad f \in \operatorname{lin}\{e_n\}_{n=0}^{\infty},$$

then $Se_n = \sqrt{1 + q + \cdots + q^n}\, e_{n+1}$, $n \geqslant 0$.

This is how it usually appears after some manipulations supposing the ground state e_0 exists. Our Rellich–Dixmier like philosophy asserts as before on creating conditions under which such a state exists. However, the situation in this case is more involved, depending on q.

q-notions. For x an integer and q real, $[x]_q \overset{\text{def}}{=} (1 - q^x)(1 - q)^{-1}$ if $q \neq 1$ and $[x]_1 \overset{\text{def}}{=} x$. If x is a nonnegative integer, $[x]_q = 1 + q \cdots + q^{x-1}$ and this is the reason it usually

referred to as a *basic* or q-number. A little step further, the q-factorial behaves like the conventional one,

$$[0]_q! \stackrel{\text{def}}{=} 1 \text{ and } [n]_q! \stackrel{\text{def}}{=} [0]_q \cdots [n-1]_q [n]_q$$

and so is the q-binomial $\begin{bmatrix} m \\ n \end{bmatrix}_q \stackrel{\text{def}}{=} \frac{[m]_q!}{[m-n]_q![n]_q!}$. Thus, if $-1 \leqslant q$ and $x \in \mathbb{N}$, the basic number $[x]_q$ is nonnegative.

For arbitrary complex numbers a and q, one can always define the *Pochhammer symbols* or *shifted factorials* $(a;q)_k$ as follows:

$$(a;q)_0 \stackrel{\text{def}}{=} 1, \quad (a;q)_k \stackrel{\text{def}}{=} (1-a)(1-aq)(1-aq^2)\cdots(1-aq^{k-1}), \quad k = 1,2,3,\ldots.$$

Then for $n > 0$, one has $[n]_q! = (q,q)_n(1-q)^{-n}$. Moreover, there are (at least) two possible definitions of q-exponential functions

$$e_q(z) \stackrel{\text{def}}{=} \sum_{k=0}^{\infty} \frac{1}{(q;q)_k} z^k, \quad z \in \omega_q,$$

$$E_q(z) \stackrel{\text{def}}{=} \sum_{k=0}^{\infty} \frac{q^{\binom{k}{2}}}{(q;q)_k} z^k, \quad z \in \omega_{q^{-1}}, \quad q \neq 0,$$

where

$$\omega_q \stackrel{\text{def}}{=} \begin{cases} \{z : \ |z| < 1\} & \text{if } |q| < 1, \\ \mathbb{C} & \text{otherwise.} \end{cases}$$

These two functions are related via

$$e_q(z) = E_{q^{-1}}(-z), \quad z \in \omega_q, \quad q \neq 0.$$

2.3.1 Spatial Interpretation of $(q\text{-}\mathrm{o})$

Again, the relation $(q\text{-}\mathrm{o})$ has nothing but a symbolic meaning unless someone says something more about it; this is because unbounded solutions may happen as well. By reason of this, we distinguish two, extreme in a sense, ways of looking at the relation $(q\text{-}\mathrm{o})$:

The first meaning of $(q\text{-}\mathrm{o})$ is

$$\begin{gathered} S \text{ closable, } \mathcal{D} \text{ is dense in } \mathcal{H} \text{ and} \\ \mathcal{D} \subset \mathcal{D}(S^*\overline{S}) \cap \mathcal{D}(\overline{S}S^*),\ S^*Sf - qSS^*f = f, f \in \mathcal{D}. \end{gathered} \qquad (q\text{-}\mathrm{o}_{\mathcal{D}})$$

The other is

$$\langle Sf, Sg\rangle - q\langle S^*f, S^*g\rangle = \langle f, g\rangle, f, g \in \mathcal{D}(S) \cap \mathcal{D}(S^*) \qquad (q\text{-}\mathrm{ow})$$

and, because this is equivalent to

$$\|Sf\|^2 - q\|S^*f\|^2 = \|f\|^2, \quad f \in \mathcal{D}(S) \cap \mathcal{D}(S^*)$$

it implies for S to be closable, $(q\text{-}\mathrm{ow})$ in turn is equivalent to

$$\langle \bar{S}f, \bar{S}g \rangle - q\langle S^*f, S^*g \rangle = \langle f, g \rangle, \quad f, g \in \mathcal{D}(\bar{S}) \cap \mathcal{D}(S^*).$$

2.3.1.1 The Self-commutator Assuming $\mathcal{D} \subset \mathcal{D}(SS^*) \cap \mathcal{D}(S^*S)$, we introduce the following operator

$$C \stackrel{\text{def}}{=} I + (q-1)SS^*, \quad \mathcal{D}(C) \stackrel{\text{def}}{=} \mathcal{D}. \tag{2.3.1}$$

This operator turns out to be an important invention in the matter. In particular, there are two immediate consequences of this definition. The first says if S satisfies $(q\text{-}\mathrm{o}_{\mathcal{D}})$ with $\mathcal{D}$ invariant for both S and S^* then $\mathcal{D}$ is invariant for C as well and

$$CSf = qSCf, \quad qCS^*f = S^*Cf, \quad f \in \mathcal{D}.$$

The other is that $(q\text{-}\mathrm{o}_{\mathcal{D}})$ takes now the form

$$S^*Sf - SS^*f = Cf, \quad f \in \mathcal{D},$$

which means that C is just the *self-commutator* of S on $\mathcal{D}$.

It turns out to be crucial to know the instances when C is a positive operator. First introduce another class of operators, which is related to subnormality. Call a densely defined operator N *hyponormal* if $\mathcal{D}(N) \subset \mathcal{D}(N^*)$ and $\|Nf\| \leqslant \|N^*f\|$ for $f \in \mathcal{D}(N)$; hyponormal operators are closable, cf. (83). Just to locate this class among those on p. 74, notify that

$$(\mathrm{H}) \implies \text{hyponormality}.$$

Fact 2.3.1 (Proposition 4 in (56)) *We have the following*

(a) for $q \geqslant 1$, $C > 0$ always;
(b) for $q < 1$, $C \geqslant 0$ if and only if S is bounded and $\|S\| \leqslant (1-q)^{-1/2}$;
(c) for S satisfying $(q\text{-}\mathrm{o}_{\mathcal{D}})$, $C \geqslant 0$ if and only if S is hyponormal.

EXAMPLE **1** Any unitary U makes the operator

$$S \stackrel{\text{def}}{=} (1-q)^{-1/2}U \tag{2.3.2}$$

to satisfy $(q\text{-}\mathrm{o}_{\mathcal{D}})$ if $q < 1$. The operator S is apparently bounded and normal. Consequently (the Spectral Theorem), it may have a bunch of nontrivial reducing subspaces (even not necessarily one dimensional) or may be irreducible. This observation ought to be dedicated to all those who start too fast generating algebras from formal commutation relations. ◀

Fact 2.3.2 *For $q < 1$, the only formally normal operators satisfying* (q-o_D) *are those of the form* (2.3.2). *For $q \geqslant 1$, there is no formally normal solution of* (q-o_D).

EXAMPLE **2** An *ad hoc* illustration can be given as follows. Take a separable Hilbert space with a basis $(e_n)_{n=-\infty}^{\infty}$ and look for a bilateral (or rather *two-sided*) weighted shift T defined as $Te_n = \tau_n e_{n+1}$, $n \in \mathbb{Z}$. Then, because $T^* e_n = \overline{\tau}_{n-1} e_{n-1}$, $n \in \mathbb{Z}$, for any $\alpha \in \mathbb{C}$ and $N \in \mathbb{Z}$, we get

$$|\tau_n|^2 = \begin{cases} \alpha q^{n+N} + (1 - q^{n+N})(1-q)^{-1} = \alpha q^{n+N} + [n+N]_q, & \text{if } q \neq 1 \\ \alpha + n & \text{if } q = 1 \end{cases};$$

this is for all $n \in \mathbb{Z}$.

The only possibility for the right-hand sides to be nonnegative (and in fact positive)[53] is $\alpha \geqslant (1-q)^{-1}$ for $0 \leqslant q < 1$ and $\alpha = (1-q)^{-1}$ for $q < 0$; the latter corresponds to Example 1.

Thus, <u>the only</u> bilateral weighted shifts satisfying (qho_D), with $D = \operatorname{lin}\{e_n : n \in \mathbb{Z}\}$, are those $Te_n = \tau_n e_{n+1}$, $n \in \mathbb{Z}$ which have the weights

$$\tau_n \stackrel{\text{def}}{=} \begin{cases} \sqrt{(1-q)^{-1}}, & q \leqslant 0 \\ \sqrt{\alpha q^{n+N} + [n+N]_q}, \quad \alpha > (1-q)^{-1}, \ N \in \mathbb{Z}, & 0 \leqslant q < 1 \\ \text{none}, & 1 \leqslant q. \end{cases}$$

However, T <u>violates hyponormality</u> if $0 < q < 1$ (pick up $f = e_0$ as a sample). In addition, C defined by (2.3.1) is <u>neither positive nor negative</u> ($\langle Ce_0, e_0 \rangle = a > 0$ while $\langle Ce_{-1}, e_{-1} \rangle < 0$). Anyway, T is apparently <u>unbounded</u> if $q > 0$. The case of $q \leqslant 0$ is precisely that of Example 1. ◀

2.3.1.2 Positive Definiteness from (q-o_D). The following is a q-version of Fact 2.2.3.

Fact 2.3.3 *If S satisfies* (q-o_D) *with D being invariant for both S and S^*, then*

$$S^{*i} S^j f = \sum_{k=0}^{\infty} [k]_q! \begin{bmatrix} i \\ k \end{bmatrix}_q \begin{bmatrix} j \\ k \end{bmatrix}_q S^{j-k} C^k S^{*(i-k)} f, \quad f \in D, \ i,j = 0,1,\ldots.$$

If, moreover, $C \geqslant 0$ then

$$\sum_{i,j=0}^{p} \langle S^i f_j, S^j f_i \rangle = \sum_{k=0}^{\infty} [k]_q! \left\| \sum_{i=0}^{p} \begin{bmatrix} i \\ k \end{bmatrix}_q C^{k/2} S^{*(i-k)} f_i \right\|^2, \quad f_0, \ldots f_p \in D.$$

All this under convention $S^l = (S^*)^l = 0$ for $l < 0$ and $\left[\begin{smallmatrix} i \\ j \end{smallmatrix}\right]_q = 0$ for $j > i$.

Consequently, S satisfies the positive definiteness condition (H).

[53] We avoid weights which are not nonnegative, for instance complex, as they lead to a unitary equivalent version only.

2.3.2 Subnormality in the q-Oscillator

The most productive part of Lemma 2.2.6, (vi) $\Longrightarrow$ (vii) is needed in this section too. Its refined essence is in the following

Lemma 2.3.4 *Let $q > 0$. Consider following conditions:*

(a) *S satisfies* ($\mathrm{qho_w}$) *and $\mathcal{D}(\overline{S}) = \mathcal{D}(S^*)$;*
(b) *$\mathcal{N}(S^*) \neq \{0\}$ and for $n = 0, 1, \ldots$*

$$f \in \mathcal{N}(S^*) \implies \overline{S}^n f \in \mathcal{D}(\overline{S}),\ \overline{S}^{(n-1)} f \in \mathcal{D}(S^*)$$

$$\& \ S^* \overline{S}^{n-1} f = (n-1)\overline{S}^{n-2} f;$$

(c) *there is $f \neq 0$ such that $\overline{S}^n f \in \mathcal{D}(\overline{S})$, $n = 0, 1, \ldots$ and $\overline{S}^m f \perp \overline{S}^n$ for $m \neq n$.*

Then (a) $\Longrightarrow$ (b) $\Longrightarrow$ (c).

2.3.2.1 *The Case of S Bounded*

Proposition 2.3.5 *Suppose S is bounded and satisfies (q-$\mathrm{o}_\mathcal{D}$).*

(a) If $q < 0$ then $\|S\| \geqslant (1-q)^{-1/2}$;
(b) if $0 \leqslant q < 1$ then $\|S\| \leqslant (1-q)^{-1/2}$;
(c) if $q \geqslant 1$ then no such an S exists.

For $q < 0$, the only bounded operator S with norm $\|S\| = (1-q)^{-1/2}$ satisfying (q-$\mathrm{o}_\mathcal{D}$) is that given by (2.3.2)

The following shows how the bounded case differs from the $q = 1$ oscillator which as we know is by force unbounded.

Theorem 2.3.6 *Suppose S satisfies* ($\mathrm{qho}_\mathcal{D}$) *with $\mathcal{D}$ dense in $\mathcal{H}$. If $0 \leqslant q < 1$, then the following facts are equivalent*

(i) *S is bounded and $\|S\| \leqslant (1-q)^{-1/2}$;*
(ii) *S is bounded;*
(iii) *S is* **subnormal***;*
(iv) *S is hyponormal.*

If S is irreducible. then (i)–(iv) *are equivalent to*

(v) *there is an orthonormal basis $(e_n)_{n=0}^{\infty}$ in $\mathcal{H}$ such that*

$$Se_n = \sqrt{[n+1]_q}\, e_{n+1}, \quad n = 0, 1, \ldots .$$

Confront (v) with Sample Theorem, p. 106.

Let us take a look at the summary of what can come out in the bounded case.

		$q < 0$	$0 \leqslant q < 1$	$1 \leqslant q$
normal	in general	SOME	SOME	
	unilat. shift		SOME	
subnormal	bilat. shift	NONE	NONE	
	others	SOME	SOME	NONE
	unilat. shifts		SOME	
hyponormal	bilat. shift	NONE	NONE	
	other	SOME	SOME	

In this table as well as that which follows "SOME" and "NONE" nickname the respective quantifiers concerning the possible occurrence .

2.3.2.2 The Case of S Unbounded Let us begin with defining two notions which are weaker than minimality of cyclic type. A normal extension N in $\mathcal{K}$ of an operator S in $\mathcal{H}$ is said to be *tight* if $\mathcal{D}(\bar{S}) = \mathcal{H} \cap \mathcal{D}(N)$; it is said to be $*$-*tight* if $\mathcal{D}(S^*) = \mathcal{H} \cap \mathcal{D}(N)^*$. Tight normal extensions may exist, cf. (75), the same refers to $*$-tight extensions.

Theorem 2.3.7 *For a densely defined closable operator S in a complex Hilbert space $\mathcal{H}$, the following conditions are equivalent.*

(i) *$\mathcal{H}$ is separable and there is an orthonormal basis in it of the form $\{e_n\}_{n=0}^{\infty}$ contained in $\mathcal{D}(\bar{S})$ and such that*

$$\bar{S}e_n = \sqrt{[n+1]_q}\, e_{n+1}, \quad n = 0, 1, \dots ;$$

(ii) *S is irreducible, satisfies* ($\mathtt{qho}_{\mathcal{D}}$) *with some $\mathcal{D}$ being invariant for S and S^* and being a core of S and S is a* **subnormal** *operator having a normal extension which is both tight and $*$-tight.*

REMARK **19** Notice the difference between condition (iv) in Theorem 2.2.7, p. 98 and condition (ii) in Theorem 2.3.7. Minimality of cyclic type in Theorem 2.2.7 forces uniqueness of the minimal extension, the case $q > 1$ lacks it: see discussion in (c), p. 113. ♣

The following table is a counterpart of that on p. 111.

		$q < 0$	$0 \leqslant q < 1$	$1 \leqslant q$
normal	general			NONE
subnormal	unilat. shift			SOME
	bilat. shift			NONE
	others	NONE	NONE	
hyponormal	unilat. shifts			
	bilat. shift			NONE
	others			MAY

REMARK **20** The "white holes" in the two tables aforementioned mean nothing has been known despite the fact that q-oscillator is no longer a top-line issue. Therefore, the invitation to fill these cells up is still open. ♣

Related questions Define two linear operators M and D_q acting on functions

$$(Mf)(z) \stackrel{\text{def}}{=} zf(z), \quad (D_q f)(z) \stackrel{\text{def}}{=} \begin{cases} \frac{f(z)-f(qz)}{z-qz} & \text{if } q \neq 1 \\ f'(z) & \text{if } q = 1. \end{cases}$$

It turns out that for $a_+ = M$ and $a_- = D_q$ the commutation relation (q-o) is always satisfied. What Bargmann did in (51) was to find, for $q = 1$, a Hilbert space of entire functions such that M and D_1 are formally adjoint. This for arbitrary $q > 0$ leads to the reproducing kernel Hilbert space $\mathcal{H}_q$ of analytic functions with the kernel

$$K(z, w) \stackrel{\text{def}}{=} e_q((1-q)z\overline{w}) \quad z, w \in |1-q|^{-1/2}\omega_q,$$

where

$$\omega_q = \begin{cases} \{z : \ |z| < 1\} & \text{if } 0 < q < 1 \\ \mathbb{C} & \text{if } q > 1. \end{cases}$$

Under these circumstances, we always have

$$\langle Z^m, Z^n \rangle_{\mathcal{H}_q} = \delta_{m,n}[m]_q!$$

and the operator $S = M$ acts as a weighted shift with the weights $(\sqrt{[n+1]_q})$ as in Sample Theorem on p. 106.

Our keynote, subnormality of M now means precisely integrability of the kernel K with some μ. Here we have three qualitatively different situations:

(a) for $0 < q < 1$, the multiplication operator M is bounded and subnormal, this implies uniqueness of μ;

(b) for $q = 1$, the multiplication operator is unbounded and subnormal, it has a normal extension of cyclic type in the sense of (11), and consequently, μ is uniquely determined as well;

(c) for $q > 1$ the multiplication operator is unbounded and subnormal, it has no normal extension of cyclic type in the sense of (11) although it does plenty of those of spectral type in the sense of (11), which are not unitary equivalent [54]. Explicit example of such, based on (84), can be found in (29) (one has to replace q by q^{-1} there to get the commutation relation (q-o) satisfied). Another explicit example, this time of nonradially invariant measure μ is struck out in (57), redesigned and generalized in (56).

The case of $0 < q < 1$ was considered by Arik and Coon (85) and many, many authors afterward. Much less has been presented in the situation of $q > 1$, papers (29, 56) show how complex the circumstances are, in somehow recent paper (86) one can find the adjective "forgotten" for this case.

2.4 BACK TO "HERMICITY"—A WAY TO SEE IT

Hermitian operators in the old times of Quantum Physics were meant as (predominantly finite) matrices and even von Neumann's fundamental work placing them in the Hilbert space environment[55] has not changed the habits too much, at least operator domains are kept secret in the main. After disapproving the notion of a Hermitian operator when thought of in an infinite dimensional Hilbert space in a loose sense, and experiencing diverse answers, if any, from MPs to the question of what it really means[56] we want to continue our comments made in Remark 1.

Let $\mathcal{D}$ be a dense linear subspace of $\mathcal{H}$. Denote, just to simplify further definitions, by L($\mathcal{D}$), the set of all closable operators A in $\mathcal{H}$ such that $\mathcal{D} \subset \mathcal{D}(A)$. Suppose $A \in$ L($\mathcal{D}$) is such that there exists operator $A^{\#} \in$ L($\mathcal{D}$) satisfying

$$\langle Af, g\rangle = \langle f, A^{\#}g\rangle, \quad f, g \in \mathcal{D}. \tag{2.4.1}$$

In other words,

$$A^*|_{\mathcal{D}} \subset A^{\#} \tag{2.4.2}$$

under the proposed circumstances. Therefore, $A^{\#}$ is a particular case (notice $\mathcal{D} \subset \mathcal{D}(A)^*$ here) of a *formal adjoint* of A, cf. (2.2.12) and in general (5, p. 67).

The requirement

$$A|_{\mathcal{D}} = A^{\#}|_{\mathcal{D}}, \tag{2.4.3}$$

[54]That is, there is no unitary map between the $\mathcal{L}^2$ spaces in question, which is the identity on $\mathcal{H}_q$.

[55]For more terminological references, besides those mentioned in **Glossary**, look at (14, p. 37) and (5, p. 72)

[56]The situation pretty often resembles Woody Alen's movie title *Everything you always wanted to know about sex but were afraid to ask.*

seems to be the closest guess of what Hermicity may mean as it is equivalent, due to (2.4.2), to $A|_{\mathcal{D}} = A^{\#}|_{\mathcal{D}} = A^{*}|_{\mathcal{D}}$ (remember $\mathcal{D} \subset \mathcal{D}(A)^{*}$ here). Because this is nothing but our guess let us try to examine the situation somewhat deeper.

Condition (2.4.3) which, due to (2.4.1), is equivalent to

$$\langle Af, g\rangle = \langle f, Ag\rangle, \quad f, g \in \mathcal{D}, \tag{2.4.4}$$

is a little bit less than symmetricity of A (notice $\mathcal{D}$ is not supposed to be the domain $\mathcal{D}(A)$[57]). Let $\mathrm{L}_{\mathrm{cor}}(\mathcal{D})$ stand for those A's in $\mathrm{L}(\mathcal{D})$ such that $\mathcal{D}$ is a core of A; then for $A \in \mathrm{L}_{\mathrm{cor}}(\mathcal{D})$ condition (2.4.4) forces *symmetricity* of A. Moreover, let $\mathrm{L}^{\#}_{\mathrm{cor}}(\mathcal{D})$ stand for those A's in $\mathrm{L}_{\mathrm{cor}}(\mathcal{D})$ for which $A^{\#}$ is in $\mathrm{L}_{\mathrm{cor}}(\mathcal{D})$ too. Then for $A \in \mathrm{L}^{\#}_{\mathrm{cor}}(\mathcal{D})$ condition (2.4.1) forces $\overline{A^{\#}} = A^{*}$ while condition (2.4.1) implies *essential self-adjointness* of A.

Now although it looks like we are about to uncover mathematical meaning of the notion of a *Herminitian operator* in MP a dilemma which of the two extreme in a sense cases is closer to a particular MP's mind appears. The possibilities for them are as follows:

(a) members of $\mathrm{L}(\mathcal{D})$;

or

(b) members of $\mathrm{L}^{\#}_{\mathrm{cor}}(\mathcal{D})$.

Notice that in any case the rigorously defined $A^{\#}$ plays the role of the mystified $A^{\dagger}$.

The meaning of "non-Hermitian" is now twofold: a closable densely defined operator A may be called "non-Hermitian" if there is *no* $\mathcal{D}$ such that A is either in $\mathrm{L}(\mathcal{D})$ (case (a)) or in $\mathrm{L}^{\#}_{\mathrm{cor}}(\mathcal{D})$ (case(b)). The first case determines the smallest class of operators while the second leads to the largest one. THE CHOICE IS YOURS.

The aforementioned should be offered to mathematicians to help them understanding the slogans "Hermitian" and "non-Hermitian" populating MP literature as well as to MPs as to invigorate them in being more careful while considering operators.

REMARK **21** It is important to notice that $A^{\#}$ depends on the choice of $\mathcal{D}$. No-one can be refrained from taking $\mathcal{D} = \mathcal{D}(A)$ if A is already closed; in this case, because $\mathcal{D}(\bar{A}) \subset \mathcal{D}(A^{*})$, one can loose invariance if an operator has to be unbounded according to (9). Invariance is something which is repeatedly hidden in MP's operators. ♣

Notice $\mathrm{L}(\mathcal{D})$ is a linear space. In order to make it an algebra, we have to impose $\mathcal{D}$ is *invariant* for members of $\mathrm{L}(\mathcal{D})$ and take restrictions of the operators in question to $\mathcal{D}$ instead. The same refers to $\mathrm{L}^{\#}(\mathcal{D})$ in which case it becomes a symmetric algebra with the mapping $A \to A^{\#}$ being an involution; the study of such algebras was initiated

[57]MPs are very reluctant to disclose what is a domain of an operator they are taking into consideration; this is why $\mathcal{D}$ is chosen here with some anxiety as a hypothetical domain.

in (87) and further developed by different authors, see (88) for more references. A more relaxed attitude is taken in (89).

Memorandum One of our goals was to emphasis that Hilbert space operators cannot live with their domains not being declared explicitly, which makes the terminological discussion crucial. On the MP ground, it might look somewhat superficial as the operators customarily appearing there as they are protected by quantum mechanical reality; even in this case, some traps can be created, cf. (3). However, it became vital when $\mathcal{PT}$-symmetry was born. By the way, on may call more attention to the meaning of the operators of the quantum harmonic oscillator which is our principal goal here.

CONCLUDING REMARKS

Our choice of topics is determined by the desire of describing solutions of the basic commutation relations in the Hilbert space environment being in line with the von Neumann standpoint of Quantum Physics. This is in contrast to the prevalent number of MP papers backed by the magic of Lie group/algebra methodology. Rellich–Dixmier theorem in particular provides with existence of "ground state" ("=vacuum" depending on some-one's preferences), having it done the rest is a matter of more or less smooth presentation.

While Rellich–Dixmier theorem is rooted in the CCR surroundings, the proper backdrop for the quantum harmonic oscillator requires in this or another way the presence of subnormal operators as they are "analytic" in nature. The flavor of analycity is also present very much in coherent states. Among unbounded subnormal operators, weighted shifts are especially pleasurable. All this has been encouraging us to devote the major part of the survey to these topics.

"Other oscillators" are by necessity shrank to one: the q-oscillator. Although the latter may look as a bit outdated, it has a methodological advantage: moment problems which stay behind it are predominantly indeterminate (the $q > 1$ case in particular).

Let us recommend additional sources (90–92, 95) for some more reading.

REFERENCES

1. Saramago J. *Todos os nomes*, Editorial Caminho S.A., Lisboa 1997.
2. Schmüdgen K. *Unbounded Self-Adjoint Operators on Hilbert Space.* Dordrecht: Springer; 2012.
3. Galapon EA. Pauli's theorem and quantum canonical pairs: the consistency of a bounded, self-adjoint time operator canonically conjugate to a Hamiltonian with non-empty point spectrum. *Proc R Soc London Ser A* 2002;458:451–472. DOI: 10.1098/rspa.2001.0874.
4. Górska K, Horzela A, Szafraniec FH. Squeezing of arbitrary order: the ups and downs. *Proc R Soc London Ser A* 20140205. DOI: 10.1098/rspa.2014.0205.

5. Weidmann J. *Linear Operators in Hilbert Spaces*. Berlin: Springer-Verlag; 1987.
6. Julia G. In: Dixmier J, Hervé M, editors. *Œuvres de Gaston Julia*. Volume IV (French). Paris: Gauthier-Villars; 1970.
7. Sz.-Nagy B Prolongements des transformations de l'espace de Hilbert qui sortent de cet espace. In: Riesz F., Sz.-Nagy B, editors. *Appendice au livre "Leçons d'analyse fonctionnelle"*. Budapest: Akadémiai Kiadó; 1955. p 36 (French).
8. Mlak W. Dilations of Hilbert space operators (General theory) [Dissertationes Math] Volume 153; 1978.
9. Ôta S. Closed linear operators with domain containing their range. *Proc Edinburgh Math Soc* 1984;27:229–233, see also the MathSciNet review # MR760619 (86e:47002).
10. Mlak W. *Hilbert Spaces and Operator Theory*. Warszawa, Dordrecht: PWN–Polish Scientific Publishers, Kluwer Academic Publishers; 1991.
11. Stochel J, Szafraniec FH. On normal extensions of unbounded operators. III. Spectral properties. *Publ RIMS, Kyoto Univ* 1989;25:105–139.
12. Naĭmark MA. On a representation of additive set functions. *C R Dokl Acad Sci URSS* 1943;41:359–361.
13. Szafraniec FH. Naĭmark dilations and Naĭmark extensions in favour of moment problems. In: Hassi S, De Snoo HSV, Szafraniec FH, editors. *Operator Methods for Boundary Value Problems*. Volume 404, *London Mathematical Society Lecture Note Series*. Cambridge University Press; 2012. p. 295–308.
14. Szőkefalvi-Nagy B. *Spektraldarstellung linearer Transformationen des Hilbertschen Raumes*. Berlin: Springer-Verlag; 1942 and 1967.
15. Coddington EA. Formally normal operators having no normal extension. *Canad J Math* 1965;17:1030–1040.
16. Schmüdgen K. A formally normal operator having no normal extension. *Proc Am Math Soc* 1985;98:503–504.
17. Nelson E. Analytic vectors. *Ann Math* 1959;70:572–615.
18. Biriuk G, Coddington EA. Normal extensions of unbounded formally normal operators. *J Math Mech* 1964;13:617–637.
19. Coddington EA. *Extension Theory of Formally Normal and Symmetric Subspaces. Memoirs of the American Mathematical Society, No. 134*. Providence (RI): American Mathematical Society; 1973.
20. Birman MS, Solomjak MZ. *Spectral Theory of Self-Adjoint Operators in Hilbert Space*. Dordrecht, Boston (MA), Lancaster, Tokyo: D. Reidel Publishing Company; 1987.
21. Stochel J, Szafraniec FH. A few assorted questions about unbounded subnormal operators. *Univ Iagel Acta Math* 1991;28:163–170.
22. Stochel J, Szafraniec FH. A characterization of subnormal operators. *Oper Theory Adv Appl* 1984;14:261–263.
23. Stochel J, Szafraniec FH. On normal extensions of unbounded operators. I. *J Oper Theory* 1985;14:31–55.
24. Stochel J, Szafraniec FH. On normal extensions of unbounded operators. II. *Acta Sci Math (Szeged)* 1989;53:153–177.

25. Szafraniec FH. On normal extensions of unbounded operators. IV. A matrix construction. *Oper Theory Adv Appl* 2005;163:337–350.
26. Szafraniec FH. Normals, subnormals and an open problem. *Oper Matr* 2010;4:485–410.
27. Bishop E. Spectral theory of operators on a Banach space. *Trans Am Math Soc* 1957;86:414–445.
28. Foiaş C. Décomposition en opérateurs et vecteurs propres. I. Études de ces décompositions et leurs rapports avec les prolongements des opérateurs. *Rev Roum Math Pures Appl* 1962;7:241-282.
29. Szafraniec FH. A RKHS of entire functions and its multiplication operator. An explicit example. *Oper Theory Adv Appl* 1990;43:309–312.
30. Cichoń D, Jan Stochel N, Szafraniec FH. How much indeterminacy may fit in a moment problem. An example. *Indiana Univ Math J* 2010;59:1947–1970.
31. Halmos P. Normal dilations and extensions of operators. *Summa Brasiliensis Math* 1950;2:125–134.
32. Bram J. Subnormal operators. *Duke Math J* 1955;22:75–94.
33. Szafraniec FH. Dilations on involution semigroups. *Proc Am Math Soc* 1977;66:30–32.
34. Andô T. Matrices of normal extensions of subnormal operators. *Acta Sci Math (Szeged)* 1963;24:91–96.
35. Szafraniec FH. The Sz.-Nagy "théorème principal" extended. Application to subnormality. *Acta Sci Math (Szeged)* 1993;57:249–262.
36. Szafraniec FH. Analytic models for the quantum harmonic oscillator. *Contemp Math* 1998;212:269–276.
37. Szafraniec FH. The reproducing kernel Hilbert space and its multiplication operators. *Oper Theory Adv Appl* 2000;114:253–263.
38. Szafraniec FH. Multipliers in the reproducing kernel Hilbert space, subnormality and noncommutative complex analysis. *Oper Theory Adv Appl* 2003;143:313–331.
39. Szafraniec FH. *Przestrzenie Hilberta z jądrem reprodukującym (Reproducing kernel Hilbert spaces, in Polish)*. Kraków: Wydawnictwo Uniwersytetu Jagiellońskiego; 2004.
40. Aronszajn N. Theory of reproducing kernels. *Trans Am Math Soc* 1950;68:337–404.
41. Zaremba S. L'équation biharmonique et une class remarquable de functions fondamentales harmoniques. *Bull Int Acad Sci Cracovie* 1907;147–196; "Sur le calcul numérique des fonctions demandées dans le problème de Dirichlet et le problème hydrodynamique", ibidem 1908;125–195.
42. Szafraniec FH. Murphy's *Positive definite kernels and Hilbert C^*-modules* reorganized. *Banach Center Publ* 2010;89:275–295.
43. Szafraniec FH. Boundedness of the shift operator related to positive definite forms: an application to moment problems. *Ark Mat* 1981;19:251–259.
44. Chihara TS. *An Introduction to Orthogonal Polynomials*. New York: Gordon and Breach; 1978.
45. Eagleson GK. A duality relation for discrete orthogonal systems. *Studia Sci Math Hungarica* 1968;3:127–136.

46. Szafraniec FH. Yet Another Face of the Creation Operator. in: *Operator Theory: Advances and Applications*. Volume 80 Basel: Birkhäuser; 1995. p 266–275.
47. van Eijndhoven SLL, Meyers JLH. New orthogonality relations for the Hermite polynomials and related Hilbert spaces. *J Math Anal Appl* 1990;146:89–98.
48. Gazeau JP, Szafraniec FH. Holomorphic Hermite polynomials and the non-commutative plane. *J Phys A Math Theor* 2011;44:495201.
49. Ali ST, Górska K, Horzela A, Szafraniec FH. Squeezed states and Hermite polynomials in a complex variable. *J Math Phys* 2014;55:012107.
50. Bargmann V. On a Hilbert space of analytic functions and an associated integral transform. *Commun Pure Appl Math* 1961;14:187–214.
51. Fock V. Verallgemeinerung und Loösung der Diracschen statistischen Gleichung. *Z Phys* 1928;49:339–350.
52. Fock V. Konfigurationsraum und zwiete Quantelung. *Z Phys* 1932;75:622–647.
53. Faddev LD, Khalfin LA, Komarov IV. *V.A. Fock - Selected Works: Quantum Mechanics and Quantum Field Theory*. Chapman and Hall/CCR; 2004.
54. Hall BC. Holomorphic methods in analysis and mathematical physics. *Contemp Math* 2000;260:1–59.
55. Szafraniec FH. Operators of the q-oscillator. *Noncommutative Harmonic Analysis with Applications to Probability*. Volume 78 Warszawa: Banach Center Publications, Institute of Mathematics Polish Academy of Sciences; 2007. p 293–307.
56. Królak I. Measures connected with Bargmann's representation of the q-commutation relation for $q > 1$. In: *Quantum Probability*. Volume 43. Warsaw: Banach Center Publications, Institute of Mathematics, Polish Academy of Sciences; 1998. p 253–257.
57. Hall BC. *Quantum Theory for Mathematicians*. New York: Springer; 2013.
58. Folland GB. *Harmonic Analysis in Phase Space*. Volume 122, *Annals of Mathematics Studies*. Princeton (NJ); Princeton University Press; 1989.
59. Fischer E. Über die Differentiationsprozesse der Algebra. *J Math* 1917;148:1–78.
60. Wintner A. The unboundedness of quantum-mechanical matrices. *Phys Rev* 1947;71:738–739.
61. Wielandt H. Über die Unbeschränktheit der Schrödingerschen Operatoren der Quantenmechanik. *Math Ann* 1949;121:21.
62. Putnam CR. *Commutation Properties of Hilbert Space Operators and Related Topics*. Springer-Verlag; 1967.
63. Foiaş C, Gehér L, Sz.-Nagy B. On the permutability condition of quantum mechanics. *Acta Math (Szeged)* 1960;21:78–89.
64. Rellich F. Der Eindeutigkeitssatz für die Lösungen der quantunmechanischen Vertauschungsrelationen. *Nachr Akad Wiss Göttingen Math-Phys Klasse* 1946;107–115.
65. Dixmier J. Sur la relation $i(PQ - QP) = 1$. *Compos Math* 1958;13:263–269.
66. Stone MH. Linear transformations in Hilbert space, III: operational methods and group theory. *Proc Natl Acad Sci USA* 1930;16:172–175.

67. von Neumann J. Die Eindeutigkeit der Schrödingerschen Operatoren. *Math Ann* 1931;104:570–578; or in: von Neumann J. In: Taub, AH, editor. *Collected Works*. Volume 2. New York and Oxford: Pergamon Press, 1961. p 221–229.
68. Szafraniec FH. Subnormality in the quantum harmonic oscillator. *Commun Math Phys* 2000;210:323–334.
69. Tillmann HG. Zur Eindeutigkeit der Lösungen der quantummechanischen Vertauschungsrelationen. *Acta Sci Math (Szeged)* 1963;24:258–270.
70. Szafraniec FH. Bounded vectors for subnormality via a group of unbounded operators. *Contemp Math* 2004;341:113–118.
71. Koekoek R, Lesky PA, Swarttouw RF. *Hypergeometric Orthogonal Polynomials and their q-Analogues*. Berlin: Springer-Verlag; 2010.
72. Szafraniec FH. Charlier polynomials and translational invariance in the quantum harmonic oscillator. *Math Nachr* 2002;241:163–169.
73. Holevo AS. *Probabilistic and Statistical Aspects of Quantum Theory*. Amsterdam, New York, Oxford: North-Holland; 1982.
74. Ôta S. On strongly normal extensions of unbounded operators. *Bull Polish Acad Sci Math* 1998;46:291–301.
75. Gazeau JP. *Coherent States in Quantum Physics*. Weinheim: Wiley-VCH; 2009.
76. Ali ST, Antoine J-P, Gazeau JP. *Coherent States, Wavelets, and their Generalizations*. New York: Springer; 2014.
77. Horzela A, Szafraniec FH. A measure free approach to coherent states. *J Phys A Math Theor* 2012;45:244018.
78. Horzela A, Szafraniec FH. A measure free approach to coherent states refined. In: Bai C, Gazeau J-P, Ge M-L, editors. *Symmetries and Groups in Contemporary Physics*. Volume 11, *Nankai Series in Pure, Applied Mathematics and Theoretical Physics*; p 277.
79. Klauder J. The action option and a Feymman quantization of spinor fields in terms of ordinary *c*-numbers. *Ann Phys* 1960;11:123.
80. Szafraniec FH. How to recognize the creation operator. *Rep Math Phys* 2007;59:401–408.
81. Möller M, Szafraniec FH. Adjoints and formal adjoints of matrices of unbounded operators. *Proc Am Math Soc* 2008;136:2165–2176.
82. Stochel J, Szafraniec FH. A peculiarity of the creation operator. *Glasgow Math J* 2001;44:137–147.
83. Askey R. Ramanujan's extension of the gamma and beta functions. *Am. Math Mon* 1980;87:346–359.
84. Arik M, Coon DD. Hilbert spaces of analytic functions and generalized coherent states. *J Math Phys* 1976;17:524–527.
85. Burban IM. Arik-Coon oscillator with $q > 1$ in the framework of unified $(q; \alpha, \beta, \gamma; \nu)$-deformation. *J Phys A Math Theor* 2010;43:305204.
86. Lassner G. Topological algebras of operators. *Rep Math Phys* 1972;3:279–293.

87. Schmüdgen K. *Unbounded Operator Algebras and Representation Theory*. Basel: Birkhäuser Verlag; 1990. p 380.
88. Antoine J-P, Trapani C. *Partial Inner Product Spaces: Theory and Applications*. Volume 1986, *Lecture Notes in Mathematics*. Berlin, Heidelberg: Springer; 2009.
89. Chaichian M, Grosse H, Presnajder P. Unitary representations of the q–oscillator algebra. *J Phys A Math Gen* 1994;27:2045–2051.
90. Cigler J. Operatormethoden für q–Identitätem. *Monatsh Math* 1979;88:87–105.
91. Klimyk AU, Schempp W. Classical and quantum Heisenberg groups, their representations and applications. *Acta Appl Math* 1996;45:143–194.
92. Summers SJ. On the Stone-von Neumann uniqueness theorem and its ramifications. In: *John von Neumann and the Foundations of Quantum Physics*. Dordrecht: Kluwer Academic Publishers; 2001. p 135–152.
93. Szafraniec FH. The reproducing kernel property and its space: more or less standard examples of applications. Handbook of Operator Theory. Basel: Springer; 2015. DOI 10.1007/978-3-0348-0692-3_70-1, online version http://www.springerreference.com/docs/index.html, under construction. Accessed 2015 March 30.
94. Szafraniec FH. The reproducing kernel property and its space: the basics. Handbook of Operator Theory. Basel: Springer; 2015. DOI: 10.1007/978-3-0348-0692-3_65-1, online version http://www.springerreference.com/docs/index.html, under construction. Accessed 2015 March 30.
95. Stochel J, Szafraniec FH. The complex moment problem and subnormality; a polar decomposition approach. *J Funct Anal* 1998;159:432-491.

3

DEFORMED CANONICAL (ANTI-)COMMUTATION RELATIONS AND NON-SELF-ADJOINT HAMILTONIANS

FABIO BAGARELLO
Università di Palermo and INFN, Torino, Italy

3.1 INTRODUCTION

In the past 20 years or so, an increasing number of physicists and mathematicians started to be interested to a *generalized version* of quantum mechanics, that is, to the situation in which the hamiltonian of the model under consideration is not necessarily self-adjoint. In fact, numerical studies first and analytical results after, showed that non-self-adjoint operators do exist with real and discrete eigenvalues. This is the case, for instance, of the celebrated $H_{cub} = p^2 + ix^3$, where x and p are the self-adjoint position and momentum operators satisfying $[x, p] = i \mathbb{1}$. In Ref. (1) Carl Bender and Stefan Boettcher discussed numerical evidences of the fact that H_{cub} has (only) real eigenvalues, even if H_{cub} is manifestly non-self-adjoint. This produced a lot of interest in the physics community and triggered the work of many researchers who, with different techniques, tried to set up a rigorous analysis of H_{cub} and its eigenvalues. For instance, a perturbative approach was proposed by Mostafazadeh in Ref. (2), while a complete, nonperturbative proof of the reality of the eigenvalues of H_{cub} can be found in Ref. (3). After this first prototype, many other non-self-adjoint hamiltonians have been found, all having real eigenvalues. After some time, it appeared clear that these hamiltonians, or more generally some non-self-adjoint operators, with real eigenvalues, are somehow related to a specific functional framework, and in particular

Non-Selfadjoint Operators in Quantum Physics: Mathematical Aspects, First Edition.
Edited by Fabio Bagarello, Jean-Pierre Gazeau, Franciszek Hugon Szafraniec and Miloslav Znojil.

to biorthogonal families of vectors, to *metric operators*, which can be used to define different scalar products in the Hilbert space where the model was originally defined, and to intertwining and similarity operators. In some literature, changing scalar product quite often is not considered to be a major problem: we just move from $\mathcal{H}$ with scalar product $\langle .,. \rangle$ to the same $\mathcal{H}$ with a different scalar product $\langle .,. \rangle_1$, and in fact, this is so when our physical system lives in a finite dimensional Hilbert space. In this case, it is easy to understand that replacing $\langle .,. \rangle$ with $\langle .,. \rangle_1$ makes almost no difference: the norms associated to these different scalar products are, in fact, equivalent, so that they define the same distance in $\mathcal{H}$[1]. A completely different situation might occur when $\mathcal{H}$ is infinite dimensional, since among other things, in this case, the metric operator, or its inverse (or both), can be unbounded. As a consequence, it is not true, in general, that we work with equivalent norms. Moreover, the set of eigenstates of the hamiltonian of our system, H, and that of $H^\dagger$, are biorthogonal but they could not be bases of $\mathcal{H}$. In particular, it is rather unlikely that they are Riesz bases. In our opinion, these aspects are not clarified in the literature, where, on the contrary, it is almost always taken for granted that the eigenstates of H form (at least) a basis and that the metric operator is everywhere defined, quite often together with its inverse. This is exactly the point of view adopted, for instance, in Ref. (4), as well as in many other and more recent papers, (5). As already stated, this is not a big problem for quantum systems living in a finite dimensional Hilbert space, for which every operator is bounded, and a basis $\mathcal{F}$ can always be written as the image, via a bounded operator T, of an orthonormal basis. Therefore, as T must be invertible, T^{-1} is also bounded and the set $\mathcal{F}$ is automatically a Riesz basis. However, when the system lives in an *infinite dimensional* Hilbert space, the situation is not so simple, mainly because the relevant operators are not bounded any longer, and domain problems are inevitable, even if, sometimes, neglected, see also Ref. (6). Even some more mathematically oriented papers, see for instance (7, 8), contain as working assumptions similar requirements. However, as we discuss in the rest of this chapter, except when $dim(\mathcal{H}) < \infty$, there is in general not guarantee that the model under consideration admits a (Riesz) basis of eigenvectors for its non-self-adjoint hamiltonian. We should also mention that results in this direction have been discussed along the years mainly by Krejcirik, Siegl, and their collaborators, see Ref. (9, 10) and references therein. They have applied some deep results in operator theory in connection with non-self-adjoint operators, in order to find a priori conditions for the related eigenstates to form Riesz bases. Their results are rather interesting but, not surprisingly, cannot be applied to many physical systems.

This chapter is meant to clarify some of these problems by making use of a rather general structure, recently introduced by us, and which, in our opinion, proved to be rather *unifying*, powerful and elegant: the central ingredient of the structure is what has been called $\mathcal{D}$-pseudo-bosons ($\mathcal{D}$-PBs in the following). We say that our structure is unifying because how we will show, many examples introduced along the

[1] Two norms on $\mathcal{H}$, $\|.\|$ and $\|.\|_1$, are equivalent if two constant A and B exist, with $0 < A \leq B < \infty$, such that $A\|f\| \leq \|f\|_1 \leq B\|f\|$, for all $f \in \mathcal{H}$. In this case, it is clear that $\|f - f_n\| \to 0$ for $n \to \infty$, if and only if $\|f - f_n\|_1 \to 0$. For finite dimensional spaces, this is always the case.

years in the literature on PT-quantum mechanics and its relatives can be rewritten in terms of $\mathcal{D}$-PBs. We also say that the structure is powerful, since the mathematical consequences are clear, easily identified, and clarify some of the claims usually found in the literature on the subject. Finally, it is elegant also because is simple: most of the results can be easily deduced, without particularly sophisticated techniques.

The chapter is organized as follows: in the following section, we introduce the definition of $\mathcal{D}$-PBs and we analyze in details some of its consequences and their connections with *ordinary* bosons. Then, we show that many physical models, originally proposed by several authors, can be rewritten in terms of $\mathcal{D}$-PBs, and this alternative point of view helps in understanding many interesting, and sometimes hidden, aspects of the models. In particular, we will mainly concentrate on two very interesting models, one related to a one-dimensional shifted harmonic oscillator and the other to a two-dimensional perturbed harmonic oscillator living in a noncommutative flat space, which clarify how losing self-adjointness of the hamiltonian causes much more problems than one could probably expect. After that we consider a much simpler, but still interesting, situation, where we deform anticommutation rather than commutation canonical rules. This produces bounded metric operators, with bounded inverse, and biorthogonal bases of the automatically finite dimensional Hilbert space of the system. Then, going back to $\mathcal{D}$-PBs, we show how our original definition should be extended in order to include, in a sort of general treatment, also those non- self-adjoint hamiltonians whose eigenvalues are not linear in the relevant quantum numbers of the system. Section 3.7 contains our final remarks.

3.2 THE MATHEMATICS OF $\mathcal{D}$-PBs

Let $\mathcal{H}$ be a given Hilbert space with scalar product $\langle .,. \rangle$ and related norm $\|.\|$. Because of their essential role all along this chapter, we give here some well-known definitions on bases and complete (or total) sets, and we discuss some interesting counterexamples which will be useful later on.

3.2.1 Some Preliminary Results on Bases and Complete Sets

A set $\mathcal{E} = \{e_n \in \mathcal{H},\, n \geq 0\}$ is a (Schauder) basis for $\mathcal{H}$ if any vector $f \in \mathcal{H}$ can be written, uniquely, as an (in general) infinite linear combination of the e_n's: $f = \sum_{n=0}^{\infty} c_n(f) e_n$. Here $c_n(f)$ are complex numbers depending on the vector f we want to expand.

A particular useful type of bases are the so-called orthonormal ones. $\mathcal{E}$ is an orthonormal basis if $\langle e_n, e_m \rangle = \delta_{n,m}$, and if $\mathcal{E}$ is a basis for $\mathcal{H}$. In this particular case, we find that $c_n(f) = \langle e_n, f \rangle$. Notice that, in our notation, the scalar product is linear in its second variable.

A slightly extended version of orthonormal bases is the one given by the so-called Riesz bases: $\mathcal{F} = \{f_n \in \mathcal{H},\, n \geq 0\}$ is a Riesz basis if a bounded operator T on $\mathcal{H}$ exists, with bounded inverse, and an orthonormal basis $\mathcal{E} = \{e_n \in \mathcal{H},\, n \geq 0\}$, such that $f_n = Te_n$, for all $n \geq 0$[2]. In this case, it is clear that $\langle f_n, f_m \rangle \neq \delta_{n,m}$, in general. However, if we consider a third set of vectors $\mathcal{G} = \{g_n = (T^{-1})^\dagger e_n,\ n \geq 0\}$, it is easy to check that (i) $\mathcal{G}$ is a Riesz basis as well and (ii) $\langle f_n, g_m \rangle = \delta_{n,m}$ for all n, m: $\mathcal{F}$ and $\mathcal{G}$ are *biorthogonal*. Here the symbol $\dagger$ indicates the adjoint with respect to the *natural* scalar product $\langle .,. \rangle$ in $\mathcal{H}$. In this case, any vector $f \in \mathcal{H}$ can be expanded as follows:

$$f = \sum_{n=0}^{\infty} \langle f_n, f \rangle\, g_n = \sum_{n=0}^{\infty} \langle g_n, f \rangle\, f_n. \tag{3.2.1}$$

The same expansion is found also when $\mathcal{F}$ and $\mathcal{G}$ are biorthogonal, but not necessarily Riesz, bases. Notice that each basis $\mathcal{F}$ in $\mathcal{H}$ possesses an unique biorthogonal set $\mathcal{G}$ which is also a basis: the expansion in (3.2.1) is unique. In fact, given $\mathcal{F}$, $\mathcal{G}$ is uniquely found, see (11).

We will also use quite often the concept of *completeness* or *totality* of a given set of vectors: given a set $\mathcal{F} = \{f_n \in \mathcal{H},\ n \geq 0\}$ and a set $V \subseteq \mathcal{H}$, we will say that $\mathcal{F}$ is complete (or total) in V if, taken $f \in V$ such that $\langle f, f_n \rangle = 0$ for all $n \geq 0$, then $f = 0$. In particular, if $V = \mathcal{H}$, then we will simply say that $\mathcal{F}$ is complete. It is well known that each basis is complete, while the converse is, in general not true. However, when $\mathcal{F}$ is an orthonormal or a Riesz basis, this is also true.

To show now how losing orthonormality can be dangerous, we first consider a simple example of a set which is complete but is not a basis[3], see (11, 13): for that, consider an orthonormal basis $\mathcal{E} = \{e_n \in \mathcal{H},\ n \geq 1\}$, and let us introduce a new set $\tilde{\mathcal{E}} := \{\tilde{e}_n := e_n + e_1,\ n = 2, 3, 4, \ldots\}$. It is clear that $\tilde{\mathcal{E}}$ is no longer orthonormal, and it is easy to check that is complete (if $f \in \mathcal{H}$ is such that $\langle f, \tilde{e}_n \rangle = 0$ for all $n = 2, 3, 4, \ldots$, then $f = 0$ necessarily). However, $\tilde{\mathcal{E}}$ is not a basis, as e_1 cannot be expanded in terms of the vectors of $\tilde{\mathcal{E}}$. In fact, assume that this can be done. Then, there exist complex α_n's such that $e_1 = \sum_{n=2}^{\infty} \alpha_n \tilde{e}_n$, where the convergence is intended in the norm of $\mathcal{H}$. Now, recalling that $\langle e_m, e_1 \rangle = \delta_{m1}$, we would get $1 = \left\langle e_1, \sum_{n=2}^{\infty} \alpha_n \tilde{e}_n \right\rangle = \sum_{n=2}^{\infty} \alpha_n$ and, for all $m \geq 2$, $0 = \left\langle e_m, \sum_{n=2}^{\infty} \alpha_n \tilde{e}_n \right\rangle = \alpha_m$. These equalities are clearly incompatible. Incidentally, we also observe that the biorthogonal set of $\tilde{\mathcal{E}}$ is easily identified to be $\hat{\mathcal{G}} = \{\hat{g}_n := e_n,\ n \geq 2\}$, and $\hat{\mathcal{G}}$ is not even complete.

Another absolutely nontrivial difference between orthonormal and not orthonormal sets is the following: let $\mathcal{E} = \{e_n \in \mathcal{H},\ n \geq 1\}$ be an orthonormal set and let $\mathcal{V}$ be a dense subspace of $\mathcal{H}$. Suppose now that each vector $f \in \mathcal{V}$ can be written as follows: $f = \sum_n \langle e_n, f \rangle\, e_n$. Then, all the vectors in $\mathcal{H}$ admit a similar expansion: $\hat{f} = \sum_n \left\langle e_n, \hat{f} \right\rangle e_n$, $\forall \hat{f} \in \mathcal{H}$. Hence, $\mathcal{E}$ is an orthonormal basis for $\mathcal{H}$, (14). The proof is simple but not entirely trivial and is based on the best approximation theorem, which we state first as a lemma:

Lemma 3.2.1 *Let* $\{x_n,\ n = 1, 2, 3, \ldots\}$ *be an arbitrary sequence of vectors which are orthonormal in* $\mathcal{H}$ *and let* α_j, $j = 1, 2, \ldots, K$ *be* K *arbitrary complex numbers. Then,*

[2] Hence, every orthonormal basis is a particular Riesz basis, with $T = T^{-1} = 1\!\!1$.

[3] This, we stress again, is something it cannot happens for orthonormal sets, see Refs (12)

for all $f \in \mathcal{H}$, *we have*

$$\left\| f - \sum_{l=1}^{K} \alpha_l x_l \right\| \geq \left\| f - \sum_{l=1}^{K} \langle x_l, f \rangle \, x_l \right\| .$$

The proof of this Lemma can be found for instance in Ref. (14) and will not be given here. More interesting, and again given in Ref. (14), is the proof of our previous assertion, which we give here in the case when $dim(\mathcal{H}) = \infty$, since otherwise the claim is clearly true. As $\mathcal{V}$ is dense in $\mathcal{H}$, given $f \in \mathcal{H}$ it surely exists $f_\epsilon \in \mathcal{V}$ such that $\|f - f_\epsilon\| \leq \epsilon$, for any fixed $\epsilon > 0$. However, because of our hypothesis on $\mathcal{E}$, for f_ϵ, we can write $f_\epsilon = \sum_{n=0}^{\infty} \langle e_n, f_\epsilon \rangle \, e_n$, meaning that a $N_\epsilon > 0$ exists such that, if $N > N_\epsilon$, then

$$\left\| f_\epsilon - \sum_{n=0}^{N} \langle e_n, f_\epsilon \rangle \, e_n \right\| \leq \epsilon.$$

Then, for these N, the best approximation theorem (with α_n identified with $\langle e_n, f_\epsilon \rangle$) implies that

$$\left\| f - \sum_{n=0}^{N} \langle e_n, f \rangle \, e_n \right\| \leq \left\| f - \sum_{n=0}^{N} \langle e_n, f_\epsilon \rangle \, e_n \right\|$$
$$\leq \|f - f_\epsilon\| + \left\| f_\epsilon - \sum_{n=0}^{N} \langle e_n, f_\epsilon \rangle \, e_n \right\| \leq 2\epsilon,$$

such that $f = \sum_{n=0}^{\infty} \langle e_n, f \rangle \, e_n$.

Let us now replace the orthonormal set $\mathcal{E}$ with a second, no longer orthonormal set $\mathcal{X} = \{x_n \in \mathcal{H}, \, n \geq 0\}$. Let us again assume that every $f \in \mathcal{V}$ can be written, in an unique way, as $f = \sum_n c_n(f) x_n$, for certain coefficients $c_n(f)$ depending on f. Then, it is not true that a similar expansion also holds for general vectors in $\mathcal{H}$. To show this, let us consider the following counterexample.

Let $\mathcal{E} = \{e_n \in \mathcal{H}, \, n \geq 1\}$ be an orthonormal basis for $\mathcal{H}$ and let us define the vectors of $\mathcal{X}$ as follows: $x_n = \sum_{k=1}^{n} \frac{1}{k} e_k$, $n \geq 1$. The set $\mathcal{X}$ is complete in $\mathcal{H}$. Indeed, suppose that $\langle f, x_n \rangle = 0$ for all $n \geq 1$. Then, in particular, $0 = \langle f, x_1 \rangle = \langle f, e_1 \rangle$. From this, we deduce that $0 = \langle f, x_2 \rangle = \frac{1}{2} \langle f, e_2 \rangle$, such that $\langle f, e_2 \rangle = 0$ as well. And so on: $\langle f, e_n \rangle = 0$ for all $n \geq 1$. Hence, being $\mathcal{E}$ complete, $f = 0$, and this means that $\mathcal{X}$ is also complete. The next step consists in proving that any vector of some dense $\mathcal{V}$ can be expanded uniquely in terms of the vectors of $\mathcal{X}$. Indeed, let $\mathcal{V}$ be the (finite) linear span of the e_n's. $\mathcal{V}$ is clearly dense in $\mathcal{H}$. Let now $f \in \mathcal{V}$. Then f is a finite linear combination of the e_n's, such that $c_j \in \mathbb{C}$, $j = 1, 2, \ldots, N$ exist for which $f = \sum_{n=1}^{N} c_n e_n$. It is easy to see that f can also be written in an unique way as a linear combination of

$x_1, x_2, \dots, x_N$. Indeed, it is easy to see that

$$\begin{pmatrix} x_1 \\ x_2 \\ x_3 \\ \dots \\ \dots \\ x_N \end{pmatrix} = \begin{pmatrix} 1 & 0 & 0 & 0 & \dots & \dots \\ 1 & \frac{1}{2} & 0 & 0 & \dots & \dots \\ 1 & \frac{1}{2} & \frac{1}{3} & 0 & \dots & \dots \\ \dots & \dots & \dots & \dots & \dots & \dots \\ \dots & \dots & \dots & \dots & \dots & \dots \\ 1 & \frac{1}{2} & \frac{1}{3} & \dots & \dots & \frac{1}{N} \end{pmatrix} \begin{pmatrix} e_1 \\ e_2 \\ e_3 \\ \dots \\ \dots \\ e_N \end{pmatrix},$$

which we rewrite as $X_N = T_N E_N$ with obvious notation. Considering that, for all finite N, $\det(T_N) = \prod_{n=1}^{N} \frac{1}{n} \neq 0$, T_N admits inverse and this means that, calling $\beta_{n,k}$ the matrix elements of T_N^{-1} (which is unique!), we can write $e_n = \sum_{k=1}^{N} \beta_{n,k} x_k$. Hence $f = \sum_{k=1}^{N} \gamma_k x_k$, where $\gamma_k = \sum_{n=1}^{N} c_n \beta_{n,k}$. Hence $\mathcal{X}$ is a basis for $\mathcal{M}$. The fact that $\mathcal{X}$ is not a basis for $\mathcal{H}$ is now not very surprising, as our argument is based on the fact that T_N^{-1} does exist, and this is not clear when $N \to \infty$ since $\det(T_N) \to 0$ in this limit. More explicitly, let us suppose that $\mathcal{X}$ is indeed a basis for $\mathcal{H}$: we will get a contradiction. In fact, let $f = \sum_{n=1}^{\infty} \frac{1}{n} e_n$. Of course $\|f\|^2 = \sum_{n=1}^{\infty} \frac{1}{n^2} < \infty$, so that $f \in \mathcal{H}$. As $\mathcal{X}$ is assumed to be a basis for $\mathcal{H}$, there must exist coefficients $d_n \in \mathbb{C}$ such that $f = \sum_{n=1}^{\infty} d_n x_n$. Now, taking the scalar product of both expansions of f with respect to e_n, $n \geq 1$, we should have that $\sum_{k=n}^{\infty} d_k = 1$ for all $n \geq 1$, which would imply that $d_k = 0$ for all $k \geq 1$, which is absurd. Hence, $\mathcal{X}$ cannot be a basis for $\mathcal{H}$.

Summarizing, the properties of orthonormal and nonorthonormal sets are really rather different. We refer to Refs (12, 13, 15) for more results on bases.

3.2.2 Back to $\mathcal{D}$-PBs

Let now a and b be two operators on $\mathcal{H}$, with domains $D(a)$ and $D(b)$, respectively, $a^\dagger$ and $b^\dagger$ their adjoint and let $\mathcal{D}$ be a dense subspace of $\mathcal{H}$ such that $a^\sharp \mathcal{D} \subseteq \mathcal{D}$ and $b^\sharp \mathcal{D} \subseteq \mathcal{D}$, where, given a generic operator x, $x^\sharp$ stands for x or $x^\dagger$: in other words, $\mathcal{D}$ is assumed to be stable under the action of a, b, $a^\dagger$ and $b^\dagger$. Notice that we are not requiring here that $\mathcal{D}$ coincides with, for example, $D(a)$ or $D(b)$. However owing to the fact that $a^\sharp f$ is well defined and belongs to $\mathcal{D}$ for all $f \in \mathcal{D}$, it is clear that $\mathcal{D} \subseteq D(a^\sharp)$. Analogously, $\mathcal{D} \subseteq D(b^\sharp)$.

Definition 3.2.1 *The operators (a, b) are $\mathcal{D}$-pseudo-bosonic ($\mathcal{D}$-pb) if, for all $f \in \mathcal{D}$, we have*

$$a\,b\,f - b\,a\,f = f. \tag{3.2.2}$$

Sometimes, to simplify the notation, instead of (3.2.2), we will simply write $[a, b] = \mathbb{1}$, where $\mathbb{1}$ is the identity operator on $\mathcal{H}$, having in mind that both sides of this equation have to act on a certain $f \in \mathcal{D}$.

Our working assumptions are the following:

Assumption $\mathcal{D}$-pb 1.– there exists a nonzero $\varphi_0 \in \mathcal{D}$ such that $a\,\varphi_0 = 0$.

Assumption $\mathcal{D}$-pb 2.– there exists a nonzero $\Psi_0 \in \mathcal{D}$ such that $b^\dagger\,\Psi_0 = 0$.

Then, as $\mathcal{D}$ is stable under the action of b and $a^\dagger$ in particular, it is obvious that $\varphi_0 \in D^\infty(b) := \cap_{k\geq 0} D(b^k)$ and that $\Psi_0 \in D^\infty(a^\dagger)$. Hence, the vectors

$$\varphi_n := \frac{1}{\sqrt{n!}}\, b^n \varphi_0, \qquad \Psi_n := \frac{1}{\sqrt{n!}}\, a^{\dagger n}\Psi_0, \tag{3.2.3}$$

$n \geq 0$, are well defined and they all belong to $\mathcal{D}$. We introduce the sets $\mathcal{F}_\Psi = \{\Psi_n,\, n \geq 0\}$ and $\mathcal{F}_\varphi = \{\varphi_n,\, n \geq 0\}$. Once again, since $\mathcal{D}$ is stable in particular under the action of $a^\dagger$ and b, we deduce that each φ_n and each Ψ_n belongs to $\mathcal{D}$ and, therefore, to the domains of $a^\sharp$, $b^\sharp$ and $N^\sharp$, where $N = ba$ and $N^\dagger = a^\dagger b^\dagger$. Notice that this last equality should be understood, here and in the following, on $\mathcal{D}$: $N^\dagger f = a^\dagger b^\dagger f$, $\forall f \in \mathcal{D}$.

It is now simple to deduce the following lowering and raising relations:

$$\begin{cases} b\,\varphi_n = \sqrt{n+1}\,\varphi_{n+1}, & n \geq 0, \\ a\,\varphi_0 = 0, \quad a\varphi_n = \sqrt{n}\,\varphi_{n-1}, & n \geq 1, \\ a^\dagger \Psi_n = \sqrt{n+1}\,\Psi_{n+1}, & n \geq 0, \\ b^\dagger \Psi_0 = 0, \quad b^\dagger \Psi_n = \sqrt{n}\,\Psi_{n-1}, & n \geq 1, \end{cases} \tag{3.2.4}$$

as well as the following eigenvalue equations: $N\varphi_n = n\varphi_n$ and $N^\dagger \Psi_n = n\Psi_n$, $n \geq 0$. In particular, the first and the third equations are just consequences of the definitions in (3.2.3). As for the second, equation $a\varphi_0 = 0$ is part of Assumption $\mathcal{D}$-pb 1. It is now easy to see that

$$a\varphi_1 = a(b\varphi_0) = (1\!\!1 + ba)\varphi_0 = \varphi_0,$$

where we have used the stability of $\mathcal{D}$ under the action of a and b and Definition 3.2.1, which applies as $\varphi_0 \in \mathcal{D}$. Now, let us assume that $a\varphi_n = \sqrt{n}\,\varphi_{n-1}$ for some particular n. We will now show that $a\varphi_{n+1} = \sqrt{n+1}\,\varphi_n$ as well. Hence, by induction, we will conclude that the second line in (3.2.4) is satisfied for all $n \geq 0$. Using (3.2.3) and Definition 3.2.1 we have

$$a\varphi_{n+1} = a\left(\frac{1}{\sqrt{n+1}}\, b\,\varphi_n\right) = \frac{1}{\sqrt{n+1}}\,(1\!\!1 + ba)\,\varphi_n$$

$$= \frac{1}{\sqrt{n+1}}\left(\varphi_n + b\sqrt{n}\,\varphi_{n-1}\right) = \frac{1}{\sqrt{n+1}}\,(1+n)\,\varphi_n = \sqrt{1+n}\,\varphi_n$$

Notice that these equalities surely hold because of (3.2.2), observing that φ_k belongs to $\mathcal{D}$ for all possible k. In a similar way, we could prove the fourth line in (3.2.4).

Let us now prove that $N\varphi_n = n\varphi_n$ for all $n \geq 0$. First we observe that, for $n = 0$, this equation is surely satisfied as $N\varphi_0 = b(a\varphi_0) = 0$. Now, if $n \geq 1$, using twice equation (3.2.4) we get

$$N\varphi_n = b(a\varphi_n) = \sqrt{n}\, b\,\varphi_{n-1} = n\,\varphi_n,$$

which is what we had to prove. In a similar way, we can also prove that $N^\dagger\Psi_n = n\Psi_n$.

As a consequence of these equations, choosing the normalization of φ_0 and Ψ_0 in such a way $\langle\varphi_0, \Psi_0\rangle = 1$, we deduce that

$$\langle\varphi_n, \Psi_m\rangle = \delta_{n,m}, \tag{3.2.5}$$

for all $n, m \geq 0$. The fact that these vectors are biorthogonal is a simple consequence of the equations $N\varphi_n = n\varphi_n$ and $N^\dagger\Psi_n = n\Psi_n$, $n \geq 0$:

$$n\,\langle\varphi_n, \Psi_m\rangle = \langle N\varphi_n, \Psi_m\rangle = \left\langle\varphi_n, N^\dagger\Psi_m\right\rangle = m\,\langle\varphi_n, \Psi_m\rangle,$$

which is satisfied if, and only if, $\langle\varphi_n, \Psi_m\rangle = 0$ for all $n \neq m$. To prove formula (3.2.5), let us now assume that, for a given n, $\langle\varphi_n, \Psi_n\rangle = 1$. Then we have

$$\left\langle\varphi_{n+1}, \Psi_{n+1}\right\rangle = \frac{1}{\sqrt{n+1}}\left\langle b\varphi_n, \Psi_{n+1}\right\rangle = \frac{1}{\sqrt{n+1}}\left\langle\varphi_n, b^\dagger\Psi_{n+1}\right\rangle = \langle\varphi_n, \Psi_n\rangle = 1,$$

because of the last formula in (3.2.4). Hence, eq. (3.2.5) follows. The conclusion is, therefore, that $\mathcal{F}_\varphi$ and $\mathcal{F}_\Psi$ are biorthonormal sets of eigenstates of N and $N^\dagger$, respectively. This, in principle, does not allow us to conclude anything about the fact that they are also bases for $\mathcal{H}$, or even Riesz bases. In fact, we will show in some concrete examples that this is not always the case. However, it may happen. For this reason, let us introduce for the time being the following assumption:

Assumption $\mathcal{D}$-pb 3.– $\mathcal{F}_\varphi$ is a basis for $\mathcal{H}$.

This assumption introduces, apparently, an asymmetry between $\mathcal{F}_\varphi$ and $\mathcal{F}_\Psi$, as this last is not required to be a basis as well. On the other hand, we can prove the following result:

Lemma 3.2.2 *$\mathcal{F}_\varphi$ is a basis for $\mathcal{H}$ if and only if $\mathcal{F}_\Psi$ is a basis for $\mathcal{H}$.*

The proof of this statement follows from the uniqueness of the basis biorthogonal to a given basis, (11–13). In this way, a complete symmetry between $\mathcal{F}_\varphi$ and $\mathcal{F}_\Psi$ is reintroduced. Notice also that, replacing *is a basis for $\mathcal{H}$* with *is complete* in Assumption $\mathcal{D}$-pb 3. would introduce an irremediable asymmetry between $\mathcal{F}_\varphi$ and $\mathcal{F}_\Psi$, because, as we have shown in Section 3.2.1, there exist complete sets with associated biorthogonal sets which are not complete. Moreover, requiring that $\mathcal{F}_\varphi$ is complete could not be enough. In fact, the same example shows that a complete set needs not to be a basis. Then, in principle, no resolution of the identity in $\mathcal{H}$ can be deduced. On the other hand, it is clear that any basis is complete. It is worth repeating that this difference does not arise when dealing with orthonormal sets, while it is unavoidable in our situation, due to the fact that the eigenvectors of non-self-adjoint operators, which is the interesting case for us, are not mutually orthogonal.

When $\mathcal{F}_\varphi$ and $\mathcal{F}_\Psi$ are Riesz basis for $\mathcal{H}$, we have called our $\mathcal{D}$-PBs *regular*. This assumption, in principle, is surely satisfied in finite dimensional Hilbert spaces, but not in general. However, examples in infinite dimensional Hilbert spaces also exist, (9, 10). Finite dimensional Hilbert spaces are of little use here as it is not hard to imagine that, as for ordinary CCR, which cannot be implemented in any such Hilbert space, also the $\mathcal{D}$-pseudo-bosonic rules in (3.2.2) do not admit such a representation. In fact, both $\mathcal{F}_\varphi$ and $\mathcal{F}_\Psi$ consist of infinitely many, but numerable, linearly independent vectors, all belonging to $\mathcal{D}$. Hence, $\mathcal{H}$ is surely infinite dimensional.

In view of the fact that some very simple quantum system do not admit a basis of eigenvectors for its hamiltonian H, in Ref. (16), we have introduced a weaker version of Assumption $\mathcal{D}$-pb 3, which proved to be quite useful in concrete applications and to produce interesting consequences: for that, let $\mathcal{G}$ be a suitable dense subspace of $\mathcal{H}$. Two biorthogonal sets $\mathcal{F}_\eta = \{\eta_n \in \mathcal{H},\ n \geq 0\}$ and $\mathcal{F}_\Phi = \{\Phi_n \in \mathcal{H},\ n \geq 0\}$ have been called $\mathcal{G}$-*quasi bases* if, for all $f, g \in \mathcal{G}$, the following holds:

$$\langle f, g\rangle = \sum_{n\geq 0} \langle f, \eta_n\rangle \langle \Phi_n, g\rangle = \sum_{n\geq 0} \langle f, \Phi_n\rangle \langle \eta_n, g\rangle . \tag{3.2.6}$$

Is is clear that, while Assumption $\mathcal{D}$-pb 3 implies (3.2.6), the reverse is false. However, if $\mathcal{F}_\eta$ and $\mathcal{F}_\Phi$ satisfy (3.2.6), we still have some (weak) form of resolution of the identity, and, from a physical point of view, we will still be able to deduce interesting results. Moreover, it is easy to check that, if $f \in \mathcal{G}$ is orthogonal to all the Φ_n's (or to all the η_n's), then f is necessarily zero: $\mathcal{F}_\Phi$ and $\mathcal{F}_\eta$ are both complete in $\mathcal{G}$. Indeed, using (3.2.6) with $g = f \in \mathcal{G}$, we find $\|f\|^2 = \sum_{n\geq 0} \langle f, \eta_n\rangle \langle \Phi_n, f\rangle = 0$ as $\langle \Phi_n, f\rangle = 0$ (or $\langle f, \eta_n\rangle = 0$) for all n. Therefore, as $\|f\| = 0, f = 0$.

With this in mind, we now consider a weaker form of Assumption $\mathcal{D}$-pb 3:

Assumption $\mathcal{D}$-pbw 3.– $\mathcal{F}_\varphi$ and $\mathcal{F}_\Psi$ are $\mathcal{G}$-quasi bases, for some subspace $\mathcal{G}$ dense in $\mathcal{H}$.

In particular, we will see in the following that in many explicit models $\mathcal{G}$ coincides with $\mathcal{D}$, but sometimes these sets must be taken different. It might happen that φ_n and Ψ_n belong to $\mathcal{G}$, but, more in general, they are simply vectors of $\mathcal{H}$.

3.2.3 The Operators S_φ and S_Ψ

Let us assume, for a moment, that Assumption $\mathcal{D}$-pb 3 is satisfied. Hence, for all $f \in \mathcal{H}$, we can write

$$f = \sum_n \langle \Psi_n, f\rangle\ \varphi_n = \sum_n \langle \varphi_n, f\rangle\ \Psi_n.$$

These formulas suggest to consider also, within our framework, two operators which should look like $S_\varphi f := \sum_n \langle \varphi_n, f\rangle\ \varphi_n$ and $S_\Psi h := \sum_n \langle \Psi_n, h\rangle\ \Psi_n$, for all f and h for which these series converge.

A first remark is that the set of functions on which two such operators can be introduced is rather rich. In fact, each Ψ_n belongs to the domain of S_φ, and each φ_n belongs to the domain of S_Ψ. In fact, using the biorthogonality of the sets $\mathcal{F}_\Psi$ and $\mathcal{F}_\varphi$, it is easy to check that the series for $S_\varphi \Psi_n$ and for $S_\Psi \varphi_n$ both converge and, in particular, that

$$S_\varphi \Psi_n = \varphi_n, \qquad S_\Psi \varphi_n = \Psi_n, \tag{3.2.7}$$

for all $n \geq 0$. These equalities together imply that $\Psi_n = (S_\Psi\, S_\varphi)\Psi_n$ and that $\varphi_n = (S_\varphi\, S_\Psi)\varphi_n$, for all $n \geq 0$. These formulas, in principle, cannot be extended to all of $\mathcal{H}$ except when both S_φ and S_Ψ are bounded. If this is the case, then we deduce that

$$S_\Psi\, S_\varphi = S_\varphi\, S_\Psi = \mathbb{1} \quad \Rightarrow \quad S_\Psi = S_\varphi^{-1}. \tag{3.2.8}$$

In other words, both S_Ψ and S_φ are invertible and one is the inverse of the other. We will arrive to a similar conclusion also for the unbounded situation, under suitable assumptions, later on in this section.

It is easy to check that S_φ and S_Ψ are necessarily bounded when $\mathcal{F}_\varphi$ and $\mathcal{F}_\Psi$ are biorthogonal Riesz bases, which is the case we consider first. In fact, as we have discussed in Section 3.2.1, in this case, there exists an orthonormal basis $\mathcal{F}_e = \{e_n,\, n \geq 0\}$, and a bounded operator R with bounded inverse R^{-1} such that $\varphi_n = Re_n$, $\Psi_n = (R^{-1})^\dagger e_n$, $\forall\, n$. Then

$$\langle \varphi_n, \Psi_m \rangle = \left\langle Re_n, (R^{-1})^\dagger e_m \right\rangle = \langle e_n, e_m \rangle = \delta_{n,m},$$

and the two sets are biorthonormal, as expected. Moreover, let $f \in D(S_\varphi)$, which for the moment we do not assume to be coincident with $\mathcal{H}$. Then

$$S_\varphi f := \sum_n \langle \varphi_n, f \rangle\, \varphi_n = \sum_n \langle Re_n, f \rangle\, Re_n = R\left(\sum_n \left\langle e_n, R^\dagger f \right\rangle\, e_n \right) = RR^\dagger f,$$

where we have used the facts that $\mathcal{F}_e$ is an orthonormal basis and that R is bounded and, therefore, continuous. For this reason, $RR^\dagger$ is bounded as well and the aforementioned equality can be extended to all of $\mathcal{H}$, so that $S_\varphi = RR^\dagger$. In a similar way, we can deduce that $S_\Psi = (R^\dagger)^{-1}R^{-1} = S_\varphi^{-1}$, which is also bounded. Moreover, using the C*-property in $B(\mathcal{H})$, we deduce that $\|S_\varphi\| = \|R\|^2$ and $\|S_\Psi\| = \|R^{-1}\|^2$.

It is now easy to see that both these operators are positive. In fact we have, for all $f \in \mathcal{H}$, $\left\langle f, S_\varphi f \right\rangle = \|R^\dagger f\|^2$, which is always nonnegative and can be zero only if $R^\dagger f = 0$, which implies, recalling that $R^\dagger$ is invertible, that $f = 0$. Similarly, we get $\langle f, S_\Psi f \rangle = \|R^{-1} f\|^2 > 0$ for all nonzero $f \in \mathcal{H}$. Then the positive square roots of S_Ψ and S_φ can be defined, and they are unique and bounded operators, (17), one the inverse of the other: $S_\Psi^{1/2} = S_\varphi^{-1/2}$. Now, if we define the new vectors $\hat{\varphi}_n = S_\Psi^{1/2} \varphi_n$, the related set $\mathcal{F}_{\hat{\varphi}} = \{\hat{\varphi}_n,\, n \geq 0\}$ is an orthonormal basis. In fact: $\langle \hat{\varphi}_n, \hat{\varphi}_m \rangle = \left\langle S_\Psi^{1/2} \varphi_n, S_\Psi^{1/2} \varphi_m \right\rangle = \langle S_\Psi \varphi_n, \varphi_m \rangle = \langle \Psi_n, \varphi_m \rangle = \delta_{n,m}$. Moreover, $\mathcal{F}_{\hat{\varphi}}$ is also total in $\mathcal{H}$: let $f \in \mathcal{H}$ be orthogonal to all the $\hat{\varphi}_n$: $\langle f, \hat{\varphi}_n \rangle = 0$, for all $n \geq 0$. As $\langle f, \hat{\varphi}_n \rangle = \left\langle f, S_\Psi^{1/2} \varphi_n \right\rangle = \left\langle S_\Psi^{1/2} f, \varphi_n \right\rangle$, and as $\mathcal{F}_\varphi$ is a basis, and therefore total, it follows that $S_\Psi^{1/2} f = 0$ which, in turn, implies that $f = 0$. In this way we have constructed an orthonormal basis of $\mathcal{H}$. It should be probably stressed that, in general, $\mathcal{F}_{\hat{\varphi}} \neq \mathcal{F}_e$, as, for instance, R^{-1} is not required to be self-adjoint, while, on the other hand, $S_\Psi^{1/2} = \left(S_\Psi^{1/2}\right)^\dagger$.

In the second part of this section, we consider a different, and slightly more general, situation, that is, the case in which $\mathcal{F}_\varphi$ and $\mathcal{F}_\Psi$ are $\mathcal{D}$-quasi bases. This means that we will be particularly interested here to consider the case in which they are not orthonormal or Riesz bases (otherwise our previous results apply). Therefore, as $\mathcal{F}_\varphi$ and $\mathcal{F}_\Psi$ are biorthogonal anyhow, this implies that no bounded operator with bounded inverse exists mapping some orthonormal basis $\mathcal{F}_e$ of $\mathcal{H}$ into the sets $\mathcal{F}_\varphi$ and $\mathcal{F}_\Psi$ in the way we discussed previously. For this reason, most of the previous results must be reconsidered with more care. In particular, S_Ψ and S_φ are, in general, unbounded. Then, it is necessary to define them properly by first fixing their domain. We put

$$D(S_\varphi) = \left\{ f \in \mathcal{H} : \sum_n \langle \varphi_n, f \rangle \, \varphi_n \text{ exists in } \mathcal{H} \right\}, \text{ and } S_\varphi f = \sum_n \langle \varphi_n, f \rangle \, \varphi_n$$

for all $f \in D(S_\varphi)$, and, similarly,

$$D(S_\Psi) = \left\{ h \in \mathcal{H} : \sum_n \langle \Psi_n, h \rangle \, \Psi_n \text{ exists in } \mathcal{H} \right\}, \text{ and } S_\Psi h = \sum_n \langle \Psi_n, h \rangle \, \Psi_n,$$

for all $h \in D(S_\Psi)$. As we have already seen it is clear that $\Psi_n \in D(S_\varphi)$ and $\varphi_n \in D(S_\Psi)$, for all $n \geq 0$. However, as $\mathcal{F}_\varphi$ and $\mathcal{F}_\Psi$ are not required to be bases here, it is convenient to work under the additional hypothesis that $\mathcal{D} \subseteq D(S_\Psi) \cap D(S_\varphi)$. In this way, S_Ψ and S_φ are automatically densely defined. In addition, as $\langle S_\Psi f, g \rangle = \langle f, S_\Psi g \rangle$ for all $f, g \in D(S_\Psi)$, S_Ψ is a symmetric operator, as well as S_φ: $\left\langle S_\varphi f, g \right\rangle = \left\langle f, S_\varphi g \right\rangle$ for all $f, g \in D(S_\varphi)$. Moreover, since they are positive operators, they are also semibounded, (18):

$$\left\langle S_\varphi f, f \right\rangle \geq 0, \qquad \langle S_\Psi h, h \rangle \geq 0,$$

for all $f \in D(S_\varphi)$ and $h \in D(S_\Psi)$. Hence, both these operators admit a self-adjoint (Friedrichs) extension, $\hat{S}_\varphi$ and $\hat{S}_\Psi$, (18), which are both also positive. Now, the spectral theorem ensures that we can define, as for the bounded case, the square roots $\hat{S}_\Psi^{1/2}$ and $\hat{S}_\varphi^{1/2}$, which have similar properties as in that case. In particular, they are self-adjoint and positive and, in general, unbounded.

Something more can be deduced if $\hat{S}_\Psi^{1/2}$ and $\hat{S}_\varphi^{1/2}$ leave $\mathcal{D}$ invariant, which is an useful assumption we will use later in Section 3.2.5. In this case, in fact, we can try to repeat our previous procedure and introduce an orthonormal set, $\mathcal{F}_{\hat{\varphi}}$, made of the vectors $\hat{\varphi}_n = \hat{S}_\Psi^{1/2}\varphi_n$ which, clearly, all belong to $\mathcal{D}$. However, this situation is slightly more complicated than for the bounded case, and several steps are needed to complete the job. First of all, we can prove that, if $\mathcal{F}_\varphi$ and $\mathcal{F}_\Psi$ are $\mathcal{D}$-quasi bases, and if $\mathcal{D} \subseteq D(S_\Psi) \cap D(S_\varphi)$, then

$$\hat{S}_\varphi\hat{S}_\Psi f = \hat{S}_\Psi\hat{S}_\varphi f = f, \tag{3.2.9}$$

for all $f \in \mathcal{D}$.

To prove this result, we first observe that the invariance of $\mathcal{D}$ under the action of $\hat{S}_\Psi^{1/2}$ and $\hat{S}_\varphi^{1/2}$ implies that, for $f \in \mathcal{D}$, both $\hat{S}_\varphi\hat{S}_\Psi f$ and $\hat{S}_\Psi\hat{S}_\varphi f$ are well-defined vectors in $\mathcal{D}$. Now, let $f, g \in \mathcal{D}$. Then in particular, $f \in D(S_\Psi)$ and $g \in D(S_\varphi)$, so that the two sequences

$$\sum_{n=0}^{N} \langle \Psi_n, f\rangle \, \Psi_n, \quad \text{and} \quad \sum_{n=0}^{N} \langle \varphi_n, g\rangle \, \varphi_n,$$

both converge, respectively, to $S_\Psi f$ and to $S_\varphi g$, which both belong to $\mathcal{D}$. Now, as $\mathcal{F}_\varphi$ and $\mathcal{F}_\Psi$ are $\mathcal{D}$-quasi bases, we have

$$\begin{aligned}\langle f, g\rangle &= \sum_{n=0}^{\infty} \langle f, \Psi_n\rangle \langle \varphi_n, g\rangle = \sum_{n,k=0}^{\infty} \langle f, \Psi_n\rangle \langle \Psi_n, \varphi_k\rangle \langle \varphi_k, g\rangle \\ &= \sum_{k=0}^{\infty} \left\langle \sum_{n=0}^{\infty} \langle \Psi_n, f\rangle \, \Psi_n, \varphi_k \right\rangle \langle \varphi_k, g\rangle = \sum_{k=0}^{\infty} \langle S_\Psi f, \varphi_k\rangle \langle \varphi_k, g\rangle,\end{aligned}$$

using the fact that the scalar product is continuous and that $\sum_{n=0}^{N} \langle \Psi_n, f\rangle \, \Psi_n$ converges in the uniform topology to $S_\Psi f$. Now we have, for similar reasons,

$$\langle f, g\rangle = \sum_{k=0}^{\infty} \langle S_\Psi f, \varphi_k\rangle \langle \varphi_k, g\rangle = \left\langle S_\Psi f, \sum_{k=0}^{\infty} \langle \varphi_k, g\rangle \, \varphi_k \right\rangle = \left\langle S_\Psi f, S_\varphi g\right\rangle.$$

On $\mathcal{D}$, the operators $\hat{S}_\Psi$ and $\hat{S}_\varphi$ coincide, respectively, with S_Ψ and S_φ. Then, as $\hat{S}_\varphi = \hat{S}_\varphi^\dagger$, we conclude that

$$\langle f, g\rangle = \left\langle \hat{S}_\varphi\hat{S}_\Psi f, g\right\rangle,$$

for all $f, g \in \mathcal{D}$ and, using the continuity of the scalar product, for all $f \in \mathcal{D}$, and for all $g \in \mathcal{H}$. Hence $\hat{S}_\varphi\hat{S}_\Psi f = f$. In a similar way, we deduce also that $\hat{S}_\Psi\hat{S}_\varphi f = f$. Moreover, it is clear that $\hat{S}_\varphi\hat{S}_\Psi f = f$ and that $\hat{S}_\Psi\hat{S}_\varphi g = g$, for all $f \in D(\hat{S}_\varphi\hat{S}_\Psi) = \{f \in D(\hat{S}_\Psi) : \hat{S}_\Psi f \in D(\hat{S}_\varphi)\}$ and for all $g \in D(\hat{S}_\Psi\hat{S}_\varphi) = \{g \in D(\hat{S}_\varphi) : \hat{S}_\varphi g \in D(\hat{S}_\Psi)\}$. Hence $\hat{S}_\Psi = \hat{S}_\varphi^{-1}$, as we have deduced, more easily, in the bounded case. Now, we can prove that

$$\hat{S}_\Psi(1\!\!1 + \hat{S}_\varphi)^{-1} \supset (1\!\!1 + \hat{S}_\varphi)^{-1}\hat{S}_\Psi,$$

and, consequently, $\hat{S}_\Psi$ and $\hat{S}_\varphi$ are strongly commuting: their spectral projectors commute. Hence, we conclude that, in particular,

$$\hat{S}_\varphi^{1/2}\hat{S}_\Psi^{1/2}f = \hat{S}_\Psi^{1/2}\hat{S}_\varphi^{1/2}f = f, \tag{3.2.10}$$

for all $f \in \mathcal{D}$.

Now we are ready to prove our main assertion: under the present assumptions, $\mathcal{F}_{\hat{\varphi}}$ is an orthonormal basis for $\mathcal{H}$. To prove this, we first observe that

$$\hat{\varphi}_n = \hat{S}_\Psi^{1/2}\varphi_n = \hat{S}_\Psi^{1/2}\hat{S}_\varphi\Psi_n = \hat{S}_\varphi^{1/2}\Psi_n,$$

because of (3.2.10). Now, it is clear that $\langle\hat{\varphi}_n, \hat{\varphi}_m\rangle = \delta_{n,m}$ and that $\mathcal{F}_{\hat{\varphi}}$ is complete in $\mathcal{D}$: if $f \in \mathcal{D}$ is orthogonal to $\hat{\varphi}_n$ for all n, then $f = 0$. Let now $f, g \in \mathcal{D}$. Then, using (3.2.10) and the fact that $\mathcal{F}_\varphi$ and $\mathcal{F}_\Psi$ are $\mathcal{D}$-quasi bases, we have

$$\langle f, g\rangle = \left\langle \hat{S}_\varphi^{1/2}\hat{S}_\Psi^{1/2}f, g\right\rangle = \left\langle \hat{S}_\Psi^{1/2}f, \hat{S}_\varphi^{1/2}g\right\rangle = \sum_{n=0}^{\infty}\left\langle \hat{S}_\Psi^{1/2}f, \varphi_n\right\rangle\left\langle \Psi_n, \hat{S}_\varphi^{1/2}g\right\rangle$$

$$= \sum_{n=0}^{\infty}\langle f, \hat{\varphi}_n\rangle\langle\hat{\varphi}_n, g\rangle = \left\langle \sum_{n=0}^{\infty}\langle\hat{\varphi}_n, f\rangle\,\hat{\varphi}_n, g\right\rangle.$$

This last equality is justified by the fact that, as $\mathcal{F}_{\hat{\varphi}}$ is an orthogonal family, $\sum_{n=0}^{\infty}\langle\hat{\varphi}_n, f\rangle\,\hat{\varphi}_n$ surely exists in $\mathcal{H}$. To prove that this series converges to f, it is now sufficient to extend, by continuity, the equality $\langle f, g\rangle = \left\langle\sum_{n=0}^{\infty}\langle\hat{\varphi}_n, f\rangle\,\hat{\varphi}_n, g\right\rangle$ to all $g \in \mathcal{H}$. Hence, the conclusion is that all $f \in \mathcal{D}$ can be written as $f = \sum_{n=0}^{\infty}\langle\hat{\varphi}_n, f\rangle\,\hat{\varphi}_n$. Now, as discussed in Section 3.2.1, using for instance the best approximation theorem, (14), we can deduce that $f = \sum_{n=0}^{\infty}\langle\hat{\varphi}_n, f\rangle\,\hat{\varphi}_n$ also holds for all $f \in \mathcal{H}$. Hence, $\mathcal{F}_{\hat{\varphi}}$ is an orthonormal basis for $\mathcal{H}$.

Remark:– Some authors suggest, in a similar context, that one between the metric operator and its inverse, S_Ψ and S_φ in our language, can be chosen to be bounded, (9). Their idea can be easily exported to our case, simply replacing for instance $S_\varphi f = \sum_n\langle\varphi_n, f\rangle\,\varphi_n$ with a generalized version, $S_\varphi^{(c)}f = \sum_n c_n\,\langle\varphi_n, f\rangle\,\varphi_n$, in which c_n should be chosen decreasing to zero sufficiently fast for diverging n, in order to absorb the possible divergence of $\|\varphi_n\|$. Of course, if we still want S_Ψ to be formally the inverse of S_φ, we need to have $S_\Psi^{(c^{-1})}f = \sum_n c_n^{-1}\,\langle\Psi_n, f\rangle\,\Psi_n$, for some suitable f. However, other than introducing an asymmetry between $S_\varphi^{(c)}$ and $S_\Psi^{(c^{-1})}$, this choice would destroy one of the crucial properties of these operators, that is, that of mapping exactly (i.e., without any extra normalization factor) $\mathcal{F}_\varphi$ into $\mathcal{F}_\Psi$ and vice versa. So we prefer not to use this possibility here.

Under our working assumptions ($\mathcal{F}_\varphi$ and $\mathcal{F}_\Psi$ are $\mathcal{D}$-quasi bases, and $\mathcal{D}$ is left invariant by $\hat{S}_\varphi^{1/2}$ and $\hat{S}_\Psi^{1/2}$), we can also prove the following intertwining relations:

$$\hat{S}_\Psi N f = N^\dagger\hat{S}_\Psi f \quad \text{and} \quad N\hat{S}_\varphi f = \hat{S}_\varphi N^\dagger f, \tag{3.2.11}$$

$\forall f \in \mathcal{D}$. These are related to the fact that the eigenvalues of, say, N and $N^\dagger$, coincide and that their eigenvectors are mapped ones into the others by the operators S_φ and S_Ψ as in (3.2.7), in agreement with the literature on intertwining operators, (19, 20), and on pseudo-Hermitian quantum mechanics, see Refs (21, 22) and references therein. We will find many other similar intertwining relations along the chapter.

To check the first equality in (3.2.11), we observe that $(\hat{S}_\Psi N - N^\dagger \hat{S}_\Psi)\varphi_n = 0$ for all $n \geq 0$. Then, taking $f \in \mathcal{D}$, we have

$$\begin{aligned}\left\langle (\hat{S}_\Psi N - N^\dagger \hat{S}_\Psi) f, \varphi_n \right\rangle &= \left\langle f, (\hat{S}_\Psi N - N^\dagger \hat{S}_\Psi)^\dagger \varphi_n \right\rangle \\ &= \left\langle f, (\hat{S}_\Psi N - N^\dagger \hat{S}_\Psi) \varphi_n \right\rangle = 0.\end{aligned}$$

Now, as $(\hat{S}_\Psi N - N^\dagger \hat{S}_\Psi) f \in \mathcal{D}$ and $\mathcal{F}_\varphi$ is complete in $\mathcal{D}$, our claim follows.

3.2.4 Θ-Conjugate Operators for $\mathcal{D}$-Quasi Bases

In this section, we slightly refine the structure, requiring that Assumption $\mathcal{D}$-pb 1, $\mathcal{D}$-pb 2, and $\mathcal{D}$-pbw 3 are satisfied, with $\mathcal{G} = \mathcal{D}$ (even if this is not strictly essential). In other words, we will not assume $\mathcal{D}$-pb 3, as this Assumption, even if it is very often taken for granted in the physical literature on non-self-adjoint hamiltonians, is not satisfied even in the simple extended harmonic oscillator we will discuss in Section 3.3.1.

Let us consider a self-adjoint, invertible, operator Θ, which leaves, together with Θ^{-1}, $\mathcal{D}$ invariant: $\Theta\mathcal{D} \subseteq \mathcal{D}$, $\Theta^{-1}\mathcal{D} \subseteq \mathcal{D}$. We introduce the following

Definition 3.2.2 *We will say that $(a, b^\dagger)$ are Θ−conjugate if $af = \Theta^{-1} b^\dagger \Theta f$, for all $f \in \mathcal{D}$.*

Briefly, we will write $a = \Theta^{-1} b^\dagger \Theta$, meaning that both sides of this formula must be applied to vectors of $\mathcal{D}$. Of course, the fact that $\mathcal{D}$ is stable under the action of both Θ and Θ^{-1}, makes the Definition 3.2.2 well posed, as $\mathcal{D}$ is also stable under the action of a and $b^\dagger$.

Then we can prove

Lemma 3.2.3 *The following statements are equivalent:*

1. *$(a, b^\dagger)$ are Θ−conjugate;*
2. *$(a^\dagger, b)$ are Θ^{-1}−conjugate;*
3. *$(b, a^\dagger)$ are Θ−conjugate;*
4. *$(b^\dagger, a)$ are Θ^{-1}−conjugate.*

Proof: We just prove here that 1. implies 2. The other statements can be proven in similar way. Let us assume that $(a, b^\dagger)$ are Θ−conjugate, and let $f, g \in \mathcal{D}$. Then

$$\left\langle f, a^\dagger g \right\rangle = \langle af, g \rangle = \left\langle \left(\Theta^{-1} b^\dagger \Theta\right) f, g \right\rangle = \left\langle f, \left(\Theta b \Theta^{-1}\right) g \right\rangle,$$

so that $\langle f, (a^\dagger - (\Theta b\, \Theta^{-1}))\, g\rangle = 0$. Then, recalling that the scalar product is continuous and that $\mathcal{D}$ is dense in $\mathcal{H}$, we deduce that

$$\langle \hat{f}, (a^\dagger - (\Theta b\, \Theta^{-1}))\, g\rangle = \lim_n \langle f_n, (a^\dagger - (\Theta b\, \Theta^{-1}))\, g\rangle = 0$$

for all $g \in \mathcal{D}$ and $\hat{f} \in \mathcal{H}$. Here $\{f_n\}$ is a sequence in $\mathcal{D}$ norm-converging to $\hat{f}$. Then $a^\dagger g = \Theta b\, \Theta^{-1} g$, for all $g \in \mathcal{D}$, which is what we had to prove.

■

Let us now suppose that $\Theta\, \varphi_0$ is not orthogonal to φ_0: $\langle \varphi_0, \Theta\varphi_0\rangle \neq 0$. We want to show that, if $(a, b^\dagger)$ are Θ–conjugate, then the two sets $\mathcal{F}_\varphi$ and $\mathcal{F}_\Psi$ introduced in the previous section are related by Θ. To prove this, it is convenient to assume that $\langle \varphi_0, \Theta\varphi_0\rangle = 1$. This is not a major requirement because if $(a, b^\dagger)$ are Θ–conjugate, then $(a, b^\dagger)$ are also $\hat{\Theta}$–conjugate, where $\hat{\Theta} := \frac{1}{\langle \varphi_0, \Theta\varphi_0\rangle}\, \Theta$. With this choice, in fact, $\left\langle \varphi_0, \hat{\Theta}\varphi_0 \right\rangle = 1$. Then we can safely assume the aforementioned normalization.

Proposition 3.2.3 *Assume that $\mathcal{F}_\varphi$ and $\mathcal{F}_\Psi$ are $\mathcal{D}$-quasi bases for $\mathcal{H}$. Then the operators $(a, b^\dagger)$ are Θ–conjugate if and only if $\Psi_n = \Theta\, \varphi_n$, for all $n \geq 0$.*

Proof: Let us first assume that $(a, b^\dagger)$ are Θ–conjugate. Then we can check that $\mathcal{F}_{\tilde{\varphi}} = \{\tilde{\varphi}_n := \Theta\, \varphi_n,\, n \geq 0\}$ is biorthogonal to $\mathcal{F}_\varphi$. For that, we first show that $N^\dagger(\Theta\varphi_k) = k(\Theta\varphi_k)$ for all $k \geq 0$. This is a consequence of Definition 3.2.2 and Lemma 3.2.3. Indeed, we have, using several times the stability of $\mathcal{D}$ under the action of the operators involved in our computations,

$$\begin{aligned} N^\dagger(\Theta\varphi_k) &= a^\dagger b^\dagger(\Theta\varphi_k) = a^\dagger(\Theta a\Theta^{-1})(\Theta\varphi_k) = a^\dagger(\Theta a\varphi_k) \\ &= (\Theta b\Theta^{-1})(\Theta a\varphi_k) = \Theta N\varphi_k = k(\Theta\varphi_k). \end{aligned}$$

Now, recalling that we have also $N\varphi_k = k\varphi_k$, $\forall\, k \geq 0$, a standard argument shows that $\langle \varphi_n, \Theta\varphi_k\rangle = 0$ whenever $n \neq k$. In fact, if from one side we have $\langle N\varphi_n, \Theta\varphi_k\rangle = n\,\langle \varphi_n, \Theta\varphi_k\rangle$, on the other side we get

$$\langle N\varphi_n, \Theta\varphi_k\rangle = \left\langle \varphi_n, N^\dagger\Theta\varphi_k\right\rangle = k\,\langle \varphi_n, \Theta\varphi_k\rangle\, .$$

Hence $(n-k)\,\langle \varphi_n, \Theta\varphi_k\rangle = 0$ and, as $n \neq k$, $\langle \varphi_n, \Theta\varphi_k\rangle$ is necessarily zero.

We further have $\langle \varphi_n, \Theta\varphi_n\rangle = 1$ for all n. Indeed, reminding first that we have chosen φ_0 and Θ in such a way that $\langle \varphi_0, \Theta\varphi_0\rangle = 1$, we use now induction on n. Then we assume that $\langle \varphi_n, \Theta\varphi_n\rangle = 1$. We just have to prove that $\left\langle \varphi_{n+1}, \Theta\varphi_{n+1}\right\rangle = 1$ as well. For that we use the fact that, since by assumption that $(a, b^\dagger)$ are Θ-conjugate, then $(a^\dagger, b)$ are Θ^{-1}-conjugate. We also need formulas $a\,\varphi_{n+1} = \sqrt{n+1}\,\varphi_n$ and $b\,\varphi_n = \sqrt{n+1}\,\varphi_{n+1}$, see (3.2.4):

$$\left\langle \varphi_{n+1}, \Theta\, \varphi_{n+1} \right\rangle = \frac{1}{\sqrt{1+n}} \left\langle \varphi_{n+1}, \Theta\, b\, \varphi_n \right\rangle = \frac{1}{\sqrt{1+n}} \left\langle \varphi_{n+1}, a^\dagger \Theta\, \varphi_n \right\rangle$$
$$= \frac{1}{\sqrt{1+n}} \left\langle a\, \varphi_{n+1}, \Theta\, \varphi_n \right\rangle = \langle \varphi_n, \Theta\, \varphi_n \rangle = 1,$$

as we had to prove.

Now, our Assumption $\mathcal{D}$-pbw 3 implies, as already stated, that $\mathcal{F}_\varphi$ and $\mathcal{F}_\Psi$ are both complete in $\mathcal{D}$, so that, if $f \in \mathcal{D}$ is orthogonal to all the φ_n's or to all the Ψ_n, then $f = 0$. Hence, as for all fixed k

$$\langle \tilde{\varphi}_k - \Psi_k, \varphi_n \rangle = \langle \tilde{\varphi}_k, \varphi_n \rangle - \langle \Psi_k, \varphi_n \rangle = \delta_{k,n} - \delta_{k,n} = 0,$$

for all $n \geq 0$, and since $\tilde{\varphi}_k - \Psi_k$ belongs to $\mathcal{D}$, we conclude that $\tilde{\varphi}_k = \Psi_k$ for each k. Hence $\Psi_k = \Theta\varphi_k$.

Let us now assume that $\Psi_n = \Theta\varphi_n$, for all $n \geq 0$. Then, taking f in $\mathcal{D}$,

$$\left\langle \left(\Theta\, a\, \Theta^{-1} - b^\dagger\right) f, \varphi_n \right\rangle = \left\langle f, \left(\Theta^{-1} a^\dagger\, \Theta - b\right) \varphi_n \right\rangle = \left\langle f, \Theta^{-1} a^\dagger\, \Psi_n - b\, \varphi_n \right\rangle$$
$$= \sqrt{n+1} \left\langle f, \Theta^{-1} \Psi_{n+1} - \varphi_{n+1} \right\rangle = 0,$$

for all $n \geq 0$. Hence, as $\mathcal{F}_\varphi$ is complete in $\mathcal{D}$, we conclude that $\left(\Theta\, a\, \Theta^{-1} - b^\dagger\right) f = 0$ for each $f \in \mathcal{D}$. Therefore $(b^\dagger, a)$ are Θ^{-1}-conjugate, and because of Lemma 3.2.3, our statement follows. ■

Remark:– Comparing equations $\Psi_n = \Theta\, \varphi_n$ and $\Psi_n = S_\Psi \varphi_n$ in (3.2.7) (or $\Psi_n = \hat{S}_\Psi \varphi_n$), we see how, essentially, the operator Θ introduced here is not very different from S_Ψ. Obviously, Θ^{-1} is quite close to S_φ. Therefore, our Definition 3.2.2 looks essentially as a different way of saying that a and $b^\dagger$ are similar and that the similarity maps are exactly given by S_φ and S_Ψ (or by $\hat{S}_\varphi$ and $\hat{S}_\Psi$). In the rest of this chapter, we sometimes call Θ a *metric operator*.

The positivity of Θ is guaranteed by the following result:

Proposition 3.2.4 *If* $(a, b^\dagger)$ *are* Θ*–conjugate then* $\langle f, \Theta f \rangle > 0$ *for all nonzero* $f \in \mathcal{D}$.

Proof: As both f and Θf belong to $\mathcal{D}$, and as $\mathcal{F}_\varphi$ and $\mathcal{F}_\Psi$ are $\mathcal{D}$-quasi bases for $\mathcal{H}$, the following expansion holds

$$\langle f, \Theta f \rangle = \sum_n \langle f, \Psi_n \rangle \langle \varphi_n, \Theta f \rangle = \sum_n \langle f, \Psi_n \rangle \langle \Theta \varphi_n, f \rangle =$$
$$= \sum_n \langle f, \Psi_n \rangle \langle \Psi_n, f \rangle = \sum_n |\langle f, \Psi_n \rangle|^2,$$

which is surely strictly positive if $f \neq 0$, due to the fact that $\mathcal{F}_\Psi$ is complete in $\mathcal{D}$. ■

Another interesting consequence of our definitions is that $Nf = \Theta^{-1}N^\dagger\Theta f$, for all $f \in \mathcal{D}$, which shows that N and $N^\dagger$ are also related by Θ. This is a weak form of an intertwining equation, relating N and $N^\dagger$ via Θ, and for this reason, it is not a big surprise to discover that N and $N^\dagger$ have the same eigenvalues and that their respective eigenvectors are related exactly by the intertwining operator Θ itself. We refer to Ref. (19) for some bibliography on the subject.

It is clear by the definition that $(a, b^\dagger)$ are Θ−conjugate if and only if $(a, b^\dagger)$ are Θ_k−conjugate, where $\Theta_k := k\Theta$, for all possible choices of nonzero real k. Here, reality of k is needed to ensure that Θ_k is self-adjoint. Notice that k could also be negative, in principle. This could seem to be in contradiction with Propositions 3.2.3 and 3.2.4. In fact, this is not so, as these results are deduced under the assumption that $\langle\varphi_0, \Theta\varphi_0\rangle = 1$, which of course fixes the value of the constant k.

In order to use the results obtained so far, in concrete models it is necessary to check if the families $\mathcal{F}_\varphi$ and $\mathcal{F}_\Psi$ are $\mathcal{G}$-quasi bases or not, for some $\mathcal{G}$ dense in $\mathcal{H}$. We know that a simple criterion ensuring that they are, for instance, Riesz bases (and therefore, automatically $\mathcal{H}$-quasi bases[4]), is to find an orthonormal basis $\mathcal{E} = \{e_n,\ n \geq 0\}$ and a bounded operator T, with bounded inverse, such that $\varphi_n = Te_n$. In this case, the biorthogonal set is $\Psi_n = (T^{-1})^\dagger e_n$, and both $\mathcal{F}_\varphi$ and $\mathcal{F}_\Psi$ are Riesz bases. If, on the other hand, T and/or T^{-1} are unbounded, the situation changes drastically. First of all, $\mathcal{F}_\varphi$ and $\mathcal{F}_\Psi$ cannot be Riesz bases. But still, there exists the possibility that they are bases for $\mathcal{H}$. However, this is not so evident. The reason is that, when trying to prove this statement, one is usually forced to exchange T or T^{-1} with some infinite series, and this is quite a dangerous operation. Actually, there are strong indications that this is not true at all: apparently, if T or T^{-1} are unbounded, then $\mathcal{F}_\varphi$ and $\mathcal{F}_\Psi$ are not bases.

However, we are left with the possibility of having $\mathcal{F}_\varphi$ and $\mathcal{F}_\Psi$ $\mathcal{G}$-quasi bases, also when T or T^{-1} are unbounded. In fact, this is the main result in Proposition 3.2.5 below. More in details, let $\mathcal{E} = \{e_n \in \mathcal{H}, n \geq 0\}$ be an orthonormal basis of $\mathcal{H}$ and let us consider a self-adjoint, invertible operator T, such that $e_n \in D(T) \cap D(T^{-1})$ for all n. Here we are considering the possibility that T or T^{-1}, or both, are unbounded. Of course $D(T)$ and $D(T^{-1})$ are, at least, dense in $\mathcal{H}$, while they both coincide with $\mathcal{H}$ when $T, T^{-1} \in B(\mathcal{H})$. Furthermore, we also assume that $\tilde{\mathcal{D}} := D(T) \bigcap D(T^{-1})$ is dense in $\mathcal{H}$. Under our assumptions, the vectors $\varphi_n = Te_n$ and $\Psi_n = T^{-1}e_n, n \geq 0$, are well-defined vectors belonging to $\mathcal{H}$. A simple consequence of these definitions is that $\varphi_n \in D(T^{-1})$, $T^{-1}\varphi_n = e_n$, and $\Psi_n \in D(T)$, $T\Psi_n = e_n$, $n \geq 0$. In addition, $\Psi_n \in D(T^2)$ and $\varphi_n \in D(T^{-2})$: $T^2\Psi_n = \varphi_n$ and $T^{-2}\varphi_n = \Psi_n$.

We can now prove the following

Proposition 3.2.5 *Under the above-mentioned assumptions: (i) the sets $\mathcal{F}_\varphi = \{\varphi_n\}$ and $\mathcal{F}_\Psi = \{\Psi_n\}$ are biorthogonal; (ii) if $f \in D(T)$ is orthogonal to all the φ_n, then $f = 0$; (iii) if $f \in D(T^{-1})$ is orthogonal to all the Ψ_n, then $f = 0$; (iv) $\forall f, g \in \tilde{\mathcal{D}}$ we have*

[4]This simply means that formula (3.2.6) holds for all $f, g \in \mathcal{H}$.

$$\langle f,g\rangle = \sum_{n=0}^{\infty} \langle f,\varphi_n\rangle \langle \Psi_n,g\rangle = \sum_{n_0=0}^{\infty} \langle f,\Psi_n\rangle \langle \varphi_n,g\rangle .$$

Therefore, $\mathcal{F}_\varphi$ and $\mathcal{F}_\Psi$ are $\tilde{\mathcal{D}}$-quasi bases; (v) if T^{-1} is bounded, then any $f \in D(T)$ can be written as $f = \sum_{n=0}^{\infty} \langle \varphi_n,f\rangle \Psi_n$. Moreover, if $\hat{g} \in \mathcal{H}$, $\langle f,\hat{g}\rangle = \sum_{n=0}^{\infty} \langle f,\varphi_n\rangle \langle \Psi_n,\hat{g}\rangle$; (vi) if T is bounded, then any $f \in D(T^{-1})$ can be written as $f = \sum_{n=0}^{\infty} \langle \Psi_n,f\rangle \varphi_n$. Moreover, if $\hat{g} \in \mathcal{H}$, $\langle f,\hat{g}\rangle = \sum_{n=0}^{\infty} \langle f,\Psi_n\rangle \langle \varphi_n,\hat{g}\rangle$.

Proof: The proofs of (*i*), (*ii*) and (*iii*) are trivial and will not be given here. To prove (*iv*) we first observe that if $f,g \in \tilde{\mathcal{D}}$, then both Tf and $T^{-1}g$ are well-defined vectors in $\mathcal{H}$. Hence, recalling that $\mathcal{E}$ is an orthonormal basis and using the definitions of φ_n and Ψ_n, we get

$$\langle f,g\rangle = \left\langle Tf,T^{-1}g\right\rangle = \sum_{n=0}^{\infty} \langle Tf,e_n\rangle \left\langle e_n,T^{-1}g\right\rangle = \sum_{n=0}^{\infty} \langle f,\varphi_n\rangle \langle \Psi_n,g\rangle .$$

Analogously,

$$\langle f,g\rangle = \left\langle T^{-1}f,Tg\right\rangle = \sum_{n=0}^{\infty} \left\langle T^{-1}f,e_n\right\rangle \langle e_n,Tg\rangle = \sum_{n=0}^{\infty} \langle f,\Psi_n\rangle \langle \varphi_n,g\rangle .$$

(*v*) If $f \in D(T)$, we can write $Tf = \sum_{n=0}^{\infty} \langle e_n,Tf\rangle e_n = \sum_{n=0}^{\infty} \langle \varphi_n,f\rangle e_n$. Now

$$\left\| f - \sum_{n=0}^{N} \langle \varphi_n,f\rangle \Psi_n \right\| = \left\| T^{-1}\left(Tf - \sum_{n=0}^{N} \langle \varphi_n,f\rangle e_n \right)\right\|$$
$$\leq \|T^{-1}\| \left\| Tf - \sum_{n=0}^{N} \langle \varphi_n,f\rangle e_n \right\| ,$$

which goes to zero when N diverges. The other statement can be proved similarly to (*iv*).

(*vi*) The proof is similar to (*v*).

■

The outcome of this proposition is that we do not really need $\mathcal{F}_\varphi$ and $\mathcal{F}_\Psi$ to be Riesz bases in order to allow a *natural* decomposition of *most* vectors of $\mathcal{H}$, that is, of all those vectors belonging to a certain dense subspace of $\mathcal{H}$. This is possible also if one between T and T^{-1} is unbounded, at least if the assumptions under which Proposition 3.2.5 is stated are satisfied. Of course, when both T and T^{-1} are bounded, then $\mathcal{F}_\varphi$ and $\mathcal{F}_\Psi$ are Riesz bases. However, in the most general case, $\mathcal{F}_\varphi$ and $\mathcal{F}_\Psi$ turn out to be $\tilde{\mathcal{D}}$-quasi bases.

Let us remember that, as we have discussed in Section 3.2.1, for orthonormal sets, being a basis in a dense set implies being a basis in $\mathcal{H}$ itself. Here, see (*v*) earlier, we proved that, if T^{-1} is bounded, then any $f \in D(T)$ can be written as $f = \sum_{n=0}^{\infty} \langle \varphi_n, f \rangle \Psi_n$. This means that $\mathcal{F}_\Psi$ is a basis in $D(T)$, which is dense in $\mathcal{H}$. So one might wonder if a similar decomposition can be extended to any vector of $\mathcal{H}$. In other words: is it true that, if T^{-1} is bounded, $\hat{f} = \sum_{n=0}^{\infty} \langle \varphi_n, \hat{f} \rangle \Psi_n$ holds for $\hat{f} \in \mathcal{H}$? However, as we have explicitly seen in Section 3.2.1, this is not ensured in general. For this reason, without extra assumptions, we cannot say more than what stated in Proposition 3.2.5.

3.2.4.1 An Historical Interlude: from PBs to $\mathcal{D}$-PBs A natural remark concerning our discussion so far is that $\mathcal{D}$ plays an essential role in the construction of the mathematical framework related to the pseudo-bosonic operators a and b. For completeness, we should mention that our original definition of pseudo-bosons, (23), was slightly different and probably *less friendly* than the one we have adopted later, that is, the one considered in this chapter, exactly because the set $\mathcal{D}$ had no particular role, at that stage. We believe that it could be interesting to sketch here the main motivations for such a change in the definition.

In Ref. (23), we proposed the following strategy: let us consider a pair of operators, a and b, acting on $\mathcal{H}$ and satisfying (formally) the commutation rule $[a, b] = 1\!\!1$. The working assumptions were the following:

Assumption 1.– there exists a nonzero $\varphi_0 \in \mathcal{H}$ such that $a\varphi_0 = 0$, and $\varphi_0 \in D^\infty(b)$.

Assumption 2.– there exists a nonzero $\Psi_0 \in \mathcal{H}$ such that $b^\dagger \Psi_0 = 0$, and $\Psi_0 \in D^\infty(a^\dagger)$.

Assumption 3.– $\mathcal{F}_\varphi = \left\{ \varphi_n = \frac{1}{\sqrt{n!}} b^n \varphi_0 \right\}$ and $\mathcal{F}_\Psi = \left\{ \Psi_n = \frac{1}{\sqrt{n!}} a^{\dagger n} \Psi_0 \right\}$ are bases for $\mathcal{H}$.

Assumption 4.– $\mathcal{F}_\Psi$ and $\mathcal{F}_\varphi$ are Riesz bases for $\mathcal{H}$.

In particular, this last requirement was not crucial in our treatment, even if, not surprisingly, it was useful to simplify the proofs of several results.

Our first remark here concerns the word *formally* we have used earlier: to give a rigorous meaning to formula $[a, b] = 1\!\!1$, we first need some dense subset of $\mathcal{H}$ on which both ab and ba can be applied. However, this is just the beginning of the story. In fact, while it is clear, by the assumptions themselves that $\varphi_n \in D(b)$ and that $\Psi_n \in D(a^\dagger)$, it is not so evident that, for instance, $\varphi_n \in D(a)$ for n larger than zero, or that, for these same values, $\Psi_n \in D(b^\dagger)$. Consequently, it is not clear a priori that $\varphi_n \in D(N)$ and that $\Psi_n \in D(N^\dagger)$, if $n \geq 1$. These properties must be checked, while they are automatic for $\mathcal{D}$-PBs, as $\mathcal{D}$ was left invariant by the action of the main operators of the framework. In addition, the operators S_Ψ and S_φ again admit self-adjoint extensions, and the square roots of these extensions can be surely computed. However, again, it seems quite useful to have some common domain left unchanged by the action of N, $N^\dagger$, S_Ψ, S_φ and their square roots. Otherwise some of the intertwining relations we deduced before, as well as the quite useful notion of Θ-conjugatness, need to be considered with more care and with different eyes. Of course, a natural choice for this common domain could be $D^\infty(b)\cap D^\infty(a^\dagger)$, but this is possibly not enough, as, for instance, if f belongs to this intersection, there is no reason for f to belong also to $D(a)$. Hence, the choice of the domain becomes more and more complicated: $D^\infty(b)\cap D^\infty(a^\dagger)\cap D^\infty(b^\dagger)\cap D^\infty(a)$. In addition, this might not be enough, as Θ and Θ^{-1} need to be added to the game as well!

Needless to say, it is much more natural to adopt our more recent point of view, in which a stable set $\mathcal{D}$, dense in $\mathcal{H}$, is introduced from the very beginning as in Definition 3.2.1. Of course, this is possible since as we will see in the second part of this chapter, many physical examples fit these assumptions. Otherwise, ours would be just a long mathematical exercise, and this was not our original aim.

We should also mention that $\mathcal{D}$-PBs allow, with our present understanding of the situation, a simpler and probably more natural comprehension of the relations between operators satisfying (in some sense) $[a,b]=1\!\!1$ and operators satisfying $[c,c^\dagger]=1\!\!1$. This is exactly the content of the following section.

3.2.5 $\mathcal{D}$-PBs Versus Bosons

In Ref. (24), we have discussed the relation existing between *standard* bosonic operators and pseudo-bosons in the sense of Section 3.2.4.1. Here we show how those results can be extended and adapted to the new, more flexible, context of $\mathcal{D}$-PBs. In the attempt to clarify the result, we divide our treatment in two parts: in Theorem 3.2.4 we show how bosons *produce*, under suitable assumptions, $\mathcal{D}$-PBs. In Theorem 3.2.5 we discuss the reverse construction, that is, how from $\mathcal{D}$-PBs we can recover ordinary bosons. Notice that, in what follows, we will focus mainly on the more general situation, that is, the one in which our construction produces sets of vectors that are not bases or even Riesz bases. Of course, when this happens, not surprisingly the results can be stated in a simpler way.

Theorem 3.2.4 *Let c be an operator densely defined on a set $\mathcal{D}\subset\mathcal{H}$ which, together with $c^\dagger$, leaves $\mathcal{D}$ invariant and such that $[c,c^\dagger]f=f$, for all $f\in\mathcal{D}$. Let us assume that a nonzero vector $\hat\varphi_0$ exists in $\mathcal{D}$ such that $c\,\hat\varphi_0=0$[5]. Suppose now that a self-adjoint, positive and invertible operator T exists such that T,T^{-1}: $\mathcal{D}\to\mathcal{D}$. Then, defining on $\mathcal{D}$ the operators a and b as follows*

$$af=T\,c\,T^{-1}f,\qquad bf=T\,c^\dagger T^{-1}f, \tag{3.2.12}$$

for all $f\in\mathcal{D}$, we have that $a^\sharp$, $b^\sharp$ leave $\mathcal{D}$ invariant; $[a,b]f=f$, $\forall f\in\mathcal{D}$; $\exists\varphi_0\in\mathcal{D}$, $\varphi_0\neq 0$, such that $a\,\varphi_0=0$; $\exists\Psi_0\in\mathcal{D}$, $\Psi_0\neq 0$, such that $b^\dagger\,\Psi_0=0$.

[5]Notice that this is exactly what happens for $c=\frac{1}{\sqrt{2}}\left(x+\frac{d}{dx}\right)$, just taking $\mathcal{D}=\mathcal{S}(\mathbb{R})$, the set of those C^∞ functions which decrease to zero, together with their derivatives, faster than any inverse power. Of course $\hat\varphi_0(x)=\frac{1}{\pi^{1/4}}\,e^{-x^2/2}$ belongs to $\mathcal{S}(\mathbb{R})$ and satisfies $c\,\hat\varphi_0=0$.

Let now put $\mathcal{F}_\varphi := \{\varphi_n,\ n \geq 0\}$ and $\mathcal{F}_\Psi := \{\Psi_n,\ n \geq 0\}$, where φ_n and Ψ_n are constructed as in (3.2.3). Then, if T and T^{-1} are bounded, $\mathcal{F}_\varphi$ and $\mathcal{F}_\Psi$ are biorthogonal Riesz bases. If, on the other hand, T or T^{-1}, or both, are unbounded, then $\mathcal{F}_\varphi$ and $\mathcal{F}_\Psi$ are biorthogonal $\mathcal{D}$-quasi bases.

Finally, we introduce the operators S_φ and S_Ψ as in Section 3.2.3: we define

$$D(S_\varphi) = \{f \in \mathcal{H} \,:\, \textstyle\sum_{n=0}^{\infty} \langle \varphi_n, f\rangle\, \varphi_n \text{ exists in } \mathcal{H}\},$$

$$D(S_\Psi) = \{g \in \mathcal{H} \,:\, \textstyle\sum_{n=0}^{\infty} \langle \Psi_n, g\rangle\, \Psi_n \text{ exists in } \mathcal{H}\},$$

and, for $f \in D(S_\varphi)$ and $g \in D(S_\Psi)$, we put

$$S_\varphi f = \sum_{n=0}^{\infty} \langle \varphi_n, f\rangle\, \varphi_n, \qquad S_\Psi g = \sum_{n=0}^{\infty} \langle \Psi_n, g\rangle\, \Psi_n. \tag{3.2.13}$$

Then, for all $f \in D(S_\varphi) \cap \mathcal{D}$ and for all $g \in D(S_\Psi) \cap \mathcal{D}$, we have

$$S_\varphi f = T^2 f, \qquad S_\Psi g = T^{-2} g, \tag{3.2.14}$$

and $(a, b^\dagger)$ are T^{-2}-conjugate: $af = T^2 b^\dagger T^{-2} f$, $f \in \mathcal{D}$. Finally, if $\mathcal{D} \subseteq D(S_\varphi) \cap D(S_\Psi)$, then $(a, b^\dagger)$ are also S_Ψ-conjugate: $af = S_\Psi^{-1} b^\dagger S_\Psi f$, $f \in \mathcal{D}$.

Remark:– As we see from the statement of the theorem, we are considering the possibility of T, T^{-1} to be bounded and the case in which this does not happen. In the first case, for instance, it is clear that $D(S_\Psi) = D(S_\varphi) = \mathcal{H}$, and equation (3.2.14) simply becomes $S_\varphi = T^2$, and $S_\Psi = T^{-2}$. Notice also that, in order for (3.2.14) to be of some use, $D(S_\varphi) \cap \mathcal{D}$ and $D(S_\Psi) \cap \mathcal{D}$ must be *rich* sets, which is surely what happens if, as in Section 3.2.3, we have $\mathcal{D} \subseteq D(S_\varphi) \cap D(S_\Psi)$.

Proof:

First we observe that, because of (3.2.12) and using the stability of $\mathcal{D}$ under T and T^{-1}, $af, bf \in \mathcal{D}$ for all $f \in \mathcal{D}$.

Next, we can check that

$$a^\dagger f = T^{-1} c^\dagger T f, \qquad b^\dagger f = T^{-1} c\, T f, \tag{3.2.15}$$

for all $f \in \mathcal{D}$. Indeed, for instance, taking $f \in \mathcal{D}$ and $g \in \mathcal{D} \subseteq D(a)$, we have

$$\langle a^\dagger f, g\rangle = \langle f, a\, g\rangle = \langle f, T c\, T^{-1} g\rangle = \langle T^{-1} c^\dagger T f, g\rangle.$$

This is because $\mathcal{D}$ is stable under the action of all the operators involved in this equation. Using now the fact that $\mathcal{D}$ is dense in $\mathcal{H}$, it is clear that each $\hat{g} \in \mathcal{H}$ can be approximated by a sequence of vectors in $\mathcal{D}$: $g_n \to \hat{g}$, for some $\{g_n\} \subset \mathcal{D}$. Then

$$\langle a^\dagger f, \hat{g}\rangle = \lim_n \langle a^\dagger f, g_n\rangle = \lim_n \langle T^{-1} c^\dagger T f, g_n\rangle = \langle T^{-1} c^\dagger T f, \hat{g}\rangle,$$

so that the first equality in (3.2.15) follows. Similarly, we can deduce the second equation in (3.2.15). Together, they also imply that $a^\dagger$ and $b^\dagger$ leave $\mathcal{D}$ invariant.

Now, checking that $[a,b]f = f$, for all $f \in \mathcal{D}$, is quite simple, as $[a,b]f = a(bf) - b(af) = T[c,c^\dagger](T^{-1}f) = T(T^{-1}f) = f$. Hence, a and b satisfy the pseudo-bosonic commutation relations on $\mathcal{D}$.

At this stage, as usual, we introduce the vectors $\hat{\varphi}_n = \frac{1}{\sqrt{n!}}(c^\dagger)^n\hat{\varphi}_0$, which all belong to $\mathcal{D}$, clearly. The set $\mathcal{F}_{\hat{\varphi}} = \{\hat{\varphi}_n,\, n \geq 0\}$ is an orthonormal basis for $\mathcal{H}$. Now, let us define two new vectors of $\mathcal{D}$ as follows: $\varphi_0 = T\hat{\varphi}_0$, and $\Psi_0 = T^{-1}\hat{\varphi}_0$. Then

$$a\varphi_0 = (TcT^{-1})(T\hat{\varphi}_0) = Tc\,\hat{\varphi}_0 = 0,$$

and

$$b^\dagger\Psi_0 = (T^{-1}cT)(T^{-1}\hat{\varphi}_0) = T^{-1}c\,\hat{\varphi}_0 = 0.$$

Let us now assume that $\varphi_n = T\hat{\varphi}_n$, for some given n. We want to show that $\varphi_{n+1} = T\hat{\varphi}_{n+1}$ as well. In fact, as $\varphi_{n+1} = \frac{1}{\sqrt{n+1}}\,b\,\varphi_n$ and $\hat{\varphi}_{n+1} = \frac{1}{\sqrt{n+1}}\,c^\dagger\,\hat{\varphi}_n$, for all $n \geq 0$, we see that

$$T\hat{\varphi}_{n+1} = \frac{1}{\sqrt{n+1}}\,T\,c^\dagger\,\hat{\varphi}_n = \frac{1}{\sqrt{n+1}}\,T\,c^\dagger\,T^{-1}\varphi_n = \frac{1}{\sqrt{n+1}}\,b\,\varphi_n = \varphi_{n+1}.$$

Then, by induction, $\varphi_n = T\hat{\varphi}_n$ for all $n \geq 0$. Here we have used (3.2.15) applied to φ_n, which is surely a vector in $\mathcal{D}$. Hence, we conclude that $\mathcal{F}_\varphi$ is the image, through T, of the orthonormal basis $\mathcal{F}_{\hat{\varphi}}$. Analogously, we can check that $\mathcal{F}_\Psi$ is the image, through T^{-1}, of $\mathcal{F}_{\hat{\varphi}}$. Then, if T and T^{-1} are bounded operators, $\mathcal{F}_\varphi$ and $\mathcal{F}_\Psi$ are biorthogonal Riesz bases, as claimed earlier.

On the other hand, let us assume that T, T^{-1} or both are unbounded. Then $\mathcal{F}_\varphi$ and $\mathcal{F}_\Psi$ cannot be Riesz bases. However, because of Proposition 3.2.5, they are still $\mathcal{D}$-quasi bases.

In order to show that $(a, b^\dagger)$ are T^{-2}-conjugate, we first remind that $af = Tc\,T^{-1}f$ and $b^\dagger h = T^{-1}c\,Th$, for all $f, h \in \mathcal{D}$. Then, as $b^\dagger h$ and $Tb^\dagger h$ both belong to $\mathcal{D}$, we deduce that $T^2b^\dagger h = TcTh = TcT^{-1}T^2h, \forall\, h \in \mathcal{D}$. Now, calling $f = T^2h$, we have $h = T^{-2}f$. Of course, as h is arbitrary, f is also arbitrary. Hence $T^2b^\dagger T^{-2}f = T\,c\,T^{-1}f$ and, as a consequence, $af = TcT^{-1}f = T^2b^\dagger T^{-2}f, \forall f \in \mathcal{D}$, which is what we had to prove.

As for the equalities in (3.2.14), they are well known when T and T^{-1} are bounded. In this case, as already stated in the Remark earlier, we can extend them to all of $\mathcal{H}$ and we get, analogously to what shown in Section 3.2.3, $S_\varphi f = T^2 f, \forall f \in \mathcal{H}$. In other words, in this case, we conclude that $S_\varphi = T^2$ and that $S_\Psi = T^{-2}$.

Let us now consider the case in which T and T^{-1} are not bounded. In this case, as we have seen, there is no reason a priori for $D(S_\varphi)$ to coincide with all of $\mathcal{H}$. Let $f \in D(S_\varphi)$. Then $\sum_{n=0}^N \langle\varphi_n, f\rangle\,\varphi_n$ converges to a vector in $\mathcal{H}$, which we call $S_\varphi f$.

Hence, taken $g \in \mathcal{H}$, we have

$$\left|\left\langle S_{\varphi}f, g\right\rangle - \sum_{n=0}^{N} \langle f, \varphi_n\rangle \langle \varphi_n, g\rangle\right| = \left|\left\langle S_{\varphi}f - \sum_{n=0}^{N} \langle f, \varphi_n\rangle \varphi_n, g\right\rangle\right|$$
$$\leq \left\|S_{\varphi}f - \sum_{n=0}^{N} \langle f, \varphi_n\rangle \varphi_n\right\| \|g\| \to 0,$$

when N diverges. Then, for all $f \in D(S_{\varphi})$ and $g \in \mathcal{H}$, we deduce that $\left\langle S_{\varphi}f, g\right\rangle = \sum_{n=0}^{\infty} \langle f, \varphi_n\rangle \langle \varphi_n, g\rangle$. Let us now take $f, g \in \mathcal{D}$. Then, as $T^2 f \in \mathcal{D}$ and as $\mathcal{F}_{\hat{\varphi}}$ is an orthonormal basis, we have

$$\left\langle T^2 f, g\right\rangle = \langle Tf, Tg\rangle = \sum_{n=0}^{\infty} \langle Tf, \hat{\varphi}_n\rangle \langle \hat{\varphi}_n, Tg\rangle = \sum_{n=0}^{\infty} \langle f, \varphi_n\rangle \langle \varphi_n, g\rangle .$$

Therefore, we conclude that $\left\langle S_{\varphi}f, g\right\rangle = \left\langle T^2 f, g\right\rangle$ for all $g \in \mathcal{D}$ and for all $f \in D(S_{\varphi}) \cap \mathcal{D}$. However, using the continuity of the scalar product and the density of $\mathcal{D}$ in $\mathcal{H}$, we can extend this equality as follows: $\left\langle S_{\varphi}f, \hat{g}\right\rangle = \left\langle T^2 f, \hat{g}\right\rangle$ for all $\hat{g} \in \mathcal{H}$ and for all $f \in D(S_{\varphi}) \cap \mathcal{D}$, so that $S_{\varphi}f = T^2 f$ for all such f, as stated earlier. A similar argument shows that $S_{\Psi}h = T^{-2}h$ for all $h \in D(S_{\Psi}) \cap \mathcal{D}$.

Our last claim follows from this result, from the equality $af = T^2 b^{\dagger} T^{-2} f, f \in \mathcal{D}$, and from the assumption that $\mathcal{D} \subseteq D(S_{\varphi})$, as in this case, $D(S_{\varphi}) \cap \mathcal{D} = \mathcal{D}$.

■

We now prove the *inverse* of Theorem 3.2.4. More explicitly, we give conditions under which $\mathcal{D}$-PBs can be used to construct ordinary bosons.

Theorem 3.2.5 *Let a and b be operators on $\mathcal{H}$ and $\mathcal{D}$ a dense subset of $\mathcal{H}$, invariant under the action of $a^{\sharp}$, $b^{\sharp}$. Let us assume that $[a, b]f = f, \forall f \in \mathcal{D}$ and that two nonzero vectors φ_0 and Ψ_0 exist in $\mathcal{D}$ such that $a\varphi_0 = 0$ and $b^{\dagger}\Psi_0 = 0$. Let us introduce $\mathcal{F}_{\varphi} := \{\varphi_n, n \geq 0\}$ and $\mathcal{F}_{\Psi} := \{\Psi_n, n \geq 0\}$, where φ_n and Ψ_n are given in (3.2.3), and let S_{φ} and S_{Ψ} the related operators, see (3.2.13), and $\hat{S}_{\varphi}$ and $\hat{S}_{\Psi}$ be their (Friedrichs) self-adjoint extensions. Assume that their square roots $\hat{S}_{\varphi}^{1/2}$ and $\hat{S}_{\Psi}^{1/2}$ leave $\mathcal{D}$ invariant and that $(a, b^{\dagger})$ are $\hat{S}_{\Psi}$-conjugate, that is, that $af = \hat{S}_{\Psi}^{-1} b^{\dagger} \hat{S}_{\Psi} f$, $\forall f \in \mathcal{D}$.*

Then, a densely defined operator c exists such that $[c, c^{\dagger}]f = f, \forall f \in \mathcal{D}$. Moreover, for these f's, $cf = \hat{S}_{\Psi}^{1/2} a \hat{S}_{\varphi}^{1/2} f$. In addition, a nonzero vector $\hat{\varphi}_0$ exists in $\mathcal{D}$ such that $c\,\hat{\varphi}_0 = 0$. Calling $\hat{\varphi}_n = \frac{1}{\sqrt{n!}} (c^{\dagger})^n \hat{\varphi}_0$, $n \geq 0$, they form an orthonormal basis for $\mathcal{H}$, $\mathcal{F}_{\hat{\varphi}}$, with $\hat{\varphi}_n \in \mathcal{D}$ and $\hat{\varphi}_n = \hat{S}_{\Psi}^{1/2} \varphi_n$, for all $n \geq 0$.

Proof: As, by assumption, $\hat{S}_{\Psi}^{1/2}$ and $\hat{S}_{\varphi}^{1/2}$ leave $\mathcal{D}$ invariant, it is possible to define an operator c, mapping also $\mathcal{D}$ into $\mathcal{D}$, as

$$cf = \hat{S}_{\Psi}^{1/2} a \hat{S}_{\varphi}^{1/2} f, \tag{3.2.16}$$

for all $f \in \mathcal{D}$. Its adjoint $c^\dagger$ also maps $\mathcal{D}$ into $\mathcal{D}$ and, in particular, it is possible to show that

$$c^\dagger f = \hat{S}_\varphi^{1/2} a^\dagger \hat{S}_\Psi^{1/2} f = \hat{S}_\Psi^{1/2} b\, \hat{S}_\varphi^{1/2} f, \tag{3.2.17}$$

for all $f \in \mathcal{D}$. In fact, we first notice that, $\forall f, g \in \mathcal{D}$,

$$\left\langle c^\dagger f, g\right\rangle = \langle f, c\, g\rangle = \left\langle f, \hat{S}_\Psi^{1/2} a \hat{S}_\varphi^{1/2} g\right\rangle = \left\langle \hat{S}_\varphi^{1/2} a^\dagger \hat{S}_\Psi^{1/2} f, g\right\rangle.$$

Then, using the continuity of the scalar product and the density of $\mathcal{D}$ in $\mathcal{H}$, we can extend this equality as follows: $\left\langle c^\dagger f, \hat{g}\right\rangle = \left\langle \hat{S}_\varphi^{1/2} a^\dagger \hat{S}_\Psi^{1/2} f, \hat{g}\right\rangle$, for all $f \in \mathcal{D}$ and $\hat{g} \in \mathcal{H}$. Then, the first equality in (3.2.17) follows. As for the second equality, we use the assumption that, as $(a, b^\dagger)$ are $\hat{S}_\Psi$-conjugate, then $af = \hat{S}_\Psi^{-1} b^\dagger \hat{S}_\Psi f$, $\forall f \in \mathcal{D}$. This is equivalent, in turns, to the fact that $(a^\dagger, b)$ are $\hat{S}_\Psi^{-1}$-conjugate, so that $a^\dagger h = \hat{S}_\Psi b \hat{S}_\Psi^{-1} h$, $\forall\, h \in \mathcal{D}$. Then, if we take $h = \hat{S}_\Psi^{1/2} f, f \in \mathcal{D}$, we recover the second equality in (3.2.17).

Now, a straightforward computation shows that $[c, c^\dagger]f = f$, for all $f \in \mathcal{D}$. In fact, using (3.2.16) and (3.2.17), we find that $[c, c^\dagger]f = \hat{S}_\Psi^{1/2}[a, b] S_\varphi^{1/2} f = \hat{S}_\Psi^{1/2} \hat{S}_\varphi^{1/2} f = f$. This last equality is a consequence of formula (3.2.10), which was deduced in the same assumptions.

Let now $\hat{\varphi}_0 = \hat{S}_\Psi^{1/2} \varphi_0$ be a nonzero vector, clearly belonging to $\mathcal{D}$. Then

$$c\, \hat{\varphi}_0 = (\hat{S}_\Psi^{1/2} a\, \hat{S}_\varphi^{1/2})(\hat{S}_\Psi^{1/2} \varphi_0) = \hat{S}_\Psi^{1/2} a\, \varphi_0 = 0.$$

Let us further define the vectors $\hat{\varphi}_n = \frac{1}{\sqrt{n!}}(c^\dagger)^n \hat{\varphi}_0$, $n \geq 0$, all belonging to $\mathcal{D}$. The set $\mathcal{F}_{\hat{\varphi}}$ of all these vectors is an orthonormal basis for $\mathcal{H}$, and a standard computation shows that $c\, \hat{\varphi}_n = \sqrt{n} \hat{\varphi}_{n-1}$, if $n \geq 1$, and, as it is clear, $c\, \hat{\varphi}_n = 0$ if $n = 0$. Moreover, $c^\dagger \hat{\varphi}_n = \sqrt{n+1}\, \hat{\varphi}_{n+1}$, $n \geq 0$.

The last point we have to prove is that $\hat{\varphi}_n = \hat{S}_\Psi^{1/2} \varphi_n$, for all $n \geq 0$. As usual, we proceed by induction on n. First we observe that this claim is true for $n = 0$, by definition. Now, assuming that for a fixed n we have $\hat{\varphi}_n = \hat{S}_\Psi^{1/2} \varphi_n$, we have, using the second equality in (3.2.17),

$$\begin{aligned}
\hat{\varphi}_{n+1} &= \frac{1}{\sqrt{n+1}} c^\dagger\, \hat{\varphi}_n = \frac{1}{\sqrt{n+1}} c^\dagger\, \hat{S}_\Psi^{1/2} \varphi_n \\
&= \frac{1}{\sqrt{n+1}} (S_\Psi^{1/2} b\, S_\varphi^{1/2})\, (\hat{S}_\Psi^{1/2} \varphi_n) = \frac{1}{\sqrt{n+1}} S_\Psi^{1/2} b\, \varphi_n = S_\Psi^{1/2} \varphi_{n+1},
\end{aligned}$$

as we had to prove.

■

The conclusion of our analysis is therefore the following: under suitable assumptions, operators (a, b) satisfying pseudo-bosonic commutation rules are, in a certain sense, not very different from an operator c which, together with its adjoint, satisfies the canonical commutation relations. Notice however that this similarity is strongly related to the fact that the assumptions discussed here are satisfied. If they are not, in fact, there is no reason a priori to imagine that c can be defined (in some reasonable way) out of a and b, and analogously, there is no reason why c should produce *real* $\mathcal{D}$-PBs, that is, operators a and b satisfying all the assumptions introduced in Section 3.2.

Remark:– For the completeness' sake we should mention that, in recent years, together with Camillo Trapani and Atsushi Inoue, we have considered a slightly different approach to pairs of operators satisfying pseudo-bosonic commutation rules in some suitable sense. The main idea was to adopt in this context a partial algebraic approach, which several times proved to be useful when unbounded operators have to be considered. We refer to Ref. (8, 25) for results in this direction.

This concludes the *more mathematical* part of the chapter, the one devoted to the construction and to the analysis of the general framework. Some easier mathematics will appear later, in connection with pseudo-fermions. In the following two sections, we will be mainly concerned with the analysis of some physical models, proposed along the years, in which our assumptions can be checked explicitly. We will see how interesting conclusions can be deduced adopting our framework.

3.3 $\mathcal{D}$-PBs IN QUANTUM MECHANICS

In this section, we review two recent models showing that many of the claims we have discussed so far have some *physical justification*, meaning with this that quantum mechanical systems do not necessarily behave as we would like them to do! In particular, we will show that:

1. also for very easy models, which are simple extensions of the one-dimensional harmonic oscillator, the eigenvectors of the related non-self-adjoint hamiltonians H and $H^\dagger$ are not biorthogonal (Riesz) bases. This occurs also for its two-dimensional, noncommutative, version.
2. For the same models, the metric operators turn out to be unbounded, invertible, and also their inverses are unbounded.
3. The biorthogonal sets, in both cases, are $\mathcal{D}$-quasi bases.

Despite of the negative results (points 1 and 2 aforementioned), point 3 produces interesting consequences, as it suggests that our previous mathematical results can be used in concrete physical situations.

3.3.1 The Harmonic Oscillator: Losing Self-adjointness

We begin with a detailed analysis of the *shifted* harmonic oscillator and of some of its possible non-self-adjoint extensions, (26). It may be worth stressing that in the literature several such extensions exist, in one or more spatial dimensions, see Ref. (27–31) and references therein for some examples.

The original ingredients of our analysis are the self-adjoint position and momentum operators x and $p = -i\frac{d}{dx}$, satisfying (in the sense of unbounded operators) $[x, p] = i\mathbb{1}$, and the standard annihilation and creation operators $a = \frac{1}{\sqrt{2}}(x + ip)$ and $a^\dagger = \frac{1}{\sqrt{2}}(x - ip)$ constructed out of them, which obey (of course, again in the sense of unbounded operators) $[a, a^\dagger] = \mathbb{1}$. Here $\mathbb{1}$ is the identity operator on $\mathcal{H} = \mathcal{L}^2(\mathbb{R})$.

Let us fix $k \in \mathbb{R}$, $\alpha, \beta \in \mathbb{C}$, and let us introduce the operators

$$c = a + k, \qquad A = a + \alpha, \qquad B = a^\dagger + \overline{\beta}, \tag{3.3.1}$$

as well as their adjoints $c^\dagger = a^\dagger + k$, $A^\dagger = a^\dagger + \overline{\alpha}$ and $B^\dagger = a + \beta$. The commutation rules

$$[c, c^\dagger] = [A, A^\dagger] = [B, B^\dagger] = \mathbb{1}, \tag{3.3.2}$$

suggest that $(c, c^\dagger)$, $(A, A^\dagger)$ and $(B, B^\dagger)$ are bosonic operators (in the usual sense) for all choices of k, α, and β. However, as we have also

$$[c, A^\dagger] = [c, B] = [A, c^\dagger] = [A, B] = [B, c] = [B, A] = \mathbb{1}, \tag{3.3.3}$$

we could also conclude that, for instance, the pairs $(A, c^\dagger)$, (A, B), or (B, c) are (at least formally), pseudo-bosonic operators. Therefore, we could introduce several self-adjoint, and non-self-adjoint, number operators, like $\hat{n} = c^\dagger c$, $N_A = A^\dagger A$, $N_B = BB^\dagger$, $N = BA$ and $N^\dagger = A^\dagger B^\dagger$, and so on. The main output of our analysis will allow us to conclude that, while the eigenstates of $\hat{n} = c^\dagger c$ produce an orthonormal basis for $\mathcal{H}$, the eigenstates of N (or those of $N^\dagger$) are not even a basis. For that we will make use of the unitary displacement operator

$$D(z) = e^{\overline{z}a - za^\dagger} = e^{-iz_i z_r} e^{-i\sqrt{2}z_i x} e^{i\sqrt{2}z_r p}, \tag{3.3.4}$$

where $z = z_r + iz_i$. The role of $D(z)$ is important, as we can write

$$c = D(k)a\,D^{-1}(k), \quad A = D(\alpha)a\,D^{-1}(\alpha), \quad B = D(\beta)a^\dagger D^{-1}(\beta).$$

Hence, calling $\hat{n}_0 := a^\dagger a$, it is clear that

$$\hat{n} = D(k)\hat{n}_0 D^{-1}(k), \quad N = D(\beta)a^\dagger D^{-1}(\beta)D(\alpha)aD^{-1}(\alpha),$$

and

$$N^\dagger = D(\alpha)a^\dagger D^{-1}(\alpha)D(\beta)aD^{-1}(\beta).$$

In particular, we see that $N = N^\dagger$ if $\alpha = \beta$, but not in general.

3.3.1.1 The Self-adjoint Shifted Harmonic Oscillator In coordinate representation the normalized vacuum of a, $a\, e_0(x) = 0$, is $e_0(x) = \frac{1}{\pi^{1/4}}\, e^{-x^2/2}$, while the other eigenstates of $\hat{n}_0$ can be written as

$$e_n(x) = \frac{1}{\sqrt{n!}}(a^\dagger)^n\, e_0(x) = \frac{1}{\sqrt{2^n n! \sqrt{\pi}}}\, H_n(x)\, e^{-x^2/2}.$$

Here $H_n(x)$ is the n-th Hermite polynomial. The eigenstates $\Phi_n(x)$ of $\hat{n}$ can be easily deduced, both with a direct computation, or from the previous ones, simply because $c\,\Phi_0(x) = 0$ is solved, choosing properly the normalization, by the function $\Phi_0(x) = D(k)e_0(x)$. This relation can be established also for the other functions of the two sets, $\mathcal{F}_e = \{e_n(x), n \geq 0\}$ and $\mathcal{F}_\Phi = \{\Phi_n(x), n \geq 0\}$. Indeed we find, using also the last equality in (3.3.4),

$$\Phi_n(x) = D(k)e_n(x) = e^{i\sqrt{2}kp}\, e_n(x) = e_n(x + \sqrt{2}k), \tag{3.3.5}$$

for all $n \geq 0$ and for all real choices of k. As $\hat{n}\Phi_n(x) = n\Phi_n(x)$, we conclude that the eigenstates of the self-adjoint operator $\hat{n}$ are just the translated version of those of $\hat{n}_0$. They are clearly orthonormal, complete, and span all the Hilbert space. These properties could be easily deduced from the fact that $D(k)$ is unitary. In fact, for instance, if $f \in \mathcal{H}$ is orthogonal to all the $\Phi_n(x)$'s, then for all $n \geq 0$, we have

$$0 = \langle f, \Phi_n \rangle = \langle f, D(k)e_n \rangle = \left\langle D^{-1}(k)f, e_n \right\rangle \quad \Rightarrow \quad D^{-1}(k)f = 0,$$

as $\mathcal{F}_e$ is complete. Hence, $f = 0$, which implies that $\mathcal{F}_\Phi$ is complete as well. To prove that $\mathcal{F}_\Phi$ is also a basis for $\mathcal{H}$, we use the fact that, $\forall g \in \mathcal{H}$, $g = \sum_{n\geq 0} \langle e_n, g \rangle\, e_n$. Then we have, using the continuity of $D(k)$ and the relation between e_n and Φ_n,

$$\begin{aligned} f &= D(k)\left(D^{-1}(k)f\right) = D(k)\left(\sum_{n\geq 0} \left\langle e_n, D^{-1}(k)f \right\rangle e_n\right) \\ &= \sum_{n\geq 0} \langle D(k)e_n, f \rangle\, D(k)e_n = \sum_{n\geq 0} \langle \Phi_n, f \rangle\, \Phi_n, \end{aligned}$$

for all $f \in \mathcal{H}$. It is important to stress that what we have done here is correct because $D(k)$ and $D^{-1}(k)$ are both bounded (and unitary). Otherwise, for instance, in the proof of the completeness of $\mathcal{F}_\Phi$, we should have taken f in the domain of $D^{-1}(k)$, and this would not allow us to conclude, in general. In addition, in the previous equation, $D(k)$ could not be *taken inside* the infinite sum on n, as there is no guarantee that the series $\sum_{n=0}^{N} \langle D(k)e_n, f \rangle\, D(k)e_n$ converges, in this case. These are exactly the kind of problems that we find when we lose self-adjointness.

3.3.1.2 The Non-self-adjoint Shifted Harmonic Oscillator Among the possible generalizations of the number operators $\hat{n}_0$ and $\hat{n}$, we could consider $A^\dagger A$ or $BB^\dagger$. However, as these appear both (at least formally) self-adjoint, not many differences are expected with respect to what we have seen in Section 3.3.1.1. For instance, if we act with powers of $A^\dagger$ on the vacuum of A, φ_0, again we get an orthonormal basis for $\mathcal{H}$, whose n-th vector can be written as $D(\alpha)e_n(x)$, and which satisfies the eigenvalue equation $(A^\dagger A)\left(D(\alpha)e_n(x)\right) = n\left(D(\alpha)e_n(x)\right)$. For our purposes, in order to find eigenstates of $N = BA$, it is more interesting to act on φ_0 with powers of B, and this is what we will do in details here.

First, let us observe that, choosing a suitable normalization, $A\varphi_0 = 0$ if $\varphi_0 = D(\alpha)e_0$. Then we have, see Section 3.2,

$$\varphi_n = \frac{1}{\sqrt{n!}}B^n\varphi_0 = \frac{1}{\sqrt{n!}}D(\beta)(a^\dagger)^n D^{-1}(\beta)D(\alpha)\,e_0 = V(\alpha,\beta)\,e_n, \tag{3.3.6}$$

where we have introduced the operator $V(\alpha,\beta) = e^{\frac{1}{2}\alpha(\overline{\beta}-\overline{\alpha})}e^{\overline{\beta}a-\alpha a^\dagger}$. Of course, each $\varphi_n \in \mathcal{H}$ since it is just a polynomial times a shifted gaussian, see below.

It is important to stress that $V(\alpha,\alpha) = D(\alpha)$, which means that, for some values of, say, β the operator $V(\alpha,\beta)$ is unitary and, therefore, bounded with bounded inverse. However, if $\alpha \neq \beta$, we will show later that $V(\alpha,\beta)$ is unbounded. Nevertheless, due to (3.3.6), each e_n belongs to its domain, which is dense in $\mathcal{H}$ as it contains the set of all the finite linear combinations of the e_n's.

As discussed in Section 3.2, the biorthogonal set $\{\Psi_n\}$ is constructed by first looking for the vacuum of $B^\dagger$: $B^\dagger\Psi_0 = 0$. This is satisfied if $a\,(D^{-1}(\beta)e_0) = 0$ and then we deduce that $\Psi_0 = \mu(\alpha,\beta)D(\beta)e_0$, where $\mu(\alpha,\beta) := e^{\frac{1}{2}(|\alpha|^2+|\beta|^2)-\beta\overline{\alpha}}$ is a suitable normalization, see below, introduced to have $\langle\varphi_0,\Psi_0\rangle = 1$. If we act on Ψ_0 with powers of B, we construct an orthogonal basis for $\mathcal{H}$ of eigenstates of $BB^\dagger$, which is not what we want here. Indeed, what is more interesting for us is finding the eigenstates of $N^\dagger = A^\dagger B^\dagger$. Then, following Section 3.2, we construct the new vectors

$$\Psi_n = \frac{1}{\sqrt{n!}}(A^\dagger)^n\Psi_0 = \frac{\mu(\alpha,\beta)}{\sqrt{n!}}D(\alpha)(a^\dagger)^n D^{-1}(\alpha)D(\beta)\,e_0$$
$$= \mu(\alpha,\beta)V(\beta,\alpha)\,e_n, \tag{3.3.7}$$

where it might be worth noticing that $V(\beta,\alpha)$ appears rather than $V(\alpha,\beta)$, see (3.3.6). It is clear that $\Psi_n \in \mathcal{H}$, for all $n \geq 0$. A simple computation shows that, with our previous choice of normalization for φ_0 and Ψ_0,

$$\langle\varphi_0,\Psi_0\rangle = \langle V(\alpha,\beta)e_0, \mu(\alpha,\beta)V(\beta,\alpha)e_0\rangle = \mu(\alpha,\beta)e^{-\frac{1}{2}(|\alpha|^2+|\beta|^2)+\beta\overline{\alpha}} = 1,$$

and therefore $\langle\varphi_n,\Psi_m\rangle = \delta_{n,m}$, for all $n,m \geq 0$. The vectors of the sets $\mathcal{F}_\varphi$ and $\mathcal{F}_\Psi$ are, respectively, eigenstates of N and $N^\dagger$, and they are biorthogonal. This does not automatically imply that $\mathcal{F}_\varphi$ and $\mathcal{F}_\Psi$ are bases for $\mathcal{H}$. We will now prove that, on the contrary, neither $\mathcal{F}_\varphi$ nor $\mathcal{F}_\Psi$ can be basis for $\mathcal{H}$.

We start proving that $\lim_{n\to\infty}\|\varphi_n\| = \infty$. In fact, with a little algebra, we have

$$\|\varphi_n\|^2 = \langle V(\alpha,\beta)\, e_n, V(\alpha,\beta)\, e_n\rangle = \|e^{(\overline{\beta}-\overline{\alpha})a}e_n\|^2$$
$$\geq 1 + |\overline{\beta}-\overline{\alpha}|^2 n = 1 + |\beta-\alpha|^2 n,$$

which clearly diverges with n diverging, if $\alpha \neq \beta$. The above-mentioned inequality follows from the fact that, $\forall \gamma \in \mathbb{C}$,

$$e^{\gamma a}\, e_n = \sum_{k=0}^{n} \frac{(\gamma a)^k}{k!}\, e_n = e_n + \gamma\sqrt{n}e_{n-1} + \cdots + \frac{\gamma^n}{\sqrt{n!}}\, e_0,$$

and from the orthogonality of the different e_k's. In a similar way, we can also prove that $\lim_{n\to\infty}\|\Psi_n\| = \infty$: the different choice of normalization, in fact, does not produce any difference, at least under this aspect.

Remark:– A crucial point to stress here is that the divergence of both $\|\varphi_n\|$ and $\|\Psi_n\|$ occurs only if $\alpha \neq \beta$, which is exactly what we expect because, if $\alpha = \beta$, $\mathcal{F}_\varphi$ and $\mathcal{F}_\Psi$ both coincide with the set $\mathcal{F}_\Phi$ previously introduced, with $k = \alpha = \beta$. Then, as its vectors are the image, via an unitary operator, of normalized vectors, they are normalized as well.

A consequence of these results is that $V(\alpha,\beta)$, if $\alpha \neq \beta$, is necessarily unbounded. The reason is simple: suppose that this is not so, and let M be the (finite) norm of $V(\alpha,\beta)$. Then, as $\|\varphi_n\| = \|V(\alpha,\beta)\, e_n\| \leq M$, we would get a contradiction. An immediate consequence of this fact is that neither $\mathcal{F}_\varphi$ nor $\mathcal{F}_\Psi$ can be Riesz basis for $\mathcal{H}$, because a Riesz basis is the image of an orthonormal basis via a bounded operator, with bounded inverse, see Section 3.2.1. However, this would not prevent, in principle, $\mathcal{F}_\varphi$ or $\mathcal{F}_\Psi$, or both, to be bases. Nevertheless, we will now show that this is also impossible, at least if $\alpha \neq \beta$.

In fact, let us assume for the moment that $\mathcal{F}_\varphi$ is a basis for $\mathcal{H}$. Hence, each $f \in \mathcal{H}$ can be written as $f = \sum_{n=0}^{\infty}\langle\Psi_n, f\rangle\,\varphi_n = \sum_{n=0}^{\infty} P_n(f)$, where $P_n(f) := \langle\Psi_n, f\rangle\,\varphi_n$. Now, we can check that $\|P_n\| = \|\varphi_n\|\|\Psi_n\|$. Indeed we have, using the Schwarz inequality,

$$\|P_n\| = \sup_{\|f\|=1}\left\|\langle\Psi_n, f\rangle\,\varphi_n\right\| \leq \sup_{\|f\|=1}\|\Psi_n\|\,\|f\|\,\|\varphi_n\| = \|\Psi_n\|\,\|\varphi_n\|.$$

On the other hand, as $\Psi_n/\|\Psi_n\|$ has norm one, we also have

$$\|P_n\| = \sup_{\|f\|=1}\left\|\langle\Psi_n, f\rangle\,\varphi_n\right\| \geq \left\|\left\langle\Psi_n, \frac{\Psi_n}{\|\Psi_n\|}\right\rangle\varphi_n\right\| = \|\Psi_n\|\,\|\varphi_n\|,$$

so that our claim follows. Then $\|P_n\| = \|\varphi_n\|\|\Psi_n\| \to \infty$, so that $\sup_n\|P_n\| = \infty$. As a consequence, the above-mentioned expansion for f cannot converge for all vectors f. Hence, $\mathcal{F}_\varphi$ cannot be a basis for $\mathcal{H}$. In a similar way, we can conclude that $\mathcal{F}_\Psi$ cannot be a basis for $\mathcal{H}$. Nevertheless, we will see in the next section that they still produce some useful weak forms of the resolution of the identity, since they will be proved to be $\mathcal{D}$-quasi bases, for a certain $\mathcal{D}$.

Summarizing we have that, in our very simple model, *the biorthogonal sets of eigenstates of N and $N^\dagger$ are not bases for $\mathcal{H}$.* This suggests that some of the claims which one can find in the physical literature on this subject, where the non-self-adjoint hamiltonians are by far more complicated than the number operators N and $N^\dagger$ considered here, are wrong or, at least, need to be justified in more details.

We should also stress that, even if they are not bases, both $\mathcal{F}_\varphi$ and $\mathcal{F}_\Psi$ are complete in $\mathcal{L}^2(\mathbb{R})$. This is because both these sets are made of polynomials times a shifted gaussian. More in details, for instance, $\varphi_n(x) = p_n(x)\, e^{-(x-\gamma)^2/2}$, where $p_n(x)$ is a polynomial of degree n and γ is some fixed shift parameter. Then $\mathcal{F}_\varphi$ is complete, (32). Hence, our results show explicitly that, when orthonormality is lost, a complete set needs not to be a basis! The same conclusion, for a slightly different model, again related to the harmonic oscillator, can be found in Ref. (31).

We end this section with some similarity relations, which can be proved explicitly, by using simple well-known formulas for a and $a^\dagger$:

$$V^{-1}(\alpha,\beta)NV(\alpha,\beta) = V^{-1}(\beta,\alpha)N^\dagger V(\beta,\alpha) = \hat{n}_0,$$

and

$$T^{-1}(\alpha,\beta)NT(\alpha,\beta) = N^\dagger,$$

where

$$T(\alpha,\beta) = V(\alpha,\beta)V^{-1}(\beta,\alpha) = e^{\frac{1}{2}(\alpha\overline{\beta}-\beta\overline{\alpha}+2|\beta|^2-2|\alpha|^2)}e^{a(\overline{\beta}-\overline{\alpha})+a^\dagger(\beta-\alpha)}.$$

Notice that all these identities cannot be defined in all of $\mathcal{H}$, as the operators involved are unbounded. However, they are surely satisfied on some dense domain.

3.3.1.3 More Results on the Non-self-adjoint Harmonic Oscillator In what we have done so far, the role of the various assumptions introduced in Section 3.2 is not completely evident. In this section, we consider exactly this aspect. To begin with, among all the possible choices of formal pseudo-bosonic operators, we will consider here only the pair (A, B), as in Section 3.3.1. We have, see (3.3.1): $A = \frac{1}{\sqrt{2}}\left(x + \frac{d}{dx} + \sqrt{2}\,\alpha\right)$ and $B = \frac{1}{\sqrt{2}}\left(x - \frac{d}{dx} + \sqrt{2}\,\overline{\beta}\right)$. The vacua of A and $B^\dagger$ are

$$\varphi_0(x) = N_\varphi \exp\left\{-\left(\frac{x^2}{2} + \sqrt{2}\,\alpha\, x\right)\right\},$$

and

$$\Psi_0(x) = N_\Psi \exp\left\{-\left(\frac{x^2}{2} + \sqrt{2}\,\beta\, x\right)\right\}.$$

Here N_φ and N_Ψ must satisfy the equality: $\overline{N_\varphi}\,N_\Psi = \frac{1}{\sqrt{\pi}}\,e^{-(\beta+\overline{\alpha})^2/2}$, which is needed to have $\langle\varphi_0, \Psi_0\rangle = 1$. Of course, N_φ and N_Ψ could be related to $\mu(\alpha,\beta)$ introduced before, but this is not relevant for us. Both vacua belong to the set

$$\mathcal{D} = \{f(x) \in \mathcal{S}(\mathbb{R}) \,:\, e^{kx}f(x) \in \mathcal{S}(\mathbb{R}),\ \forall k \in \mathbb{R}\}. \tag{3.3.8}$$

This set is dense in $\mathcal{L}^2(\mathbb{R})$, as it contains $D(\mathbb{R})$, the set of all C^∞ functions with compact support, and is stable under the action of $A^\sharp$ and $B^\sharp$. Assumptions $\mathcal{D}$-pb 1 and $\mathcal{D}$-pb 2 are satisfied. As for Assumption $\mathcal{D}$-pb 3, our previous results show that this does not hold. However, it is possible to show that, on the other hand, Assumption $\mathcal{D}$-pbw 3 is satisfied, with $\mathcal{G} \equiv \mathcal{D}$. Indeed let us take $f, g \in \mathcal{D}$. Then, as $\mathcal{F}_e$ is an orthonormal basis and as $\mathcal{D} \subseteq D(V^\dagger(\alpha,\beta)) \cap D(V^{-1}(\alpha,\beta))$, we have

$$\begin{aligned}\langle f, g\rangle &= \left\langle V^\dagger(\alpha,\beta)f, V^{-1}(\alpha,\beta)g\right\rangle = \sum_n \left\langle V^\dagger(\alpha,\beta)f, e_n\right\rangle \left\langle e_n, V^{-1}(\alpha,\beta)g\right\rangle \\ &= \sum_n \langle f, \varphi_n\rangle \langle \Psi_n, g\rangle,\end{aligned} \tag{3.3.9}$$

because of the equalities $\varphi_n = V(\alpha,\beta)e_n$ and $\Psi_n = (V^{-1}(\alpha,\beta))^\dagger e_n$. Equation (3.3.9) shows that, as required by Assumption $\mathcal{D}$-pbw 3, $\mathcal{F}_\varphi$ and $\mathcal{F}_\Psi$ produce a weak form of the resolution of the identity[6].

Let us now put together $\varphi_n = V(\alpha,\beta)\,e_n$ and $\Psi_n = \mu(\alpha,\beta)V(\beta,\alpha)\,e_n$. Then we deduce that

$$\Psi_n = \Theta(\alpha,\beta)\varphi_n, \tag{3.3.10}$$

where

$$\Theta(\alpha,\beta) = \mu(\alpha,\beta)V(\beta,\alpha)V^{-1}(\alpha,\beta) = e^{-\frac{1}{2}|\alpha+\beta|^2+2|\alpha|^2}\,e^{a(\overline{\alpha}-\overline{\beta})+a^\dagger(\alpha-\beta)}, \tag{3.3.11}$$

which can also be written as $\Theta(\alpha,\beta) = e^{|\alpha|^2-|\beta|^2}e^{a(\overline{\alpha}-\overline{\beta})}e^{a^\dagger(\alpha-\beta)}$. We see that $\Theta(\alpha,\beta)$ is self-adjoint, leaves $\mathcal{D}$ invariant together with its inverse and that $\Theta(\alpha,\alpha) = 1\!\!1$. Moreover, a simple computation shows that $\Theta^{-1}(\alpha,\beta)B^\dagger\Theta(\alpha,\beta)f = Af$, for all $f \in \mathcal{D}$, so that $(A, B^\dagger)$ are $\Theta(\alpha,\beta)$-conjugate. This, in view of (3.3.10), is exactly the content of Proposition 3.2.3. Then, a simple consequence of this is, for instance, that $\Theta^{-1}(\alpha,\beta)N^\dagger\Theta(\alpha,\beta)f = Nf$, for all $f \in \mathcal{D}$. In addition, for each nonzero $f \in \mathcal{D}$, we deduce that $\langle f, \Theta(\alpha,\beta)f\rangle = e^{|\alpha|^2-|\beta|^2}\,\|e^{a^\dagger(\alpha-\beta)}f\|^2$, which is strictly positive, as expected because of Proposition 3.2.4. Finally, it is interesting to see that $\mathcal{D}$ is also stable under the action of $\Theta^{1/2}(\alpha,\beta)$ and of its inverse, which was one of the working assumption of Theorem 3.2.5, identifying $\Theta(\alpha,\beta)$ with $\hat{S}_\Psi$. Hence, we recover, for our simple model, the general structure and results discussed in Section 3.2.

[6]Equation (3.3.9) is nothing that Proposition 3.2.5 , point (iv), explicitly adapted to the present situation.

Once again, we stress that $\Theta(\alpha, \beta)$ is what in the literature is usually called *the metric operator*, and an estimate similar to that used earlier for $V(\alpha, \beta)$ allows us to conclude that both $\Theta(\alpha, \beta)$ and its inverse are unbounded operators, at least if $\alpha \neq \beta$. This result contradicts what is often assumed in the literature, that is, that the metric operator, its inverse, or both, are bounded. This seems to be not so automatic and need to be checked even in very simple systems[7]. This might appear as a very strong (and strange) result, as here α and β in (3.3.1) could be taken as small as we like and still, if $\alpha \neq \beta$, we are forced to admit that the eigenstates of N and $N^\dagger$ are not bases and that Θ and Θ^{-1} are unbounded. However, noticing that

$$N = BA = a^\dagger a + \left(\overline{\beta}\, a + \alpha a^\dagger + \alpha\overline{\beta}\, 1\!\!1\right) = \hat{n}_0 + \delta\hat{n}_0,$$

and that $\delta\hat{n}_0$ is an unbounded operator even when α and β are almost zero, we can easily understand why going from $\hat{n}_0$ to N these major differences are produced: α and β can be small, but the perturbation $\delta\hat{n}_0$ is not small at all!

3.3.2 A Two-dimensional Model in a Flat noncommutative space

The model we are going to describe now is a sort of linearly perturbed two-dimensional harmonic oscillator, defined in a noncommutative space, (33). The hamiltonian of the model is

$$\hat{H} = \frac{1}{2}(\hat{p}_1^2 + \hat{x}_1^2) + \frac{1}{2}(\hat{p}_2^2 + \hat{x}_2^2) + i\left[A(\hat{x}_1 + \hat{x}_2) + B(\hat{p}_1 + \hat{p}_2)\right], \tag{3.3.12}$$

where A and B are real constants. The self-adjoint operators $\hat{x}_j$ and $\hat{p}_k$ are assumed to satisfy the following commutation rules:

$$[\hat{x}_j, \hat{p}_k] = i\delta_{j,k} 1\!\!1, \quad [\hat{x}_j, \hat{x}_k] = i\theta\epsilon_{j,k} 1\!\!1, \tag{3.3.13}$$

the other commutators being zero. Here θ is a small parameter, which measure the noncommutativity of the system, and we have $\epsilon_{j,j} = 0$, $\epsilon_{1,2} = -\epsilon_{2,1} = 1$.

Instead of the perturbative approach adopted in Refs (29, 34), we will see under which conditions we can diagonalize exactly $\hat{H}$. For that, we will adopt the representation of $(\hat{x}_j, \hat{p}_j)$ used also in Ref. (35). If we introduce two pairs of self-adjoint, canonically conjugate operators, $(\tilde{x}_j, \tilde{p}_j)$, $j = 1, 2$ ($[\tilde{x}_j, \tilde{p}_k] = i\delta_{j,k} 1\!\!1$, $[\tilde{x}_j, \tilde{x}_k] = [\tilde{p}_j, \tilde{p}_k] = 0$), we can recover (3.3.13) if we assume that

$$\hat{x}_1 = \tilde{x}_1 - \frac{\theta}{2}\tilde{p}_2, \quad \hat{x}_2 = \tilde{x}_2 + \frac{\theta}{2}\tilde{p}_1, \quad \hat{p}_1 = \tilde{p}_1, \quad \hat{p}_2 = \tilde{p}_2, \tag{3.3.14}$$

[7] As already discussed, since we prefer to have $\Psi_n = \Theta(\alpha, \beta)\varphi_n$, without any extra multiplication factor, our metric operator is defined without using any rapidly decreasing sequence $\{c_n\}$, see the final Remark in Section 3.2.3.

which is usually referred to as the Bopp shift. Let us further introduce the pairs

$$x_j = \frac{\tilde{x}_j}{\sqrt[4]{1+\frac{\theta^2}{4}}}, \qquad p_j = \tilde{p}_j \sqrt[4]{1+\frac{\theta^2}{4}},$$

$j = 1, 2$. They still satisfy the same commutation rules: $[x_j, p_k] = i1\!\!1\,\delta_{j,k}$. After some algebra we find that $\hat{H} = \gamma^2 H$, where $\gamma = \sqrt[4]{1+\frac{\theta^2}{4}}$ depends[8] on θ and

$$\begin{cases} H = H_1 + H_2 + H_3, \\ H_1 = \frac{1}{2}(p_1^2 + x_1^2) + \frac{1}{2}(p_2^2 + x_2^2), \\ H_2 = \frac{\theta}{2\gamma^2}(p_1 x_2 - p_2 x_1), \\ H_3 = \frac{i}{\gamma}\left[A(x_1 + x_2) + \frac{1}{\gamma^2}\left(p_1\left(B + \theta\frac{A}{2}\right) + p_2\left(B - \theta\frac{A}{2}\right)\right)\right]. \end{cases} \tag{3.3.15}$$

In order to write H in terms of pseudo-bosonic number operators, we introduce the operators a_j and b_j as follows

$$\begin{aligned} a_1 &= \frac{1}{2}(x_1 + ip_1 + ix_2 - p_2 + \alpha_1 1\!\!1), \\ a_2 &= \frac{1}{2}(-ix_1 + p_1 - x_2 - ip_2 + \alpha_2 1\!\!1), \\ b_1 &= \frac{1}{2}(x_1 - ip_1 - ix_2 - p_2 + \beta_1 1\!\!1), \\ b_2 &= \frac{1}{2}(ix_1 + p_1 - x_2 + ip_2 + \beta_2 1\!\!1), \end{aligned} \tag{3.3.16}$$

where α_j and β_j are complex quantities to be determined. These operators satisfy, among the others, the formal two-dimensional pseudo-bosonic commutation rules:

$$[a_j, b_k] = i1\!\!1\,\delta_{j,k}, \quad [a_j, a_k] = [b_j, b_k] = 0. \tag{3.3.17}$$

A simple consequence of (3.3.16) is the following: $b_j = a_j^\dagger$ if and only if $\overline{\beta_j} = \alpha_j$, $j = 1, 2$. In other words, if for some reason we take the complex constants in (3.3.16) in such a way that this relation is satisfied, our (formal) pseudo-bosonic operators become (truly) bosonic creation and annihilation operators. But, on the other hand, if for some reason, we have $\overline{\beta_1} \neq \alpha_1$ or $\overline{\beta_2} \neq \alpha_2$, or both, then the best we can have is that (a_j, b_j) behave as $\mathcal{D}$-PBs, for some suitable dense $\mathcal{D}$ in $\mathcal{H}$, still to be identified. Moreover, we also observe that, as for instance $b_1 - \beta_1 = a_1^\dagger - \overline{\alpha_1}$, we have

[8]To simplify the notation, we drop here the dependence of γ on θ.

$[a_1, a_1^\dagger] = 1\!\!1$. Hence, the pair (a_1, b_1) is pseudo-bosonic while $(a_1, a_1^\dagger)$ is truly bosonic. So, the natural question to ask is: why not to use the friendly $(a_1, a_1^\dagger)$ rather than the more complicated pair (a_1, b_1)? The answer is simple: we cannot really choose the operators we have to use because, essentially, this choice is driven by the hamiltonian itself. And, indeed, we will see that this is the situation, for our H in (3.3.15), which turns out to be diagonal using $\mathcal{D}$-PBs, but not otherwise.

Let us now introduce, as in Section 3.2, the number operators $N_j = b_j a_j$, $j = 1, 2$. Then our hamiltonian H can be written in terms of N_1, N_2 and of the identity operator $1\!\!1$ as follows:

$$H = k_1 N_1 + k_2 N_2 + k_0 1\!\!1, \tag{3.3.18}$$

where a direct comparison between the two sides of this equation allows us to fix, first of all, $k_1 = 1 + \frac{\theta}{2\gamma^2}$ and $k_2 = 1 - \frac{\theta}{2\gamma^2}$. In the same way, we also deduce that α_j, β_j and k_0 must be fixed, in terms of A, B and θ (and γ, k_1 and k_2, which however can also be rewritten only in terms of θ), as follows:

$$\begin{cases} \alpha_1 = -\frac{(1-i)}{2\gamma^3 k_1}\left(2iB + A(2\gamma^2 + \theta)\right), \\ \alpha_2 = \frac{1}{2\gamma^3 k_2}\left(2(1+i)B + 2(1-i)A\gamma^2 - (1-i)A\theta\right), \\ \beta_1 = \frac{(1+i)}{2\gamma^3 k_1}\left(-2iB + A(2\gamma^2 + \theta)\right), \\ \beta_2 = \frac{(1+i)}{2\gamma^3 k_2}\left(2iB + A(-2\gamma^2 + \theta)\right), \\ k_0 = k_1(2 - \alpha_1\beta_1) + k_2(2 - \alpha_2\beta_2). \end{cases} \tag{3.3.19}$$

Once the explicit expressions of the constants α_j and β_j have been fixed, it is important to notice that, except for the very particular case in which $A = B = 0$, α_1 is not the complex conjugate of β_1. As a consequence, if we want to rewrite H in terms of number operators, we are forced, as already claimed, to use pseudo-bosons, since bosons *are not enough*, except when $\hat{H}$ is self-adjoint ($A = B = 0$).

Remark:– as for the shifted harmonic oscillator it is worth to stress that, even if A and B are very small, the *perturbation* $i\left[A(\hat{x}_1 + \hat{x}_2) + B(\hat{p}_1 + \hat{p}_2)\right]$ in (3.3.12) is unbounded. This is most probably the reason why moving away from the hamiltonian $\hat{H}_0 = \frac{1}{2}(\hat{p}_1^2 + \hat{x}_1^2) + \frac{1}{2}(\hat{p}_2^2 + \hat{x}_2^2)$ we deduce similar major changes here, as those found in Section 3.3.1. This will be clear in the following.

The conclusion is therefore that, with the definitions we have given earlier, the hamiltonian $\hat{H}$ in (3.3.12) can be rewritten as follows:

$$\hat{H} = \gamma^2\left(k_1 N_1 + k_2 N_2 + k_0 1\!\!1\right). \tag{3.3.20}$$

We are now going to derive the eigenvalues and the eigenvectors of $\hat{H}$ and to check that the assumptions of Section 3.2 are indeed satisfied.

3.3.2.1 Checking the Pseudo-bosonic Assumptions We start introducing the operators

$$\begin{cases} A_1 = \frac{1}{2}\left(x_1 + ip_1 + ix_2 - p_2\right), \\ A_2 = \frac{1}{2}\left(-ix_1 + p_1 - x_2 - ip_2\right), \end{cases} \tag{3.3.21}$$

which, together with their adjoint, satisfy the canonical commutation relations. Indeed, we have $[A_j, A_k^\dagger] = \delta_{j,k} \, 1\!\!1$ and $[A_j, A_k] = 0$, $j, k = 1, 2$. Then, introducing for future convenience the complex quantities $\gamma_j = \frac{\alpha_j}{2}$ and $\nu_j = \frac{\overline{\beta_j}}{2}$, as well as the unitary operators

$$D_j(z_j) = \exp\left\{\overline{z_j} A_j - z_j A_j^\dagger\right\}, \qquad D(\underline{z}) = D_1(z_1)D_2(z_2),$$

we see that

$$\begin{cases} a_j = A_j + \gamma_j = D(\underline{\gamma})A_j D^{-1}(\underline{\gamma}), \\ b_j = A_j^\dagger + \overline{\nu_j} = D(\underline{\nu})A_j^\dagger D^{-1}(\underline{\nu}), \end{cases} \tag{3.3.22}$$

$j = 1, 2$. An orthonormal basis for $\mathcal{H} = \mathcal{L}^2(\mathbb{R}^2)$ can be easily constructed: let $e_{0,0} = e_{\underline{0}}$ be the vacuum of A_1 and A_2: $A_j e_{\underline{0}} = 0$, $j = 1, 2$. Then, as usual, we introduce

$$e_{n_1,n_2} = e_{\underline{n}} = \frac{1}{\sqrt{n_1! n_2!}} (A_1^\dagger)^{n_1} (A_2^\dagger)^{n_2} e_{\underline{0}},$$

and the related set $\mathcal{F}_e = \{e_{\underline{n}},\ n_1, n_2 \geq 0\}$. Of course, if $\hat{n}_j = A_j^\dagger A_j$, $\hat{n}_j e_{\underline{n}} = n_j e_{\underline{n}}$.

In order to apply and check the results of Section 3.2, we first look for the vacuum of a_j: $a_1 \varphi_{\underline{0}} = a_2 \varphi_{\underline{0}} = 0$ if, and only if, $A_j(D^{-1}(\underline{\gamma})\varphi_{\underline{0}}) = 0$, $j = 1, 2$. This implies that, choosing properly the normalization, $\varphi_{\underline{0}} = D(\underline{\gamma}) e_{\underline{0}}$. Notice that, being $D(\underline{\gamma})$ unitary, $\varphi_{\underline{0}}$ is surely well defined and normalized.

Before analyzing in more details $\varphi_{\underline{0}}$, and the vectors which arise out of $\varphi_{\underline{0}}$, let us look for the vacuum of $b_j^\dagger$, as required in Section 3.2. We have $b_1^\dagger \Psi_{\underline{0}} = b_2^\dagger \Psi_{\underline{0}} = 0$ if, and only if, $A_j(D^{-1}(\underline{\nu})\Psi_{\underline{0}}) = 0$, $j = 1, 2$. The solution is $\Psi_{\underline{0}} = N_\Psi D(\underline{\nu}) e_{\underline{0}}$, which is again well defined for the same reason. Here N_Ψ is a normalization needed to have $\left\langle \varphi_{\underline{0}}, \Psi_{\underline{0}} \right\rangle = 1$. It turns out that

$$N_\Psi = \exp\left\{\frac{1}{2}\left(|\gamma_1|^2 + |\gamma_2|^2 + |\nu_1|^2 + |\nu_2|^2 - 2\overline{\gamma_1}\nu_1 - 2\overline{\gamma_2}\nu_2\right)\right\}.$$

Remark:– These results could also be found in the coordinate representation quite easily. For instance, $a_1 \varphi_{\underline{0}} = a_2 \varphi_{\underline{0}} = 0$ are equivalent to

$$\left(x_1 + \frac{\partial}{\partial x_1} + ix_2 + i\frac{\partial}{\partial x_2} + \alpha_1\right) \varphi_{\underline{0}}(x_1, x_2) = 0$$

and

$$\left(-ix_1 - i\frac{\partial}{\partial x_1} - x_2 - \frac{\partial}{\partial x_2} + \alpha_2\right)\varphi_{\underline{0}}(x_1,x_2) = 0,$$

whose solution is proportional to $e^{-\frac{1}{2}(x_1^2+x_2^2)-\frac{1}{2}(\alpha_1+i\alpha_2)x_1+\frac{1}{2}(\alpha_2+i\alpha_1)x_2}$. Similarly, $\Psi_{\underline{0}}(x_1,x_2)$ is proportional to $e^{-\frac{1}{2}(x_1^2+x_2^2)-\frac{1}{2}(\beta_1+i\beta_2)x_1+\frac{1}{2}(\beta_2+i\beta_1)x_2}$. We see that both these functions belong to the following set:

$$\mathcal{D} = \left\{f(x_1,f_2) \in \mathcal{S}(\mathbb{R}^2),\ \text{such that } e^{k_1x_1+k_2x_2}f(x_1,x_2) \in \mathcal{S}(\mathbb{R}^2),\ \forall k_1,k_2 \in \mathbb{C}\right\},$$

where $\mathcal{S}(\mathbb{R}^2)$ is the set of the C^∞ functions which decrease to zero, together with their derivatives, faster than any inverse power of x_1 and x_2. It is clear that $\mathcal{D}$ is dense in $\mathcal{H}$, as it contains $D(\mathbb{R}^2)$. We also observe that this is just a two-dimensional version of the set introduced in (3.3.8) for the shifted harmonic oscillator.

Following our general framework, we are now interested in getting the usual vectors $\varphi_{\underline{n}} = \frac{1}{\sqrt{n_1!n_2!}}b_1^{n_1}b_2^{n_2}\varphi_{\underline{0}}$ and $\Psi_{\underline{n}} = \frac{1}{\sqrt{n_1!n_2!}}(a_1^\dagger)^{n_1}(a_2^\dagger)^{n_2}\Psi_{\underline{0}}$. Notice that, because of the stability of $\mathcal{D}$ under the action of b_j and $a_j^\dagger$, and recalling that $\varphi_{\underline{0}}, \Psi_{\underline{0}} \in \mathcal{D}$, both $\varphi_{\underline{n}}$ and $\Psi_{\underline{n}}$ necessarily belong to $\mathcal{D}$, for all $\underline{n}$. If we now introduce the operators

$$V_j(\gamma_j,\nu_j) = e^{\frac{1}{2}\gamma_j(\overline{\nu_j}-\overline{\gamma_j})}e^{(\overline{\nu_j}A_j-\gamma_jA_j^\dagger)}, \qquad V(\underline{\gamma},\underline{\nu}) = V_1(\gamma_1,\nu_1)V_2(\gamma_2,\nu_2),$$

we deduce that

$$\varphi_{\underline{n}} = V(\underline{\gamma},\underline{\nu})e_{\underline{n}}, \qquad \Psi_{\underline{n}} = N_\Psi V(\underline{\nu},\underline{\gamma})e_{\underline{n}}. \tag{3.3.23}$$

In complete analogy with what discussed for the shifted harmonic oscillator, we see that, while $V(\underline{\gamma},\underline{\gamma}) = D(\underline{\gamma})$, which is unitary and, as a consequence, bounded with bounded inverse, if $\underline{\gamma} \neq \underline{\nu}$ then $V(\underline{\gamma},\underline{\nu})$ is unbounded, as well as its inverse. Then the two sets $\mathcal{F}_\varphi = \{\varphi_{\underline{n}}\}$ and $\mathcal{F}_\Psi = \{\Psi_{\underline{n}}\}$ cannot be Riesz bases. In fact, they are both related to the orthonormal basis $\mathcal{F}_e$ by unbounded operators as in (3.3.23). More than this: they are not even bases, while they are both complete in $\mathcal{H}$. The proofs of these claims are just a two-dimensional version of those given in the previous example and will not be repeated here.

Formulas (3.3.22) show how (a_j, b_j) are related to $(A_j, A_j^\dagger)$ by the unitary operators $D(\underline{\gamma})$ and $D(\underline{\nu})$. We can further see that

$$\begin{cases} A_j = V^{-1}(\underline{\gamma},\underline{\nu})a_jV(\underline{\gamma},\underline{\nu}), \\ A_j^\dagger = V^{-1}(\underline{\gamma},\underline{\nu})b_jV(\underline{\gamma},\underline{\nu}), \end{cases} \tag{3.3.24}$$

which only involve a single (unbounded) operator in both equations, rather than the two (bounded) operators used in (3.3.22). In addition, a direct computation also shows that

$$\begin{cases} A_j = V^{-1}(\underline{\nu}, \underline{\gamma}) b_j^\dagger V(\underline{\nu}, \underline{\gamma}), \\ A_j^\dagger = V^{-1}(\underline{\nu}, \underline{\gamma}) a_j^\dagger V(\underline{\nu}, \underline{\gamma}). \end{cases} \tag{3.3.25}$$

A consequence of these formulas is the following relation between the various number operators: $\hat{n}_j = V^{-1}(\underline{\gamma}, \underline{\nu}) N_j V(\underline{\gamma}, \underline{\nu}) = V^{-1}(\underline{\nu}, \underline{\gamma}) N_j^\dagger V(\underline{\nu}, \underline{\gamma})$, which in turns implies that

$$N_j = T(\underline{\gamma}, \underline{\nu}) N_j^\dagger T^{-1}(\underline{\gamma}, \underline{\nu}), \tag{3.3.26}$$

where $T(\underline{\gamma}, \underline{\nu}) = V(\underline{\gamma}, \underline{\nu}) V^{-1}(\underline{\nu}, \underline{\gamma})$. Needless to say, all these formulas are well defined on $\mathcal{D}$, but not on the whole $\mathcal{H}$. Incidentally, we also observe that $T(\underline{\gamma}, \underline{\gamma}) = \mathbb{1}$.

As in Ref. (26), we can prove that $\mathcal{F}_\varphi$ and $\mathcal{F}_\Psi$ are $\mathcal{D}$-quasi bases. In fact, repeating almost the same steps, we deduce that for instance, $\forall f, g \in \mathcal{D}$,

$$\langle f, g \rangle = \sum_{\underline{n}} \left\langle f, \varphi_{\underline{n}} \right\rangle \left\langle \Psi_{\underline{n}}, g \right\rangle = \sum_{\underline{n}} \left\langle f, \Psi_{\underline{n}} \right\rangle \left\langle \varphi_{\underline{n}}, g \right\rangle.$$

We now introduce the operator $\Theta(\underline{\gamma}, \underline{\nu}) = N_\Psi T(\underline{\nu}, \underline{\gamma})$. It is possible to show that $\Theta(\underline{\gamma}, \underline{\nu})$ is self-adjoint, invertible, and leaves $\mathcal{D}$ invariant. Moreover, $\Theta(\underline{\gamma}, \underline{\gamma}) = 1$, and

$$\Theta(\underline{\gamma}, \underline{\nu}) = \prod_{j=1}^{2} e^{|\gamma_j|^2 - |\nu_j|^2} e^{A_j(\overline{\gamma}_j - \overline{\nu}_j)} e^{A_j^\dagger(\gamma_j - \nu_j)},$$

which implies also that $< f, \Theta(\underline{\gamma}, \underline{\nu}) f >> 0$ for all nonzero vectors $f \in \mathcal{D}$. This is in agreement with the facts that (i) $\Psi_{\underline{n}} = \Theta(\underline{\gamma}, \underline{\nu}) \varphi_{\underline{n}}, \forall \underline{n}$, and (ii) $(a_j, b_j^\dagger)$ are Θ-conjugate: $a_j f = \Theta^{-1}(\underline{\gamma}, \underline{\nu}) b_j^\dagger \Theta(\underline{\gamma}, \underline{\nu}) f$, for all $f \in \mathcal{D}$. We conclude also that, again for all $f \in \mathcal{D}$,

$$N_j f = \Theta^{-1}(\underline{\gamma}, \underline{\nu}) N_j^\dagger \Theta(\underline{\gamma}, \underline{\nu}) f.$$

Finally, the stability of $\mathcal{D}$ under the action of the square root of $\Theta(\underline{\gamma}, \underline{\nu})$ and of its inverse is also guaranteed such that Theorem 3.2.5 applies. Then, also for this model, we recover exactly the same general results we deduced at the end of Section 3.2.

3.3.2.2 Back to the Hamiltonian We can now return to our original problem, that is, that of deducing the eigenvalues and the eigenvectors for $\hat{H}$ in (3.3.12). As we have shown, this can be rewritten as in (3.3.20), where γ, k_1, and k_2 are real quantities. As for k_0, it turns out that k_0 is real as well. In fact, because of the explicit forms of α_j and β_j, it is possible to see that both $\alpha_1\beta_1$ and $\alpha_2\beta_2$ are real, so that, see (3.3.19), k_0 is also real.

It is clear that

$$\hat{H} \varphi_{\underline{n}} = \epsilon_{\underline{n}} \varphi_{\underline{n}}, \tag{3.3.27}$$

where we have introduced $\epsilon_{\underline{n}} = \gamma^2(k_1n_1 + k_2n_2 + k_0)$, which is real for all $\underline{n} = (n_1, n_2)$. Our results in Section 3.2 show that, introducing $H^\dagger = \gamma^2 \left(k_1N_1^\dagger + k_2N_2^\dagger + k_0 1\!\!1 \right)$, the eigensystem of this operator can be easily deduced. In fact, we get

$$\hat{H}^\dagger \Psi_{\underline{n}} = \epsilon_{\underline{n}} \Psi_{\underline{n}}. \tag{3.3.28}$$

The conclusion is identical to that deduced for the shifted quantum harmonic oscillator: two biorthogonal sets of eigenstates of an hamiltonian and of its adjoint, need not to be automatically (Riesz) bases, even when they are complete! This is exactly the case here: $\mathcal{F}_\varphi$ and $\mathcal{F}_\Psi$ are biorthogonal, complete, eigenstates of $\hat{H}$ and $\hat{H}^\dagger$, respectively, but neither $\mathcal{F}_\varphi$ nor $\mathcal{F}_\Psi$ are bases for $\mathcal{H}$. However, interestingly enough, they are still $\mathcal{D}$-quasi bases, and this produces good consequences and the properties we have explicitly checked for our model.

Related to this aspect, we also deduce that the metric operator and its inverse are unbounded. This could suggest that *well, then the model is simply unphysical.* However, our analysis proves that a rigorous treatment of the system is possible, with results (reality of the eigenvalues, eigenvectors producing a weak resolution of the identity, existence of a stable dense domain where all the relevant operators are well defined, and so on) which suggest that, also in *extreme* situations like these, some interesting physics can be extracted out of some strange models.

3.4 OTHER APPEARANCES OF $\mathcal{D}$-PBs IN QUANTUM MECHANICS

Along the years, $\mathcal{D}$-PBs appeared also in other physical models. In this section, we briefly consider other examples where this was observed recently.

3.4.1 The Extended Quantum Harmonic Oscillator

The first example we want to consider was introduced, in a pseudo-bosonic context, in Ref. (36). The hamiltonian of this model, considered first in Ref. (37), is the manifestly non-self-adjoint operator $H_\nu = \frac{\nu}{2}\left(p^2 + x^2\right) + i\sqrt{2}p$, where ν is a strictly positive parameter and $[x, p] = i1\!\!1$.

Introducing the standard bosonic operators $a = \frac{1}{\sqrt{2}}\left(x + \frac{d}{dx}\right)$, $a^\dagger = \frac{1}{\sqrt{2}}\left(x - \frac{d}{dx}\right)$, $[a, a^\dagger] = 1\!\!1$, and the related operators $A_\nu = a - \frac{1}{\nu}$, and $B_\nu = a^\dagger + \frac{1}{\nu}$, we can write $H_\nu = \nu\left(B_\nu A_\nu + \gamma_\nu 1\!\!1\right)$, where $\gamma_\nu = \frac{2+\nu^2}{2\nu^2}$. It is clear that, for all $\nu > 0$, $A_\nu^\dagger \neq B_\nu$ and $[A_\nu, B_\nu] = 1\!\!1$. Hence, we have to do, at least formally, with pseudo-bosonic operators.

This model can be seen as a particular case of the shifted harmonic oscillator considered in Section 3.3.1, where we considered, see formula (3.3.1), the shifted operators $A = a + \alpha$ and $B = a^\dagger + \overline{\beta}$. Hence, all the results deduced there can be applied here, just identifying α with $-\frac{1}{\nu}$ and $\overline{\beta}$ with $\frac{1}{\nu}$. Notice that, with these choices, we surely have $\alpha \neq \beta$: we are not recovering orthonormal basis, then.

In particular, we find that the sets $\mathcal{F}_\varphi = \{\varphi_n^{(\nu)}(x)\}$ and $\mathcal{F}_\Psi = \{\Psi_n^{(\nu)}(x)\}$ of eigenstates of H_ν and $H_\nu^\dagger$ are not biorthogonal bases, but still they are $\mathcal{D}$-quasi bases. In Ref. (36), we have deduced, among other results, the following expressions for these eigenvectors

$$\varphi_n^{(\nu)}(x) = \frac{1}{\sqrt{n!}} B_\nu^n \varphi_0^{(\nu)}(x) = \frac{e^{-1/\nu^2}}{\pi^{1/4}\sqrt{2^n n!}}\left(x - \frac{d}{dx} + \frac{\sqrt{2}}{\nu}\right)^n e^{-\frac{1}{2}(x-\sqrt{2}/\nu)^2},$$

and

$$\Psi_n^{(\nu)}(x) = \frac{1}{\sqrt{n!}} A_\nu^{\dagger n} \Psi_0^{(\nu)}(x) = \frac{e^{1/\nu^2}}{\pi^{1/4}\sqrt{2^n n!}}\left(x - \frac{d}{dx} - \frac{\sqrt{2}}{\nu}\right)^n e^{-\frac{1}{2}(x+\sqrt{2}/\nu)^2}.$$

Notice that the normalization chosen here is slightly different from the one adopted in Ref. (36). This is useful to unify the notation of this Chapter. They both correspond to the same eigenvalue, $\nu(n+\gamma_\nu)$, for H_ν and $H_\nu^\dagger$, respectively. The operator Θ of Section 3.2 can be identified using (3.3.11):

$$\Theta_\nu := \Theta\left(-\frac{1}{\nu}, \frac{1}{\nu}\right) = e^{2/\nu^2} e^{-\frac{2}{\nu}(a+a^\dagger)}.$$

Obviously we have, adapting the results of Section 3.3.1 to this particular situation,

$$\Psi_n^{(\nu)}(x) = \Theta_\nu \varphi_n^{(\nu)}(x), \qquad A_\nu f = \Theta_\nu^{-1} B_\nu^\dagger \Theta_\nu f,$$

for all $f \in \mathcal{D}$, where $\mathcal{D}$ is the set defined as in (3.3.8). In addition,

$$\langle f, \Theta_\nu f\rangle \geq 0, \text{ and } N_\nu f = \Theta_\nu^{-1} N_\nu^\dagger \Theta_\nu f,$$

for all $f \in \mathcal{D}$. The conclusion is that, not surprisingly, the model fits perfectly the general scheme of Section 3.2.

3.4.2 The Swanson Model

The starting point is the non-self-adjoint hamiltonian,

$$H_\theta = \frac{1}{2}\left(p^2 + x^2\right) - \frac{i}{2}\tan(2\theta)\left(p^2 - x^2\right),$$

where θ is a real parameter taking value in $\left(-\frac{\pi}{4}, \frac{\pi}{4}\right) \setminus \{0\} =: I$, (37). As before, $[x,p] = i1\!\!1$. Of course, $\theta = 0$ is excluded from I just to avoid going back to the standard, self-adjoint, harmonic oscillator, and not for any other deeper reason. Notice also that H_θ can be rewritten as

$$H_\theta = \frac{1}{2\cos(2\theta)}\left(p^2 e^{-2i\theta} + x^2 e^{2i\theta}\right) = \frac{e^{-2i\theta}}{2\cos(2\theta)}\left(p^2 + x^2 e^{4i\theta}\right),$$

which has, except for an unessential overall complex constant, the same form considered in (30), $H = -\frac{d^2}{dx^2} + z^4x^2$, $z \in \mathbb{C}$, taking $z = e^{i\theta}$.

Introducing now the annihilation and creation operators $a, a^\dagger$, and their linear combinations

$$\begin{cases} A_\theta = \cos(\theta)\, a + i\sin(\theta)\, a^\dagger = \frac{1}{\sqrt{2}}\left(e^{i\theta}x + e^{-i\theta}\frac{d}{dx}\right), \\ B_\theta = \cos(\theta)\, a^\dagger + i\sin(\theta)\, a = \frac{1}{\sqrt{2}}\left(e^{i\theta}x - e^{-i\theta}\frac{d}{dx}\right), \end{cases}$$

we can write $H_\theta = \omega_\theta\left(B_\theta A_\theta + \frac{1}{2}\mathbb{1}\right)$, where $\omega_\theta = \frac{1}{\cos(2\theta)}$ is well defined because $\cos(2\theta) \neq 0$ for all $\theta \in I$. It is clear that, for θ in this set, $A_\theta^\dagger \neq B_\theta$ and $[A_\theta, B_\theta] = \mathbb{1}$. The two vacua of A_θ and $B_\theta^\dagger$ are $\varphi_0^{(\theta)}(x) = N_1 \exp\left\{-\frac{1}{2}e^{2i\theta}x^2\right\}$, and $\Psi_0^{(\theta)}(x) = N_2 \exp\left\{-\frac{1}{2}e^{-2i\theta}x^2\right\}$, where N_1 and N_2 are suitable normalization constants, which, because of $\left\langle \varphi_0^{(\theta)}, \Psi_0^{(\theta)}\right\rangle = 1$, must satisfy the condition $\overline{N_1}\, N_2 = \frac{e^{-i\theta}}{\sqrt{\pi}}$.

Notice that, as $\Re(e^{\pm 2i\theta}) = \cos(2\theta) > 0$ for all $\theta \in I$, both $\varphi_0^{(\theta)}(x)$ and $\Psi_0^{(\theta)}(x)$ belong to $\mathcal{L}^2(\mathbb{R})$. The functions of the sets $\mathcal{F}_\varphi^{(\theta)}$ and $\mathcal{F}_\Psi^{(\theta)}$ have been found in Ref. (36):

$$\begin{cases} \varphi_n^{(\theta)}(x) = \frac{N_1}{\sqrt{2^n\, n!}} H_n\left(e^{i\theta}x\right) \exp\left\{-\frac{1}{2}e^{2i\theta}x^2\right\}, \\ \Psi_n^{(\theta)}(x) = \frac{N_2}{\sqrt{2^n\, n!}} H_n\left(e^{-i\theta}x\right) \exp\left\{-\frac{1}{2}e^{-2i\theta}x^2\right\}, \end{cases}$$

where $H_n(x)$ is the n-th Hermite polynomial. We see from these formulas that, for all $n \geq 0$, $\frac{1}{N_1}\varphi_n^{(\theta)}(x) = \frac{1}{N_2}\Psi_n^{(-\theta)}(x)$. Moreover, they all belong to $\mathcal{L}^2(\mathbb{R})$, which is a clear indication that $\varphi_0^{(\theta)}(x) \in D^\infty(B_\theta)$ and $\Psi_0^{(\theta)}(x) \in D^\infty(A_\theta^\dagger)$. Before identifying the set $\mathcal{D}$, we try to see whether $\mathcal{F}_\varphi = \{\varphi_n^{(\theta)}(x)\}$ and $\mathcal{F}_\Psi = \{\Psi_n^{(\theta)}(x)\}$ are (Riesz) bases or not. For that we use the following result, (36):

$$\left\|\varphi_n^{(\theta)}\right\|^2 = |N_1|^2 \sqrt{\frac{\pi}{\cos(2\theta)}}\, P_n\left(\frac{1}{\cos(2\theta)}\right),$$

where $P_n(x)$ is the n-th Legendre Polynomial. As for $\left\|\Psi_n^{(\theta)}\right\|$, we have

$$\left\|\Psi_n^{(\theta)}\right\| = \left|\frac{N_2}{N_1}\right| \left\|\varphi_n^{(-\theta)}\right\| = \left|\frac{N_2}{N_1}\right| \left\|\varphi_n^{(\theta)}\right\|.$$

In particular, if $\theta = 0$, we get

$$\frac{\left\|\varphi_n^{(0)}\right\|^2}{|N_1|^2} = \frac{\left\|\Psi_n^{(0)}\right\|^2}{|N_2|^2} = \sqrt{\pi}\, P_n(1) = \sqrt{\pi},$$

which is in agreement with the well-known normalization of the eigenfunctions of the quantum harmonic oscillator. In fact, in this case,

$$\varphi_n^{(0)}(x) = \Psi_n^{(0)}(x) = e_n(x) = \frac{1}{\sqrt{2^n\, n!\, \sqrt{\pi}}} H_n(x) e^{-x^2/2}.$$

Let now take $\theta \in I$. Then $(\cos(2\theta))^{-1} > 1$ and, see Ref. (38), $P_n\left(\frac{1}{\cos(2\theta)}\right) \to \infty$ when $n \to \infty$. Therefore, both $\left\|\varphi_n^{(\theta)}\right\|$ and $\left\|\Psi_n^{(\theta)}\right\|$ diverge with n. Then, using the same arguments as in Section 3.3.1, we conclude that $\mathcal{F}_\varphi = \{\varphi_n^{(\theta)}(x)\}$ and $\mathcal{F}_\Psi = \{\Psi_n^{(\theta)}(x)\}$ are not Riesz bases, and not even bases. Now, in order to prove that they are $\mathcal{G}$-quasi bases, for a suitable $\mathcal{G}$, we first recall that, as $\left\langle \varphi_0^{(\theta)}, \Psi_0^{(\theta)} \right\rangle = 1$, we automatically have $\left\langle \varphi_n^{(\theta)}, \Psi_m^{(\theta)} \right\rangle = \delta_{n,m}$, for all $n, m \geq 0$.

We observe that we have, in general, to identify two sets, $\mathcal{D}$ and $\mathcal{G}$, which in the examples discussed before turned out to coincide. Here, on the contrary, we will need to consider two different dense subspaces of $\mathcal{L}^2(\mathbb{R})$. We now show why.

First, in order to identify $\mathcal{D}$, we start noticing that, for instance, $\varphi_0^{(\theta)}(x)$, $\Psi_0^{(\theta)}(x) \in \mathcal{S}(\mathbb{R})$, which is also stable under the action of $A_\theta^\sharp$ and $B_\theta^\sharp$. For this reason, taking $\mathcal{D} = \mathcal{S}(\mathbb{R})$ is a natural choice. However, it is not difficult to show that $\mathcal{F}_\varphi$ and $\mathcal{F}_\Psi$ are not necessarily also $\mathcal{D}$-quasi bases.

To show this, let us introduce the following self-adjoint, invertible operator

$$T_\theta = e^{i\frac{\theta}{2}\left(a^2 - a^{\dagger 2}\right)} = e^{i\frac{\theta}{2}\left(x\frac{d}{dx} + \frac{d}{dx}x\right)}.$$

For sufficiently regular functions we have that $(T_\theta f)(x) = e^{i\frac{\theta}{2}} f(e^{i\theta}x)$. It is clear that $(T_\theta^{-1} f)(x) = e^{-i\frac{\theta}{2}} f(e^{-i\theta}x)$: $T_\theta^{-1} = T_{-\theta}$. It is further clear that $\mathcal{S}(\mathbb{R})$ is not stable under the action of T_θ. Even more: not any function of $\mathcal{S}(\mathbb{R})$ belongs to the domain of T_θ. In fact, let $f(x) = e^{-x^4}$. This is clearly in $\mathcal{S}(\mathbb{R})$. However, $f \notin D(T_\theta)$, as $(T_\theta f)(x) = e^{-e^{4i\theta}x^4}$, which is not square integrable for certain values of $\theta \in I$, and in particular for those values for which we have $\cos(4\theta) < 0$. Despite these difficulties, T_θ and its inverse play a very important role in our analysis. In fact, let $\mathcal{L}_e$ be the linear span of the functions $e_n(x)$ introduced above. As $\mathcal{F}_e = \{e_n(x)\}$ is an orthonormal basis for $\mathcal{H} = \mathcal{L}^2(\mathbb{R})$, $\mathcal{L}_e$ is dense in $\mathcal{H}$. Each function in $\mathcal{L}_e$ belongs to the domain of T_θ and T_θ^{-1}, and in particular we have

$$\begin{cases} \varphi_n^{(\theta)}(x) = N_1\, \pi^{1/4} e^{-i\frac{\theta}{2}} \left(T_\theta\, e_n\right)(x), \\ \Psi_n^{(\theta)}(x) = N_2\, \pi^{1/4} e^{i\frac{\theta}{2}} \left(T_\theta^{-1} e_n\right)(x). \end{cases} \tag{3.4.1}$$

Hence T_θ (essentially) maps $\mathcal{F}_e$ into $\mathcal{F}_\varphi$, while T_θ^{-1} maps $\mathcal{F}_e$ into $\mathcal{F}_\Psi$.

Now, let us take $f, g \in \mathcal{L}_e$. Then, as $\mathcal{F}_e$ is an orthonormal basis,

$$\langle f, g\rangle = \left\langle T_\theta f, T_\theta^{-1} g\right\rangle = \sum_n \langle T_\theta f, e_n\rangle \left\langle e_n, T_\theta^{-1} g\right\rangle = \sum_n \langle f, T_\theta e_n\rangle \left\langle T_\theta^{-1} e_n, g\right\rangle$$
$$= \frac{e^{i\theta}}{\sqrt{\pi} N_1 \overline{N_2}} \sum_n \left\langle f, \varphi_n^{(\theta)}\right\rangle \left\langle \Psi_n^{(\theta)}, g\right\rangle = \sum_n \left\langle f, \varphi_n^{(\theta)}\right\rangle \left\langle \Psi_n^{(\theta)}, g\right\rangle,$$

where we have used equation (3.4.1) and the equality $\overline{N_1}\, N_2 = \frac{e^{-i\theta}}{\sqrt{\pi}}$. Then we conclude that $\mathcal{F}_\varphi$ and $\mathcal{F}_\Psi$ are $\mathcal{L}_e$-quasi bases. They are also complete, (36). Notice that, since for all $\theta \in I$, $\varphi_0^{(\theta)}(x)$, $\Psi_0^{(\theta)}(x) \notin \mathcal{L}_e$, while they both belong to $\mathcal{S}(\mathbb{R})$ $\mathcal{L}_e$ could not be identified with the set $\mathcal{D}$ for this model. So, as already stated, the two spaces $\mathcal{D}$ and $\mathcal{G}$ introduced in Section 3.2 are not the same, for the Swanson model[9]. Notice also that, while it is essential in our framework that $\mathcal{D}$ is stable under the action of the pseudo-bosonic operators and of their adjoint, it could be not so important to have a similar stability of $\mathcal{G}$ under the action of the metric operators Θ and Θ^{-1}, even if we have assumed it in Section 3.2. In fact, even if this stability is lost, we will see now that most of the general results deduced in that section can still be recovered.

First of all, it is now possible to check that both T_θ and T_θ^{-1} are unbounded. The argument is identical to that used in Section 3.3.1: suppose this is not so, and let, for instance, $M < \infty$ be the (finite) value of $\|T_\theta\|$. Then, because of the first equality in (3.4.1), we have $\|\varphi_n^{(\theta)}\| \leq |N_1|\, \pi^{1/4} \|T_\theta\| \|e_n\| = |N_1|\, \pi^{1/4} M$, for all $n \geq 0$, which is not compatible with the fact that, as we have previously deduced, $\|\varphi_n^{(\theta)}\| \to \infty$ with n. Therefore, T_θ cannot be bounded. Analogously, as $\|\Psi_n^{(\theta)}\| \to \infty$ with n, also T_θ^{-1} is unbounded. This, of course, could also be deduced from the fact that neither T_θ nor T_θ^{-1} are everywhere defined in $\mathcal{H}$.

Now, it is easy to see that, taking $\Theta := \frac{1}{\sqrt{\pi}\, |N_1|^2}\, T_\theta^{-2}$, each $\varphi_n^{(\theta)} \in D(\Theta)$, and $\Psi_n^{(\theta)} = \Theta\, \varphi_n^{(\theta)}$, for all $n \geq 0$. For the sake of simplicity we restrict here to $\theta \in I_r := \left]-\frac{\pi}{8}, \frac{\pi}{8}\right[\setminus \{0\}$. In this case, $\mathcal{L}_e \subseteq D(\Theta) \cap D(\Theta^{-1})$, and we deduce that $\langle f, \Theta f\rangle = \frac{1}{\sqrt{\pi}\, |N_1|^2} \|T_\theta^{-1} f\| > 0$ for all nonzero $f \in \mathcal{L}_e$. We also get, with a direct computation, that $A_\theta f = \Theta^{-1} B_\theta^\dagger \Theta f$, for all $f \in \mathcal{L}_e$: $(A_\theta, B_\theta^\dagger)$ are Θ-conjugate. Finally, H_θ is similar, on $\mathcal{L}_e$, to a self-adjoint hamiltonian: $H_\theta f = T_\theta\, h_\theta\, T_\theta^{-1} f$, where $h_\theta = \omega_\theta \left(a^\dagger a + \frac{1}{2}\, 1\!\!1\right)$.

3.4.3 Generalized Landau Levels

We consider now a model, originally introduced in Ref. (39), which extends the well known model for the Landau levels. We refer to Ref. (39) for the details of our derivation. Here we just skip most of the technicalities.

[9] This does not exclude that other different, and maybe more convenient, choices could be done.

The essential idea is that we have a non-self-adjoint hamiltonian acting on $\mathcal{H} = \mathcal{L}^2(\mathbb{R}^2)$, which with a suitable choice of variables, can be written as $h' = B'A' - \frac{1}{2}1\!\!1$, where

$$A' = \alpha' \left(\partial_x - i\partial_y + \frac{x}{2}(1+2k_2) - \frac{iy}{2}(1-2k_1) \right)$$

and

$$B' = \gamma' \left(-\partial_x - i\partial_y + \frac{x}{2}(1-2k_2) + \frac{iy}{2}(1+2k_1) \right),$$

for suitable complex constants α' and γ', and for $k_j \in \left]-\frac{1}{2}, \frac{1}{2}\right[$. This hamiltonian commutes with a second, again non-self-adjoint, operator $h = BA - \frac{1}{2}1\!\!1$, a second hamiltonian, where $A = \alpha\left(-i\partial_x + \partial_y - \frac{ix}{2}(1+2k_2) + \frac{y}{2}(1-2k_1)\right)$ and $B = \gamma\left(-i\partial_x - \partial_y + \frac{ix}{2}(1-2k_2) + \frac{y}{2}(1+2k_1)\right)$, with $\alpha, \gamma \in \mathbb{C}$ chosen properly, (39). We have discussed in (39) in which sense h and h' extend the ordinary two-dimensional hamiltonian of the Landau levels to a non-self-adjoint situation. In particular, we go back to Landau levels simply when $k_1 = k_2 = 0$ and $\alpha = \alpha' = \gamma = \gamma' = \frac{1}{\sqrt{2}}$.

The vacua of A, A' and of $B^\dagger$, $B'^\dagger$ are found to be

$$\varphi_{0,0}(x,y) = N_\varphi\, e^{\left\{-\frac{x^2}{4}(1+2k_2) - \frac{y^2}{4}(1-2k_1)\right\}}$$

and

$$\Psi_{0,0}(x,y) = N_\Psi\, e^{\left\{-\frac{x^2}{4}(1-2k_2) - \frac{y^2}{4}(1+2k_1)\right\}},$$

where N_φ and N_Ψ are normalization constants chosen in such a way that $\langle \varphi_{0,0}, \Psi_{0,0} \rangle = 1$.

In Ref. (39), it is shown that the vectors

$$\varphi_{n,l}(x,y) = \frac{B'^n\, B^l}{\sqrt{n!\, l!}}\, \varphi_{0,0}(x,y), \text{ and } \Psi_{n,l}(x,y) = \frac{(A'^\dagger)^n\, (A^\dagger)^l}{\sqrt{n!\, l!}}\, \Psi_{0,0}(x,y),$$

$n, l = 0, 1, 2, 3, \ldots$, are related to the vectors $e_{n,l}(x,y)$ of an orthonormal basis of $\mathcal{L}^2(\mathbb{R}^2)$, in a reasonably simple way. Here the functions $e_{n,l}(x,y)$ are those deduced by, for instance, $\varphi_{n,l}(x,y)$ by taking $k_1 = k_2 = 0$ and $\alpha = \alpha' = \gamma = \gamma' = \frac{1}{\sqrt{2}}$, which define an orthonormal basis $\mathcal{F}_e = \{e_{n,l},\ n, l \geq 0\}$ of our Hilbert space. In particular, we have shown that $\varphi_{n,l}(x, y) = Te_{n,l}(x, y)$, while $\Psi_{n,l}(x, y) = T^{-1}e_{n,l}(x, y)$, $n, l \geq 0$, with $T = \sqrt{2\pi}\, N_\varphi\, e^{-\frac{x^2}{2}k_2+\frac{y^2}{2}k_1}$, a simple multiplication operator. We now have

$$D(T) = \left\{ f(x, y) \in \mathcal{L}^2(\mathbb{R}^2) : e^{-\frac{x^2}{2}k_2+\frac{y^2}{2}k_1} f(x, y) \in \mathcal{L}^2(\mathbb{R}^2) \right\},$$

and a similar definition can be introduced for $D(T^{-1})$. We stress again that $k_j \in \left]-\frac{1}{2}, \frac{1}{2}\right[$ here. These two sets are dense in $\mathcal{L}^2(\mathbb{R}^2)$, together with their intersection, since they both contain $D(\mathbb{R}^2)$, the set of C^∞ functions in two variables with compact support. Then, Proposition 3.2.5 implies that $\mathcal{F}_\varphi = \{\varphi_{n,l}(x, y),\ n, l \geq 0\}$ and $\mathcal{F}_\Psi = \{\Psi_{n,l}(x, y),\ n, l \geq 0\}$ are $D(T) \cap D(T^{-1})$-quasi bases. They are also complete in $\mathcal{L}^2(\mathbb{R}^2)$, (39). This is not the end of the story. In fact, $\mathcal{F}_\varphi$ and $\mathcal{F}_\Psi$ cannot be Riesz bases for $\mathcal{L}^2(\mathbb{R}^2)$, since T and T^{-1} are both unbounded. In addition, it is possible to argue, by computing the norm of the operator $P_{n,l}$ defined as $P_{n,l}(f) = \left\langle \Psi_{n,l}, f \right\rangle \varphi_{n,l}$, that they cannot even be bases. This claim can be proved easily for particular (non both zero) values of k_1 and k_2, and focusing on the norm of $P_{n,0}$, $\|P_{n,0}\| = \|\varphi_{n,0}\|\,\|\Psi_{n,0}\|$, rather than on the generic $P_{n,l}$. To fix the ideas, we further restrict here to $\alpha = \alpha' = \gamma = \gamma' = \frac{1}{\sqrt{2}}$. With these choices we have, for instance, that $\varphi_{n,0}(x, y) = \frac{1}{\sqrt{2^n n!}}(x + iy)^n \varphi_{0,0}(x, y)$, so that

$$\|\varphi_{n,0}\|^2 \geq |N_\varphi|^2 s^{n+1} \frac{\Gamma\left(\frac{n+1}{2}\right)^2}{\Gamma(n+1)},$$

where $s = \sqrt{\frac{4}{(1+2k_2)(1-2k_1)}}$, which is surely larger than one. A similar estimate can be deduced for $\|\Psi_{n,0}\|^2$, and the asymptotic analysis of $\|P_{n,0}\|$ shows that there exist values of k_1 and k_2 in the interval $\left]-\frac{1}{2}, \frac{1}{2}\right[$, with $k_1 = k_2 = 0$ excluded, such that $\|P_{n,0}\| \to \infty$ when n diverges. It is enough that the inequalities $(1 + 2k_2)(1 - 2k_1) < 1$ and $(1 - 2k_2)(1 + 2k_1) < 1$ are both satisfied. $k_1 = -\frac{1}{4}$ and $k_2 = -\frac{1}{3}$ is one possible solution, but it is not hard to find other pairs of (k_1, k_2) also satisfying these inequalities. We believe that, refining these estimates, the same conclusion can be extended to all the allowed values of k_1 and k_2.

Concerning Assumptions $\mathcal{D}$-pb 1 and 2, a set $\mathcal{D}$ with the required properties can be easily identified: in fact, it is enough to take $\mathcal{D} \equiv \mathcal{S}(\mathbb{R}^2)$. It is clear that both $\varphi_{0,0}(x, y)$ and $\Psi_{0,0}(x, y)$ belong to $\mathcal{S}(\mathbb{R}^2)$, which is dense in $\mathcal{L}^2(\mathbb{R}^2)$ and that $A^\sharp$, $A'^\sharp$, $B^\sharp$ and $B'^\sharp$, all leave this space stable. However, this is not so for T and T^{-1}. As for the Swanson model, to bypass this difficulty, we could look for different choices of $\mathcal{D}$ which, at the same time, are also stable under the action of $T^{\pm 1}$. As we have already seen, the space $D(\mathbb{R}^2)$ appear to be a nice choice, as it is indeed stable under the action of all the operators involved into the game. However, this choice has a major drawback (the same we met in Section 3.4.2): the functions $\varphi_{0,0}(x,y)$ and $\Psi_{0,0}(x,y)$ do not belong to this space. Therefore, also for this model, it seems to be necessary to use two different dense subspaces of $\mathcal{L}^2(\mathbb{R}^2)$, rather than just one as we did in Sections 3.3.1 and 3.3.2: one for the definition of the $\mathcal{D}$-PBs, and the other to check that $\mathcal{F}_\varphi$ and $\mathcal{F}_\Psi$ are quasi bases *somewhere*, that is, for some suitable dense subset of $\mathcal{H}$. In other words: also in this model $\mathcal{D} \neq \mathcal{G}$, because $\mathcal{D} = \mathcal{S}(\mathbb{R}^2)$ while $\mathcal{G} = D(T) \cap D(T^{-1})$.

3.4.4 An Example by Bender and Jones

The next example we want to consider was originally introduced by Carl Bender and Hugh Jones in Ref. (27) and then considered further in Ref. (28). The starting point is the following, manifestly non-self-adjoint, hamiltonian:

$$H = (p_1^2 + x_1^2) + (p_2^2 + x_2^2 + 2ix_2) + 2\epsilon x_1 x_2, \tag{3.4.2}$$

where ϵ is a real constant, with $\epsilon \in]-1,1[$. Here the following commutation rules are assumed: $[x_j, p_k] = i\delta_{j,k}\mathbb{1}$, $\mathbb{1}$ being the identity operator on $\mathcal{L}^2(\mathbb{R}^2)$, $j,k = 1,2$. All the other commutators are zero.

In order to rewrite H in a more convenient form we perform few changes of variables, (28):

1. first of all we introduce the *capital* operators $P_j, X_j, j = 1,2$, via

$$P_1 := \frac{1}{2a}(p_1 + \xi p_2), \quad P_2 := \frac{1}{2b}(p_1 - \xi p_2),$$
$$X_1 := a(x_1 + \xi x_2), \qquad X_2 := b(x_1 - \xi x_2),$$

   where ξ can be ± 1, while a and b are real, nonzero, arbitrary constants. These operators satisfy the same canonical commutation rules as the original ones: $[X_j, P_k] = i\delta_{j,k}\mathbb{1}$.

2. Secondly, we introduce the operators

$$\Pi_1 = P_1, \quad \Pi_2 = P_2, \quad q_1 = X_1 + i\frac{a\xi}{1+\epsilon\,\xi}, \quad q_2 = X_2 - i\frac{b\xi}{1-\epsilon\,\xi}.$$

   The first clear fact is that $\Pi_j^\dagger = \Pi_j$, while $q_j^\dagger \neq q_j, j = 1,2$. However, the commutation rules are preserved: $[q_j, \Pi_k] = i\delta_{j,k}\mathbb{1}$.

3. The third step consists in introducing the following operators:

$$\begin{cases} a_1 = \frac{a}{\sqrt[4]{1+\epsilon\,\xi}}\left(i\Pi_1 + \frac{\sqrt{1+\epsilon\,\xi}}{2a^2}\,q_1\right), \\ a_2 = \frac{a}{\sqrt[4]{1-\epsilon\,\xi}}\left(i\Pi_2 + \frac{\sqrt{1-\epsilon\,\xi}}{2b^2}\,q_2\right), \end{cases} \tag{3.4.3}$$

and

$$\begin{cases} b_1 = \frac{a}{\sqrt[4]{1+\epsilon\,\xi}}\left(-i\Pi_1 + \frac{\sqrt{1+\epsilon\,\xi}}{2a^2}\,q_1\right), \\ b_2 = \frac{a}{\sqrt[4]{1-\epsilon\,\xi}}\left(-i\Pi_2 + \frac{\sqrt{1-\epsilon\,\xi}}{2b^2}\,q_2\right). \end{cases} \tag{3.4.4}$$

It may be worth remarking that $b_j \neq a_j^\dagger$, the reason being that the q_j's are not self-adjoint. These operators satisfy, at least formally, the pseudo-bosonic commutation rules

$$[a_j, b_k] = \delta_{j,k}\,1\!\!1, \tag{3.4.5}$$

the other commutators being zero.

Going back to H, and introducing the operators $N_j := b_j a_j$, we can write

$$\begin{aligned} H &= H_1 + H_2 + \frac{1}{1-\epsilon^2}\,1\!\!1, \\ H_1 &= \sqrt{1+\epsilon\,\xi}(2N_1 + 1\!\!1), \quad H_2 = \sqrt{1-\epsilon\,\xi}(2N_2 + 1\!\!1). \end{aligned} \tag{3.4.6}$$

We are now ready to check if the general framework described in Section 3.2 applies to the present model. In other words, we want to check if Assumptions $\mathcal{D}$-pb 1, $\mathcal{D}$-pb 2 and $\mathcal{D}$-pb 3 (or $\mathcal{D}$-pbw 3) hold true or not here.

For that, the first thing to do is to rewrite the operators a_j and b_j in terms of the original x_j and p_j, appearing in (3.4.2):

$$\begin{cases} a_1 = \frac{1}{2\sqrt[4]{1+\epsilon\,\xi}}\left((ip_1 + \sqrt{1+\epsilon\,\xi}\,x_1) + \xi(ip_2 + \sqrt{1+\epsilon\,\xi}\,x_2) + i\,\frac{\xi}{\sqrt{1+\epsilon\,\xi}}\right), \\ a_2 = \frac{1}{2\sqrt[4]{1-\epsilon\,\xi}}\left((ip_1 + \sqrt{1-\epsilon\,\xi}\,x_1) - \xi(ip_2 + \sqrt{1-\epsilon\,\xi}\,x_2) - i\,\frac{\xi}{\sqrt{1-\epsilon\,\xi}}\right), \\ b_1 = \frac{1}{2\sqrt[4]{1+\epsilon\,\xi}}\left((-ip_1 + \sqrt{1+\epsilon\,\xi}\,x_1) + \xi(-ip_2 + \sqrt{1+\epsilon\,\xi}\,x_2) + i\,\frac{\xi}{\sqrt{1+\epsilon\,\xi}}\right), \\ b_2 = \frac{1}{2\sqrt[4]{1-\epsilon\,\xi}}\left((-ip_1 + \sqrt{1-\epsilon\,\xi}\,x_1) - \xi(-ip_2 + \sqrt{1-\epsilon\,\xi}\,x_2) - i\,\frac{\xi}{\sqrt{1-\epsilon\,\xi}}\right). \end{cases}$$

We now have to find a dense subspace $\mathcal{D}$ of $\mathcal{L}^2(\mathbb{R}^2)$ which is stable under the action of a_j, b_j and their adjoints. Moreover $\mathcal{D}$ must also contain the two vacua of a_j and $b_j^\dagger$, if they exist. Hence, from a practical point of view, it is convenient to look first for a solution of the equations $a_1\varphi_{0,0}(x_1,x_2) = a_2\varphi_{0,0}(x_1,x_2) = 0$ and $b_1^\dagger\Psi_{0,0}(x_1,x_2) = b_2^\dagger\Psi_{0,0}(x_1,x_2) = 0$. Using $p_j = -i\frac{\partial}{\partial x_j}$, these are simple two-dimensional differential equations which can be easily solved, and the results are

$$\begin{cases} \varphi_{0,0}(x_1,x_2) = N\exp\left\{-\frac{1}{2}\alpha_+(x_1^2+x_2^2) - k_-x_1 - k_+x_2 - \xi\alpha_-x_1x_2\right\}, \\ \Psi_{0,0}(x_1,x_2) = N'\exp\left\{-\frac{1}{2}\alpha_+(x_1^2+x_2^2) + k_-x_1 + k_+x_2 - \xi\alpha_-x_1x_2\right\}, \end{cases} \tag{3.4.7}$$

where we have introduced the following constants:

$$\alpha_\pm = \frac{1}{2}\left(\sqrt{1+\epsilon\,\xi} \pm \sqrt{1-\epsilon\,\xi}\right), \quad k_- = \frac{-i\xi\alpha_-}{\sqrt{1-\epsilon^2}}, \quad k_+ = \frac{i\alpha_+}{\sqrt{1-\epsilon^2}}.$$

N and N' in (3.4.7) are normalization constants, (partially) fixed requiring that $\langle \varphi_{0,0}, \Psi_{0,0} \rangle = 1$. This scalar product is finite since, being $\alpha_+ > 0$, $\varphi_{0,0}(x_1, x_2)$, $\Psi_{0,0}(x_1, x_2) \in \mathcal{L}^2(\mathbb{R}^2)$. Of course, there is more than this: both $\varphi_{0,0}(x_1, x_2)$ and $\Psi_{0,0}(x_1, x_2)$ belong to the set $\mathcal{D}$ of Section 3.3.2. This will be our choice also here. In fact, other than having $\varphi_{0,0}(x_1, x_2)$, $\Psi_{0,0}(x_1, x_2) \in \mathcal{D}$, $\mathcal{D}$ is also stable under the action of a_j, b_j and of their adjoints. It remains open the possibility that the set $\mathcal{G}$ in Assumption $\mathcal{D}$-pbw 3 coincides or not with $\mathcal{D}$. We will return to this later.

At this point we can construct, as usual, the functions

$$\begin{cases} \varphi_{n_1,n_2}(x_1, x_2) = \frac{1}{\sqrt{n_1!\,n_2!}}\, b_1^{n_1} b_2^{n_2} \varphi_{0,0}(x_1, x_2), \\ \Psi_{n_1,n_2}(x_1, x_2) = \frac{1}{\sqrt{n_1!\,n_2!}}\, a_1^{\dagger n_1} a_2^{\dagger n_2} \Psi_{0,0}(x_1, x_2), \end{cases}$$

and the related sets $\mathcal{F}_\varphi = \{\varphi_{n_1,n_2}(x_1, x_2), n_j \geq 0\}$, $\mathcal{F}_\Psi = \{\Psi_{n_1,n_2}(x_1, x_2), n_j \geq 0\}$. It is clear that both $\varphi_{n_1,n_2}(x_1, x_2)$ and $\Psi_{n_1,n_2}(x_1, x_2)$ differ from $\varphi_{0,0}(x_1, x_2)$ and $\Psi_{0,0}(x_1, x_2)$ for some polynomial in x_1 and x_2. Hence they are still functions in $\mathcal{D}$, as expected.

Now we have to check whether $\mathcal{F}_\varphi$ and $\mathcal{F}_\Psi$ are bases for $\mathcal{H}$ or not. This is not evident, in principle. What is much easier to check is that these sets are both complete in $\mathcal{H}$, but we know that completeness of a certain nonorthogonal set does not imply that this set is also a basis. Following Ref. (27), we define an unbounded, self-adjoint and invertible operator $T = e^{\frac{1}{1-\epsilon^2}(p_2 - \epsilon p_1)}$. Then, simple computations show that

$$T H T^{-1} = (p_1^2 + x_1^2) + (p_2^2 + x_2^2) + 2\epsilon x_1 x_2 + \frac{1}{1-\epsilon^2} =: h. \tag{3.4.8}$$

It is clear that, contrarily to H, $h = h^\dagger$ (at least formally). This hamiltonian can be diagonalized essentially with the same changes of variables as before. In particular, we can again introduce the capital operators P_j, X_j as before, and then the operators

$$A_1 = \frac{a}{\sqrt[4]{1+\epsilon\,\xi}}\left(i\Pi_1 + \frac{\sqrt{1+\epsilon\,\xi}}{2a^2} X_1\right), \; A_2 = \frac{b}{\sqrt[4]{1-\epsilon\,\xi}}\left(i\Pi_2 + \frac{\sqrt{1-\epsilon\,\xi}}{2b^2} X_2\right),$$

and their adjoints $A_j^\dagger$. These are *true* bosonic operators: $[A_j, A_k^\dagger] = \delta_{j,k}\mathbb{1}$, in terms of which we can write $h = h_1 + h_2 + \frac{1}{1-\epsilon^2}\,\mathbb{1}$, with

$$h_1 = \sqrt{1+\epsilon\,\xi}\,(2\hat{N}_1 + \mathbb{1}), \quad h_2 = \sqrt{1-\epsilon\,\xi}\,(2\hat{N}_2 + \mathbb{1}).$$

Here $\hat{N}_j := A_j^\dagger A_j$ is a bosonic number operator.

Now, if $e_{0,0}$ is the vacuum of A_j, $A_1 e_{0,0} = A_2 e_{0,0} = 0$, we can construct, *more solito*, the set $\mathcal{F}_e := \{e_{n_1,n_2},\ n_j \geq 0\}$, where $e_{n_1,n_2} = \frac{1}{\sqrt{n_1!\,n_2!}} A_1^{\dagger n_1} A_2^{\dagger n_2} e_{0,0}$. $\mathcal{F}_e$ is an orthonormal basis for $\mathcal{H}$, and the $e_{n_1,n_2}(x_1, x_2)$ can be factorized as follows:

$$e_{n_1,n_2}(x_1, x_2) = e_{n_1}(x_1) e_{n_2}(x_2),$$

where $e_n(x)$ is the usual eigenstate of a one-dimensional harmonic oscillator: $e_n(x) = \frac{1}{\sqrt{2^n\, n!\sqrt{\pi}}} H_n(x) e^{-\frac{1}{2}x^2}$. Each $e_{n_1,n_2}(x_1, x_2)$ belongs to $\mathcal{D}$, which is left invariant also by T and T^{-1} (and by their square roots). In particular we can check that $Te_{n_1,n_2}(x_1, x_2) = e_{n_1,n_2}(x_1 + \delta_1, x_2 + \delta_2)$, where $\delta_1 = \frac{i}{1-\epsilon^2}(a\epsilon - \xi)$ and $\delta_2 = \frac{i}{1-\epsilon^2}(b\epsilon + \xi)$. Needless to say, $T^{-1} e_{n_1,n_2}(x_1, x_2) = e_{n_1,n_2}(x_1 - \delta_1, x_2 - \delta_2)$.

Let us now show that, for all n_1 and n_2, $Te_{n_1,n_2}(x_1, x_2) = \Psi_{n_1,n_2}(x_1, x_2)$ and $T^{-1} e_{n_1,n_2}(x_1, x_2) = \varphi_{n_1,n_2}(x_1, x_2)$. For that it is convenient to recall that the following equations must all be satisfied: $h\, e_{n_1,n_2} = E_{n_1,n_2}\, e_{n_1,n_2}$, $H\, \varphi_{n_1,n_2} = E_{n_1,n_2}\, \varphi_{n_1,n_2}$, $THT^{-1} = h$, and that

$$E_{n_1,n_2} = \sqrt{1 + \epsilon\, \xi}(2n_1 + 1) + \sqrt{1 - \epsilon\, \xi}(2n_2 + 1) + \frac{1}{1 - \epsilon^2}.$$

If $\epsilon \neq 0$, each E_{n_1,n_2} is not degenerate. It is convenient here to work in this assumption, even because, if $\epsilon = 0$, the original hamiltonian H simplifies a lot and becomes less interesting for us. As $e_{n_1,n_2} \in D(T^{-1})$, equation $he_{n_1,n_2} = E_{n_1,n_2} e_{n_1,n_2}$ can be rewritten as: $H(T^{-1} e_{n_1,n_2}) = E_{n_1,n_2}(T^{-1} e_{n_1,n_2})$. Therefore, $T^{-1} e_{n_1,n_2}$ must be proportional to φ_{n_1,n_2}. For similar reasons, we can check that Te_{n_1,n_2} must be proportional to Ψ_{n_1,n_2}, since $H^\dagger \Psi_{n_1,n_2} = E_{n_1,n_2} \Psi_{n_1,n_2}$. These proportionality constants can be taken all equal to one. We are in the conditions of Proposition 3.2.5; therefore we can easily conclude that $\mathcal{F}_\varphi$ and $\mathcal{F}_\Psi$ are both $\mathcal{D}$-quasi bases for $\mathcal{H}$. This means that Assumption $\mathcal{D}$-pbw 3 is also satisfied, with $\mathcal{G} \equiv \mathcal{D}$. In principle, we could hope to have something more. For example, we could ask ourselves whether $\mathcal{F}_\varphi$ and $\mathcal{F}_\Psi$ are also Riesz bases. The answer is negative, since T and T^{-1} are unbounded. It remains open, in principle, the possibility that they are bases for $\mathcal{H}$. However, again, this is not so. Indeed we have, first of all,

$$\Psi_{n_1,n_2}(x_1, x_2) = e_{n_1,n_2}(x_1 + i\hat{\delta}_1, x_2 + i\hat{\delta}_2) = \overline{\varphi_{n_1,n_2}(x_1, x_2)},$$

where we have introduced the real quantities $\hat{\delta}_1 = \frac{a\epsilon - \xi}{1-\epsilon^2}$ and $\hat{\delta}_2 = \frac{b\epsilon + \xi}{1-\epsilon^2}$. Therefore, $\|\Psi_{n_1,n_2}\| = \|\varphi_{n_1,n_2}\|$. Moreover

$$\|\Psi_{n_1,n_2}\|^2 = \int_{\mathbb{R}} |e_{n_1}(x_1 + i\hat{\delta}_1)|^2\, dx_1 \int_{\mathbb{R}} |e_{n_2}(x_2 + i\hat{\delta}_2)|^2\, dx_2,$$

and we have, for instance,

$$\int_{\mathbb{R}} |e_{n_1}(x_1 + i\hat{\delta}_1)|^2\, dx_1 = \frac{e^{\hat{\delta}_1^2}}{2^{n_1}\, n_1! \sqrt{\pi}} \int_{\mathbb{R}} H_{n_1}(x_1 - i\hat{\delta}_1) H_{n_1}(x_1 + i\hat{\delta}_1)\, e^{-x_1^2}\, dx_1.$$

This integral can be computed by means of the following result,

$$\int_{\mathbb{R}} e^{-x^2} H_m(x+y)H_n(x+z)\,dx = 2^n\sqrt{\pi}\,m!\,z^{n-m}L_m^{n-m}(-2yz),$$

which holds for all $m \leq n$, (41), Formula 7.377. Here L_m^k is a Laguerre polynomial. Hence we deduce that

$$\|\Psi_{n_1,n_2}\|^2 = e^{\hat{\delta}_1^2+\hat{\delta}_2^2}L_{n_1}(-2\hat{\delta}_1^2)L_{n_2}(-2\hat{\delta}_2^2),$$

where $L_n(x) = L_n^0(x)$. The asymptotic behavior of the Laguerre polynomials for negative values of their argument can be found in Ref. (38), Theorem 8.22.3, and the conclusion is that, for large values of n_1 and n_2, and for nonzero $\hat{\delta}_j$,

$$\|\Psi_{n_1,n_2}\|^2 \simeq \frac{1}{4\pi(4n_1n_2\hat{\delta}_1^2\hat{\delta}_2^2)^{1/4}}\,e^{\sqrt{8(n_1\hat{\delta}_1^2+n_2\hat{\delta}_2^2)}},$$

which is clearly divergent for n_j diverging. Therefore, calling as usual P_{n_1,n_2} the operator defined as $P_{n_1,n_2}(f) = \left\langle \Psi_{n_1,n_2}, f\right\rangle \varphi_{n_1,n_2}$, we conclude that $\|P_{n_1,n_2}\| = \|\varphi_{n_1,n_2}\|\,\|\Psi_{n_1,n_2}\| \to \infty$, for $n_j \to \infty$: $\mathcal{F}_\varphi$ and $\mathcal{F}_\Psi$ cannot be bases for $\mathcal{L}^2(\mathbb{R}^2)$.

Let us now take $\Theta := T^2$. It is clear that Θ^{-1} exists and that, together with Θ, leaves $\mathcal{D}$ invariant. Moreover $\Psi_{n_1,n_2} = \Theta\,\varphi_{n_1,n_2}$ so that, as discussed in Section 3.2, $(a_j, b_j^\dagger)$ turn out to be Θ-conjugate and Θ is positive. The intertwining relation $N_j f = \Theta^{-1}N_j^\dagger\Theta f$, $f \in \mathcal{D}$, can also be established. The conclusion is therefore that also this model fits into the general structure described in Section 3.2.

3.4.5 A Perturbed Harmonic Oscillator in $d = 2$

In this section we consider a different quantum mechanical model, originally introduced, in our knowledge, in Ref. (29). The starting point is the following manifestly non-self-adjoint hamiltonian,

$$H = \frac{1}{2}(p_1^2 + x_1^2) + \frac{1}{2}(p_2^2 + x_2^2) + i\left[A(x_1 + x_2) + B(p_1 + p_2)\right], \tag{3.4.9}$$

where A and B are real constants, while x_j and p_j are the self-adjoint position and momentum operators, satisfying $[x_j, p_k] = i\delta_{j,k}\mathbb{1}$. Notice that this is a commutative version (i.e., with $[x_1, x_2] = 0$) of the hamiltonian considered in Section 3.3.2.

Let us introduce the shifted operators

$$P_1 = p_1 + iB,\quad P_2 = p_2 + iB,\quad X_1 = x_1 + iA,\quad X_2 = x_2 + iA,$$

and then

$$a_j = \frac{1}{\sqrt{2}}(X_j + iP_j),\qquad b_j = \frac{1}{\sqrt{2}}(X_j - iP_j), \tag{3.4.10}$$

$j = 1, 2$. It is easy to check that $[X_j, P_k] = i\delta_{j,k}1\!\!1$, $[a_j, b_k] = \delta_{j,k}1\!\!1$, and that, since (if $A \neq 0$ or $B \neq 0$) $X_j^\dagger \neq X_j$ and $P_j^\dagger \neq P_j$, $b_j \neq a_j^\dagger$. Introducing further $N_j = b_j a_j$ we can rewrite H as follows: $H = N_1 + N_2 + (A^2 + B^2 + 1)1\!\!1$.

The eigenstates of H and $H^\dagger$ can now be easily constructed if assumptions $\mathcal{D}$-pb 1 and $\mathcal{D}$-pb 2 are satisfied. If assumption $\mathcal{D}$-pb 3 is also satisfied, then the sets of these eigenstates are biorthogonal bases for $\mathcal{H} = \mathcal{L}^2(\mathbb{R}^2)$. If, rather than this, assumption $\mathcal{D}$-pbw 3 holds true, then they are $\mathcal{G}$-quasi bases, for some $\mathcal{G}$ to be identified.

Rewriting

$$a_j = \frac{1}{\sqrt{2}}(x_j + ip_j + C), \qquad b_j = \frac{1}{\sqrt{2}}(x_j - ip_j + D),$$

$j = 1, 2$, where $C = iA - B$ and $D = iA + B$, the two vacua of a_j and $b_j^\dagger$ are found to be

$$\varphi_{0,0}(x_1, x_2) = Ne^{-\frac{1}{2}(x_1^2+x_2^2)-C(x_1+x_2)}, \;\; \Psi_{0,0}(x_1, x_2) = N'e^{-\frac{1}{2}(x_1^2+x_2^2)-\overline{D}(x_1+x_2)},$$

where N and N' are normalization constant chosen in such a way $\left\langle \varphi_{0,0}, \Psi_{0,0} \right\rangle = 1$. Also for this example we observe that both $\varphi_{0,0}(x_1, x_2)$ and $\Psi_{0,0}(x_1, x_2)$ belong to $\mathcal{S}(\mathbb{R}^2)$, which we could take as the space $\mathcal{D}$ for our $\mathcal{D}$-PBs. However, to guarantee the stability of $\mathcal{D}$ also under the action of other useful operators (T and T^{-1} below), it is more convenient to define $\mathcal{D}$ as in Section 3.3.2:

$$\mathcal{D} = \left\{ f(x_1, f_2) \in \mathcal{S}(\mathbb{R}^2), \;\text{ such that } e^{k_1x_1+k_2x_2}f(x_1, x_2) \in \mathcal{S}(\mathbb{R}^2), \;\forall k_1, k_2 \in \mathbb{C} \right\}.$$

Of course, $\varphi_{0,0}, \Psi_{0,0} \in \mathcal{D}$. It is easy to see that $\varphi_{n_1,n_2}(x_1, x_2)$ can be factorized. In fact we have

$$\varphi_{n_1,n_2}(x_1, x_2) = \frac{N}{\sqrt{n_1!n_2!2^{n_1+n_2}}} \left[\left(x_1 - \frac{\partial}{\partial x_1} + D \right)^{n_1} e^{-\frac{1}{2}x_1^2 - Cx_1} \right] \times$$
$$\times \left[\left(x_2 - \frac{\partial}{\partial x_2} + D \right)^{n_2} e^{-\frac{1}{2}x_2^2 - Cx_2} \right],$$

while $\Psi_{n_1,n_2}(x_1, x_2)$ can be deduced from $\varphi_{n_1,n_2}(x_1, x_2)$ simply replacing C with $\overline{D}$ and vice versa everywhere. Incidentally we observe that, as expected, $\varphi_{n_1,n_2}(x_1, x_2)$ and $\Psi_{n_1,n_2}(x_1, x_2)$ are all in $\mathcal{D}$.

We next prove that both

$$\mathcal{F}_\varphi = \{\varphi_{n_1,n_2}(x_1, x_2), n_j \geq 0\}, \quad \mathcal{F}_\Psi = \{\Psi_{n_1,n_2}(x_1, x_2), n_j \geq 0\}$$

are $\mathcal{D}$-quasi bases for $\mathcal{H}$. Then, for this model, $\mathcal{G}$ can be taken coincident with $\mathcal{D}$ itself. Again, we will show that Proposition 3.2.5 is useful for this task. In fact, let us introduce the following unbounded, self-adjoint, invertible operator T:

$$T = e^{-A(p_1+p_2)+B(x_1+x_2)}.$$

T and T^{-1} both leave $\mathcal{D}$ stable. It is possible to see that $H = T\tilde{h}T^{-1}$, where $\tilde{h} = \frac{1}{2}(p_1^2 + x_1^2) + \frac{1}{2}(p_2^2 + x_2^2) + (A^2 + B^2)1\!\!1$. Therefore, if we introduce the standard bosonic operators $c_j = \frac{1}{\sqrt{2}}(x_j + ip_j)$, together with their adjoints, we can write $\tilde{h} = c_1^\dagger c_1 + c_2^\dagger c_2 + (A^2 + B^2 + 1)1\!\!1$. The eigenvalues of $\tilde{h}$ are $E_{n_1,n_2} = n_1 + n_2 + A^2 + B^2 + 1$, and the related eigenvectors are constructed in a standard way: given $e_{0,0}(x_1, x_2) \in \mathcal{H}$ such that $c_j e_{0,0} = 0$, $j = 1, 2$, the set of eigenstates of $\tilde{h}$ are obtained using the raising operators $c_j^\dagger$: $e_{n_1,n_2} := \frac{1}{\sqrt{n_1!n_2!}}(c_1^\dagger)^{n_1}(c_2^\dagger)^{n_2} e_{0,0}$, $n_j \geq 0$. The set $\mathcal{F}_e = \{e_{n_1,n_2}, n_j \geq 0\}$ is an orthonormal basis for $\mathcal{H}$, and it is a simple exercise to check not only that $e_{n_1,n_2} \in \mathcal{D}$ but also that $\varphi_{n_1,n_2} = Te_{n_1,n_2}$ and $\Psi_{n_1,n_2} = T^{-1}e_{n_1,n_2}$. We are in the conditions of Proposition 3.2.5, so that our claim follows: $\mathcal{F}_\varphi$ and $\mathcal{F}_\Psi$ are $\mathcal{D}$-quasi bases. As in the previous example, we can also easily understand that $\mathcal{F}_\varphi$ and $\mathcal{F}_\Psi$ are not Riesz bases. Moreover, with arguments similar to those of the previous example, we can also check that they are not even bases. This is a consequence of the fact that we can prove that $\|\Psi_{n_1,n_2}\| = \|\varphi_{n_1,n_2}\| \to \infty$ when n_1 or n_2 diverge.

The metric operator Θ is now $\Theta = T^{-2} = e^{2A(p_1+p_2)-2B(x_1+x_2)}$, which maps $\mathcal{D}$ into itself, together with their square roots. This suggests that we can apply Theorem 3.2.5.

3.4.6 A Last Perturbative Example

The last example we want to consider here is the same noncommutative model considered in Section 3.3.2, with the same hamiltonian as in (3.3.12):

$$\hat{H} = \frac{1}{2}(\hat{p}_1^2 + \hat{x}_1^2) + \frac{1}{2}(\hat{p}_2^2 + \hat{x}_2^2) + i\left[A(\hat{x}_1 + \hat{x}_2) + B(\hat{p}_1 + \hat{p}_2)\right], \tag{3.4.11}$$

and where, as before, A and B are real constants. The difference stands in the fact that the self-adjoint operators $\hat{x}_j$ and $\hat{p}_k$ are now assumed to satisfy the following commutation rules:

$$[\hat{x}_j, \hat{p}_k] = i\delta_{j,k}1\!\!1, \quad [\hat{x}_j, \hat{x}_k] = i\theta\epsilon_{j,k}1\!\!1, \quad [\hat{p}_j, \hat{p}_k] = i\tilde{\theta}\epsilon_{j,k}1\!\!1. \tag{3.4.12}$$

Here θ and $\tilde{\theta}$ are two small parameters: the noncommutativity is no longer restricted to the space variables, but involves also the momentum operators. In other words, this model returns the one in Section 3.3.2, if $\tilde{\theta} = 0$.

To deal with this hamiltionan, it is convenient to set up a perturbative approach: in what follows we will only keep the terms which are, at most, linear in θ and $\tilde{\theta}$, neglecting all the quadratic, cubic, ...terms, as in Ref. (29).

With this in mind, if we introduce two pairs of canonically conjugate operators, $(x_j, p_j), j = 1, 2$ (i.e. satisfying $[x_j, p_k] = i\delta_{j,k}\mathbb{1}$, $[x_j, x_k] = [p_j, p_k] = 0$), we can recover (3.4.12) if we assume that

$$\begin{aligned} \hat{x}_1 &= x_1 - \frac{1}{2}\theta p_2, \quad \hat{x}_2 = x_2 + \frac{1}{2}\theta p_1, \\ \hat{p}_1 &= p_1 + \frac{1}{2}\tilde{\theta} x_2, \quad \hat{p}_2 = p_2 - \frac{1}{2}\tilde{\theta} x_1. \end{aligned} \tag{3.4.13}$$

Then $\hat{H}$ can be rewritten, up to corrections which are quadratic in θ and $\tilde{\theta}$, as follows

$$\hat{H} = \frac{1}{2}(p_1^2 + x_1^2) + \frac{1}{2}(p_2^2 + x_2^2) + i\left[A(x_1 + x_2) + B(p_1 + p_2)\right] + \frac{1}{2}(\theta + \tilde{\theta})(p_1 x_2 - p_2 x_1)$$

$$+ i\left[\frac{A\theta}{2}(p_1 - p_2) - \frac{B\tilde{\theta}}{2}(x_1 - x_2)\right] \tag{3.4.14}$$

Defining now new, non-self-adjoint, operators $P_j = p_j + iB_j$, $X_j = x_j + iA_j$, $j = 1, 2$, we observe that $[X_j, P_k] = i\delta_{j,k}\mathbb{1}$, while $[X_j, X_k] = [P_j, P_k] = 0$. Here we have introduced

$$A_1 = A + \frac{1}{2}\theta B, \quad A_2 = A - \frac{1}{2}\theta B, \quad B_1 = B - \frac{1}{2}\tilde{\theta} A, \quad B_2 = B + \frac{1}{2}\tilde{\theta} A.$$

We introduce further the following formally pseudo-bosonic operators:

$$\begin{cases} a_1 = \frac{1}{2}\left(X_1 + iP_1 + iX_2 - P_2\right), & a_2 = \frac{1}{2}\left(-iX_1 + P_1 - X_2 - iP_2\right), \\ b_1 = \frac{1}{2}\left(X_1 - iP_1 - iX_2 - P_2\right), & b_2 = \frac{1}{2}\left(iX_1 + P_1 - X_2 + iP_2\right). \end{cases} \tag{3.4.15}$$

We see that $[a_j, b_k] = \delta_{j,k}\mathbb{1}$, while all the other commutators are zero. Moreover, $b_j \neq a_j^\dagger$. We have used here the word *formally* since we still have to check whether Assumptions $\mathcal{D}$-pb 1, $\mathcal{D}$-pb 2 and $\mathcal{D}$-pb 3, are satisfied or not. In terms of these operators $\hat{H}$ can be written as

$$\hat{H} = (N_1 + N_2 + \mathbb{1}) + \frac{1}{2}(\theta + \tilde{\theta})(N_1 - N_2) + (A^2 + B^2)\mathbb{1}. \tag{3.4.16}$$

Let us rewrite the operators a_j and b_j in terms of the variables x_j and p_j. We find:

$$\begin{cases} a_1 = \frac{1}{2}(x_1 + ip_1 + ix_2 - p_2 + k_1), & a_2 = \frac{1}{2}(-ix_1 + p_1 - x_2 - ip_2 + k_2), \\ b_1 = \frac{1}{2}(x_1 - ip_1 - ix_2 - p_2 + \tilde{k}_1), & b_2 = \frac{1}{2}(ix_1 + p_1 - x_2 + ip_2 + \tilde{k}_2), \end{cases}$$

where

$$\begin{cases} k_1 = A\left(1-\frac{\tilde{\theta}}{2}\right)(i-1) + B\left(\frac{\theta}{2}-1\right)(i+1), \\ k_2 = A\left(1+\frac{\tilde{\theta}}{2}\right)(1-i) + B\left(\frac{\theta}{2}+1\right)(i+1), \\ \tilde{k}_1 = A\left(1-\frac{\tilde{\theta}}{2}\right)(i+1) + B\left(\frac{\theta}{2}-1\right)(i-1), \\ \tilde{k}_2 = -A\left(1+\frac{\tilde{\theta}}{2}\right)(i+1) + B\left(\frac{\theta}{2}+1\right)(i-1). \end{cases}$$

The functions annihilated by a_1 and a_2 and by $b_1^\dagger$ and $b_2^\dagger$ can be found easily, solving two systems of two coupled differential equations. The result is the following:

$$\begin{cases} \varphi_{0,0}(x_1,x_2) = Ne^{-\frac{1}{2}(x_1^2+x_2^2)-\alpha_1 x_1-\alpha_2 x_2}, \\ \Psi_{0,0}(x_1,x_2) = N'e^{-\frac{1}{2}(x_1^2+x_2^2)+\alpha_1 x_1+\alpha_2 x_2}, \end{cases}$$

where $\alpha_1 = \frac{k_1+ik_2}{2}$, $\alpha_2 = \frac{k_1-ik_2}{2i}$, while N and N' are normalization constants, chosen in the usual way (i.e. requiring that $\left\langle \varphi_{0,0}, \Psi_{0,0}\right\rangle = 1$). Also in this model, it looks natural at this stage to take $\mathcal{D} \equiv \mathcal{S}(\mathbb{R}^2)$. In fact, with this choice, $\mathcal{D}$ would be stable under the action of a_j, b_j, and their adjoints. Moreover, $\varphi_{0,0}(x_1,x_2)$ and $\Psi_{0,0}(x_1,x_2)$ both belong to $\mathcal{D}$. However, for future convenience, to ensure its stability under the action of other useful operators, it is better to take $\mathcal{D}$ as in the previous model, i.e.

$$\mathcal{D} = \left\{f(x_1,f_2) \in \mathcal{S}(\mathbb{R}^2), \text{ such that } e^{k_1x_1+k_2x_2}f(x_1,x_2) \in \mathcal{S}(\mathbb{R}^2), \forall k_1,k_2 \in \mathbb{C}\right\}.$$

The functions $\varphi_{n_1,n_2}(x_1,x_2)$ and $\Psi_{n_1,n_2}(x_1,x_2)$ are constructed as usual, and they all belong to $\mathcal{D}$. In conclusion, Assumptions $\mathcal{D}$-pb 1 and $\mathcal{D}$-pb 2 are both satisfied.

For what concerns Assumption $\mathcal{D}$-pbw 3, the idea is again to look for an orthonormal basis which is mapped into $\mathcal{F}_\varphi$ and $\mathcal{F}_\Psi$ by some suitable operators. For that, it is convenient to introduce two bosonic operators, $c_j = \frac{1}{\sqrt{2}}(x_j + ip_j)$, $j = 1,2$, and their adjoints $c_j^\dagger$. Moreover, following Ref. (42), we introduce now two new bosonic lowering and raising operators $c_g = \frac{1}{\sqrt{2}}(c_1 + ic_2)$ and $c_d = \frac{-i}{\sqrt{2}}(c_1 - ic_2)$. They satisfy the following:

$$[c_g, c_g^\dagger] = [c_d, c_d^\dagger] = 1\!\!1,$$

all the other commutators being zero. The vacuum of c_g, c_d, $\chi_{0,0}$, coincides clearly with that of c_1, c_2, $\Phi_{0,0}$: in other words, if $c_1\Phi_{0,0} = c_2\Phi_{0,0} = 0$, then, calling $\chi_{0,0} = \Phi_{0,0}$, we automatically have $c_d\chi_{0,0} = c_g\chi_{0,0} = 0$. Introducing now the vectors $\chi_{n_d,n_g} = \frac{1}{\sqrt{n_d!\,n_g!}}(c_d^\dagger)^{n_d}(c_g^\dagger)^{n_g}\chi_{0,0}$, $n_d, n_g \geq 0$, the set $\mathcal{F}_\chi = \{\chi_{n_d,n_g}\}$ of all these vectors is an orhonormal basis for $\mathcal{H}$, (42). Let us further define the unbounded, self-adjoint and invertible operator T as

$$T = \exp\left\{-\frac{1}{2}\left(k_1 c_g^\dagger + k_2 c_d^\dagger + \overline{k_1} c_g + \overline{k_2} c_d\right)\right\}.$$

$\mathcal{D}$ is stable under T and T^{-1}. We find that

$$\begin{cases} T c_g T^{-1} = c_g + \frac{k_1}{2} = a_1, & T c_d T^{-1} = c_d + \frac{k_2}{2} = a_2, \\ T c_g^\dagger T^{-1} = c_g^\dagger - \frac{\overline{k_1}}{2} = b_1, & T c_d^\dagger T^{-1} = c_d^\dagger - \frac{\overline{k_2}}{2} = b_2. \end{cases}$$

Now, except at most for an unessential normalization, we can check that $\varphi_{0,0} = T\chi_{0,0}$ and that $\Psi_{0,0} = T^{-1}\chi_{0,0}$. This follows, for instance, from the fact that $c_g(T\chi_{0,0}) = -\frac{k_1}{2}(T\chi_{0,0})$ and $c_d(T\chi_{0,0}) = -\frac{k_2}{2}(T\chi_{0,0})$. These equalities can be now easily extended to all the vectors: $\varphi_{n_1,n_2} = T\chi_{n_1,n_2}$ and $\Psi_{n_1,n_2} = T^{-1}\chi_{n_1,n_2}$, for all $n_j \geq 0$. This allow us to use Proposition 3.2.5 so that we can conclude that $\mathcal{F}_\varphi$ and $\mathcal{F}_\Psi$ are $\mathcal{D}$-quasi bases, but, again, they are not Riesz bases. Of course, the metric operator Θ can be introduced, mapping $\mathcal{F}_\Psi$ into $\mathcal{F}_\varphi$ and vice versa, and the same results as in the previous models can be deduced.

3.5 A MUCH SIMPLER CASE: PSEUDO-FERMIONS

So far we have considered operators which, for their intrinsic nature, are often unbounded. Therefore, they cannot be defined on all of $\mathcal{H}$, and their domains must be determined and properly considered. This is, in general, not an easy task. On the other hand, similar problems simply do not exist for fermions, which are described by bounded operators, and this makes the whole story much simpler, but still absolutely nontrivial. This section is devoted to show how the canonical anticommutation relations (CAR) can be modified along the same lines we considered for $\mathcal{D}$-PBs, and why this extension is surely *safer*, and mathematically simpler than that considered in Section 3.2.

The starting point is a modification of the CAR $\{c, c^\dagger\} = c\,c^\dagger + c^\dagger\,c = 1\!\!1$, $\{c, c\} = \{c^\dagger, c^\dagger\} = 0$, between two operators, c and $c^\dagger$, acting on a two-dimensional Hilbert space $\mathcal{H} = \mathbb{C}^2$. The CAR are replaced here by the following rules, (43):

$$\{a, b\} = 1\!\!1, \quad \{a, a\} = 0, \quad \{b, b\} = 0, \tag{3.5.1}$$

where the relevant situation is when $b \neq a^\dagger$. Following what we did for $\mathcal{D}$-PBs, the first two assumptions we might need to require are the following:

- **p1.** a nonzero vector φ_0 exists in $\mathcal{H}$ such that $a\,\varphi_0 = 0$,
- **p2.** a nonzero vector Ψ_0 exists in $\mathcal{H}$ such that $b^\dagger\,\Psi_0 = 0$.

However, these two requirements are automatically satisfied, as a consequence of (3.5.1) so that we do not need to require anything, at least at this stage. In fact, in $\mathcal{H}$, it is easy to check that the only nontrivial possible choices of a and b satisfying (3.5.1) are the following:

$$a(1) = \begin{pmatrix} 0 & 1 \\ 0 & 0 \end{pmatrix}, \quad b(1) = \begin{pmatrix} \beta & -\beta^2 \\ 1 & -\beta \end{pmatrix},$$

$$a(2) = \begin{pmatrix} \alpha & 1 \\ -\alpha^2 & -\alpha \end{pmatrix}, \quad b(2) = \begin{pmatrix} 0 & 0 \\ 1 & 0 \end{pmatrix},$$

with nonzero α and β, or, maybe more interestingly,

$$a(3) = \begin{pmatrix} \alpha_{11} & \alpha_{12} \\ -\alpha_{11}^2/\alpha_{12} & -\alpha_{11} \end{pmatrix}, \quad b(3) = \begin{pmatrix} \beta_{11} & \beta_{12} \\ -\beta_{11}^2/\beta_{12} & -\beta_{11} \end{pmatrix}, \tag{3.5.2}$$

with

$$2\alpha_{11}\beta_{11} - \frac{\alpha_{11}^2\beta_{12}}{\alpha_{12}} - \frac{\beta_{11}^2\alpha_{12}}{\beta_{12}} = -\frac{1}{\alpha_{12}\beta_{12}}\left(\alpha_{11}\beta_{12} - \alpha_{12}\beta_{11}\right)^2 = 1. \tag{3.5.3}$$

Other possibilities also exist, but are those in which a and b exchange their roles or those in which a and b are standard fermion operators. Notice also that $a(1)$ and $b(1)$ appear to be a special case of $a(3)$ and $b(3)$, with $\alpha_{11} = 0$, $\alpha_{12} = 1$, $\beta_{11} = \beta$ and $\beta_{12} = -\beta^2$, choice which satisfies the above-mentioned condition on the matrix entries. On the other hand, $a(2)$ and $b(2)$ cannot be recovered from $a(3)$ and $b(3)$ for any choice of parameters[10].

For all these choices, it is easy to show that the two nonzero vectors φ_0 and Ψ_0 of **p1** and **p2** do automatically exist. This is not surprising, as, because of (3.5.1), $\det(a) = \det(b^\dagger) = 0$, necessarily. For instance, if we take $\alpha_{11} = \frac{1}{3}$, $\beta_{11} = \frac{2}{3}$, and $\alpha_{12} = -\beta_{12} = -i$, we find:

$$a(3) = \begin{pmatrix} 1/3 & -i \\ -i/9 & -1/3 \end{pmatrix}, \; b(3) = \begin{pmatrix} 2/3 & i \\ 4i/9 & -2/3 \end{pmatrix},$$

with

$$\varphi_0 = \alpha \begin{pmatrix} 1 \\ -i/3 \end{pmatrix}, \quad \Psi_0 = \beta \begin{pmatrix} 1 \\ -3i/2 \end{pmatrix}.$$

They satisfy the normalization condition $\langle \varphi_0, \Psi_0 \rangle = 1$ if $\overline{\alpha}\,\beta = \frac{2}{3}$.

[10] As a matter of fact, we can recover $a(2)$ and $b(2)$ out of $a(3)$ and $b(3)$ but only considering a suitable limit: we take $\alpha_{11} = \alpha$, $\alpha_{12} = 1$, $\beta_{11} = x$, $\beta_{12} = -x^2$, and then we send x to zero.

It is now possible to recover similar results as those for $\mathcal{D}$-PBs. In particular, we first introduce the following nonzero vectors

$$\varphi_1 := b\,\varphi_0, \quad \Psi_1 = a^\dagger \Psi_0, \tag{3.5.4}$$

as well as the non-self-adjoint operators

$$N = ba, \quad N^\dagger = a^\dagger b^\dagger. \tag{3.5.5}$$

Of course, it makes no sense to consider $b^n\,\varphi_0$ or ${a^\dagger}^n \Psi_0$ for $n \geq 2$, as all these vectors are automatically zero. Let us now introduce the self-adjoint operators S_φ and S_Ψ via their action on a generic $f \in \mathcal{H}$:

$$S_\varphi f = \sum_{n=0}^{1} \langle \varphi_n, f\rangle\, \varphi_n, \quad S_\Psi f = \sum_{n=0}^{1} \langle \Psi_n, f\rangle\, \Psi_n. \tag{3.5.6}$$

In the present settings there is no problem now with the existence of S_φ and S_Ψ, and with their domains, as the sums in (3.5.6) are finite. This is extremely different from what we had to do for $\mathcal{D}$-PBs, see Section 3.2.3. Now it is very easy to get the following results, similar in part to those for $\mathcal{D}$-PBs. The proofs of our claims are straightforward and will not be given here:

1. $$a\varphi_1 = \varphi_0, \quad b^\dagger \Psi_1 = \Psi_0. \tag{3.5.7}$$

2. $$N\varphi_n = n\varphi_n, \quad N^\dagger \Psi_n = n\Psi_n, \tag{3.5.8}$$

   for $n = 0, 1$.
3. If the normalizations of φ_0 and Ψ_0 are chosen in such a way that $\langle \varphi_0, \Psi_0\rangle = 1$, then

   $$\langle \varphi_k, \Psi_n\rangle = \delta_{k,n}, \tag{3.5.9}$$

   for $k, n = 0, 1$.
4. S_φ and S_Ψ are bounded, strictly positive, self-adjoint, and invertible. They satisfy

   $$\|S_\varphi\| \leq \|\varphi_0\|^2 + \|\varphi_1\|^2, \quad \|S_\Psi\| \leq \|\Psi_0\|^2 + \|\Psi_1\|^2, \tag{3.5.10}$$

   $$S_\varphi \Psi_n = \varphi_n, \qquad S_\Psi \varphi_n = \Psi_n, \tag{3.5.11}$$

   for $n = 0, 1$, as well as $S_\varphi = S_\Psi^{-1}$ and the following intertwining relations

   $$S_\Psi N = N^\dagger S_\Psi, \qquad S_\varphi N^\dagger = N S_\varphi. \tag{3.5.12}$$

All these formulas show that (i) N and $N^\dagger$ behave as fermionic number operators, having eigenvalues 0 and 1; (ii) their related eigenvectors are respectively the vectors in $\mathcal{F}_\varphi = \{\varphi_0, \varphi_1\}$ and $\mathcal{F}_\Psi = \{\Psi_0, \Psi_1\}$; (iii) a and $b^\dagger$ are lowering operators for $\mathcal{F}_\varphi$ and $\mathcal{F}_\Psi$ respectively; (iv) b and $a^\dagger$ are raising operators for $\mathcal{F}_\varphi$ and $\mathcal{F}_\Psi$ respectively; (v) the two sets $\mathcal{F}_\varphi$ and $\mathcal{F}_\Psi$ are biorthonormal; (vi) the *very well-behaved* operators S_φ and S_Ψ map $\mathcal{F}_\varphi$ in $\mathcal{F}_\Psi$ and vice versa; (vii) S_φ and S_Ψ intertwine between operators which are not self-adjoint, in the very same way as they do for $\mathcal{D}$-PBs.

It is clear that we do not need to add any condition on the possibility of computing, for instance, $b\,\varphi_0$, as we had to do for $\mathcal{D}$-PBs. In fact, we can always act on a two-dimensional vector with a two-by-two matrix: no problem with the domains! In addition, we do not need to check (or to ask for) Assumption 3, as this is automatically satisfied: being biorthogonal, the vectors of both $\mathcal{F}_\varphi$ and $\mathcal{F}_\Psi$ are linearly independent. Hence φ_0 and φ_1 are two linearly independent vectors in a two-dimensional Hilbert space. Then $\mathcal{F}_\varphi$ is a basis for $\mathcal{H}$. The same argument obviously can be used for $\mathcal{F}_\Psi$. We will show in a moment that both these sets are also Riesz bases, which is the best we can have after orthonormal bases. This will appear to be a consequence of the properties listed earlier for S_φ and S_Ψ, but we need first a preliminary result.

We have analyzed in Section 3.2.5 the relations existing between $\mathcal{D}$-PBs and ordinary bosons. In particular, we have seen how slightly more complicated methods of functional analysis are needed when the operators S_φ and S_Ψ are unbounded. Therefore, it is not surprising that this kind of difficulties do not appear here since as the inequalities in (3.5.10) show, S_φ and S_Ψ are both bounded operators. More explicitly, we get:

Theorem 3.5.1 *Let c and $T = T^\dagger$ be two operators on $\mathcal{H}$ such that $\{c, c^\dagger\} = 1\!\!1$, $c^2 = 0$, and $T > 0$. Then, defining*

$$a = T\,c\,T^{-1}, \quad b = T\,c^\dagger\,T^{-1}, \tag{3.5.13}$$

these operators satisfy (3.5.1).

Vice versa, given two operators a and b acting on $\mathcal{H}$, satisfying (3.5.1), it is possible to define two operators, c and T, such that $\{c, c^\dagger\} = 1\!\!1$, $c^2 = 0$, $T = T^\dagger$ is strictly positive, and (3.5.13) holds.

Proof: The proof of the first part of the theorem is trivial and will not be given here. As for the second part, we start with the following remark: as S_Ψ is positive and invertible, the operators $S_\Psi^{\pm 1/2}$ are both well defined, self-adjoint and can be chosen, in an unique way, strictly positive. Hence we can define, using $\mathcal{F}_\varphi$ and $S_\Psi^{1/2}$, another family $\mathcal{F}_e = \{e_0, e_1\}$ of vectors $e_n := S_\Psi^{1/2}\varphi_n$, $n = 0, 1$. $\mathcal{F}_e$ is an orthonormal basis for $\mathcal{H}$ and we can naturally define an operator c via its action on e_0 and e_1: we put $c\,e_0 = 0$ and $c\,e_1 = e_0$. Hence, $c^\dagger e_0 = e_1$ and $c^\dagger e_1 = 0$. It is now easy to check that $\{c, c^\dagger\} = 1\!\!1$ and that $c^2 = 0$. Moreover, as $S_\Psi^{-1/2} c\, S_\Psi^{1/2}\varphi_0 = 0$ and $S_\Psi^{-1/2} c\, S_\Psi^{1/2}\varphi_1 = \varphi_0$, we deduce that $a = S_\Psi^{-1/2} c\, S_\Psi^{1/2}$. Analogously we find that $b = S_\Psi^{-1/2} c^\dagger\, S_\Psi^{1/2}$, and therefore we can now identify T in (3.5.13) with $S_\Psi^{-1/2}$, which is strictly self-adjoint and positive.

■

A first consequence of this theorem is that, as $\mathcal{F}_\varphi$ is the image of the orthonormal basis $\mathcal{F}_e$ via a bounded operator, with bounded inverse, $S_\Psi^{-1/2}$, $\mathcal{F}_\varphi$ is a Riesz basis, as we claimed before. Incidentally, it is obvious that the sets $\mathcal{F}_\varphi$ and $\mathcal{F}_\Psi$ are $\mathcal{H}$-quasi bases. A second consequence is that, introducing the self-adjoint number operator for the fermionic operators, $N_0 := c^\dagger c$, this can be related to both N and $N^\dagger$:

$$N = S_\Psi^{-1/2} N_0 \, S_\Psi^{1/2}, \quad N^\dagger = S_\Psi^{1/2} N_0 \, S_\Psi^{-1/2}, \tag{3.5.14}$$

which can also be written in form of intertwining relations: $S_\Psi^{1/2} N_0 = N^\dagger S_\Psi^{1/2}$, $S_\Psi^{1/2} N = N_0 S_\Psi^{1/2}$. Putting together these equations we also recover (3.5.12).

3.5.1 A First Example from the Literature

In 2007, in Ref. (44), an effective non-self-adjoint hamiltonian describing a two-level atom interacting with an electromagnetic field was analyzed in connection with pseudo-hermitian systems. We will discuss here how this model can be very naturally rewritten in terms of pseudo-fermionic operators, and that the structure previously described naturally arises. The starting point is the Schrödinger equation

$$i\dot{\Phi}(t) = H_{eff}\Phi(t), \qquad H_{eff} = \frac{1}{2}\begin{pmatrix} -i\delta & \overline{\omega} \\ \omega & i\delta \end{pmatrix}. \tag{3.5.15}$$

Here δ is a real quantity, related to the decay rates for the two levels, while the complex parameter ω characterizes the radiation-atom interaction. We refer to Ref. (44) for further details. It is clear that $H_{eff} \neq H_{eff}^\dagger$. It is convenient to write $\omega = |\omega| e^{i\theta}$. Then, we introduce the operators

$$a = \frac{1}{2\Omega}\begin{pmatrix} -|\omega| & -e^{-i\theta}(\Omega + i\delta) \\ e^{i\theta}(\Omega - i\delta) & |\omega| \end{pmatrix}$$

and

$$b = \frac{1}{2\Omega}\begin{pmatrix} -|\omega| & e^{-i\theta}(\Omega - i\delta) \\ -e^{i\theta}(\Omega + i\delta) & |\omega| \end{pmatrix}.$$

Here $\Omega = \sqrt{|\omega|^2 - \delta^2}$, which we will assume here to be real and strictly positive. A direct computation shows that $a \neq b^\dagger$ (we are taking here $\delta \neq 0$), $\{a, b\} = \mathbb{1}$, $a^2 = b^2 = 0$. Hence a and b are pseudo-fermionic operators. Moreover, H_{eff} can be written in terms of these operators as $H_{eff} = \Omega\left(ba - \frac{1}{2}\mathbb{1}\right)$.

Remark:– Notice that a and b aforementioned have exactly the same expressions as $a(3)$ and $b(3)$ in (3.5.2), with $\alpha_{11} = \beta_{11} = -\frac{|\omega|}{2\Omega}$, $\alpha_{12} = -\frac{1}{2\Omega} e^{-i\theta}(\Omega + i\delta)$, and $\beta_{12} = \frac{1}{2\Omega} e^{-i\theta}(\Omega - i\delta)$. Notice also that this choice satisfies condition (3.5.3).

To recover the pseudo-fermionic structure we first need to find a nonzero vector φ_0 annihilated by a and a second nonzero vector Ψ_0 annihilated by $b^\dagger$. These vectors are

$$\varphi_0 = k\begin{pmatrix} 1 \\ -\frac{e^{i\theta}(\Omega-i\delta)}{|\omega|} \end{pmatrix}, \qquad \Psi_0 = k'\begin{pmatrix} 1 \\ -\frac{e^{i\theta}(\Omega+i\delta)}{|\omega|} \end{pmatrix},$$

where k and k' are normalization constants, partially fixed by the requirement that $\langle \varphi_0, \Psi_0 \rangle = \overline{k}\,k'\left(1 + \frac{1}{|\omega|^2}(\Omega + i\delta)^2\right) = 1$. We now also introduce the vectors

$$\varphi_1 = b\varphi_0 = k\begin{pmatrix} \frac{i\delta-\Omega}{|\omega|} \\ -e^{i\theta} \end{pmatrix}, \qquad \Psi_1 = a^\dagger \Psi_0 = k'\begin{pmatrix} \frac{-i\delta-\Omega}{|\omega|} \\ -e^{i\theta} \end{pmatrix}.$$

It is now easy to check that $\mathcal{F}_\varphi$ and $\mathcal{F}_\Psi$ are biorthonormal bases of $\mathcal{H}$, and we can also check that

$$H_{eff}\varphi_0 = -\frac{\Omega}{2}\varphi_0, \quad H_{eff}\varphi_1 = \frac{\Omega}{2}\varphi_1, \quad H_{eff}^\dagger\Psi_0 = -\frac{\Omega}{2}\Psi_0, \quad H_{eff}^\dagger\Psi_1 = \frac{\Omega}{2}\Psi_1.$$

Therefore, H_{eff} and $H_{eff}^\dagger$ are isospectral as expected. To carry on our analysis we now compute S_φ and S_Ψ, which are found to be

$$S_\varphi = 2|k|^2\begin{pmatrix} 1 & \frac{-i\delta}{|\omega|}e^{-i\theta} \\ \frac{i\delta}{|\omega|}e^{i\theta} & 1 \end{pmatrix}, \qquad S_\Psi = \frac{|\omega|^2}{2|k|^2\Omega^2}\begin{pmatrix} 1 & \frac{i\delta}{|\omega|}e^{-i\theta} \\ \frac{-i\delta}{|\omega|}e^{i\theta} & 1 \end{pmatrix},$$

and turn out to be one the inverse of the other. They are also, under our assumption $\Omega > 0$, positive definite matrices, as they should. Using now Theorem 3.5.1, we can use $S_\varphi^{\pm 1/2}$ to define two *standard* fermion operators c and $c^\dagger$, and their related number operator $N_0 = c^\dagger c$, out of a and b. Hence we easily find that

$$H_{eff} = S_\varphi^{1/2}\,h\,S_\varphi^{-1/2},$$

where $h = \Omega\left(c^\dagger c - \frac{1}{2}\mathbb{1}\right)$ is a self-adjoint operator. This shows that the effective hamiltonian H_{eff} is similar to a self-adjoint operator, suggesting that, at least for this model, the appearance of a non-self-adjoint hamiltonian is related to the choice of a *natural but possibly not well chosen* scalar product in the Hilbert space[11] $\mathcal{H}$. In fact, replacing the standard scalar product $\langle f, g \rangle$ with a new one,

[11] This is matter of debate as, sometimes, working with non-self-adjoint hamiltonians is exactly what one wants to do in concrete applications so that there is no reason to look for a different scalar product in which the hamiltonian appears to be self-adjoint.

$\langle f,g\rangle_S=\left\langle S_\varphi^{-1/2}f,S_\varphi^{-1/2}g\right\rangle$, would make H_{eff} self-adjoint: for all $f,g\in\mathcal{H}$ indeed we find

$$\left\langle H_{eff}f,g\right\rangle_S=\left\langle S_\varphi^{-1/2}\left(S_\varphi^{1/2}h\,S_\varphi^{-1/2}\right)f,S_\varphi^{-1/2}g\right\rangle=\left\langle S_\varphi^{-1/2}f,h\,S_\varphi^{-1/2}g\right\rangle$$
$$=\left\langle S_\varphi^{-1/2}f,S_\varphi^{-1/2}\left(S_\varphi^{1/2}h\,S_\varphi^{-1/2}\right)g\right\rangle=\langle f,H_{eff}\,g\rangle_S.$$

Here, once again, we can work without having to fight with domains of operators, as they are all everywhere defined in $\mathcal{H}$. We refer to Refs (45, 46) for an extension of our framework to more pseudo-fermionic modes, and for further applications to concrete situations. In particular, in Ref. (46), pseudo-fermions have been applied to the analysis of a gain-loss electronic circuit. Rather than showing this particular application, we will now briefly discuss how several other two-by-two hamiltonians, introduced by several authors along the years in connection with PT-quantum mechanics, can be written using pseudo-fermionic operators.

3.5.2 More Examples from the Literature

As we have seen before, the choice (3.5.2) provides essentially the most general form of operators satisfying (3.5.1). For this reason, we will now concentrate on this particular form of $a:=a(3)$ and $b:=b(3)$ and use these operators to write a very general diagonalizable hamiltonian, which can be written as

$$H=\omega N+\rho 1\!\!1=\begin{pmatrix}\omega\gamma\alpha+\rho & \omega\gamma\\ -\omega\gamma\alpha\beta & -\omega\gamma\beta+\rho\end{pmatrix}, \tag{3.5.16}$$

where ω and ρ, in principle, could be complex numbers, while $\alpha=\frac{\alpha_{11}}{\alpha_{12}}$, $\beta=\frac{\beta_{11}}{\beta_{12}}$, and $\gamma=\alpha_{12}\beta_{11}-\alpha_{11}\beta_{12}=\alpha_{12}\beta_{12}(\beta-\alpha)$. Then we can rewrite

$$a=\alpha_{12}\begin{pmatrix}\alpha & 1\\ -\alpha^2 & -\alpha\end{pmatrix},\qquad b=\beta_{12}\begin{pmatrix}\beta & 1\\ -\beta^2 & -\beta\end{pmatrix},$$

while condition (3.5.3) can be written as $-\gamma^2=\alpha_{12}\beta_{12}$. This also implies that $(\alpha-\beta)\gamma=1$.

The eigensystem of H is trivially deduced: the eigenvalues are $\epsilon_0=\rho$ and $\epsilon_1=\omega+\rho$, which are real if and only if ρ and ω are both real. In this case, ϵ_0 and ϵ_1 are also the eigenvalues of $H^\dagger=\omega N^\dagger+\rho 1\!\!1$. It might be interesting to notice that, adopting the limiting procedure previously described ($\alpha_{11}=\alpha$, $\alpha_{12}=1$, $\beta_{11}=x$, $\beta_{12}=-x^2$, and $x\to 0$), we recover a diagonal matrix: $H=\rho 1\!\!1$.

The eigenvectors of N and $N^\dagger$, and of H and $H^\dagger$ as a consequence, are the following:

$$\varphi_0=N_\varphi\begin{pmatrix}1\\ -\alpha\end{pmatrix},\qquad \varphi_1=b\,\varphi_0=\frac{\gamma N_\varphi}{\alpha_{12}}\begin{pmatrix}1\\ -\beta\end{pmatrix}, \tag{3.5.17}$$

and

$$\Psi_0 = N_\Psi \begin{pmatrix} 1 \\ \frac{-1}{\beta} \end{pmatrix}, \quad \Psi_1 = a^\dagger \Psi_0 = \frac{\overline{\gamma N_\Psi}}{\overline{\beta_{11}}} \begin{pmatrix} \overline{\alpha} \\ 1 \end{pmatrix}, \tag{3.5.18}$$

where $N_\varphi \overline{N_\Psi} = \frac{\alpha_{12}\beta_{11}}{\gamma}$. This choice is dictated by the request that $\langle \Psi_0, \varphi_0 \rangle = 1$. We observe that $a\,\varphi_0 = 0$ and $b^\dagger \Psi_0 = 0$, and as $N\varphi_j = j\varphi_j$ and $N^\dagger \Psi_j = j\Psi_j$, $j = 0, 1$, we also deduce the following:

$$H\varphi_j = \epsilon_j \varphi_j, \qquad H^\dagger \Psi_j = \epsilon_j \Psi_j, \tag{3.5.19}$$

$j = 0, 1$.

It is a straightforward computation to check that $\mathcal{F}_\varphi$ and $\mathcal{F}_\Psi$ produce, together, a resolution of the identity. Then, as expected, they are biorthogonal bases for $\mathcal{H}$.

The operators S_φ and S_Ψ can be written as

$$S_\varphi = |N_\varphi|^2 \begin{pmatrix} 1 + \left|\frac{\gamma}{\alpha_{12}}\right|^2 & -\overline{\alpha} - \overline{\beta} \left|\frac{\gamma}{\alpha_{12}}\right|^2 \\ -\alpha - \beta \left|\frac{\gamma}{\alpha_{12}}\right|^2 & |\alpha|^2 + \left|\frac{\gamma\beta}{\alpha_{12}}\right|^2 \end{pmatrix} \tag{3.5.20}$$

and

$$S_\Psi = |N_\Psi|^2 \begin{pmatrix} 1 + \left|\frac{\alpha\gamma}{\beta_{11}}\right|^2 & \frac{1}{\beta} + \overline{\alpha} \left|\frac{\gamma}{\beta_{11}}\right|^2 \\ \frac{1}{\overline{\beta}} + \alpha \left|\frac{\gamma}{\beta_{11}}\right|^2 & \left|\frac{1}{\beta}\right|^2 + \left|\frac{\gamma}{\beta_{11}}\right|^2 \end{pmatrix}, \tag{3.5.21}$$

which are both clearly self-adjoint. Using, for instance, the Sylvester's criterion, it is possible to check explicitly that, if $\alpha \neq \beta$, both S_φ and S_Ψ are positive definite. This can also be deduced looking at the eigenvalues of the two matrices, or just using the definition: $\langle f, S_\varphi f \rangle$ and $\langle f, S_\Psi f \rangle$ are both strictly positive for any nonzero $f \in \mathcal{H}$, if $\alpha \neq \beta$. Interestingly enough, $\alpha = \beta$ implies that condition (3.5.3) cannot be satisfied, and this means, in turn, that we are loosing the pseudo-fermionic structure described at the beginning of this section. In fact, a and b cannot satisfy any longer the anticommutation rules in (3.5.1). Therefore, it is not surprising that S_φ and S_Ψ do not admit inverse, contrarily to what happens when (3.5.1) are satisfied.

Because of their positivity, there exist unique square root matrices $S_\varphi^{1/2}$ and $S_\Psi^{1/2}$, which are also positive and self-adjoint. They have a rather involved expression, which can be found in Ref. (47), together with other related results. In particular, we have reconsidered several older models, introduced along the years, and we have shown how their hamiltonians are all particular forms of the one in (3.5.16), for different choices of the constants. This is what happens, for instance, for

$$H_{DG} = \begin{pmatrix} r\,e^{i\theta} & s\,e^{i\phi} \\ t\,e^{-i\phi} & r\,e^{-i\theta} \end{pmatrix},$$

where r, s, t, θ and ϕ are all real quantities, originally considered in Ref. (48), for

$$H_{GMM} = \begin{pmatrix} \sigma_1 - i\Gamma_1 & \nu_0 \\ \nu_0 & \sigma_2 - i\Gamma_2 \end{pmatrix},$$

where Γ_1 and Γ_2 are positive quantities, σ_1 and σ_2 are reals, and ν_0 is complex-valued, introduced in Ref. (49), and

$$H_{MO} = E \begin{pmatrix} \cos\theta & e^{-i\phi}\sin(\theta) \\ e^{i\phi}\sin(\theta) & -\cos\theta \end{pmatrix},$$

where $\theta, \phi \in \mathbb{C}$, $\Re(\theta) \in [0, \pi)$, and $\Re(\phi) \in [0, \pi)$, first introduced in Ref. (50). Of course, here the interesting situation is when $E \neq 0$ and $\theta \neq 0$. Finally, we have also considered, see Refs (51, 52),

$$H_{rel} = \begin{pmatrix} mc^2 & cp_x + v \\ cp_x - v & -mc^2 \end{pmatrix},$$

which, when $c^2p_x^2 \neq v^2$, can be seen as a particular case of the hamiltonian H_{MO}. On the other hand, when $c^2p_x^2 = v^2 \neq 0$, H_{rel} and H_{MO} are different for any possible choice of the parameters. This is easy to understand: while, in this case, only one nondiagonal matrix element in H_{rel} can be different from zero, the analogous elements in H_{MO} are both zero or both not zero. Nevertheless, even this hamiltonian H_{rel}, fixing properly the parameters, coincides with H in (3.5.16), (47): our H covers also this other example, then.

In conclusion, our analysis shows that PFs provide an unifying language for several interesting models introduced along the years.

3.6 A POSSIBLE EXTENSION: NONLINEAR $\mathcal{D}$-PBS

As we have seen, $\mathcal{D}$-PBs appear quite often in many physical quantum models proposed in the literature. Actually, our claim is the following:
suppose that the non-self-adjoint hamiltonian H of a quantum system with N degrees of freedom has eigenvalues $E_{n_1,\ldots,n_N}$ which depends linearly on the quantum numbers n_j, $E_{n_1,\ldots,n_N} = \sum_{j=1}^N \omega_j n_j + \omega_0$, for some real[12] *$\omega_j$, $j = 0, 1, 2, \ldots, N$. Then we can introduce N pairs of $\mathcal{D}$-pseudo-bosonic operators a_j and b_j such that the hamiltonian can be written as $H = \sum_{j=1}^N \omega_j b_j a_j + \omega_0 \mathbb{1}$.*

This is what we have found in all the models proposed in the literature so far, but we have still no rigorous proof of this claim. As one can see, there is a strong limitation concerning the kind of hamiltonians which can be rewritten in terms of

[12] As a matter of fact, we believe that this claim can be generalized. In fact, there is no apparent mathematical reason a priori to restrict to real values of the ω_j. We think that this same claim can also be stated when $\omega_j \in \mathbb{C}$, even if, in this case, the energy operator would have (possibly) unphysical complex eigenvalues.

$\mathcal{D}$-PBs: they must necessarily have *linear* eigenvalues, like the $E_{n_1,\ldots,n_N}$'s above[13]. This suggests that these hamiltonians should have some relations with the quantum harmonic oscillator (in dimension N). This is indeed what happens for the Swanson's model, for the quantum Hall effect, and so on. Our claim also suggests that, when the eigenvalues of a certain hamiltonian are **not** linear in the quantum numbers, $\mathcal{D}$-PBs are not expected to appear, in general. For this reason, we have recently introduced a slightly extended definition, which seems to provide a natural settings for this kind of hamiltonians: let us consider a strictly increasing sequence $\{\epsilon_n\}$: $0 = \epsilon_0 < \epsilon_1 < \cdots < \epsilon_n < \cdots$. Then, given two operators a and b on $\mathcal{H}$, and a set $\mathcal{D} \subset \mathcal{H}$ which is dense in $\mathcal{H}$, and which is stable under the action of $a^\sharp$ and $b^\sharp$.

Definition 3.6.1 *We will say that the triple $(a, b, \{\epsilon_n\})$ is a family of $\mathcal{D}$-nonlinear pseudo-bosons ($\mathcal{D}$-NLPBs) if the following properties hold:*

- **p1.** *a nonzero vector Φ_0 exists in $\mathcal{D}$ such that $a\,\Phi_0 = 0$;*
- **p2.** *a nonzero vector η_0 exists in $\mathcal{D}$ such that $b^\dagger\,\eta_0 = 0$;*
- **p3.** *Calling*

$$\Phi_n := \frac{1}{\sqrt{\epsilon_n!}}\, b^n\,\Phi_0, \qquad \eta_n := \frac{1}{\sqrt{\epsilon_n!}}\, a^{\dagger n}\,\eta_0, \tag{3.6.1}$$

we have, for all $n \geq 0$,

$$a\,\Phi_n = \sqrt{\epsilon_n}\,\Phi_{n-1}, \qquad b^\dagger\eta_n = \sqrt{\epsilon_n}\,\eta_{n-1}. \tag{3.6.2}$$

- **p4.** *The set $\mathcal{F}_\Phi = \{\Phi_n,\ n \geq 0\}$ is a basis for $\mathcal{H}$.*

Remarks:– (1) As $\mathcal{D}$ is stable under the action of b and $a^\dagger$, it follows that $\Phi_n, \eta_n \in \mathcal{D}$, for all $n \geq 0$.

(2) $\mathcal{D}$-PBs are recovered choosing $\epsilon_n = n$.

(3) In the literature, if $\mathcal{F}_\Phi$ is a Riesz basis for $\mathcal{H}$, the $\mathcal{D}$-NLPBs are called *regular*.

(4) The set $\mathcal{F}_\eta = \{\eta_n,\ n \geq 0\}$ is automatically a basis for $\mathcal{H}$ as well. This follows from the fact that, calling $M = ba$, we have $M\Phi_n = \epsilon_n\Phi_n$ and $M^\dagger\eta_n = \epsilon_n\eta_n$. Therefore, choosing the normalization of η_0 and Φ_0 in such a way $\langle\eta_0, \Phi_0\rangle = 1$, $\mathcal{F}_\eta$ is biorthogonal to the basis $\mathcal{F}_\Phi$. Then, it is possible to check that $\mathcal{F}_\eta$ is the unique basis which is biorthogonal to $\mathcal{F}_\Phi$.

Also in this context it is possible to deduce interesting intertwining relations. We just consider here the simple situation in which the two bases are related by a suitable self-adjoint, invertible and, in general, unbounded operator Θ which, together with

[13]Of course, we are not considering here trivial situations as for instance, the case in which H is just, say, the square of some operator with linear eigenvalues, for instance $H = (ba)^2$ for a $\mathcal{D}$-pseudo-bosonic pair (a, b). In this case, the eigenvalues are quadratic, $E_n = n^2$, but the underlying structure is identical to that in Section 3.2.

Θ^{-1}, leaves $\mathcal{D}$ invariant. More explicitly, we require that $\eta_n = \Theta\,\Phi_n, \forall\, n$. In this case, we easily get

$$\left(M^{\dagger}\Theta - \Theta M\right)\Phi_n = 0,$$

for all n. As already stressed in Refs (53–55), this could be relevant in discussing physical systems described by some hamiltonian which is not self-adjoint, but crypto-hermitian or PT-symmetric, and with eigenvalues ϵ_n which are not necessarily linear in the quantum number n.

As in Section 3.2, it could be reasonable to replace **p4.** with some weaker requirement. In particular, also in view of our previous discussion, it is natural to assume that

- **pw4.** $\mathcal{F}_\Phi$ and $\mathcal{F}_\eta$ are $\mathcal{G}$-quasi bases for some $\mathcal{G} \subset \mathcal{H}$, dense in $\mathcal{H}$.

The consequences of this extension have not yet been analyzed, and this is, indeed, part of our future projects.

3.7 CONCLUSIONS

In this chapter we have discussed two possible extensions of the canonical commutation and anticommutation relations, showing that, in both cases, an interesting functional structure arises: orthonormal bases are replaced by biorthogonal sets, non-self-adjoint operators appear having discrete and real eigenstates, intertwining operators are naturally defined, and some of them map non-self-adjoint to self-adjoint operators. What is also interesting, and somehow unexpected at the beginning of this analysis, is that several models proposed during the last twenty years in connection with what is now called PT-quantum mechanics, perfectly fit in our scheme. Therefore, what at the beginning was just a mathematical curiosity, generated soon also a physically oriented line of research. Research which is still rich of open questions. We want to end this chapter with a short list of *things to do*. Here they are, listed in random order:

1. In Ref. (23), we defined some sort of coherent states, as eigenstates of a and $b^{\dagger}$, where a and b were pseudo-bosonic operators in the sense of Section 3.2.4.1. This is something to be considered in more details, both from a mathematical side, and also in view of the possibility of applying these *bicoherent states* in connection with quantization procedure, (56). A first step in this direction is already discussed in Ref. (57)
2. The mathematical aspects of $\mathcal{D}$-NLPBs have not been yet well explored in details. And not many nontrivial examples of the general settings have been analyzed, so far. These are gaps which we need to fill up.
3. Finite dimensional models quite often are easily handled and under control. They are also useful in concrete situations, like in quantum optics or in the analysis of electronic circuits. The structure proposed in Section 3.5 has already proven to be useful in several such situations, and we believe more applications

can be easily found. There are, in fact, some indications that this approach can also be useful in connection with exceptional points. Now the natural problem to consider is therefore the following: how much of this connection survives when going from PFs to $\mathcal{D}$-PBs?

4. All along the chapter, the role of $\mathcal{G}$-quasi bases appeared to be quite often essential. We believe that there is much more to discover about these bases, both from a mathematical and from a physical point of view.

As you can see, there is still work to do!

3.8 ACKNOWLEDGMENTS

The material discussed here is also the result of several interesting and useful discussions I had, along the years, with several colleagues and friends. My warm thanks go in particular to Camillo Trapani, Christopher Heil, Atsushi Inoue and Petr Siegl, which I disturbed several times, asking for hints, references, comments, etc.... And, surprisingly, they still reply to me!

REFERENCES

1. Bender CM, Boettcher S. Real spectra in non-Hermitian Hamiltonians having PT-symmetry. *Phys Rev Lett* 1998;80:5243–5246.
2. Mostafazadeh A. Metric operator in pseudo-Hermitian quantum mechanics and the imaginary cubic potential. *J Phys A Math Gen* 2006;39:10171–10188.
3. Dorey P, Dunning C, Tateo R. Spectral equivalences, Bethe ansatz equations, and reality properties in PT-symmetric quantum mechanics. *J Phys A Math Gen* 2001;34:5679–5704.
4. (a) Faisal FHM, Moloney JV. Time-dependent theory of non-Hermitian Schrodinger equation: application to multiphoton-induced ionisation decay of atoms. *J Phys B At Mol Phys* 1981;14:3603–3620; (b) Dattoli G, Mignani R, Torre A. Geometrical phase in the cyclic evolution of non-Hermitian systems. *J Phys A Math Gen* 1990;23:5795–5806; (c) Bender CM, Mannheim PD. Exactly solvable PT-symmetric Hamiltonian having no Hermitian counterpart. *Phys. Rev. D* 2008;78:025022.
5. (a) Maleki Y. Para-grassmannian coherent and squeezed states for pseudo-Hermitian q-oscillator and their entanglement. *SIGMA* 2011;7:084, 20; (b) Ghatak A, Mandal BP. Comparison of different approaches of finding the positive definite metric in pseudo-Hermitian theories. *Commun Theor Phys* 2013;59:533–539.
6. (a) Bebiano N, Da Providencia J, Da Providencia JP. Classes of non-Hermitian operators with real eigenvalues. *Electron J Linear Algebra* 2010;21:98–109; (b) Govaerts J, Bwayi CM, Mattelaer O. The Klauder-Daubechies construction of the phase space path integral and the harmonic oscillator. *J Phys A Math Theor* 2009;42:445304.

7. (a) Mostafazadeh A. Metric operators for quasi-Hermitian Hamiltonians and symmetries of equivalent Hermitian Hamiltonians. *J Phys A Math Theor* 2008;41:244017; (b) Mostafazadeh A. Non-Hermitian Hamiltonians with a real spectrum and their physical applications. *Pramana-J Phys* 2009;73:269–277; (d) Mostafazadeh A. Pseudo-Hermiticity versus PT-symmetry: the necessary condition for the reality of the spectrum of a non-Hermitian Hamiltonian. *J Math Phys* 2002;43:205–214.
8. Bagarello F, Inoue A, Trapani C. Non-self-adjoint Hamiltonians defined by Riesz bases. *J Math Phys* 2014;55:033501.
9. (a) Siegl P. PT-symmetric square well-perturbations and the existence of metric operator. *Int J Theor Phys* 2011;50:991–996.; (b) Krejcirik D, Siegl P, Zelezny J. *On the Similarity of Sturm-Liouville Operators with Non-Hermitian Boundary Conditions to Self-Adjoint and Normal Operators*. Complex Analysis and Operator Theory 2014;8:255–281.
10. Krejcirik D, Siegl P. On the metric operator for the imaginary cubic oscillator. *Phys. Rev. D* 2012;86:121702(R).
11. Young RM. On complete biorthogonal bases. *Proc Am Math Soc* 1981;83(3):537–540.
12. Christensen O. *An Introduction to Frames and Riesz Bases*. Boston (MA): Birkhäuser; 2003.
13. Heil C. *A Basis Theory Primer: Expanded Edition*. New York: Springer-Verlag; 2010.
14. Hansen VL. *Functional Analysis: Entering Hilbert Space*. Singapore: World Scientific; 2006.
15. Gohberg IC, Krein MG. *Introduction to the Theory of Linear Nonselfadjoint Operators*. Providence (RI): American Mathematical Society; 1969.
16. Bagarello F. More mathematics for pseudo-bosons. *J Math Phys* 2013;54:063512.
17. Reed M, Simon B. *Methods of Modern Mathematical Physics I: Functional Analysis*. New York: Academic Press; 1980.
18. Pedersen GK. *Analysis Now*. New York: Springer-Verlag; 1989.
19. (a) Kuru S, Tegmen A, Vercin A. Intertwined isospectral potentials in an arbitrary dimension. *J Math Phys* 2001;42(8):3344–3360; (b) Kuru S, Demircioglu B, Onder M, Vercin A. Two families of superintegrable and isospectral potentials in two dimensions. *J Math Phys* 2002;43(5):2133–2150; (c) Samani KA, Zarei M. Intertwined Hamiltonians in two-dimensional curved spaces. *Ann Phys* 2005;316:466–482.
20. (a) Bagarello F. Extended SUSY quantum mechanics, intertwining operators and coherent states. *Phys Lett A* 2008. DOI: 10.1016/ j.physleta. 2008.08.047; (b) Bagarello F. Vector coherent states and intertwining operators. *J Phys A* 2009. DOI: 10.1088/1751-8113/42/7/075302; (c) Bagarello F. Intertwining operators between different Hilbert spaces: connection with frames. *J Math Phys* 2009;50:043509, 13pp. DOI: 10.1063/1.3094758.
21. Mostafazadeh A. Pseudo-Hermitian quantum mechanics. *Int J Geom Methods Mod Phys* 2010;7:1191–1306.
22. Bender C. Making sense of non-Hermitian Hamiltonians. *Rep Progr Phys* 2007;70: 947–1018.

23. Bagarello F. Pseudo-bosons, Riesz bases and coherent states. *J Math Phys* 2010;50: 023531.

24. Bagarello F. (Regular) pseudo-bosons versus bosons. *J Phys A* 2011;44:015205.

25. (a) Bagarello F, Inoue A, Trapani C. Weak commutation relations of unbounded operators and applications. *J Math Phys* 2011;52:113508; (b) Bagarello F, Inoue A, Trapani C. Weak commutation relations of unbounded operators: nonlinear extensions. *J Math Phys* 2012;53:123510.

26. Bagarello F. From self to non self-adjoint harmonic oscillators: physical consequences and mathematical pitfalls. *Phys Rev A* 2013;88:032120.

27. Bender CM, Jones HF. Interactions of Hermitian and non-Hermitian Hamiltonians. *J Phys A* 2008;41:244006.

28. Li J-Q, Li Q, Miao Y-G. Investigation of PT-symmetric Hamiltonian systems from an alternative point of view. *Commun Theor Phys* 2012;58:497.

29. Li J-Q, Miao Y-G, Xue Z. Algebraic method for pseudo-Hermitian Hamiltonians, arXiv:1107.4972 [quant-ph].

30. Davies EB, Kuijlaars BJ. Spectral asymptotics of the non-self-adjoint harmonic oscillator. *J Lond Math Soc* 2004;70:420–426.

31. Davies EB. Pseudospectra, the harmonic oscillator and complex resonances. *Proc R Soc London Ser A* 1999;455:585–599.

32. Kolmogorov A, Fomine S. Eléments de la théorie des fonctions et de lanalyse fonctionelle, Mir; 1973.

33. Bagarello F, Fring A. A non self-adjoint model on a two dimensional noncommutative space with unbound metric, *Phys Rev A* 2013;88. DOI: 10.1103/PhysRevA.88.042119.

34. Bagarello F, Lattuca M. $\mathcal{D}$ pseudo-bosons in quantum models. *Phys Lett A* 2013;377(44):3199–3204.

35. Bagarello F, Ali ST, Gazeau JP. Extended pseudo-fermions from non commutative bosons. *J Math Phys* 2013;54:073516.

36. Bagarello F. Examples of pseudo-bosons in quantum mechanics. *Phys Lett A* 2010;374:3823–3827.

37. da Providência J, Bebiano N, da Providência JP. Non Hermitian operators with real spectrum in quantum mechanics. *Electron J Linear Algebra* 2010;21:98–109.

38. Szegö G. *Orthogonal Polynomials*. Providence (RI): American Mathematical Society; 1939.

39. Ali ST, Bagarello F, Gazeau J-P. Modified Landau levels, damped harmonic oscillator and two-dimensional pseudo-bosons. *J Math Phys* 2010;51:123502.

40. Prudnikov AP, Bryehkov YuA, Marichev OI. *Integrals and Series*. Volume 2, *Special functions*. Amsterdam: Opa; 1986.

41. Gradshteyn IS, Ryzhik IM. *Table of Integrals, Series and Products*. 7th ed. San Diego (CA): Academic Press; 2007.

42. Messiah A. *Quantum Mechanics*. Volume 1. Amsterdam: North Holland Publishing Company; 1961.

43. Bagarello F. Linear pseudo-fermions. *J Phys A Math Theor* 2012;45:444002.
44. Cherbal O, Drir M, Maamache M, Trifonov DA. Fermionic coherent states for pseudo-Hermitian two-level systems. *J Phys A* 2007;40:1835–1844.
45. Bagarello F. Damping and pseudo-fermions. *J Math Phys* 2013;54:023509.
46. Bagarello F, Pantano G. Pseudo-fermions in an electronic loss-gain circuit. *Int J Theor Phys* 2013;52:4507–4518.
47. Bagarello F, Gargano F. Model pseudofermionic systems: connections with exceptional points. *Phys Rev A* 2014;89:032113.
48. Das A, Greenwood L. An alternative construction of the positive inner product for pseudo-Hermitian Hamiltonians: examples. *J Math Phys* 2010;51(4):042103.
49. Gilary I, Mailybaev AA, Moiseyev N. Time-asymmetric quantum-state-exchange mechanism. *Phys Rev A* 2013;88:010102(R).
50. Mostafazadeh A, Özcelik S. Explicit realization of pseudo-Hermitian and quasi-Hermitian quantum mechanics for two-level systems. *Turk J Phys* 2006;30:437–443.
51. Mandal BP, Gupta S. Pseudo-Hermitian interactions in Dirac theory: examples. *Mod Phys Lett A* 2010;25:1723.
52. Ghatak A, Mandal BP. Comparison of different approaches of finding the positive definite metric in pseudo-hermitian theories. *Commun Theor Phys* 2013;59:533–539.
53. Bagarello F. Non linear pseudo-bosons. *J Math Phys* 2011;52:063521.
54. Bagarello F, Znojil M. Non linear pseudo-bosons versus hidden Hermiticity. *J Phys A* 2011;44:415305.
55. Bagarello F, Znojil M. Non linear pseudo-bosons versus hidden Hermiticity. II: the case of unbounded operators. *J Phys A* 2012;45:115311.
56. Gazeau JP *Coherent States in Quantum Physics*. Berlin: Wiley-VCH; 2009.
57. Bagarello F, Triolo S. Some invariant biorthogonal sets with an application to coherent states. *J Math Anal Appl* 2014;415:462–476.

4

CRITERIA FOR THE REALITY OF THE SPECTRUM OF $\mathcal{PT}$-SYMMETRIC SCHRÖDINGER OPERATORS AND FOR THE EXISTENCE OF $\mathcal{PT}$-SYMMETRIC PHASE TRANSITIONS

EMANUELA CALICETI AND SANDRO GRAFFI
Dipartimento di Matematica, Università di Bologna, Bologna, Italy

4.1 INTRODUCTION

As illustrated in this book, a major mathematical problem in $\mathcal{PT}$-symmetric quantum mechanics (see e.g., 1–7) is to determine whether or not the spectrum of any given non-self-adjoint but $\mathcal{PT}$-symmetric Schrödinger operator is real. In this case, the $\mathcal{PT}$-symmetry is called *proper* (8). Clearly, in this connection, an equally important issue is the spontaneous violation of the $\mathcal{PT}$-symmetry, which might occur in a $\mathcal{PT}$-symmetric operator family. The spontaneous violation of the $\mathcal{PT}$-symmetry is defined as the transition from real values of the spectrum to complex ones at the variation of the parameter labeling the family. Its occurrence is referred to also as the $\mathcal{PT}$-symmetric phase transition.

This chapter intends to present a review of the recent results concerning these two mathematical points, within the standard notions of spectral theory for Hilbert space operators. The main technical instrument is represented by perturbation theory.

Non-Selfadjoint Operators in Quantum Physics: Mathematical Aspects, First Edition.
Edited by Fabio Bagarello, Jean-Pierre Gazeau, Franciszek Hugon Szafraniec and Miloslav Znojil.

The plan of the presentation is as follows: in Section 2, we formulate some general conditions under which perturbation theory may give global information about the spectrum in this $\mathcal{PT}$-symmetric situation. In the framework of one-dimensional, $\mathcal{PT}$-symmetric Schrödinger operators, in Section 3, we review the criteria for the reality of the discrete spectrum, and in Section 4, we describe a criterion for the reality of the continuous band spectrum for $\mathcal{PT}$-symmetric periodic perturbations of periodic potentials. In Section 5, a class of one-dimensional anharmonic oscillators is described where the existence of the $\mathcal{PT}$-symmetric phase transition can be explicitly proved. Finally, in Section 6, we show, in the case of $\mathcal{PT}$-symmetric Schrödinger operators on the torus, how the method of the convergent quantum normal form can be successfully applied to prove the reality of the spectrum for a class of $\mathcal{PT}$-symmetric nonlocal perturbations.

Conventions and notation

The $\mathcal{PT}$-symmetry operation acts in $L^2(\mathbb{R}^d)$, $d \geq 1$, as the following semilinear operator:

$$(\mathcal{PT}\psi)(x) = \overline{\psi}(-x), \qquad \psi \in L^2(\mathbb{R}^d).$$

Here, $\mathcal{P}$ denotes the parity operation $(\mathcal{P}\psi)(x) = \psi(-x)$, $\forall \psi \in L^2(\mathbb{R}^d)$, and $\mathcal{T}$ the complex conjugation $(\mathcal{T}\psi)(x) = \overline{\psi}(x)$, $\forall \psi \in L^2(\mathbb{R}^d)$. An operator S acting in $L^2(\mathbb{R}^d)$ is $\mathcal{PT}$-symmetric if $\mathcal{PT}[D(S)] \subset D(S)$ and $[S, \mathcal{PT}] = 0$ on $D(S)$.
Moreover, we assume:

1. The $\mathcal{PT}$-symmetric Schrödinger operators under consideration always act in $L^2(\mathbb{R}^d)$ and have the form $H = T + V$. Here, T is a self-adjoint, $\mathcal{PT}$-symmetric operator, V a closed, $\mathcal{PT}$-symmetric operator, and H is defined on the maximal domain $D(H) = D(T) \cap D(V)$;
2. Unless otherwise specified, as in the case of $\mathcal{PT}$-symmetric operators with periodic potentials in Section 4, the spectrum Spec (T) of the self-adjoint operator T is discrete.

Under these conditions, the operator H is $\mathcal{PT}$-symmetric; moreover, Spec (H) is discrete if the identity operator is compact from the graph domain of H to L^2. Let us furthermore recall the following notation:

1. $H^p(\mathbb{R}^d)$ is the Sobolev space of order $p \in \mathbb{R}$ with norm $L^2(\mathbb{R}^d)$, namely:

$$H^p(\mathbb{R}^d) := \{u \in L^2(\mathbb{R}^d) \; : \; (1 + |\xi|^2)^{p/2}\, \widehat{u}(\xi) \in L^2(\mathbb{R}^d)\}, \tag{4.1.1}$$

   Here $\widehat{u}(\xi)$ is the Fourier transform of $u(x) \in L^2(\mathbb{R}^d)$;
2. $L^2_p(\mathbb{R}^d)$ is the weighted L^2-space of order p, namely:

$$L^2_p(\mathbb{R}^d) := \{u \in L^2(\mathbb{R}^d) \; : \; (1 + |x|^2)^{p/2}\, u(x) \in L^2(\mathbb{R}^d)\}. \tag{4.1.2}$$

Examples

1. $\mathcal{PT}$-symmetric bounded perturbations of harmonic oscillators in $L^2(\mathbb{R}^d)$. Here:

$$T = -\Delta + \sum_{i=1}^{d} \omega_i^2 x_i^2, \ \omega_i \in \mathbb{R}; \qquad D(T) = H^2(\mathbb{R}^d) \cap L_2^2(\mathbb{R}^d); \tag{4.1.3}$$

   V is the maximal multiplication operator in L^2 by any bounded, continuous, imaginary valued function $V(x)$ such that

$$V(-x_1, \dots, -x_d) = -V(x_1, \dots, x_d).$$

   Thus, V is a bounded operator in L^2. Therefore, H is $\mathcal{PT}$-symmetric with discrete spectrum.
2. Odd anharmonic oscillators in $L^2(\mathbb{R}^d)$. Here, T is as above, and:

$$V(x) = iP_{2M+1}(x_1, \dots x_d), \quad M = 1, 2, \dots, \tag{4.1.4}$$

   where $P_{2M+1}(x)$ is any real-valued odd polynomial of degree $2M + 1$ in $(x_1, \dots, x_d)$. In this case, it is known (see e.g., 9) that the maximal operator H in $L^2(\mathbb{R}^d)$ defined by $T + V$ on $D(T) \cap D(V)$ has discrete spectrum.

Remark

Let $g \in \mathbb{C}$. Under the conditions of Example (1) on T and V, consider the operator family in L^2 defined as $H(g) := T + gV$ on the maximal domain $D(T) \cap D(V) = D(T)$. Then, $g \mapsto H(g)$ is a type-A holomorphic operator family in the sense of Kato (10, §VII.1) because its domain is independent of g and the scalar products $g \mapsto \langle u, G(g)u \rangle$ are holomorphic $\forall\, g \in \mathbb{C}$ for any $u \in\ D(T) = D(H(g))$. This entails, in particular, that simple isolated eigenvalues of $g \mapsto H(g)$ are locally holomorphic functions of g. Under the conditions of Example (2), the family $g \mapsto H(g)$ is holomorphic away from $g = 0$.

4.2 PERTURBATION THEORY AND GLOBAL CONTROL OF THE SPECTRUM

Unless otherwise specified, the present $\mathcal{PT}$ symmetric operators appear as perturbations of self-adjoint operators with discrete spectrum. Now a preliminary problem to be solved is the formulation of conditions under which the perturbation theory of discrete eigenvalues, typically a local one, may yield global information on the spectrum. The first general result is just an elementary remark formulated under the form of a lemma.

Lemma 4.2.1 *Let $g \mapsto H(g)$ be a type-A holomorphic family of operators defined on an open set $\Omega \subset \mathbb{C}$ symmetric with respect to the real axis. Furthermore, let $H(g)$ be $\mathcal{PT}$-symmetric for $g \in \mathbb{R}$. Let $g_0 \in \Omega \cap \mathbb{R}$ and $\lambda(g_0) \in \mathbb{R}$ be an isolated and simple eigenvalue of $H(g_0)$. Then, there is $r(g_0) > 0$ such that, for $|g - g_0| < r(g_0)$, $H(g)$ has one and only one eigenvalue $\lambda(g)$ near $\lambda(g_0)$. Moreover, $\lambda(g) \in \mathbb{R}$ for $g \in \mathbb{R}$, $|g - g_0| < r(g_0)$.*

Proof: The $\mathcal{PT}$-symmetry entails that $\lambda(g)$ is an isolated eigenvalue of $H(g), g \in \mathbb{R}$, if and only if $\overline{\lambda}(g)$ enjoys the same property. In turn, the holomorphy entails that the isolated simple eigenvalue $\lambda(g_0) \in \mathbb{R}$ is *stable* (see e.g., 10, §VIII.1), namely, there exists $r(g_0) > 0$ such that $H(g)$ admits one and only one eigenvalue $\lambda(g)$ near $\lambda(g_0)$ for $|g - g_0| < r(g_0)$, analytic at $g = g_0$. The two statements may simultaneously hold for $g \in \mathbb{R}$ if and only if $\lambda(g) = \overline{\lambda}(g)$, and this proves the lemma. ■

An easy consequence of this lemma is the following:

Corollary 4.2.2 *Consider the holomorphic operator family $H(g) = T + gV$ in $L^2(\mathbb{R}^d)$, $g \in \mathbb{C}$, $\mathcal{PT}$-symmetric for $g \in \mathbb{R}$, where T and V are as in Example (1). Assume furthermore that the frequencies $\omega_i > 0$ are rationally independent, namely, that the equation*

$$\omega_1 \nu_1 + \cdots + \omega_d \nu_d = 0, \quad \nu_i \in \mathbb{Z}, \ i = 1, \ldots, d \tag{4.2.1}$$

has only the trivial solution $\nu_1 = \cdots = \nu_d = 0$. Then, for any given $n_i \in \mathbb{Z} \cup \{0\}$, $i = 1, \ldots, d$, there is $r(n_1, \ldots, n_d) > 0$ such that $H(g)$ has one and only one eigenvalue $\lambda_{n_1,\ldots,n_d}(g)$ near the eigenvalue $\lambda_{n_1,\ldots,n_d} = \omega_1 n_1 + \cdots + \omega_d n_d$ of T, which is real for $|g| < r(n_1, \ldots, n_d)$, $g \in \mathbb{R}$.

Proof: Any eigenvalue $\lambda_{n_1,\ldots,n_d}$ of T is simple by condition (4.2.1), and thus by the holomorphy of the family, the corresponding eigenvalue $\lambda_{n_1,\ldots,n_d}(g)$ of $H(g)$ for $|g| < r(n_1, \ldots, n_d)$ is simple. Hence, the assertion follows as in Lemma 4.2.1. This proves the corollary. ■

Remark

The same assertion holds when H and V fulfill the conditions of Example (2). However, in this case, the operator family $H(g)$ is not analytic at $g = 0$ and the proof is based on the norm resolvent convergence of $H(g)$ to T as $g \to 0$ (see e.g., 11).

Let us now state and prove the general result ensuring the reality of the spectrum of the $\mathcal{PT}$-symmetric operator family $H(g), g \in \mathbb{R}$, mentioned earlier.

Theorem 4.2.3 *Let the $\mathcal{PT}$-symmetric self-adjoint operator T be bounded below (and thus positive, without loss), with compact resolvent, and simple eigenvalues $0 < \lambda_0 < \lambda_1 < \ldots \lambda_\ell < \ldots$. Let the $\mathcal{PT}$ symmetric operator V be bounded. Consider the operator family $H(g) = T + gV$ defined on $D(T)$, holomorphic for $g \in \mathbb{C}$ and $\mathcal{PT}$-symmetric for $g \in \mathbb{R}$. Assume:*

$$\delta := \frac{1}{2} \inf_{j \geq 0} |\lambda_{j+1} - \lambda_j| > 0. \tag{4.2.2}$$

Then Spec $H(g) \subset \mathbb{R}$ *if* $g \in \mathbb{R}$, $|g| < \dfrac{\delta}{\|V\|}$.

Example

Here, $d = 1$, $T = -\dfrac{d^2}{dx^2} + W(x)$, $W(x) = kx^{2m}$, $k > 0$, $m \geq 1$, $D(T) = H^2(\mathbb{R}) \cap D(W)$; $V(x) = iQ(x)$, $Q(x) \in \mathbb{R}$, $Q \in L^\infty(\mathbb{R})$, $Q(-x) = -Q(x)$. We have, as is well known:

$$\lambda_n \sim k^{\frac{1}{2m}} n^{\frac{2m}{m+1}}, \qquad n \to \infty.$$

Each eigenvalue λ_n is simple. Clearly, we may assume without loss $\delta \geq 1$. Then, $H(g) := T + gV$ has real discrete spectrum for $|g| < \|V\|_\infty^{-1} = \|Q\|_\infty^{-1}$.

Proof: We repeat here the argument of Ref. (12). Under the present assumptions, we recall that $H(g)$ is a type-A holomorphic family of operators, with compact resolvents, $\forall\, g \in \mathbb{C}$. Hence, $\mathrm{Spec}(H(g)) = \{\lambda_l(g) : l = 0, 1, \dots\}$. In particular:

(i) The eigenvalues $\lambda_l(g)$ are locally holomorphic functions of g with only algebraic singularities;

(ii) the eigenvalues $\lambda_l(g)$ are stable, namely, given any eigenvalue $\lambda_l(g_0)$ of $H(g_0)$ there is exactly one eigenvalue $\lambda_l(g)$ of $H(g)$ such that $\lim_{g \to g_0} \lambda_l(g) = \lambda_l(g_0)$;

(iii) the Rayleigh-Schrödinger perturbation expansions for the eigenprojections and the eigenvalues near any eigenvalue λ of T have convergence radius $\delta_\lambda / \|V\|$ where $\delta_\lambda := \dfrac{1}{2} \inf_{j : \lambda_j \neq \lambda} |\lambda - \lambda_j|$ is half the isolation distance of λ.

Remark that as $\delta_\lambda \geq \delta$, $\forall\, \lambda$, all series will be convergent for all $g \in \Omega_{r_0}$; $\Omega_{r_0} := \{g \in \mathbb{C} : |g| < r_0\}$, where $r_0 := \delta / \|V\|$ is a uniform lower bound for all convergence radii. To simplify the notation, we can now assume, without loss of generality, $\|V\| = 1$. By hypothesis $|\lambda_l - \lambda_{l+1}| \geq 2\delta > 0$, $\forall\, l \in \mathbb{N}$. First remark that if $g \in \mathbb{R}$, $|g| < r_0$, and $\lambda(g)$ is an eigenvalue of $H(g)$, then $|\Im \lambda(g)| < \delta$, that is, $\mathrm{Spec}(H_(g)) \cap \mathbb{C}_\delta = \emptyset$, $\mathbb{C}_\delta := \{z \in \mathbb{C} \mid |\Im z| \geq \delta\}$. Set indeed

$$R_0(z) := [T - z]^{-1}, \qquad z \notin \mathrm{Spec}(T).$$

Then, $\forall\, z \in \mathbb{C}$ such that $|\Im z| \geq \delta$, we have

$$\|gVR_0(z)\| \leq |g| \cdot \|V\| \cdot \|R_0(z)\| \leq \frac{|g|}{\mathrm{dist}[z, \mathrm{Spec}(T)]} \leq \frac{|g|}{|\Im z|} \leq \frac{|g|}{\delta} < 1. \tag{4.2.3}$$

Hence, the resolvent

$$R_g(z) := [H(g) - z]^{-1} = R_0(z)[1 + VR_0(z)]^{-1}$$

exists and is bounded if $|\Im z| \geq \delta$ because (4.2.3) entails the uniform norm convergence of the Neumann expansion for the resolvent:

$$\begin{aligned}\|R_g(z)\| &= \|[H(g)-z]^{-1}\| = \|R_0(z)\sum_{k=0}^{\infty}[-gVR_0(z)]^k\| \\ &\leq \|R_0(z)\| \sum_{k=0}^{\infty} |g^k| \|VR_0(z)]\|^k \leq \|R_0(z)\| \sum_{k=0}^{\infty} (|g|/\delta)^k \\ &\leq \|R_0(z)\| \frac{\delta}{\delta - |g|}. \end{aligned} \tag{4.2.4}$$

Now $\forall\, l \in \mathbb{N}$ let $Q_l(\delta)$ denote the open square of side 2δ centered at λ_l. As $|\lambda_l - \lambda_{l+1}| \geq 2\delta$, it follows as aforementioned that $R_g(z)$ exists and is bounded for $z \in \partial Q_l(\delta)$, the boundary of $Q_l(\delta)$. We can, therefore, according to the standard procedure (see e.g., 10, Chapter III.2), define the strong Riemann integrals

$$P_l(g) = \frac{1}{2\pi i}\int_{\partial Q_l(\delta)} R_g(z)\,dz, \qquad l = 1, 2, \ldots.$$

As is well known, P_l is the spectral projection onto the part of $\mathrm{Spec}(H(g))$ inside Q_l. As $g \mapsto H(g)$ is a holomorphic family, by well-known results (see e.g., 10, Thm. VII.2.1), the same is true for $P_l(g)$ for all $l \in \mathbb{N}$. In particular, this entails the continuity of $P_l(g)$ for $|g| < r_0$. Now $P_l(0)$ is a one-dimensional projection: hence, the same is true for $P_l(g)$. As a consequence, there is one and only one point of $\mathrm{Spec}(H(g))$ inside any Q_l. Now $\mathrm{Spec}(H(g))$ is discrete, and thus any such point is an eigenvalue; moreover, any such point is real for g real because $\mathrm{Spec}(H(g))$ is symmetric with respect to the real axis. Finally, we note (see also 13 for more detail) that if $z \in \mathbb{R}$, $z \notin \bigcup_{l=1}^{\infty}]\lambda_l - \delta, \lambda_l + \delta[$ the Neumann series (4.2.4) is convergent and the resolvent $R_g(z)$ is continuous. This concludes the proof of Theorem 4.2.3. ∎

4.3 ONE-DIMENSIONAL $\mathcal{PT}$-SYMMETRIC HAMILTONIANS: CRITERIA FOR THE REALITY OF THE SPECTRUM

The criterion to be described in this section applies to a class of $\mathcal{PT}$-symmetric anharmonic oscillators. It is based on an extension of Theorem 4.2.3 to relatively bounded perturbations.

Let $P(x) : x \in \mathbb{R}$ be a real, even polynomial of degree $2p$, $p \geq 1$, $Q(x)$ a real, odd polynomial of degree $2r-1$, and g a real number. Consider the Schrödinger operator family in $L^2(\mathbb{R})$ $H(g) = T + igQ$, $g \in \mathbb{R}$, defined on the maximal domain $D(T) \cap D(Q)$. Here, T is the differential operator in $L^2(\mathbb{R})$ generated by the differential expression $-\dfrac{d^2}{dx^2} + P(x)$ acting on the maximal domain $H^2(\mathbb{R}) \cap D(P)$ so that

$H(g)$ is the operator family defined by the differential operator $-\frac{d^2}{dx^2} + P(x) + igQ(x)$ acting on its maximal domain:

$$\begin{aligned} H(g)\psi &:= -\frac{d^2\psi}{dx^2} + P(x)\psi + igQ(x)\psi, \\ D(H(g)) &= H^2(\mathbb{R}) \cap D(P) \cap D(Q). \end{aligned} \tag{4.3.1}$$

As is well known, T is self-adjoint and positive in $L^2(\mathbb{R})$, with compact resolvent. Its eigenvalues λ_n form an increasing sequence such that $\lim_{n\to\infty} \lambda_n = +\infty$ and fulfill the estimate (see e.g., 14)

$$\lambda_n = Bn^{2p/(p+1)} + O(n^{\frac{p-1}{p+1}}), \quad n \to \infty \tag{4.3.2}$$

for some positive constant B.
$H(g)$ is $\mathcal{PT}$-symmetric and has compact resolvent (see e.g., 14) and, therefore, discrete spectrum; obviously, however, it is not self-adjoint and not even normal.

Theorem 4.3.1 *In the aforementioned notations, let* $p > 2r$. *Then, there exists* $\overline{R} > 0$ *such that* $\mathrm{Spec}(H(g)) \subset \mathbb{R}$ *if* $|g| < \overline{R}$.

Remark

For $P = (-iq)^m$, $Q = -i\sum_{j=1}^{m-1} a_j(iq)^{m-j}$, $m \geq 2$, g arbitrary, and $(j-k)a_k \geq 0\ \forall\, k$ for at least one $1 \leq j \leq m/2$, the same result has been proved by Shin (15) for all g. For $P(q) = q^2$, $Q(q) = q^{2r-1}$, by Dorey *et al.* (16) again for all g. Hence, we see that for $m = 2p$, the present result extends the previous ones.
To prove the theorem, we first establish a preliminary result:

Proposition 4.3.2 *All eigenvalues* λ_n *of* T *are stable with respect to* $H(g)$ *as* $|g| \to 0$. *suitably small; moreover, there is* $\overline{R} > 0$ *independent of* n *such that the perturbation expansion for any eigenvalue* $\lambda_n(g)$ *near* λ_n *converges for* $|g| < \overline{R}$.

Proof: T is self-adjoint with compact resolvent, and Q is relatively bounded with respect to T with relative bound zero. This means that $D(Q) \supset D(T)$ and that for any $b > 0$, there is $a > 0$ such that

$$\|Qu\| \leq b\|H(0)u\| + a\|u\|, \qquad \forall\, u \in D(H(0)). \tag{4.3.3}$$

This entails that $H(g)$ defined on the maximal domain $D(T)$ is a closed operator with compact resolvent (and hence discrete spectrum) for all $g \in \mathbb{C}$. Moreover, all eigenvalues of T, which are simple, are stable as eigenvalues of $H(g)$ for $|g|$ suitably small, that is, there is one and only one simple eigenvalue $\lambda_n(g)$ of $H(g)$ near λ_n for $|g|$ suitably small (see e.g., 17 for the formal definition). Moreover, under the present conditions, the (Rayleigh–Schrödinger) perturbation expansion near any unperturbed eigenvalue λ_n has a positive convergence radius; a lower bound for the convergence radius is given by (see 10, formula VII.2.34):

$$\overline{R}(n) = \left[\frac{2(a + b|\lambda_n|)}{d_n} + 2b + 1\right]^{-1}. \tag{4.3.4}$$

Here, b and a are the constants of the estimate (4.3.3) and d_n is half the isolation distance of the eigenvalue λ_n, namely:

$$d_n := \frac{1}{2}\min\left(|\lambda_n - \lambda_{n-1}|, |\lambda_n - \lambda_{n+1}|\right). \tag{4.3.5}$$

In this case, by (4.3.2), we have

$$d_n \sim n^{\frac{p-1}{p+1}}, \qquad \frac{\lambda_n}{d_n} \sim n, \quad n \to \infty. \tag{4.3.6}$$

Therefore, to prove the assertion, it is enough to verify that, choosing $b = b_n = \dfrac{1}{n}$ in (4.3.3), we can find $a = a_n$ such that

$$\frac{a_n}{d_n} \leq A < +\infty, \qquad n \to \infty \tag{4.3.7}$$

for a suitable constant $A > 0$. By (4.3.4), this entails that the perturbation expansions near the unperturbed eigenvalues λ_n have a common convergence circle of radius $\overline{R} > 0$ independent of n, that is, $\overline{R}(n) \geq \overline{R}$, $\forall\, n$. We have indeed:

Lemma 4.3.3 *Let $p > 2r$, and $n \in \mathbb{N}$. Then, there exist $K > 0$ and $N > 0$ such that the estimate (4.3.3) holds with $b_n = \dfrac{K}{n}$ and $a_n < Nn^{\frac{p-1}{p+1}}$.*

Proof: Introduce from now on the standard notation $Du = -i\dfrac{du}{dx}$ such that

$$D^2 u = -\frac{d^2 u}{dx^2}, \qquad T = D^2 + P(x).$$

The following quadratic estimate is well known (see e.g., 18)

$$\|D^2 u\| + \|P(x)u\| \leq \alpha\|(D^2 + P(x))u\| + \beta\|u\|, \qquad \forall\, u \in D(H(0)) \tag{4.3.8}$$

for some $\alpha > 0, \beta > 0$. Therefore, to prove (4.3.3) with the stated constants a_n and b_n, it will be enough to prove the further estimate

$$\|Qu\| \leq \frac{1}{n}\|Pu\| + a_n\|u\|, \qquad \forall\, u \in D(P) \tag{4.3.9}$$

because we then have $\|Qu\| \leq \frac{\alpha}{n}\|Tu\| + b_n\beta\|u\| + a_n\|u\|$ and the constant $b_n\beta$ can be obviously absorbed in a_n. In turn, (4.3.9) follows from

$$\|Qu\|^2 \leq \frac{1}{n^2}\|Pu\|^2 + a_n^2\|u\|^2, \qquad \forall\, u \in D(P). \tag{4.3.10}$$

Now this L^2 inequality is clearly implied by the pointwise inequality

$$\frac{1}{n^2}P(x)^2 - Q(x)^2 + a_n^2 \geq 0, \qquad \forall\, x \in \mathbb{R}. \tag{4.3.11}$$

Next we remark that, up to an additive constant, which can be absorbed in the constant β of the estimate (4.3.8), we can limit ourselves to verify this inequality for homogeneous polynomials P and Q of degree $2p$ and $2r - 1$, respectively. Then, this last inequality reads

$$x^{4r-2} \leq \frac{1}{n^2}x^{4p} + a_n^2, \qquad \forall\, x \in \mathbb{R}. \tag{4.3.12}$$

As there are only even powers, we can restrict to $x \geq 0$. Remark that the inequality

$$x^{4r-2} \leq n^{-2}x^{4p}$$

is fulfilled if

$$x \geq n^{\alpha}, \qquad \alpha = \frac{1}{2(p-r)+1}.$$

On the other hand, if $x < n^{\alpha}$ then $x^{4r-2} < n^{(4r-2)\alpha}$; hence, the inequality $x^{4r-2} < a_n^2$, which yields (4.3.12) in this case, will be fulfilled if $a_n \geq n^{\frac{2r-1}{2(p-r)+1}}$ If suffices, then to have $\frac{2r-1}{2(p-r)+1} < \frac{p-1}{p+1}$ and this inequality is *a fortiori* true if $p > \frac{4r+1}{2}$. This concludes the proof of the Lemma. ∎

An immediate consequence of this is the existence of the constant A as in (4.3.7) and consequently of a common convergence radius $\overline{R}$. This verifies Proposition 4.3.2. ∎

Proof of Theorem 4.3.1

We have only to verify that for $|g| < \overline{R}$, the eigenvalues of $H(g)$ defined by the convergent perturbation expansions are real and that $H(g)$ has no other eigenvalue. We carry out this by adopting the argument of Theorem 2.3. Under the present assumptions, $H(g)$ is a type-A holomorphic family of operators in the sense of Kato with compact resolvents $\forall\, g \in \mathbb{C}$. In particular, this entails that any eigenvalue $\lambda_n(g)$ is a locally holomorphic function of g with only algebraic singularities. Moreover, any such eigenvalue is stable.

We have seen earlier that $\forall\, n \in \mathbb{N}$, the Rayleigh-Schrödinger perturbation expansion for the eigenvalue $\lambda_n(g)$ near any eigenvalue λ_n of T is convergent for all $g \in \Omega_{\overline{R}} := \{g \in \mathbb{C} : |g| \leq \overline{R}\}$, where $\overline{R}$ is the uniform lower bound for all convergence radii.

The first part of the argument concerns localization of the eigenvalues of $H(g_0)$, $g_0 \in \mathbb{C}$. As $b_n = K/n$, by (4.3.6) and ((4.3.7) there exists $A > 0$ sufficiently large such that

$$3b_n + \frac{b_n \lambda_n}{d_n} + \frac{a_n}{d_n} \leq A, \quad \forall n \in \mathbb{N} \tag{4.3.13}$$

$$2b_n + \frac{b_n(\lambda_{n+1} - d_{n+1})}{\delta_n} + \frac{a_n}{\delta_n} \leq A, \quad \forall n \in \mathbb{N} \tag{4.3.14}$$

$$2b_1 + \frac{a_1}{|\lambda_1 - d_1|} \leq A. \tag{4.3.15}$$

Here,

$$\delta_n := \min(d_n, d_{n+1}).$$

For any $n \in \mathbb{N}$, let $\mathcal{Q}_n$ be the square centered at λ_n and of side $2d_n$ in the complex z plane, and recall that the eigenvalues $\lambda_n : n \in \mathbb{N}$ form an increasing sequence. Let $\mathcal{A} := \{z \in \mathbb{C} : \text{Re}z \geq (\lambda_1 - d_1)\} \setminus (\bigcup_{n\in\mathbb{N}} \mathcal{Q}_n)$. Let us show that this set has empty intersection with the spectrum of $H(g)$ for $|g| < 1/A$, that is, $\mathcal{A} \subset \rho(H(g)) = \mathbb{C} \setminus \text{Spec}(H(g))$. $\forall z \in \mathcal{A}$, there are indeed two possibilities:

$$a) \quad \exists\ s \in \mathbb{N} \quad s.t. \quad |\Im z| \geq d_s \quad \text{and} \quad |\Re z - \lambda_s| \leq \delta_s$$

$$b) \quad \exists\ s \in \mathbb{N} \quad s.t. \quad \lambda_s + d_s \leq \Re z \leq \lambda_{s+1} - d_{s+1}.$$

Case a): Let $R_0(z) := (T - z)^{-1}$ be the free resolvent. Then, we have:

$$\begin{aligned}
\|gQR_0(z)\| &\leq |g| \cdot \|QR_0(z)\| \\
&\leq |g|[b_s\|[T - z]R_0(z)\| + b_s|z|\,\|R_0(z)\| + a_s\,\|R_0(z)\|] \\
&\leq |g| \left[b_s + \frac{b_s|z| + a_s}{\text{dist}(z, \text{Spec}(T))}\right] \leq |g| \left[b_s + b_s\frac{\lambda_s + d_s + |\Im z|}{|\Im z|} + \frac{a_s}{|\Im z|}\right] \\
&\leq |g| \left[3b_s + b_s\frac{\lambda_s}{d_s} + \frac{a_s}{d_s}\right] \leq |g|A < 1, \quad (4.3.16)
\end{aligned}$$

if $|g| < 1/A$. This formula follows by the relative boundedness condition $\|Qu\| \leq b_n\|Tu\| + a_n\|u\|$, the fact that $\|R_0(z)\| = \dfrac{1}{\text{dist}(z, \text{Spec}(T)} \leq \dfrac{1}{|\Im z|}$, and formula (4.3.13). We now prove that the resolvent

$$R_g(z) := [H(g) - z]^{-1} = R_0(z)[1 + igQR_0(z)]^{-1}$$

exists and is bounded by the uniform norm convergence of the Neumann expansion. We have indeed, by (4.3.16):

$$\|R_g(z)\| = \|R_0(z)[1+igQR_0(z)]^{-1}\| = \|R_0(z)\sum_{k=0}^{\infty}[-igQR_0(z)]^k\| \leq$$

$$\leq \|R_0(z)\| \sum_{k=0}^{\infty} |g^k| \|QR_0(z)]\|^k \leq \frac{\|R_0(z)\|}{1-|g|A}. \tag{4.3.17}$$

Moreover, $\|R_0(z)\|$ is uniformly bounded for z in the compacts of $\rho(T)$, and this entails the same property for $R_g(z)$.
Case b) Analogous computations yield

$$\|gR_0(z)\| \leq |g|A < 1$$

provided $|g| < 1/A$. Here, we have used (4.3.14). Hence, as in *a)* $z \in \rho(H(g))$ if $|g| < 1/A$. Finally, let us prove that $\mathbb{C} \setminus \bigcup_{n\in\mathbb{N}} \mathcal{Q}_n \subset \rho(H(g))$ if $|g| < 1/A$. We only need to show that $z \in \rho(H(g))$ if $|g| < 1/A$ and $\Re z \leq \lambda_1 - d_1$. In fact, once more we have:

$$\|gR_0(z)\| \leq |g|A < 1$$

for $|g| < 1/A$.
The results so far obtained allow us to assert that if $\lambda(g_0)$ is an eigenvalue of $H(g_0)$ with $g_0 \in \mathbb{R}$, $|g_0| < 1/A$, then $\lambda(g_0) \in \bigcup_{n\in\mathbb{N}} \mathcal{Q}_n$. As the open squares $\mathcal{Q}_n$ are disjoint, there exists $n_0 \in \mathbb{N}$ such that $\lambda(g_0) \in \mathcal{Q}_{n_0}$. We can assume, without loss, $g_0 > 0$. Moreover, if $g \mapsto \lambda(g)$ is a continuous function defined on any open subset D of the circle $\{g : |g| < 1/A\}$ containing g_0, then $\lambda(g) \in \mathcal{Q}_{n_0}$. Now, as $\lambda(g_0)$ is stable, it generates a continuous function $\lambda(g)$ defined in a real left neighborhood of g_0 (see 19 for details), which can be continued to a holomorphic function $\lambda(g)$ defined on $]-1/A, 1/A[$, taking values in $\mathcal{Q}_{n_0}$, such that $\lambda(g)$ is an eigenvalue of $H(g)$. In particular, $\lambda(0)$ coincides with λ_{n_0}, which is the only eigenvalue of $H(0)$ inside $\mathcal{Q}_{n_0}$. As λ_{n_0} is simple, $\lambda(g)$ is the only eigenvalue of $H(g)$ close to λ_{n_0} for g small. Thus, $\lambda(g)$ must be real, because if it is complex also its conjugate $\overline{\lambda}(g)$ enjoys the same property, which is ruled out by the stability. Moreover, $\lambda(g)$ is the sum of the perturbation expansion around λ_{n_0} and, therefore, is a holomorphic function for $|g| < \overline{R}$. Let from now on $|g| < g_1 := \min(\overline{R}, 1/A)$. Then, the holomorphy implies that the real valuedness for $|g|$ small extends to all $|g| < g_1$. This concludes the proof of Theorem 4.3.1.

4.4 $\mathcal{PT}$-SYMMETRIC PERIODIC SCHRÖDINGER OPERATORS WITH REAL SPECTRUM

We now proceed to deal with the problem of the reality of the spectrum of $\mathcal{PT}$-symmetric Schrödinger operators in the context of periodic potentials on $\mathbb{R}$, already considered in Ref. (20–25). Without loss, the period is assumed to be 2π. If the potential V is periodic and real valued, it is well known (see e.g., 14) that, under mild regularity assumptions, the spectrum of $H = T + V$ is absolutely continuous on $\mathbb{R}$ and band-shaped.

It is then natural to ask whether or not there exist classes of $\mathcal{PT}$ symmetric, complex periodic potentials generating Schrödinger operators that are non-self-adjoint but with real band spectrum.

We describe here an explicit criterion isolating a class of operators enjoying the aforementioned properties.

Denote $\tau(u)$ the nonnegative quadratic form in $L^2(\mathbb{R})$ with domain $H^1(\mathbb{R})$ defined by the kinetic energy:

$$\tau(u) := \int_{\mathbb{R}} |u'|^2 \, dx, \qquad u \in H^1(\mathbb{R}). \tag{4.4.1}$$

Let q be a real-valued, tempered distribution. Assume:

(1) q is a 2π-periodic, $\mathcal{P}$ symmetric distribution belonging to $H^{-1}_{loc}(\mathbb{R})$;
(2) $W : \mathbb{R} \to \mathbb{C}$ belongs to $L^\infty(\mathbb{R})$ and is $\mathcal{PT}$-symmetric, that is, $\overline{W(-x)} = -W(x)$, $\forall x \in \mathbb{R}$;
(3) q generates a real quadratic form $\mathcal{Q}(u)$ in $L^2(\mathbb{R})$ with domain $H^1(\mathbb{R})$;
(4) $\mathcal{Q}(u)$ is relatively bounded with respect to $\tau(u)$ with relative bound $b < 1$, that is, there are $b < 1$ and $a > 0$ such that

$$\mathcal{Q}(u) \leq b\tau(u) + a\|u\|^2, \quad \forall\, u \in H^1(\mathbb{R}). \tag{4.4.2}$$

Under these assumptions, the real quadratic form

$$\mathcal{H}_0(u) := \tau(u) + \mathcal{Q}(u), \quad u \in H^1(\mathbb{R}) \tag{4.4.3}$$

is closed and bounded below in $L^2(\mathbb{R})$. We denote $H(0)$ the corresponding self-adjoint operator. This is the self-adjoint realization of the formal differential expression (note the abuse of notation)

$$H(0) = -\frac{d^2}{dx^2} + q(x).$$

Under these circumstances, it is known (see e.g., 26) that the spectrum of $H(0)$ is continuous and band-shaped. For $n = 0, 1, 2, \ldots$, we denote

$$B_{2n} := [\alpha_{2n}, \beta_{2n}], \quad B_{2n+1} := [\beta_{2n+1}, \alpha_{2n+1}]$$

the bands of $H(0)$, and $\Delta_{2n} =]\beta_{2n}, \beta_{2n+1}[$, $\Delta_{2n+1} =]\alpha_{2n+1}, \alpha_{2n+2}[$, the gaps between the bands. Here:

$$0 \le \alpha_0 \le \beta_0 \le \beta_1 \le \alpha_1 \le \alpha_2 \le \beta_2 \le \beta_3 \le \alpha_3 \le \alpha_4 \le \dots .$$

The maximal multiplication operator by W is continuous in L^2, and, therefore, so is the quadratic form $\langle u, Wu \rangle$. It follows that the quadratic form family

$$\mathcal{H}_g(u) := \tau(u) + \mathcal{Q}(u) + g\langle u, Wu \rangle, \quad u \in H^1(\mathbb{R}) \tag{4.4.4}$$

is closed and sectorial in $L^2(\mathbb{R})$ for any $g \in \mathbb{C}$.
We denote $H(g)$ as the uniquely associated $m-$sectorial operator family in $L^2(\mathbb{R})$. This is the realization of the formal differential operator family

$$H(g) = -\frac{d^2}{dx^2} + q(x) + gW(x).$$

By definition, $H(g)$ is a holomorphic family of operators of type B in the sense of Kato for $g \in \mathbb{C}$; by (2) it is also $\mathcal{PT}$ symmetric for $g \in \mathbb{R}$. Then, the criterion is:

Theorem 4.4.1 *Let all gaps and bands of $H(0)$ be open, namely: $\alpha_{2n} < \beta_{2n} < \beta_{2n+1} < \alpha_{2n+1} < \alpha_{2n+2}$, $\forall n = 0, 1, \dots$, and assume that*

$$d := \frac{1}{2} \inf_{n \in \mathbb{N}} |\Delta_n| > 0. \tag{4.4.5}$$

Here $|\Delta_n|$ denotes the width of the gap Δ_n. Then, if

$$|g| < \frac{d^2}{2(1+d)\|W\|_\infty} := \overline{g} \tag{4.4.6}$$

there exist

$$0 \le \alpha_0(g) < \beta_0(g) < \beta_1(g) < \alpha_1(g) < \alpha_2(g) < \beta_2(g) < \dots$$

such that

$$\sigma(H(g)) = \left(\bigcup_{n \in \mathcal{N}} B_{2n}(g) \right) \bigcup \left(\bigcup_{n \in \mathcal{N}} B_{2n+1}(g) \right) \tag{4.4.7}$$

where, as for $g = 0$:

$$B_{2n}(g) := [\alpha_{2n}(g), \beta_{2n}(g)], \quad B_{2n+1}(g) := [\beta_{2n+1}(g), \alpha_{2n+1}(g)].$$

Remark

The theorem states that for $|g|$ being small enough, the spectrum of the non-self-adjoint operator $H(g)$ remains real and band-shaped. The proof is critically dependent on the validity of the lower bound (4.4.5). Therefore, it cannot apply to smooth potentials $q(x)$, in which case the gaps vanish as $n \to \infty$. Actually, we have the following:

Example: $\mathcal{PT}$-symmetric perturbations of the Kronig–Penney potential

A locally $H^{-1}(\mathbb{R})$ distribution $q(x)$ fulfilling the aforementioned conditions is

$$q(x) = \sum_{n\in\mathbb{Z}} \delta(x - 2\pi n)$$

the periodic δ function. Here, we have:

$$Q(u) = \int_{\mathbb{R}} q(x)|u(x)|^2\,dx = \sum_{n\in\mathbb{Z}} |u(2\pi n)|^2, \quad u \in H^1(\mathbb{R}).$$

This example is the well-known Kronig–Penney model in the one-electron theory of solids.

Let us verify that condition (4.4.2) is satisfied. As is known, this follows from the inequality (see e.g., 10, §VI.4.10):

$$|u(2\pi n)|^2 \le \epsilon \int_{2\pi n}^{2\pi(n+1)} |u'(y)|^2\,dy + \eta \int_{2\pi n}^{2\pi(n+1)} |u(y)|^2\,dy,$$

where ϵ can be chosen arbitrarily small for η large enough. In fact, if $u \in H^1(\mathbb{R})$, this inequality yields:

$$\begin{aligned}\int_{\mathbb{R}} q(x)|u(x)|^2\,dx = \sum_{n\in\mathbb{Z}} |u(2\pi n)|^2 &\le \epsilon \int_{\mathbb{R}} |u'(y)|^2\,dy + \eta \int_{R} |u(y)|^2\,dy \\ &= \epsilon\tau(u) + \eta\|u\|^2, \qquad \forall\, u \in H^1(\mathbb{R}),\end{aligned}$$

which in turn entails the closedness of $\tau(u) + Q(u)$ defined on H^1 by the standard Kato criterion. The closedness and sectoriality of $\mathcal{H}_g(u)$ defined on $H^1(\mathbb{R})$ are an immediate consequence of the continuity of W as a maximal multiplication operator in L^2. For the verification of (4.4.5), see for example, Ref. (26). Hence, any bounded $\mathcal{PT}$-symmetric periodic perturbation of the Kronig-Penney potential has real spectrum for $g \in \mathbb{R}$, $|g| < \overline{g}$, where $\overline{g}$ is defined by (4.4.6).

To prove Theorem 4.4.1, let us first recall that if $f \in L^\infty(\mathbb{R};\mathbb{C})$ is 2π periodic, $f(x + 2\pi) = f(x)$, $x \in \mathbb{R}$, and $\mathcal{PT}$ symmetric, $\overline{f}(-x) = f(x), x \in \mathbb{R}$, then its Fourier coefficients

$$f_n = \frac{1}{2\pi}\int_0^{2\pi} f(x)e^{-inx}\,dx$$

are such that $\overline{f}_n = f_n$, $\forall n \in \mathbb{Z}$. Moreover, by the Floquet–Bloch theory (see e.g., 14, 27), $\lambda \in \text{Spec}(H(g))$ if and only if the equation $H(g)\psi = \lambda\psi$ has a nonconstant bounded solution. In turn, all bounded solutions have the (Bloch) form

$$\psi_p(x; \lambda, g) = e^{ipx}\phi_p(x; \lambda, g), \tag{4.4.8}$$

where $p \in]-1/2, 1/2] := B$ (the Brillouin zone) and ϕ_p is 2π-periodic. It is indeed immediately verified that $\psi_p(x; \lambda, g)$ solves $H(g)\psi = \lambda\psi$ if and only if $\phi_p(x; \lambda, g)$ is a solution of

$$H_p(g)\phi_p(x; \lambda, g) = \lambda\phi_p(x; \lambda, g).$$

Here, $H_p(g)$ is the operator in $L^2(0, 2\pi)$ given by

$$H_p(g)u = \left(-i\frac{d}{dx} + p\right)^2 u + qu + gWu, \quad u \in D(H_p(g)) \tag{4.4.9}$$

with periodic boundary conditions; its realization will be now recalled. More precisely, denote S^1 the one-dimensional torus, that is, the interval $[-\pi, \pi]$ with the endpoints identified. By Assumptions (1) and (2), the restriction of q to S^1, still denoted q by a standard abuse of notation, belongs to $H^{-1}(S^1)$ and generates a real quadratic form $\mathcal{Q}_p(u)$ in $L^2(S^1)$ with domain $H^1(S^1)$. By Assumption (3), $\mathcal{Q}_p(u)$ is relatively bounded, with relative bound zero, with respect to

$$\tau_p(u) := \int_{-\pi}^{\pi} [-iu' + pu][i\overline{u}' + p\overline{u}]\, dx, \quad D(\tau_p(u)) = H^1(S^1) \tag{4.4.10}$$

such that the real semibounded form $\mathcal{H}_p^0(u) := \tau_p(u) + \mathcal{Q}_p(u)$ defined on $H^1(S^1)$ is closed. The corresponding self-adjoint operator in $L^2(S^1)$ is the self-adjoint realization of the formal differential expression (note again the abuse of notation)

$$H_p(0) = -\frac{d^2}{dx^2} - 2ip\frac{d}{dx} + p^2 + q.$$

As mentioned earlier, the form $\mathcal{H}_p(g)(u) := \mathcal{H}_p^0(u) + \langle u, Wu\rangle$ defined on $H^1(S^1)$ is closed and sectorial in $L^2(S^1)$. Let $H_p(g)$ be the associated m-sectorial operator in $L^2(S^1)$. On $u \in D(H_p(g))$, the action of the operator $H_p(g)$ is specified by (4.4.9); moreover, $H_p(g)$ has compact resolvent. Let

$$\text{Spec}(H_p(g)) := \{\lambda_n(g; p) : n = 0, 1, \ldots\}$$

denote the spectrum of $H_p(g)$, with $p \in [-1/2, 1/2]$, $|g| < \overline{g}$. By the above remarks, we have

$$\text{Spec}(H(g)) = \bigcup_{p \in]-1/2, 1/2]} \sigma(H_p(g)) = \bigcup_{p,n} \lambda_n(g; p).$$

Then, $\mathrm{Spec}(H(g)) \subset \mathbb{R}$, for $|g| < \overline{g}$, if and only if $\lambda_n(g;p) \in \mathbb{R} \, : \, n = 0, 1, \dots ; p \in [-1/2, 1/2]$.

To prove this, let us further recall the construction of the bands for $g = 0$: it can be proved that, under the present conditions, all eigenvalues $\lambda_n(0;p)$ are simple $\forall p \in [-1/2, 1/2]$; the functions $\lambda_n(0;p)$ are continuous and even in $[-1/2, 1/2]$ with respect to p so that one can restrict to $p \in [0, 1/2]$; the functions $\lambda_{2k}(0;p)$ are strictly increasing on $[0, 1/2]$ while the functions $\lambda_{2k+1}(0;p)$ are strictly decreasing, $k = 0, 1 \dots$, (see e.g., 14). Set:

$$\alpha_k = \lambda_k(0, 0), \quad \beta_k = \lambda_k(0, 1/2).$$

Then

$$\alpha_0 < \beta_0 < \beta_1 < \alpha_1 < \alpha_2 < \beta_2 < \beta_3 < \cdots .$$

The intervals $[\alpha_{2n}, \beta_{2n}]$ and $[\beta_{2n+1}, \alpha_{2n+1}]$ coincide with the range of $\lambda_{2n}(0, p)$, $\lambda_{2n+1}(0, p)$, respectively, and represent the bands of $\mathrm{Spec}(H(0))$; the intervals

$$\Delta_{2n} :=]\beta_{2n}, \beta_{2n+1}[, \quad \Delta_{2n+1} :=]\alpha_{2n+1}, \alpha_{2n+2}[$$

the gaps between the bands.
The monotonicity of the functions $\lambda_n(0, p)$ and assumption (4.4.5) entail

$$\inf_n \min_{p \in [0,1/2]} |\lambda_n(0, p) - \lambda_{n+1}(0, p)| \geq 2d. \tag{4.4.11}$$

Let us now state the following preliminary result

Proposition 4.4.2

1. *Let $g \in \overline{D}$, where $\overline{D}$ is the disk $\{g \in \mathbb{C} \, : \, |g| < \overline{g}\}$. For any n, there is a function $\lambda_n(g, p) \, : \, \overline{D} \times [0, 1/2] \to \mathbb{C}$, holomorphic in g and continuous in p, such that $\lambda_n(g, p)$ is a simple eigenvalue of $H_p(g)$ for all $(g, p) \in \overline{D} \times [0, 1/2]$.*
2. $$\sup_{g \in \overline{D}, p \in [0,1/2]} |\lambda_n(g, p) - \lambda_n(0, p)| < \frac{d}{2}.$$
3. *If $g \in \mathbb{R} \cap \overline{D}$ all eigenvalues $\lambda_n(g, p)$ are real;*
4. *If $g \in \mathbb{R} \cap \overline{D}$ then $\sigma(\mathcal{H}_p(g)) \equiv \{\lambda_n(g, p)\}_{n=0}^{\infty}$.*

Assuming the validity of this proposition, the proof of Theorem 4.4.1 is immediate.
Proof of Theorem 4.4.1
As the functions $\lambda_n(g;p)$ are real and continuous for $g \in \mathbb{R} \cap \overline{D}$, $p \in [0, 1/2]$, we can define:

$$\alpha_{2n}(g) := \min_{p \in [0,1/2]} \lambda_{2n}(g;p); \quad \beta_{2n}(g) := \max_{p \in [0,1/2]} \lambda_{2n}(g;p) \tag{4.4.12}$$

$$\alpha_{2n+1}(g) := \max_{p \in [0,1/2]} \lambda_{2n+1}(g;p); \quad \beta_{2n+1}(g) := \min_{p \in [0,1/2]} \lambda_{2n+1}(g;p). \tag{4.4.13}$$

Then

$$\text{Spec}(H(g)) = \bigcup_{n=0}^{\infty} B_n(g),$$

where the bands $B_n(g)$ are defined, in analogy with the $g = 0$ case, by:

$$B_{2n}(g) := [\alpha_{2n}(g), \beta_{2n}(g)], \quad B_{2n+1}(g) := [\beta_{2n+1}(g), \alpha_{2n+1}(g)].$$

By Assertion 2 of Proposition 4.4.2 we have, $\forall\, n = 0, 1, \ldots, \forall\, g \in \overline{D}$:

$$\begin{aligned}\alpha_{2n}(g) - \frac{d}{2} &= \lambda_{2n}(0,0) - \frac{d}{2} \le \lambda_{2n}(0,p) - \frac{d}{2} \\ &\le \lambda_{2n}(g,p) \le \lambda_{2n}(0,1/2) + \frac{d}{2} = \beta_{2n} + \frac{d}{2}\end{aligned}$$

whence

$$\alpha_{2n}(g) - \frac{d}{2} \le \lambda_{2n}(g,p) \le \beta_{2n} + \frac{d}{2}, \quad \forall\, n,\ \forall\, g \in \overline{D}.$$

This yields (through an analogous argument for the second inclusion):

$$\begin{aligned}B_{2n}(g) &\subset \left[\alpha_{2n} - \frac{d}{2}, \beta_{2n} + \frac{d}{2}\right], \\ B_{2n+1}(g) &\subset \left[\beta_{2n+1} - \frac{d}{2}, \alpha_{2n+1} + \frac{d}{2}\right].\end{aligned}$$

Therefore, the bands are pairwise disjoint, because the gaps

$$\Delta_{2n}(g) :=]\beta_{2n}(g), \beta_{2n+1}(g)[, \quad \Delta_{2n+1}(g) :=]\alpha_{2n+1}(g), \alpha_{2n+2}(g)[$$

are all open and their width is not smaller than d. In fact, by (4.4.5) we have:

$$|\alpha_n - \alpha_{n+1}| \ge 2d, \quad |\beta_n - \beta_{n+1}| \le 2d.$$

This concludes the proof of Theorem 4.4.1.
Proof of Proposition 4.4.2
Assertions 1 and 2. As the maximal multiplication operator by W is continuous in $L^2(S^1)$ with norm $\|W\|_\infty$, the operator family $H_p(g)$ is a type-A holomorphic family with respect to $g \in \mathbb{C}$, uniformly with respect to $p \in [0, 1/2]$. Hence, we can directly apply regular perturbation theory (see e.g., 10): the perturbation expansion near any eigenvalue $\lambda_n(0;p)$ of $H_p(0)$ exists and is convergent for $g \in \overline{D}$ to a simple eigenvalue $\lambda_n(g;p)$ of $H_p(g)$:

$$\lambda_n(g;p) = \lambda_n(0,p) + \sum_{s=1}^{\infty} \lambda_n^s(0;p) g^s, \quad g \in \overline{D}. \tag{4.4.14}$$

The convergence radius $r_n(p)$ is not smaller than $\overline{g}$. Hence, $\overline{g}$ represents a lower bound for $r_n(p)$ independent of n and p. Moreover, $\lambda_n^s(0;p)$ is continuous for all $p \in [0, 1/2]$, and hence the same is true for the sum $\lambda_n(g;p)$. This proves Assertion 1.
To prove Assertion 2, recall that the coefficients $\lambda_n^s(0;p)$ fulfill the majorization (see 10, §II.3)

$$|\lambda_n^s(0;p)| \leq \left(\frac{2\|W\|_\infty}{\inf_k \min_{p\in[0,1/2]} |\lambda_k(0,p) - \lambda_{k\pm1}(0,p)|} \right)^s \leq \left(\|W\|_\infty/d\right)^s. \quad (4.4.15)$$

Therefore, by (4.4.14):

$$|\lambda_n(g;p) - \lambda_n(0,p)| \leq \frac{|g| \left(\|W\|_\infty/d\right)}{1 - \left(2|g|\|W\|_\infty/d\right)} = \frac{|g|\|W\|_\infty}{d - 2|g|\|W\|_\infty} < \frac{d}{2}$$

whence the stated majorization on account of (4.4.6).
Assertion 3. As is known, and very easy to verify, the $\mathcal{PT}$ symmetry entails that the eigenvalues of a $\mathcal{PT}$-symmetric operator are either real or complex conjugate. By standard regular perturbation theory (see e.g., 10, §VII.2) any eigenvalue $\lambda_n(0;p)$ of $H_p(0)$ is stable with respect to $H_p(g)$; as $\lambda_n(0,p)$ is simple, for g suitably small there is one and only one eigenvalue $\lambda_n(g,p)$ of $H_p(g)$ near $\lambda_n(0,p)$, and $\lambda_n(g,p) \to \lambda_n(0,p)$ as $g \to 0$. This excludes the existence of the complex conjugate eigenvalue $\overline{\lambda}_n(g,p)$ distinct from $\lambda_n(g,p)$. Thus, for $g \in \mathbb{R}$, $|g|$ suitably small, $\lambda_n(g,p)$ is real. This entails the reality of series expansion (4.4.14) for g small and hence $\forall\, g \in \overline{\mathcal{D}}$. This in turn implies the reality of $\lambda_n(g,p)\,, \forall\, g \in \overline{\mathcal{D}}$.
Assertion 4. We directly refer the reader to the argument of Ref. (28), which is in turn just a repetition of the argument introduced in Ref. (12, 13).

4.5 AN EXAMPLE OF $\mathcal{PT}$-SYMMETRIC PHASE TRANSITION

As explained in, for example, Refs (1–7), an important issue in $\mathcal{PT}$-symmetric quantum mechanics is the spontaneous violation of the $\mathcal{PT}$ symmetry, namely, the transition from real values of the spectrum to complex ones at the variation of the parameter labeling the family. Its occurrence is referred to also as the $\mathcal{PT}$-symmetric phase transition. A basic example is represented (8) by the $\mathcal{PT}$-symmetric Schrödinger operator family in $L^2(\mathbb{R})$ defined by the maximal action of the differential operator

$$H(g)u = -\frac{d^2u}{dx^2} + x^2(ix)^g u, \quad g \in \mathbb{R}.$$

It has been long known that the spectrum of $H(g)$ is empty for $g = -1$ (29). More recently, its reality has been proved for $g = 1$, (15, 16). There is strong evidence (8) of a $\mathcal{PT}$-symmetric phase transition at $g = 0$, but a full mathematical proof is to our knowledge still missing. More generally, to our knowledge, no example is still known in $L^2(\mathbb{R})$ where the occurrence of the $\mathcal{PT}$-phase transition has been proved (see, however, Remark (3) for $\mathcal{PT}$-symmetric operators with different boundary conditions at infinity).

We describe here a class of such examples recently isolated in Ref. (30). They consist in the analytic continuation up to imaginary coupling constant of a class of self-adjoint even anharmonic oscillators perturbed by an odd potential breaking the parity symmetry. Consider the polynomial of degree $2M$ of the form

$$V_{2M}(x;g) := P_{2M}(x) + gP_{2q-1}(x), \quad M = 2,4,\dots;\ q = 1,2\dots,M;\ g \in \mathbb{C},$$

where $P_{2M}(x)$ is an even polynomial of degree $2M$ and $P_{2q-1}(x)$ an odd polynomial of degree $2q-1 \le 2M-1$. Let now $H(g)$ be the Schrödinger operator family in $L^2(\mathbb{R})$ whose domain and action are defined as

$$\begin{aligned} D(H(g)) &= H^2(\mathbb{R}) \cap L^2_{2M}(\mathbb{R}), \\ H(g)u &= -\frac{d^2u}{dx^2} + V_{2M}(x;g)u, \quad u \in D(H(g)),\ g \in \mathbb{C}. \end{aligned} \tag{4.5.1}$$

Then, (see 30 for details) $H(g)$ is a type-A entire holomorphic operator family in the sense of Kato (10, §VII.2) with compact resolvents. $H(g), g \in \mathbb{R}$, is self-adjoint and bounded below, while $H(ig)$ is $\mathcal{PT}$ symmetric. Denote now $E_n(g), n = 0, 1, \dots,$ the eigenvalues of $H(g)$, $g \in \mathbb{R}$. The stated holomorphy of $H(g)$ entails that each eigenvalue $E_n(g)$ is holomorphic near $g \in \mathbb{R}$.

We can now state the result concerning the existence of a $\mathcal{PT}$-symmetric phase transition.

Theorem 4.5.1 *Let*

$$P_{2M} = \frac{x^{2M}}{2M}, \qquad P_{2q-1} = -\frac{x^{M-1}}{M-1}, \ (q = M/2)$$

M even. Let $E_n(g)$ be any eigenvalue of $H(g)$, $n = 0, 1, \dots$. Then, there is $R(n) > 0$ such that

1. *$E_n(g)$ is holomorphic for $g \in \mathbb{R}$ and in the disk $D_n := \{g \in \mathbb{C} : |g| < R(n)\}$. If $g \in D_n \cap \mathbb{R}$ then $E_n(ig) = E_n(-ig) \in \mathbb{R}$.*
2. *There is $R_1(n) > R(n) > 0$ such that $E_n(g)$ admits two analytic continuations, one from $g \in \mathbb{R}_+$ and one from $g \in \mathbb{R}_-$ to the sectors*

$$\mathcal{F}^+_M := \left\{g \in \mathbb{C} : |g| > R_1(n)\ ;\ |\arg g| < \frac{M+1}{M-1}\frac{\pi}{2}\right\}, \tag{4.5.2}$$

$$\mathcal{F}^-_M := \left\{g \in \mathbb{C} : |g| > R_1(n)\ ;\ |\arg g - \pi| < \frac{M+1}{M-1}\frac{\pi}{2}\right\}, \tag{4.5.3}$$

respectively, across the imaginary axis.

3. *If* $g \in \mathbb{R}$, $|g| > R_1(n)$, $E_n(ig)$ *bifurcates in two complex conjugate eigenvalues* $E'_n(ig)$ *and* $E''_n(ig)$. *Namely:*
4. *Let* $g_0 > R_1(n)$. *Then*

$$\begin{aligned} E''_n(ig_0) = \overline{E'_n(ig_0)} &= \lim_{\substack{g\to ig_0 \\ \arg g<\pi/2}} E_n(g) \\ &= \lim_{\substack{g\to ig_0 \\ \arg g>\pi/2}} \overline{E_n(g)} = \lim_{\substack{g\to -ig_0 \\ \arg g<3\pi/2}} E_n(g) = \lim_{\substack{g\to -ig_0 \\ \arg g>3\pi/2}} \overline{E_n(g)}. \end{aligned} \tag{4.5.4}$$

$$\begin{aligned} \lim_{\substack{g\to ig_0 \\ \arg g>\pi/2}} \Im E_n(g) = \Im E'_n(ig_0) &= - \lim_{\substack{g\to ig_0 \\ \arg g<\pi/2}} \Im E_n(g) \\ &= -\Im E''_n(ig_0) \neq 0. \end{aligned} \tag{4.5.5}$$

The symmetric statements hold for $g_0 < -R_1(n)$.

Thus, the $\mathcal{PT}$ symmetry is violated and the $\mathcal{PT}$ phase transition takes place.

Remarks

1. If $H(g)\psi_n(x,g) = E_n(g)\psi_n(x,g)$ then $H(-g)\psi_n(-x,g) = E_n(g)\psi_n(-x,g)$, $H(\overline{g})\overline{\psi}_n(x,g) = \overline{E_n(g)}\overline{\psi}_n(x,g)$. This yields, $\forall\, g \in \mathbb{C}$, $\forall\, n = 0, 1, \ldots$:

$$\begin{cases} E_n(g) = E_n(-g) \\ \overline{E}_n(g) = E_n(\overline{g}) \end{cases} \implies E_n(-\overline{g}) = \overline{E}_n(g). \tag{4.5.6}$$

Therefore, in discussing the analyticity properties of $E_n(g)$, we can henceforth restrict our considerations to the half-plane $\Re g > 0$. Thus, $\forall\, g \in \mathbb{C} \setminus i\mathbb{R}$:

$$E_n(g) = E_n(-g) = \overline{E_n(\overline{g})} = \overline{E_n(-\overline{g})}. \tag{4.5.7}$$

2. Numerical evidence for the present results is presented in Ref. (31), together with a different proof, valid in the semiclassical regime.
3. In Refs (16, 32) and (33), the existence problem for the $\mathcal{PT}$-symmetric phase transitions has been examined for the class of Schrödinger operators defined by the formal differential operator

$$\mathcal{H} = -\frac{d^2}{dx^2} - (ix)^{2M} - g(ix)^{M-1} + \frac{\lambda^2 - 1/4}{x^2} \tag{4.5.8}$$

where $M \in \mathbb{R}$, $g \in \mathbb{R}$, $\lambda \in \mathbb{R}$; a spectral problem with discrete spectrum is generated out of the differential equation $\mathcal{H}\psi = E\psi$ imposing suitable boundary conditions at infinity along a contour C in the complex plane. For $M = 3$, the breakdown of the $\mathcal{PT}$-symmetry was proved, and a strong numerical evidence for it was found for $M \neq 3$. On the other hand, there are no values of M for which $H(g)$ and $\mathcal{H}$ define the same eigenvalue problem.

Let us now briefly outline the main steps needed to obtain the proof of Theorem 4.5.1.

1. The Rayleigh-Schrödinger perturbation theory generates the Taylor expansion near $g = 0$ for any eigenvalue $E_n(g)$, with convergence radius $R(n) > 0$: $E_n(g) = \sum_{\ell=0}^{\infty} A_{\ell,n} g^\ell$. Then, $A_{2\ell+1,n} = 0$. This makes E_n a function of $\beta = g^2$:

$$\begin{aligned} E_n(g) = F_n(\beta) &:= \sum_{\ell=0}^{\infty} B_{\ell,n}\beta^\ell, \\ B_{\ell,n} := A_{2\ell,n}, &\quad |\beta| < r(n) := R(n)^2. \end{aligned} \tag{4.5.9}$$

Hence, $E_n(ig) \in \mathbb{R}$, $g \in \mathbb{R}$, $|g| < R(n)$, as $A_{\ell,n} \in \mathbb{R}$.

2. Let $\mu := g^{-1/2} > 0$, and let $\{L(\mu) : \mu = g^{-1/2} > 0\}$ be the self-adjoint operator family in $L^2(\mathbb{R})$ defined as

$$\begin{aligned} &D(L(\mu)) = H^2(\mathbb{R}) \cap L^2_{2M}(\mathbb{R}); \\ &(L(\mu)f)(x) = -f''(x) + a_2 x^2 f(x) + \left[\sum_{k=3}^{2M} a_k\, \mu^{k-2} x^k\right] f(x), \\ &a_k = \frac{(2M-k+1)\cdots(2M-1) - (M-2)\cdots(M-k)}{k!}. \end{aligned} \tag{4.5.10}$$

Denote $\Lambda_n(\mu), n = 0, 1, \ldots,$ its eigenvalues. Then, Λ_n is a function of $\mu^2 = g^{-1} := \sigma$, because its Rayleigh-Schrödinger perturbation expansion at $\mu = 0$ near the eigenvalue $\Lambda_n(0) = \sqrt{a_2}(2n+1)$ of the harmonic oscillator is Borel summable and its odd terms vanish. We denote $\Lambda_n(\mu) = G_n(g^{-1}) = G_n(\sigma)$. Then, the following formula holds:

$$E_n(g) = g^{(M-1)/(M+1)} G_n(g^{-1}) - \frac{M+1}{2M(M-1)}\, g^{\frac{2M}{M+1}}, \quad g \in \mathbb{R}. \tag{4.5.11}$$

3. There is $R_2(n) > 0$ such that $G_n(\sigma)$ is a holomorphic function in the sector (on a $M+1$-sheeted Riemann surface)

$$\mathcal{D}_M := \{\sigma \in \mathbb{C} : 0 < |\sigma| < R_2(n);\ |\arg\sigma| < \frac{M+1}{M-1}\frac{\pi}{2}\} \tag{4.5.12}$$

and a stable eigenvalue as $|\sigma| \to 0$, $\sigma \in \mathcal{D}_M$.

4. By (4.5.11) and step 3, any eigenvalue $E_n(g)$ admits a holomorphic continuation from $g \in \mathbb{R}$ to the sector

$$\begin{aligned} &\mathcal{F}_M := \left\{g \in \mathbb{C} :\ |g| > R_1(n);\ |\arg g| < \frac{M+1}{M-1}\frac{\pi}{2}\right\}, \\ &R_1(n) := R_2(n)^{-1} \end{aligned} \tag{4.5.13}$$

enjoying the properties listed in Theorem 4.5.1.

Let us begin by summing up the relevant properties of the operator family in $L^2(\mathbb{R})$: $g \in \mathbb{C} \mapsto H(g)$, referring the reader to Ref. (30) for the simple details.

1. $H(g)$ is a type-A entire holomorphic operator family (in the sense of Kato) with compact resolvents, self-adjoint and bounded below for $g \in \mathbb{R}$. The real eigenvalues $E_n(g) : g \in \mathbb{R}$ are simple.
2. $H(ig) : g \in \mathbb{R}$ is $\mathcal{PT}$-symmetric.

In the particular case of Theorem 4.5.1, $V_{2M}(x,g) = \dfrac{x^{2M}}{2M} - g\dfrac{x^{M-1}}{M-1}$, we get:

$$\begin{aligned} V_{2M}(x,g) = &-\frac{M+1}{2M(M-1)}\, g^{\frac{2M}{M+1}} + \\ &+ \sum_{k=2}^{2M} a_k\, g^{(2M-k)/(M+1)}(x - g^{1/(M+1)})^k. \end{aligned} \tag{4.5.14}$$

where

$$-\frac{M+1}{2M(M-1)}\, g^{\frac{2M}{M+1}} = V_{2M}(x_0, g), \quad x_0 := g^{1/(M+1)}$$

and the coefficients $a_k : k = 2, 3, \ldots$ are given by (4.5.10).
Introduce now the unitary scaling operator $U(\lambda)$ in $L^2(\mathbb{R})$:

$$(U(\lambda)f)(x) := \lambda^{1/4} f(\lambda^{1/2}x) \qquad \lambda > 0, \quad U(\lambda)^{-1} = U(\lambda^{-1}). \tag{4.5.15}$$

The action of the operator family $U(\lambda)H(g)U(\lambda)^{-1}$ unitarily equivalent to $H(g)$ is

$$U(\lambda)H(g)U(\lambda)^{-1} = \lambda^{-1}[-\frac{d^2}{dx^2} + \lambda P_{2M}(\lambda^{1/2}x) + \lambda g P_{2q-1}(\lambda^{1/2}x)]. \tag{4.5.16}$$

In the particular case $P_{2M} = \frac{x^{2M}}{2M}$, $P_{2q-1}(x) = -\frac{x^{M-1}}{M-1}$, $U(\lambda)H(g)U(\lambda)^{-1}$ becomes:

$$U(\lambda)H(g)U(\lambda)^{-1} = \lambda^{-1}[-\frac{d^2}{dx^2} + \lambda^{M+1}\frac{x^{2M}}{2M} - g\lambda^{(M+1)/2}\frac{x^{M-1}}{M-1}]. \tag{4.5.17}$$

Moreover, we have the following:

Lemma 4.5.2 *Let $g > 0$. Then, $H(g)$ is unitarily equivalent to the operator*

$$K(g) = g^{(M-1)/(M+1)} L(g^{-1/2}) - \frac{M+1}{2M(M-1)}\, g^{\frac{2M}{M+1}}. \tag{4.5.18}$$

Here, $L(g^{-1/2})$ is the self-adjoint operator family in $L^2(\mathbb{R})$ defined as in (4.5.10), *namely:*

$$D(L(g^{-1/2})) = H^2(\mathbb{R}) \cap L^2_{2M}(\mathbb{R});$$

$$(L(g^{-1/2})f)(x) = -f''(x) + a_2x^2f(x) + \left[\sum_{k=3}^{2M} a_k\, g^{1-k/2}x^k\right] f(x). \tag{4.5.19}$$

Proof: By (4.5.14), $H(g)$ takes the form:

$$H(g) = -\frac{d^2}{dx^2} - \frac{M+1}{2M(M-1)}\, g^{\frac{2M}{M+1}} + \\ + \sum_{k=2}^{2M} a_k\, g^{(2M-k)/(M+1)}(x - g^{1/(M+1)})^k. \tag{4.5.20}$$

The translation $x \mapsto x + g^{1/(M+1)}$ makes $H(g)$ unitarily equivalent to the operator family defined on the same domain by the differential expression

$$T(g) = -\frac{d^2}{dx^2} - \frac{M+1}{2M(M-1)}\, g^{\frac{2M}{M+1}} + \sum_{k=2}^{2M} a_k\, g^{(2M-k)/(M+1)}x^k. \tag{4.5.21}$$

Performing now on $T(g)$, the unitary dilation $x \mapsto \lambda^{1/2}x$, $\lambda := g^{-\frac{M-1}{M+1}}$, $H(g)$ becomes unitarily equivalent also to the operator family

$$K(g) := U(\lambda)T(g)U(\lambda)^{-1} = \\ g^{(M-1)/(M+1)}\left[-\frac{d^2}{dx^2} + a_2x^2 + \sum_{k=3}^{2M} a_k\, g^{1-k/2}x^k\right] - \frac{M+1}{2M(M-1)}\, g^{\frac{2M}{M+1}}$$

and this proves (4.5.18) as well as (4.5.19). ■

Consider now the eigenvalues $E_n(g)$ of the operator $H(g)$ and the eigenvalues $\Lambda_n(\mu)$ of the operator $L(\mu)$. Recall that the *isolation distance* $d_n(g)$ of any eigenvalue $E_n(g)$ is

$$d_n(g) := \inf_{m \neq n} |E_m(g) - E_n(g)|. \tag{4.5.22}$$

Let us furthermore set:

$$\alpha_M := \min_{x\in\mathbb{R}}[P_{2M}(x) - P_{2q-1}(x)]. \tag{4.5.23}$$

$$\overline{r}(n) := \frac{1}{2}\left[\frac{|\alpha_M| + E_n(0)}{d_n(0)} + 1\right]^{-1}. \tag{4.5.24}$$

Proposition 4.5.3 *Consider any eigenvalue $E_n(g)$, $n = 0, 1, \ldots$ of $H(g)$. Then:*

1. *The Rayleigh-Schrödinger perturbation expansion for $E_n(g)$ near $E_n(0)$:*

$$E_n(g) = \sum_{l=0}^{\infty} A_{\ell,n} g^{\ell}, \quad A_{0,n} = E_n(0) \tag{4.5.25}$$

has radius of convergence $R(n) \geq \overline{r}(n)$.

2. *$A_{\ell,n} \in \mathbb{R}$; $A_{2\ell+1,n} = 0, \ell = 0, 1, \ldots$. Setting $\beta = g^2$, $B_{\ell,n} := A_{2\ell,n}$ then:*

$$E_n(g) = F_n(\beta), \quad F_n(\beta) := \sum_{\ell=0}^{\infty} B_{\ell,n} \beta^{\ell}. \tag{4.5.26}$$

3. *The function $\beta \mapsto F_n(\beta)$ has a holomorphic continuation from the disk $|\beta| < r(n) = R(n)^2$ to a neighborhood of $]-r(n), +\infty[$.*

Remark

Under the present conditions, it is proved (10, §VIII.2) that an eigenvalue $E_n(g)$ may develop a singularity at a point $g_0 \in \mathbb{C}$ if and only if there is a crossing at g_0, that is, there is $m \neq n$ such that $E_m(g_0) = E_n(g_0)$.

Proof:

(1) First recall that the existence and convergence of the Rayleigh–Schrödinger perturbation expansion for $E_n(g)$ near $E_n(0)$ follow from the holomorphy of the family $H(g)$ (see again 10, §VII.2). To obtain the lower bound (4.5.24) for its convergence radius, remark that (4.5.23) entails $|P_{2q-1}(x)| \leq |P_{2M}(x)| + |\alpha_M|$ whence, $\forall\, u \in H^1(\mathbb{R}) \cap L^2_{2M}$,

$$\begin{aligned}\int_{\mathbb{R}} |P_{2q-1}(x)|\, |u(x)|^2\, dx &\leq \int_{\mathbb{R}} |P_{2M}(x)|\, |u(x)|^2\, dx + |\alpha_M| \int_{\mathbb{R}} |u(x)|^2\, dx \\ &\leq \int_{\mathbb{R}} |u'(x)|^2\, dx + \int_{\mathbb{R}} |P_{2M}(x)|\, |u(x)|^2\, dx + |\alpha_M| \int_{\mathbb{R}} |u(x)|^2\, dx.\end{aligned}$$

In other words, P_{2q-1} is relatively form-bounded (10, §VI.1)with respect to $H(0)$ with relative bound 1 so that $H(g)$ can be considered also a holomorphic family of type B (see 10, §VII-3). Therefore, we can directly apply the lower bound VII-(4.47) for the convergence radius, in which we have to make $\epsilon = 1$, $b = 1$, $\lambda = E_n$, $a = |\alpha_M|$, $d = d_n$. This proves assertion (1). Assertion (2) is an immediate consequence of the $\mathcal{PT}$-symmetry, which entails that $\overline{E_n(ig)}$ is an eigenvalue if $E_n(ig)$ is, and of the holomorphy, which entails the stability of the eigenvalues of $H(0)$ for g small, so that $E_n(ig) = \overline{E_n(ig)} \in \mathbb{R}$, which can be true if and only if $A_{2\ell+1,n} = 0, \ell = 0, 1, \ldots$.

Finally, Assertion (3) is simply the transcription in terms of the variable $\sigma = g^2$ of the analytic properties of $E_n(g)$: holomorphy for $|g| < R(n)$ (by the convergence of the perturbation series) and in a neighborhood of the real axis (by the holomorphy, self-adjointness, and simplicity of the spectrum of the operator family). This concludes the proof of the proposition. ∎

4.5.1 Holomorphy and Borel Summability at Infinity

Consider again the operator family $L(\mu)$ defined by (4.5.10) or, equivalently, by (4.5.19) with $\mu = g^{-1/2}$, with eigenvalues $\Lambda_n(\mu), n = 0, 1, \ldots$. The functions $\Lambda_n(\mu), n = 0, 1, \ldots$ actually depend only on $\mu^2 := \sigma$. Hence, denoting $L(\mu) = L(\sqrt{\sigma})$, as earlier, we also denote $G_n(\sigma)$ the eigenvalues $\Lambda_n(\mu)$. Concerning these eigenvalues, the following statement holds.

Proposition 4.5.4 *Consider any eigenvalue $G_n(\sigma)$: $n = 0, 1, \ldots$ of $L(\sqrt{\sigma})$. Then:*

1. *Any path of analytic continuation of the eigenvalue $G_n(\sigma)$, $\sigma > 0$, lies on a $M + 1$-sheeted Riemann surface;*
2. *there is $R_2(n) > 0$ such that the eigenvalue $G_n(\sigma)$ is analytic in the sector $\mathcal{D}_M$ (on a $M + 1$-sheeted Riemann surface), and*

$$\lim_{\sigma \to 0,\, \sigma \in \mathcal{D}_M} G_n(\sigma) = \sqrt{a_2}(2n + 1). \tag{4.5.27}$$

Proof: The unitary image of $L(\sqrt{\sigma})$ under the unitary dilation $U(\lambda)$ defined by (4.5.15) with $\lambda = \sigma^{-(M-1)/(M+1)}$ is the operator family $S(\sigma)$ with domain $H^2(\mathbb{R}) \cap L^2_{2M}(\mathbb{R})$ and action

$$S(\sqrt{\sigma})u(x) = \sigma^{(M-1)/(M+1)} \Big[-u''(x) + +a_2\sigma^{-2(M-1)/(M+1)}x^2u(x)+ \\ +\left(\sum_{k=3}^{2M-1} a_k\sigma^{-(2M-k)/(M+1)}x^k\right)u(x) + a_{2M}x^{2M}u(x)\Big]. \tag{4.5.28}$$

Therefore, after any closed path of analytic continuation starting from $\sigma_0 \in \mathbb{R}_+$ and making exactly $M + 1$ turns around the origin, $G_n(\sigma)$ ends up with the initial value $G_n(\sigma_0)$, while this does not occur after any other closed path making ℓ turns, $1 \le \ell \le M$. This proves Assertion (1). To see Assertion (2), we follow the argument introduced by Simon (18), §II.10. We scale out the phase of σ in the operator family $L(\sqrt{\sigma})$, that is, we consider the similarity $U(\lambda)$ in L^2 with $\lambda := e^{-i[(M-1)/(M+1)]\arg\sigma}$. Then, $L(\sqrt{\sigma})$ is similar to the operator $J(\sqrt{\sigma})$ in $L^2(\mathbb{R})$ with domain $H^2(\mathbb{R}) \cap L^2_{2M}(\mathbb{R})$ and action

$$J(\sqrt{\sigma})u(x) = e^{i(M-1)/(M+1)\arg\sigma}[-u''(x) + a_2e^{-2i[(M-1)/(M+1)]\arg\sigma}x^2u(x) \\ + \sum_{k=3}^{2M-1} a_ke^{-i[(M-1)(1+k/2)/(M+1)]\arg\sigma}\,|\sigma|^{k/2}\,x^ku(x) + a_{2M}\,|\sigma|^Mx^{2M}u(x).$$

We denote $L_\gamma(0)$ the harmonic oscillator operator with complex frequency $\sqrt{a_2\gamma}$ away from the nonpositive real axis, that is,

$$\begin{aligned} L_\gamma(0)u &= -u''(x) + a_2\gamma x^2 u(x), \\ D(L_\gamma(0)) &= H^2(\mathbb{R}) \cap L^2_2(\mathbb{R}), \quad \gamma = e^{-2i[(M-1)/(M+1)]\arg\sigma}. \end{aligned} \tag{4.5.29}$$

Now we can apply directly the results of Ref. (18), §III.1: $J(\sqrt{\sigma})$ converges to $L_\gamma(0)$ in the norm resolvent sense for $|\sigma| \to 0$ as long as $|2[(M-1)/(M+1)]\arg\sigma| < \pi$, that is, as long as

$$|\arg\sigma| < \frac{M+1}{M-1}\frac{\pi}{2}.$$

This norm resolvent convergence entails (see again 18) the stated Assertion (2) on the eigenvalues. This concludes the proof of the Proposition. ■

Let us now turn to the examination of the perturbation theory of the eigenvalues $G_n(\sigma)$.

Proposition 4.5.5 *Consider any eigenvalue $G_n(\sigma)$: $n = 0, 1, \ldots$ of the operator family $L(\sqrt{\sigma})$.*

1. *The Rayleigh-Schrödinger perturbation expansion of $\Lambda_n(\mu)$ near $\Lambda_n(0) = \sqrt{a_2}(2n+1)$ exists to all orders and admits only the even powers of $\mu = \sqrt{\sigma}$:*

$$\Lambda_n(\mu) \sim \sum_{\ell=0}^{\infty} C_{\ell,n}\mu^\ell, \; C_{0,n} = \Lambda_n(0) = \sqrt{a_2}(2n+1), \; C_{2\ell+1} = 0; \tag{4.5.30}$$

$$\Lambda_n(\mu) = G_n(\sigma) \sim \sum_{\ell=0}^{\infty} D_{\ell,n}\sigma^\ell, \quad D_{\ell,n} = C_{2\ell,n}. \tag{4.5.31}$$

2. *The series* (4.5.31) *diverges as $\ell!$, that is, there are $0 < \Gamma_1(M,n) < \Gamma_2(M,n)$ such that*

$$\Gamma_1(M,n)^\ell\,\ell! < D_{\ell,n} < \Gamma_2(M,n)^\ell\,\ell!, \quad \forall \ell = 1, 2, \ldots \tag{4.5.32}$$

3. *The series* (4.5.31) *is a uniform asymptotic expansion to all orders for $G_n(\sigma)$ as $|\sigma| \to 0$, $\sigma \in \mathcal{D}_M$, and is Borel summable to $G_n(\sigma)$ for $\sigma \in \mathcal{E}_M$, where*

$$\mathcal{E}_M := \left\{\sigma \in \mathbb{C} : 0 < |\sigma| < R_2(n), \; |\arg\sigma| < \frac{\pi}{M-1}\right\}.$$

Proof: The perturbation term in $L(g^{-1/2}) = L(\mu)$ is the sum of $2M-2$ terms:

$$V = \sum_{k=1}^{2(M-1)} V^{(k)}\mu^k; \quad V^{(k)} := a_k x^{k+2}.$$

Let us apply formula II.(2.30) of Ref. (10), which reads:

$$
\begin{aligned}
C_{\ell,n} &= -\frac{(-1)^{\ell}}{2\pi i} \sum_{\nu_1+\ldots+\nu_p=\ell} \operatorname{Tr} \oint_{\Gamma_n} V^{(\nu_1)}R_{\gamma}(z)V^{(\nu_2)}\cdots R_{\gamma}(z)V^{(\nu_p)}R_{\gamma}(z)\,dz \\
&= -\frac{(-1)^{\ell}}{2\pi i} \oint_{\Gamma_n} \sum_{\nu_1+\ldots+\nu_p=\ell} \langle \psi_n, V^{(\nu_1)}R_{\gamma}(z)V^{(\nu_2)}\cdots R_{\gamma}(z)V^{(\nu_p)}R_{\gamma}(z)\psi_n\rangle\,dz. \quad (4.5.33)
\end{aligned}
$$

where $R_{\gamma}(z) = [L_{\gamma}(0) - z]^{-1}$. Remark that the vector

$$
V^{(\nu_1)}R_{\gamma}(z)V^{(\nu_2)}\cdots R_{\gamma}(z)V^{(\nu_p)}R_{\gamma}(z)\psi_n
$$

exists for all ℓ because the polynomial $V^{(\nu_s)}$ leaves the linear span of the eigenvectors ψ_n of $L_{\gamma}(0)$ invariant. Now $V^{(\nu_s)}$ is even or odd according to the parity of ν_s. Then, we can easily prove that

$$
\begin{aligned}
&\mathcal{P}[V^{(\nu_1)}R_{\gamma}(z)V^{(\nu_2)}\cdots R_{\gamma}(z)V^{(\nu_p)}R_{\gamma}(z)\psi_n] \\
&= (\pm 1)^{\nu_1+\ldots+\nu_p+n}[V^{(\nu_1)}R_{\gamma}(z)V^{(\nu_2)}\cdots R_{\gamma}(z)V^{(\nu_p)}R_{\gamma}(z)\psi_n] \\
&= (\pm 1)^{\ell+n}[V^{(\nu_1)}R_{\gamma}(z)V^{(\nu_2)}\cdots R_{\gamma}(z)V^{(\nu_p)}R_{\gamma}(z)\psi_n].
\end{aligned}
$$

because $\mathcal{P}\psi_n = (-1)^n\psi_n$. Therefore ψ_n and $V^{(\nu_1)}R_{\gamma}(z)\cdots R_{\gamma}(z)V^{(\nu_p)}R_{\gamma}(z)\psi_n$ have opposite parities if ℓ is odd. Hence, as above, the scalar product in (4.5.33) vanishes if ℓ is odd and $C_{\ell,n} = 0$ if ℓ is odd. This proves Assertion (1).

Let us now turn to Assertions (2) and (3). To simplify the notation, we drop the dependence on the eigenvalue index n, with the provision that all constants in the estimates to be obtained actually depend on n. Let $P_{\Lambda}(\mu)$ be the projection operator on the eigenspace corresponding to $\Lambda(\mu)$. The norm resolvent convergence of $J(\sqrt{\sigma}) = J(\mu)$ to $L_{\gamma}(0)$ of Proposition 4.5.4 entails that $\Lambda(\mu)$ is a stable eigenvalue. Hence, there is $\varepsilon > 0$ independent of μ such that the resolvent $[L(\mu) - z]^{-1}$ is bounded uniformly in $\mu \in \mathcal{D}_M$ for $z \in \{z \in \mathbb{C} : 0 < |z - \lambda| = \varepsilon\}$; moreover, $\forall\, \mu \in \mathcal{D}_M$, there exists a circumference Γ independent of (μ, γ) of radius smaller than ε encircling $\Lambda(\mu)$ and no other eigenvalue of $L(\mu)$ such that:

$$
\begin{aligned}
&P_{\Lambda}(\mu)u = \frac{1}{2\pi i}\oint_{\Gamma}[L(\mu) - z]^{-1}u\,dz \quad \forall\, u \in L^2(\mathbb{R}) \\
&\lim_{\mu\in\mathcal{D}_M\to 0} \|P_{\Lambda}(\mu) - P_{\Lambda}(0)\| = 0.
\end{aligned}
$$

$P_{\Lambda}(0)$ is the projection on the one-dimensional eigenspace of $\Lambda(0)$. This implies the one-dimensionality also of $P_{\Lambda}(\mu)$. By well-known arguments (see e.g., 34), (2), and

(3) follow by the estimates:

$$\sup_{z\in\Gamma} \| \sum_{v_1+\ldots+v_p=\ell} V^{(v_1)}R_\gamma(z)V^{(v_2)}\cdots R_\gamma(z)V^{(v_p)}R_\gamma(z)P_\Lambda(0)\|$$
$$\leq C\sigma^\ell(\sqrt{\ell})^\ell. \tag{4.5.34}$$

for suitable $C > 0$, $\sigma > 0$. To prove (4.5.34), we follow again the argument of Ref. (34). Remark that as the eigenfunctions of $L_\gamma(0)$ have the form

$$\phi_n = c_n H_n(\sqrt{a_2\gamma}x)e^{-\sqrt{a_2\gamma}x^2/2},$$

where $H_n(x)$ is the n-th Hermite polynomial, we can write

$$\|e^{\alpha x^2}P_\Lambda(0)\| < +\infty, \qquad 0 < \alpha := \frac{1}{2}\Re\sqrt{a_2\gamma}.$$

Now define:

$$f(x) := \alpha x^2; \quad R_{\gamma,f}(z) = e^f R_\gamma(z)e^{-f}. \tag{4.5.35}$$

It is known that with this choice of f the operator $x^2R_{\gamma,sf}$ is bounded for any $0 \leq s \leq 1$. Write now:

$$V^{(v_1)}R_\gamma(z)V^{(v_2)}\cdots R_\gamma(z)V^{(v_p)}R_\gamma(z)P_\Lambda(0) = V^{(v_1)}e^{-f/\ell}R_{\gamma f/\ell}(z)$$
$$\times V^{(v_2)}e^{-f/\ell}R_{\gamma,2f/\ell}V^{(v_3)}e^{-f/\ell}\cdots R_{\gamma,(\ell-1)/\ell}(z)V^{(v_p)}e^{-f/\ell}e^fP_\Lambda(0). \tag{4.5.36}$$

Now the boundedness of $x^2R_{\gamma,sf}$ entails that there is $C_1 > 0$ independent of γ and $1 \leq s \leq \ell$ such that $\sup_{z\in\Gamma}\|x^2R_{\gamma,sf/\ell}(z)\| \leq C_1$. As $V^{(v)} = a_v x^{v+2}$, we can write:

$$\|V^{(v_1)}e^{-f/\ell}R_{\gamma f/\ell}(z)V^{(v_2)}e^{-f/\ell}R_{\gamma,2f/\ell}V^{(v_3)}e^{-f/\ell}\cdots R_{\gamma,(\ell-1)/\ell}(z)V^{(v_p)}e^{-f/\ell}e^fP_\Lambda(0)\|$$
$$= |a_{v_1}\cdots a_{v_p}|\|x^{v_1}e^{-f/\ell}x^2R_{\gamma f/\ell}(z)x^{v_2}e^{-f/\ell}x^2R_{\gamma,2f/\ell}\cdots x^2R_{\gamma,(\ell-1)/\ell}(z)x^{v_p}e^{-f/\ell}e^fP_\Lambda(0)\|$$
$$\leq |a_{v_1}\cdots a_{v_p}|\cdot C_1^p\|x^{v_1+\ldots+v_p}e^{-\ell f}e^fP_\Lambda(0)\| \leq |a_{v_1}\cdots a_{v_p}|\cdot C_1^\ell(\sqrt{\ell})^\ell \sup_{t\in\mathbb{R}}(te^{-t^2})^\ell\|e^fP_\Lambda(0)\|$$
$$\leq \rho^\ell(\sqrt{\ell})^\ell$$

for some $\rho > 0$. Therefore:

$$\sup_{z\in\Gamma} \| \sum_{v_1+\ldots+v_p=\ell} V^{(v_1)}R_\gamma(z)V^{(v_2)}\cdots R_\gamma(z)V^{(v_p)}R_\gamma(z)P_\Lambda(0)\|$$
$$\leq \sum_{v_1+\ldots+v_p=\ell} \sup_{z\in\Gamma}\|V^{(v_1)}R_\gamma(z)V^{(v_2)}\cdots R_\gamma(z)V^{(v_p)}R_\gamma(z)P_\Lambda(0)\|$$
$$\leq \rho^\ell(\sqrt{\ell})^\ell \leq \rho^\ell(\sqrt{\ell})^\ell \max_{1\leq p\leq\ell}\binom{\ell}{p} \leq C\sigma^\ell(\sqrt{\ell})^\ell$$

for suitable $C > 0$, $\sigma > 0$, which is (4.5.34). As $D_{2\ell+1} = 0$, this estimate entails the upper bound (4.5.32). We omit the proof of the lower bound. ■

4.5.2 Analytic Continuation of the Eigenvalues and Proof of the Theorem

We are now in position to prove Theorem 4.5.1. Before turning to this point, we examine the behavior of the eigenvalues $G_n(\sigma)$ as $\sigma \to \infty$, that is, as $\sigma^{-1} = g \to 0$. As the arguments in this section do not depend on the particular eigenvalue, to simplify the notation, we drop the n-dependence. First we remark that since $\sigma = g^{-1}$, formula (4.5.11) and Proposition 4.5.3(2) yield, for $|\sigma| > r(n)^{-1}$:

$$G(\sigma) = \sigma^{(M-1)/(M+1)} \sum_{\ell=0}^{\infty} B_\ell \sigma^{-2\ell} + \frac{M+1}{2M(M-1)}\sigma^{-1} \tag{4.5.37}$$

as $|\sigma| \to \infty$, $|\arg \sigma| < \dfrac{M+1}{M-1}\dfrac{\pi}{2}$.

Proof of Theorem 4.5.1
Assertion (1) is already proved. Assertion (2) is a direct consequence of Proposition 4.5.4, by (4.5.11), because $\sigma = g^{-1}$.

As far as Assertions (3) and (4) are concerned, consider again (4.5.11) in the form

$$E(g) = g^{(M-1)/(M+1)} \left[G(g^{-1}) - \frac{M+1}{2M(M-1)}\, g \right]. \tag{4.5.38}$$

Now set $\mathcal{G} := \{g \in \mathbb{C} : |g| < R(n)\} \cup \mathbb{R}$. Then, by Proposition 4.5.4, the eigenvalue $E(g)$ is analytic in $\mathcal{G}$, and by Proposition 4.5.3(2) starting from $\mathcal{G}^+ := \{g \in \mathbb{C} : |g| < R(n)\} \cup \mathbb{R}_+$, it can be analytically continued throughout $\mathcal{F}_M^+$. Since by (4.5.6) $E(-g) = E(g)$, the eigenvalue $E(g)$ is holomorphic in $\mathcal{G}^- := \{g \in \mathbb{C} : |g| < R(n)\} \cup \mathbb{R}_-$ and can be analytically continued through

$$\mathcal{F}_M^- := \left\{ g \in \mathbb{C};\ |g| > R_1(n);\ |\arg g - \pi| < \frac{M+1}{M-1}\frac{\pi}{2} \right\}.$$

We now prove that the limits of $E(g)$ as g approaches the imaginary axis from $\mathcal{F}_M^+$ and from $\mathcal{F}_M^-$, respectively, are complex conjugate to each other with nonvanishing imaginary part. This entails the actual many-valuedness of the analytic continuation of $E(g)$ across the imaginary axis from iR_1 to $i\infty$ ($-iR_1$ to $-i\infty$) on the $M+1$-sheeted Riemann surface. The natural cut contains the aforementioned subset of the imaginary axis, and the first Riemann sheet contains $\mathcal{G}^+ \cup \mathcal{G}^- \cup \{g \in \mathbb{C};\ |g| > R_1(n);\ |\arg g - \pi| < \frac{\pi}{2}\} \cup \{g \in \mathbb{C};\ |g| > R_1(n);\ |\arg g| < \frac{\pi}{2}\}$.

We have indeed, by Remark (1) after Theorem 4.5.1, namely, $E(-g) = E(g)$,

$$E(|g|e^{i(\pi/2+\epsilon)}) = E(|g|e^{i(-\pi/2+\epsilon)}), \quad \forall\, \epsilon > 0.$$

Therefore, for $g_0 > R_1(n)$:

$$E'(ig_0) := \lim_{g\to ig_0; g\in \mathcal{F}_M^-} E(g) = \lim_{\varepsilon\downarrow 0} E(g_0 e^{i(\pi/2+\varepsilon)})$$
$$= \lim_{g\to -ig_0; g\in \mathcal{F}_M^+} E(g) = \lim_{\varepsilon\downarrow 0} E(g_0 e^{i(-\pi/2+\varepsilon)}).$$

Similarly:

$$E''(ig_0) := \lim_{g\to ig_0; g\in \mathcal{F}_M^+} E(g) = \lim_{\varepsilon\downarrow 0} E(g_0 e^{i(\pi/2-\varepsilon)})$$
$$= \lim_{g\to -ig_0; g\in \mathcal{F}_M^-} E(g) = \lim_{\varepsilon\downarrow 0} E(g_0 e^{i(-\pi/2-\varepsilon)}).$$

On the other hand, once more by Remark (1) after Theorem 4.5.1, equation (4.5.7), we have

$$E(g_0 e^{i(\pi/2+\varepsilon)}) = \overline{E}(g_0 e^{i(\pi/2-\varepsilon)})$$

and consequently

$$E'(ig_0) = \overline{E''}(ig_0).$$

We now prove that $\Im E'(ig_0) = -\Im E''(ig_0) \neq 0$. We have

$$\Im E'(ig_0) = \lim_{\varepsilon\downarrow 0} E(g_0 e^{i(\pi/2+\varepsilon)}) =$$
$$\Im[(g_0 e^{i\pi/2})^{(M-1)/(M+1)}[G(g_0^{-1} e^{-i\pi/2}) - \frac{M+1}{2M(M-1)} g_0 e^{i\pi/2}]] \tag{4.5.39}$$
$$= -\frac{M+1}{2M(M-1)} g_0^{2M/(M+1)} \left[\sin\left(\frac{2M}{M+1}\frac{\pi}{2}\right) + O(g_0^{-1})\right] \neq 0$$

for g_0 large enough. The same argument works for $g_0 < -R_1(n)$.

To sum up: let $E(i\delta) \in \mathbb{R}$ be any eigenvalue of $H(i\delta)$ for $\delta \in \mathbb{R}$, $|\delta| < R(n)$. Then there exist two analytic continuation paths $\Gamma_- \subset \mathcal{F}_M^-$ and $\Gamma_+ \subset \mathcal{F}_M^+$ originating from δ and ending at g_0, $|g_0| > R_1(n)$, such that, setting

$$\lim_{g\to ig_0; g\in \mathcal{F}_M^-} E(g) = E'(ig_0); \quad \lim_{g\to ig_0 \in \mathcal{F}_M^+} E(g) = E''(ig_0),$$

then $E'(ig_0)$ and $E''(ig_0)$ are nonreal eigenvalues of $H(ig_0)$ with the property

$$\overline{E'(ig_0)} = E''(ig_0),$$

with the imaginary part given by (4.5.39).

Therefore, any eigenvalue of $H(ig)$, $g \in \mathbb{R}$, $|g| < R_1(n)$ bifurcates for $|g| > R(n)$ large enough and this proves assertions (3) and (4). The proof of Theorem 4.5.1 is now complete.

4.6 THE METHOD OF THE QUANTUM NORMAL FORM

If a given $\mathcal{PT}$-symmetric operator can be conjugated to a self-adjoint one through a similarity transformation, of course its spectrum is real. The possibility of such similarity has been extensively studied (relevant references are Refs (3, 5–7, 35), and (36); see also (37, 38), and (39) for its examination in an abstract setting). Quite recently, a complete characterization has been obtained of $\mathcal{PT}$-symmetric quadratic Schrödinger operators similar to a self-adjoint one (40).

Here, we construct such a similarity transformation with the techniques of the quantum normal form (QNF) (see e.g., 41, 42), and provide a class of $\mathcal{PT}$-symmetric operators for which the procedure works. Namely, the QNF of the given $\mathcal{PT}$-symmetric Schrödinger operator is real and convergent, uniformly with respect to $\hbar \in [0, 1]$. The convergence of the QNF not only provides the similarity with a self-adjoint operator but also has the following straightforward consequences:

1) It yields an *exact* quantization formula for the eigenvalues;
2) As the QNF reduces to the classical normal form (CNF) for $\hbar = 0$, the CNF is convergent as well, and the corresponding classical system is therefore integrable.

Not surprisingly, we are able to prove a result so much stronger than simple similarity with a self-adjoint operator only for a very restricted class of operators, namely, a class of holomorphic, $\mathcal{PT}$-symmetric perturbations of the quantization of the linear diophantine flow over the torus $\mathbb{T}^l$.

Consider indeed a classical Hamiltonian family, defined in the phase space $\mathbb{R}^l \times \mathbb{T}^l, l = 1, 2. \dots,$ expressed in the action-angle variables (ξ, x), $\xi \in \mathbb{R}^l$, $x \in \mathbb{T}^l$:

$$\mathcal{H}_g(\xi, x) = \mathcal{L}_\omega(\xi) + g\mathcal{M}(\xi, x), \quad g \in \mathbb{R}, \tag{4.6.1}$$

where $\mathcal{L}_\omega(\xi) := \langle \omega, \xi \rangle$, $\omega := (\omega_1, \dots, \omega_l) \in \mathbb{R}^l$, is the Hamiltonian generating the linear quasiperiodic flow $x_i \mapsto x_i + \omega_i t$, $\forall i = 1, \dots, l$, with frequencies ω_i over $\mathbb{T}^l$, and $\mathcal{M}$ is an a priori complex-valued holomorphic function of (ξ, x), assumed to be $\mathcal{PT}$-symmetric. Namely, if $\mathcal{P} : x \to -x$ denotes the parity operation, that is, $(\mathcal{P}f)(\xi, x) = f(\xi, -x)$, $\forall f \in L^2(\mathbb{R}^l \times \mathbb{T}^l)$ and $\mathcal{T} : f \to \overline{f}$, the complex conjugation in $L^2(\mathbb{R}^l \times \mathbb{T}^l)$, then

$$((\mathcal{PT})\mathcal{M})(\xi, x) := \overline{\mathcal{M}}(\xi, -x) = \mathcal{M}(\xi, x), \quad \forall (\xi, x) \in \mathbb{R}^l \times \mathbb{T}^l.$$

Writing $\mathcal{M}$ through its uniformly convergent Fourier expansion:

$$\mathcal{M}(\xi, x) = \sum_{q \in \mathbb{Z}^l} \mathcal{M}_q(\xi) e^{i\langle q, x \rangle}; \quad \mathcal{M}_q(\xi) = (2\pi)^{-l/2} \int_{\mathbb{T}^l} \mathcal{M}(\xi, x) e^{-i\langle q, x \rangle}\, dx \tag{4.6.2}$$

the equivalent formulation of the $\mathcal{PT}$ symmetry in terms of the Fourier coefficients is immediately seen:

$$\mathcal{M}_q(\xi) = \overline{\mathcal{M}_q(\xi)}, \qquad \forall\,(\xi, q) \in \mathbb{R}^l \times \mathbb{Z}^l. \tag{4.6.3}$$

Moreover, we assume that

$$\mathcal{M}_{-q}(\xi) = -\mathcal{M}_q(\xi); \qquad \mathcal{M}_q(-\xi) = \mathcal{M}_q(\xi), \qquad \forall\,(\xi, q) \in \mathbb{R}^l \times \mathbb{Z}^l, \tag{4.6.4}$$

which ensures that the potential $\mathcal{M}(\xi, x)$ is even in the variable ξ and odd in the variable x:

$$\mathcal{M}(-\xi, x) = \mathcal{M}(\xi, x), \qquad \mathcal{M}(\xi, -x) = -\mathcal{M}(\xi, x), \qquad \forall\,(\xi, x) \in \mathbb{R}^l \times \mathbb{T}^l.$$

We denote V the operator in $L^2(\mathbb{T}^l)$ generated by the Weyl quantization of the symbol $\mathcal{M}$ (see Appendix A.2), namely, the operator acting on $L^2(\mathbb{T}^l)$ in the following way:

$$(Vf)(x) := \int_{\mathbb{R}^l} \sum_{q\in\mathbb{Z}^l} \widehat{\mathcal{M}}_q(p) e^{i(\langle q,x\rangle + \langle p,q\rangle \hbar/2)} f(x + p\hbar)\, dp, \quad \forall f \in L^2(\mathbb{T}^l), \tag{4.6.5}$$

where

$$\widehat{\mathcal{M}}_q(p) := (2\pi)^{-l/2} \int_{\mathbb{R}^l} \mathcal{M}_q(\xi) e^{-i\langle p,\xi\rangle}\, d\xi$$

is the Fourier transform of the Fourier coefficient $\mathcal{M}_q(\xi)$.

Then, the quantization of $\mathcal{H}_g$ is the $\mathcal{PT}$-symmetric (verification below), non-self-adjoint operator in $L^2(\mathbb{T}^l)$ acting as

$$H(\omega, g) = i\hbar\langle \omega, \nabla\rangle + gV = L(\omega, \hbar) + gV, \quad L(\omega, \hbar) := i\hbar\langle \omega, \nabla\rangle. \tag{4.6.6}$$

The Schrödinger operator $H(\omega, g)$ thus represents a perturbation of the self-adjoint operator $L(\omega, \hbar)$ in $L^2(\mathbb{T}^l)$, whose spectrum obviously consists of the eigenvalues $\lambda_{n,\omega} = \hbar\langle \omega, n\rangle$, $n = (n_1, \ldots, n_l) \in \mathbb{Z}^l$, with corresponding normalized eigenfunctions $\phi_n(x) = (2\pi)^{-l/2} e^{i\langle n,x\rangle}$.

Remark

By the assumptions to be specified below V will represent a regular perturbation of $L(\omega, \hbar)$. However, the spectrum of $L(\omega, \hbar)$, although pure point, is *dense* in $\mathbb{R}$. Therefore, the standard (Rayleigh–Schrödinger) perturbation theory of quantum mechanics *cannot* be applied here because no eigenvalue is isolated, and the approach through the Normal Form is therefore necessary, insofar as it represents an alternative method that serves the purpose.

The statement of the result will profit in clarity by first sketching the construction of the QNF (see e.g., (41, 42), and in this particular context (43)). Its purpose in this connection is to construct a similarity transformation $U(g)$ in $L^2(\mathbb{R}^l)$, generated by a continuous operator $W(g)$, $U(g) = e^{iW(g)/\hbar}$, such that

$$U(g)H(\omega, g)U(g)^{-1} = e^{iW(g)/\hbar}(L(\omega, \hbar) + gV)e^{-iW(g)/\hbar} = S(g), \tag{4.6.7}$$

where the similar operator $S(g)$ is self-adjoint. The procedure is as follows:

1. Look for that particular similarity transformation $U(g) = e^{iW(g)/\hbar}$, such that the transformed operator $S(g)$ assumes the form

$$S(g) = L(\omega, \hbar) + \sum_{k=1}^{\infty} g^k B_k(\hbar) \tag{4.6.8}$$

under the additional conditions

$$[B_k, L] = 0, \qquad B_k = B_k^*, \qquad \forall k = 1, 2, \dots, \tag{4.6.9}$$

where $B_k := B_k(\hbar)$, $\forall k$, and $L := L(\hbar, \omega)$. If it can be proved that the series (4.6.8) (under the additional conditions (4.6.9)) has a positive convergence radius g^*, then obviously $S(g)$ is self-adjoint for $|g| < g^*$ so that its spectrum is real; moreover, $S(g)$ is diagonal on the eigenvector basis of $L(\hbar, \omega)$. The series (4.6.8), assuming the validity of conditions (4.6.9), is called the *operator quantum normal form (O-QNF)*.

2. To determine the O-QNF, we first construct the QNF for the symbols (S-QNF). That is, we first construct, for any $k = 1, 2, \dots$, the symbol $\mathcal{B}_k(\xi; x; \hbar)$ of the self-adjoint operator B_k. The symbol $\mathcal{B}_k$ turns out to be a function only of ξ (depending parametrically on $\hbar$), so that the application of the Weyl quantization formula (see Appendix A.2) specifies the action of B_k:

$$B_k f = \mathcal{B}_k(i\hbar\langle\omega, \nabla\rangle)f = \mathcal{B}_k(L_\omega)f, \quad \forall f \in L^2(\mathbb{T}^l), \quad L_\omega := L = L(\hbar, \omega).$$

Hence, $[B_k, L_\omega] = 0$, $\forall k$, and the eigenvalues of B_k are simply $\mathcal{B}_k(n\hbar, \hbar)$, $n \in \mathbb{Z}^l$. Then, the symbol of $S(g)$ is

$$\Sigma(\xi, g, \hbar) = \mathcal{L}_\omega(\xi) + \sum_{k=1}^{\infty} \mathcal{B}_k(\xi, \hbar)g^k$$

provided the series has a nonzero convergence radius. In that case, the eigenvalues of $S(g)$, and hence of $H(\omega, g)$, are clearly given by the following *exact* quantization formula:

$$\lambda_n(\varepsilon, g) = \langle\omega, n\rangle\hbar + \sum_{k=1}^{\infty} \mathcal{B}_k(n\hbar, \hbar)g^k, \tag{4.6.10}$$

that is, by the symbol $\Sigma(\xi, g, \hbar)$ evaluated at the *quantized* values $n\hbar$ of the classical actions $\xi \in \mathbb{R}^l$. Moreover, the spectrum of $S(g)$, that is, of $H(\omega, g)$, is real if $S(g)$ is self-adjoint, namely if B_k is self-adjoint $\forall\, k = 1, \ldots$; again by the Weyl quantization formula (Appendix A.2), this is true if $\mathcal{B}_k(\xi; \hbar)$ is real and bounded $\forall\, k = 1, 2, \ldots$.

3. By construction, each coefficient $\mathcal{B}_k(\xi, \hbar), k = 1, \ldots,$ of the S-QNF turns out to be a smooth function of $\hbar$ near $\hbar = 0$, and $\mathcal{B}_k(\xi, 0) := \mathcal{B}_k(\xi)$ is just the k-term of the CNF generated by canonical perturbation theory applied to the classical Hamiltonian $\mathcal{H}_g(\xi, x)$. More precisely:

$$\mathcal{H}_g(\xi, x) \sim \mathcal{L}_\omega(\xi) + \sum_{k=1}^{\infty} \mathcal{B}_k(\xi) g^k, \tag{4.6.11}$$

where $\sim$ denotes canonical equivalence. Therefore, if the convergence of the S-QNF is *uniform* with respect to $\hbar \in [0, 1]$ the CNF (4.6.11) is also convergent, and therefore, the classical hamiltonian $\mathcal{H}_g(\xi, x)$ is integrable because the equivalent hamiltonian depends only on the actions.

We can now proceed to the precise statement of the results. First we describe the assumptions. Consider again the operator

$$L(\omega, \hbar)\psi = i\hbar\langle\omega, \nabla\rangle\psi = -i\hbar\left[\omega_1\partial_{x_1} + \ldots + \omega_l\partial_{x_l}\right]\psi, \quad D(L_\omega) = H^1(\mathbb{T}^l);$$
$$H^1(\mathbb{T}^l) := \{\psi = \sum_{n\in\mathbb{Z}^l} \psi_n e^{i\langle n,x\rangle} \in L^2(\mathbb{T}^l) \;:\; \sum_{n\in\mathbb{Z}^l} |n|^2\, |\psi_n|^2 < +\infty\}.$$

The first assumption is:
(A1) *The frequencies* $\omega = (\omega_1, \ldots, \omega_l)$ *are rationally independent (see* (4.2.1)*), and* diophantine, *that is,* $\exists \gamma > 0,\ \tau > l$ *such that:*

$$|\langle\omega, q\rangle|^{-1} \le \gamma |q|^\tau, \quad q \in \mathbb{Z}^l,\ q \neq 0. \tag{4.6.12}$$

Remark that (4.6.12) entails that all the eigenvalues $\lambda_{n,\omega} = \langle n, \omega\rangle\hbar$ of $L(\omega, \hbar)$ are simple.

Let now $(t, x) \mapsto \mathcal{M}(t, x)$ be a complex-valued smooth function defined on $\mathbb{R} \times \mathbb{T}^l$, that is, $\mathcal{M} \in C^\infty(\mathbb{R} \times \mathbb{T}^l; \mathbb{C})$. Write its Fourier expansion:

$$\mathcal{M}(t, x) = \sum_{q\in\mathbb{Z}^l} \mathcal{M}_q(t) e^{i\langle q,x\rangle}, \quad \mathcal{M}_q(t) := (2\pi)^{-l/2} \int_{\mathbb{T}^l} \mathcal{M}(t, x) e^{-i\langle q,x\rangle}\, dx \tag{4.6.13}$$

and define the functions $\mathcal{M}_\omega(\xi, x) : \mathbb{R}^l \times \mathbb{T}^l \to \mathbb{C}$ in the following way:

$$\mathcal{M}_\omega(\xi, x) := \mathcal{M}(\langle\omega, \xi\rangle, x) = \sum_{q\in\mathbb{Z}^l} \mathcal{M}_{\omega,q}(\xi) e^{i\langle q,x\rangle}, \; \mathcal{M}_{\omega,q}(\xi) := \mathcal{M}_q(\langle\omega, \xi\rangle). \tag{4.6.14}$$

Now consider the space Fourier transform of $\mathcal{M}_q(t), q \in \mathbb{Z}^l$:

$$\widehat{\mathcal{M}}_q(p) := \frac{1}{\sqrt{2\pi}} \int_{\mathbb{R}} \mathcal{M}_q(t) e^{-ipt}\, dt, \quad p \in \mathbb{R}.$$

Then (see formula (A.2.1)), the Weyl quantization of $\mathcal{M}_\omega(\xi, x)$ is the operator in $L^2(\mathbb{T}^l)$ acting as follows:

$$(V_\omega f)(x) = \int_{\mathbb{R}} \sum_{q \in \mathbb{Z}^l} \widehat{\mathcal{M}}_q(p) e^{i(\langle q,x\rangle + \hbar p\langle \omega,q\rangle/2)} f(x + \hbar p\omega)\, dp, \quad f \in L^2(\mathbb{T}^l).$$

V_ω is actually a continuous operator in $L^2(\mathbb{T}^l)$ (see Appendix, Remark (d)) by virtue of our second assumption, namely:

(A2) *Let the diophantine constants γ and τ be such that*

$$\gamma\tau^\tau(\tau+2)^{4(\tau+2)} < \frac{1}{2}$$

and let there exist $\rho > 2$ such that

$$\|\mathcal{M}_\omega\|_\rho := \sum_{q \in \mathbb{Z}^l} e^{\rho|q|} \int_{\mathbb{R}} e^{\rho|p|} |\widehat{\mathcal{M}}_q(p)|\, dp < +\infty. \tag{4.6.15}$$

Remarks

(i) Actually, by formula (A.2.6), $\|V_\omega\|_{L^2 \to L^2} \le \|\mathcal{M}_\omega\|_\rho$. Moreover, assumption (A2) makes $\mathcal{M}_\omega$ a holomorphic function of (ξ, x) in $\mathbb{C}^{2l}_\rho := \{(\xi, x) \in \mathbb{C}^{2l} : |\mathrm{Im}\xi_i| < \rho;\ |\mathrm{Im} x_i| < \rho,\ \forall i = 1, \dots, l\}$.

(ii) As discussed in Ref. (43), $\mathcal{M}(t, x)$ must depend explicitly on t if $l > 1$ to make the problem a nontrivial one. Once more by Assumption (A2), formula (4.6.15), $\mathcal{M}(t, x)$ vanishes exponentially fast as $|t| \to \infty$ uniformly with respect to $x \in \mathbb{T}^l$.

Our third assumption concerns the $\mathcal{PT}$-symmetry and is formulated as follows (see (4.6.3) and (4.6.4)):

(A3) *The Fourier coefficients $\mathcal{M}_{\omega,q}(\xi)$ enjoy the following symmetry properties:*

$$\begin{gathered}\mathcal{M}_{\omega,q}(\xi) = \overline{\mathcal{M}_{\omega,q}(\xi)};\ \mathcal{M}_{\omega,-q}(\xi) = -\mathcal{M}_{\omega,q}(\xi);\\ \mathcal{M}_{\omega,q}(-\xi) = \mathcal{M}_{\omega,q}(\xi),\ \forall(\xi, q) \in \mathbb{R}^l \times \mathbb{T}^l.\end{gathered} \tag{4.6.16}$$

Remark

Clearly (A3) entails $\mathcal{M}_\omega(\xi, -x) = -\mathcal{M}_\omega(\xi, x)$ and

$$((\mathcal{PT})\mathcal{M}_\omega)(\xi, x) = (\mathcal{PT})\left(\sum_{q \in \mathbb{Z}^l} \mathcal{M}_{\omega,q}(\xi) e^{i\langle q,x\rangle}\right) = \mathcal{M}_\omega(\xi, x), \qquad \forall(\xi, x) \in \mathbb{R}^l \times \mathbb{T}^l,$$

that is, $\mathcal{M}_\omega(\xi,x)$ is a $\mathcal{PT}$-invariant function, odd with respect to x. Moreover from (4.6.16) one can easily obtain $\widehat{\mathcal{M}}_{\omega,q}(-p) = \widehat{\mathcal{M}}_{\omega,q}(p) \in \mathbb{R}$, $\forall p \in \mathbb{R}^l$, $\forall q \in \mathbb{Z}^l$. This entails that $V := V_\omega$ is a $\mathcal{PT}$-symmetric operator in $L^2(\mathbb{T}^l)$, that is, $[V, \mathcal{PT}] = 0$. We have indeed

$$\begin{aligned}
(\mathcal{PT})(Vf)(x) &= \int_{\mathbb{R}} \sum_{q\in\mathbb{Z}^l} \widehat{\mathcal{M}}_{\omega,q}(p) e^{i\langle q,x\rangle - i\hbar p\langle\omega,q\rangle/2} \overline{f}(-x+\hbar p\omega)\, dp \\
&= \int_{\mathbb{R}} \sum_{q\in\mathbb{Z}^l} \widehat{\mathcal{M}}_{\omega,q}(p) e^{i(\langle q,x\rangle + \hbar p\langle\omega,q\rangle/2)} \overline{f}(-x-\hbar p\omega)\, dp \\
&= \int_{\mathbb{R}} \sum_{q\in\mathbb{Z}^l} \widehat{\mathcal{M}}_{\omega,q}(p) e^{i(\langle q,x\rangle + \hbar p\langle\omega,q\rangle/2)} (\mathcal{PT}f)(x+\hbar p\omega)\, dp \\
&= V(\mathcal{PT}f)(x), \qquad \forall f \in L^2(\mathbb{T}^l),\ \forall x \in \mathbb{T}^l.
\end{aligned}$$

To sum up, the operator acting as

$$H(g) = i\hbar\langle\omega, \nabla\rangle + gV$$

and defined on $D(H(g)) = H^1(\mathbb{T}^l)$ has pure-point spectrum denoted $\mathrm{Spec}(H(g))$, and we will prove that it consists of a sequence of nonisolated eigenvalues denoted $\{l_n(\hbar, g) : n \in \mathbb{Z}^l\}$. The symbol of $H(g)$ is the Hamiltonian family defined on $\mathbb{R}^l \times \mathbb{T}^l$:

$$\mathcal{H}_g(\xi,x) = \langle\omega,\xi\rangle + g\mathcal{M}_\omega(\xi,x) = \mathcal{L}_\omega(\xi) + g\mathcal{M}_\omega(\xi,x).$$

We can now state the main result.

Theorem 4.6.1 *Under Assumptions (A1-A3), there exists $g_0 > 0$ independent of $\hbar \in [0,1]$ such that for $|g| < g_0$ the spectrum of $H(g)$ is given by the exact quantization formula:*

$$\lambda_n(\hbar, g) = \langle\omega, n\rangle\hbar + \mathcal{B}(n\hbar, \hbar; g), \quad n \in \mathbb{Z}^l \tag{4.6.17}$$

$$\mathcal{B}(n\hbar, \hbar; g) := \sum_{k=1}^{\infty} \mathcal{B}_k(n\hbar, \hbar) g^k, \tag{4.6.18}$$

where

1. *$\mathcal{B}_k(\xi, \hbar) \in C^\infty(\mathbb{R}^l \times [0,1])$ is real-valued, $k = 1, 2, \dots$;*
2. *$\mathcal{B}_{2s+1} = 0$, $s = 0, 1, \dots$;*
3. *The series* (4.6.18) *converges uniformly with respect to $(\xi, \hbar) \in \mathbb{R}^l \times [0,1]$;*
4. *$\mathcal{B}_k(n\hbar, \hbar)$ is obtained from the Weyl quantization formula applied to $\mathcal{B}_k(\xi, \hbar)$, which is the symbol of the operator B_k, the term of order k of the QNF.*

Corollary 4.6.2 *Let $|g| < g_0$. Then the operator $H(\omega, g)$ is similar to the self-adjoint operator*

$$S(g) := L(\omega, \hbar) + \sum_{k=1}^{\infty} B_k(\hbar) g^k.$$

Remark

The explicit construction of the bounded operator $W(g)$ realizing the similarity $U = U(\omega, g, \hbar) = e^{iW(g)/\hbar}$ is described in the proof of Theorem 4.6.1.

A straightforward consequence of the uniformity (with respect to $\hbar \in [0,1]$) of the convergence of the QNF is a convergence result for the corresponding CNF, valid for a class of $\mathcal{PT}$-symmetric, nonholomorphic perturbations of nonresonant harmonic oscillators. Consider indeed the inverse transformation into action-angle variables

$$C(\xi, x) = (\eta, y) := \begin{cases} \eta_i = -\sqrt{\xi_i} \sin x_i \\ y_i = \sqrt{\xi_i} \cos x_i \end{cases} \qquad i = 1, \dots, l.$$

It is defined only on $\mathbb{R}_+^l \times \mathbb{T}^l$ and does not preserve the regularity at the origin. On the other hand, C is an analytic, canonical map between $\mathbb{R}_+^l \times \mathbb{T}^l$ and $\mathbb{R}^{2l} \setminus \{0,0\}$. Then:

$$(\mathcal{H}_g \circ C^{-1})(\eta, y) = \sum_{s=1}^{l} \omega_s(\eta_s^2 + y_s^2) + g(\mathcal{M} \circ C^{-1})(\eta, y)$$
$$:= \mathcal{P}_0(\eta, y) + g\mathcal{P}_1(\eta, y),$$

where for $(\eta, y) \in \mathbb{R}^{2l} \setminus \{0,0\}$

$$\mathcal{P}_1(\eta, y) = (\mathcal{M} \circ C^{-1})(\eta, y) = \mathcal{P}_{1,R}(\eta, y) + \mathcal{P}_{1,I}(\eta, y),$$

$$\mathcal{P}_{1,R}(\eta, y) = \frac{1}{2} \sum_{k \in \mathbb{Z}^l} (\Re \mathcal{M}_k \circ C^{-1})(\eta, y) \prod_{s=1}^{l} \left(\frac{\eta_s - i y_s}{\sqrt{\eta_s^2 + y_s^2}} \right)^{k_s}$$

$$\mathcal{P}_{1,I}(\eta, y) = \frac{1}{2} \sum_{k \in \mathbb{Z}^l} (\Im \mathcal{M}_k \circ C^{-1})(\eta, y) \prod_{s=1}^{l} \left(\frac{\eta_s - i y_s}{\sqrt{\eta_s^2 + y_s^2}} \right)^{k_s}.$$

Corollary 4.6.3 *The Birkhoff normal form of $\mathcal{H}_g$ is real and uniformly convergent on any compact of $\mathbb{R}^{2l} \setminus \{0,0\}$ if $|g| < g_0$. Hence, the system is integrable.*

Proof of Theorem 4.6.1 . Under the present conditions, statements (3) and (4) are proved in Ref. (43), as well as the smoothness of $\mathcal{B}_k(\xi, \hbar)$ asserted

in (1). The assertions left to prove are therefore the reality statement (1), $B_k(\xi,\hbar)=\overline{B}_k(\xi,\hbar)$, $\forall\,(\xi,\hbar)\in\mathbb{R}^l\times[0,1]$, and the even nature of the QNF, $\mathcal{B}_{2s+1}=B_{2s+1}=0$, $\forall s=0,1,\ldots$. This requires a detailed examination of the structure of the QNF, whose construction we now recall in Section 4.6.1. In Section 4.6.2, we describe the inductive argument proving the reality assertion and the symmetry argument proving the vanishing of the odd terms.
Proof of Corollary 4.6.3. For this proof, the reader is directly referred to Ref. (44).

4.6.1 The Quantum Normal Form: the Formal Construction

(We follow Sjöstrand (41) and Bambusi *et al.* (42)).

Given $H(g)=L(\omega,\hbar)+\varepsilon V$ in $L^2(\mathbb{T}^l)$, look for a similarity transformation $U=U(\omega,g,\hbar)$, in general nonunitary ($W(g)\neq W(g)^*$):

$$U(\omega,g,\hbar)=e^{iW(g)/\hbar}\;:\;L^2(\mathbb{T}^l)\leftrightarrow L^2(\mathbb{T}^l)$$

such that

$$S(g):=UH(g)U^{-1}=L(\omega,\hbar)+gB_1+g^2B_2+\ldots=L(\omega,\hbar)+\sum_{k=1}^{\infty}B_kg^k \qquad (4.6.19)$$

under the requirement:

$$[B_k,L]=0,\qquad \forall k.$$

Recall the formal commutator expansion

$$S(g)=e^{iW(g)/\hbar}H(g)e^{-iW(g)/\hbar}=\sum_{k=0}^{\infty}H_k \qquad (4.6.20)$$

$$H_0:=H(g),\quad H_k:=\frac{[W(g),H_{k-1}]}{i\hbar k},\qquad k\geq 1$$

and look for $W(g)$ under the form of a power series expansion in g: $W(g)=gW_1+g^2W_2+\ldots$.
Then (4.6.19) becomes:

$$S(g)=\sum_{k=0}^{\infty}g^kB_k, \qquad (4.6.21)$$

where

$$B_0=L(\omega,\hbar);\quad B_k:=\frac{[W_k,L]}{i\hbar}+V_k,\qquad k\geq 1, \qquad (4.6.22)$$

$V_1 \equiv V$ and

$$V_k = \sum_{r=2}^{k} \frac{1}{r!} \sum_{\substack{j_1+\ldots+j_r=k \\ j_s \geq 1}} \frac{[W_{j_1}, [W_{j_2}, \ldots, [W_{j_r}, L] \ldots]}{(i\hbar)^r}$$
$$+ \sum_{r=1}^{k-1} \frac{1}{r!} \sum_{\substack{j_1+\ldots+j_r=k-1 \\ j_s \geq 1}} \frac{[W_{j_1}, [W_{j_2}, \ldots, [W_{j_r}, V] \ldots]}{(i\hbar)^r}. \quad (4.6.23)$$

V_k depends on $W_1, \ldots, W_{k-1}$ but not on W_k. Thus, we get the recursive homological equations:

$$\frac{[W_k, L]}{i\hbar} + V_k = B_k, \qquad [L, B_k] = 0. \quad (4.6.24)$$

To solve (4.6.24) for the two unknowns B_k, W_k, we look for their symbols and then apply the Weyl quantization formula. First recall (see e.g., (45) or (46)) that the symbol of the commutator $[F, G]/i\hbar$ of two operators F and G is the *Moyal bracket* $\{\mathcal{F}, \mathcal{G}\}_M$ of the symbols $\mathcal{F} = \mathcal{F}(\xi, x, \hbar)$ of F and $\mathcal{G} = \mathcal{G}(\xi, x, \hbar)$ of G, where $\{\mathcal{F}, \mathcal{G}\}_M$ is defined through its Fourier representation

$$\{\mathcal{F}, \mathcal{G}\}_M(\xi, x; \hbar) = \int_{\mathbb{R}^l} \sum_{q \in \mathbb{Z}^l} (\widehat{\{\mathcal{F}, \mathcal{G}\}_M})_q(p, \hbar) e^{i(\langle p, \xi \rangle + \langle q, x \rangle)} \, dp \quad (4.6.25)$$

and

$$(\widehat{\{\mathcal{F}, \mathcal{G}\}_M})_q(p, \hbar) =$$
$$\frac{2}{\hbar} \int_{\mathbb{R}^l} \sum_{q' \in \mathbb{Z}^l} \widehat{\mathcal{F}}_{q-q'}(p - p', \hbar) \widehat{\mathcal{G}}_{q'}(p', \hbar) \sin[\frac{2}{\hbar}(\langle p', q \rangle - \langle p, q' \rangle)] \, dp' . \quad (4.6.26)$$

Notice that $\{\mathcal{F}, \mathcal{G}\}_M = -\{\mathcal{G}, \mathcal{F}\}_M$. The above equations (4.6.20)–(4.6.23) become, once written for the symbols:

$$\Sigma(g) = \sum_{k=0}^{\infty} \mathcal{H}_k \quad (4.6.27)$$

$$\mathcal{H}_0 := \mathcal{L}_\omega + g\mathcal{M}, \quad \mathcal{H}_k := \frac{\{\mathcal{W}(g), \mathcal{H}_{k-1}\}_M}{k}, \ k \geq 1,$$

where $\mathcal{W}(g) = g\mathcal{W}_1 + g^2\mathcal{W}_2 + \ldots$,

$$\Sigma(g) = \sum_{k=0}^{\infty} g^k \mathcal{B}_k \quad (4.6.28)$$

and

$$\mathcal{B}_0 = \mathcal{L}_\omega = \langle \omega, \xi \rangle; \quad \mathcal{B}_k = \{\mathcal{W}_k, \mathcal{L}_\omega\}_M + \mathcal{M}_k, \ k \geq 1, \quad \mathcal{M}_1 \equiv \mathcal{M} \tag{4.6.29}$$

$$\mathcal{M}_k = \sum_{r=2}^{k} \frac{1}{r!} \sum_{\substack{j_1+\ldots+j_r=k \\ j_s \geq 1}} \{\mathcal{W}_{j_1}, \{\mathcal{W}_{j_2}, \ldots, \{\mathcal{W}_{j_r}, \mathcal{L}_\omega\}_M \ldots\}_M\}_M \tag{4.6.30}$$
$$+ \sum_{r=1}^{k-1} \frac{1}{r!} \sum_{\substack{j_1+\ldots+j_r=k-1 \\ j_s \geq 1}} \{\mathcal{W}_{j_1}, \{\mathcal{W}_{j_2}, \ldots, \{\mathcal{W}_{j_r}, \mathcal{M}\}_M \ldots\}_M\}_M. \quad k > 1$$

Therefore, the symbols $\mathcal{W}_k$ and $\mathcal{B}_k$ of W_k and B_k can be recursively found solving the homological equation:

$$\{\mathcal{W}_k, \mathcal{L}_\omega\}_M + \mathcal{M}_k = \mathcal{B}_k, \qquad k = 1, \ldots \tag{4.6.31}$$

under the condition:

$$\{\mathcal{L}_\omega, \mathcal{B}_k\}_M = 0. \tag{4.6.32}$$

Here

$$\mathcal{W}_k = \mathcal{W}_k(\xi, x; \hbar), \ \mathcal{M}_k = \mathcal{M}_k(\xi, x; \hbar), \ \mathcal{B}_k = \mathcal{B}_k(\xi, x; \hbar).$$

Notice that, in view of Theorem A.1.1 in Appendix, (4.6.32) is immediately satisfied if $\mathcal{B}_k = \mathcal{B}_k(\xi; \hbar)$ does not depend on x. Moreover, by Theorem A.1.1(2), as $\mathcal{L}_\omega = \mathcal{L}_\omega(\xi) = \langle \omega, \xi \rangle$ is linear in ξ, we have

$$\{\mathcal{W}_k, \mathcal{L}_\omega\}_M = \{\mathcal{W}_k, \mathcal{L}_\omega\} = -\langle \nabla_x \mathcal{W}_k, \omega \rangle$$

and (4.6.31) becomes

$$-\langle \nabla_x \mathcal{W}_k(\xi, x), \omega \rangle + \mathcal{M}_k(\xi, x; \hbar) = \mathcal{B}_k(\xi; \hbar). \tag{4.6.33}$$

Write now $\mathcal{W}_k(\xi, x; \hbar)$ and $\mathcal{M}_k(\xi, x; \hbar)$ under their Fourier series representation, respectively:

$$\mathcal{W}_k(\xi, x; \hbar) = \sum_{q \in \mathbb{Z}^l} \mathcal{W}_{k,q}(\xi; \hbar) e^{i\langle q, x \rangle}, \qquad \mathcal{M}_k(\xi, x; \hbar) = \sum_{q \in \mathbb{Z}^l} \mathcal{M}_{k,q}(\xi; \hbar) e^{i\langle q, x \rangle}.$$

Then (4.6.33) in turn becomes:

$$-i \sum_{q \neq 0} \langle q, \omega \rangle \mathcal{W}_{k,q}(\xi; \hbar) e^{i\langle q, x \rangle} + \sum_{q \in \mathbb{Z}^l} \mathcal{M}_{k,q}(\xi; \hbar) e^{i\langle q, x \rangle} = \mathcal{B}_k(\xi; \hbar) \tag{4.6.34}$$

whence, imposing the equality of the Fourier coefficients of both sides, we obtain the solutions

$$\mathcal{B}_k(\xi,\hbar)=\mathcal{M}_{k,0}(\xi,\hbar),\qquad \mathcal{W}_{k,q}(\xi,\hbar)=\frac{\mathcal{M}_{k,q}(\xi,\hbar)}{i\langle q,\omega\rangle},\quad \forall q\neq 0. \tag{4.6.35}$$

4.6.2 Reality of $\mathcal{B}_k$: the Inductive Argument

Denote now $\mathcal{M}_1\equiv\mathcal{M}=\mathcal{M}_\omega$. As $\mathcal{M}_{\omega,q}(\xi)$ is real $\forall q\in\mathbb{Z}^l$ by assumption, we have

$$\mathcal{B}_1(\xi,\hbar)=\mathcal{M}_{\omega,0}(\xi)\in\mathbb{R}$$

and

$$\mathcal{W}_{1,q}(\xi,\hbar)=\frac{\mathcal{M}_{\omega,q}(\xi)}{i\langle q,\omega\rangle}\in i\mathbb{R},\quad \forall q\neq 0. \tag{4.6.36}$$

Moreover, as no requirement is asked on $\mathcal{W}_{1,0}$, we can choose $\mathcal{W}_{1,0}=0$. Now assume inductively:
($\mathbf{A_1}$) $\mathcal{M}_{j,q}(\xi,\hbar)\in\mathbb{R},\quad \forall j=1,\dots,k-1,\ \forall q\in\mathbb{Z}^l$;
($\mathbf{A_2}$) we can choose $\mathcal{W}_{j,0}=0,\ \forall j=1,\dots,k-1$.
Remark that ($\mathbf{A_1}$) entails

$$\mathcal{W}_{j,q}(\xi,\hbar))=\frac{\mathcal{M}_{j,q}(\xi,\hbar)}{i\langle q,\omega\rangle}\in i\mathbb{R}\,,\qquad \mathcal{B}_j(\xi,\hbar)=\mathcal{M}_{j,0}\in\mathbb{R},\quad \forall j=1,\dots,k-1. \tag{4.6.37}$$

Then the following assertions hold:
($\mathbf{R_1}$) $\mathcal{M}_{k,q}(\xi,\hbar)\in\mathbb{R},\ \forall q\in\mathbb{Z}^l$;
($\mathbf{R_2}$), we can choose $\mathcal{W}_{k,0}=0$.
Remark that ($\mathbf{R_1}$) entails

$$\mathcal{W}_{k,q}(\xi,\hbar)=\frac{\mathcal{M}_{k,q}(\xi,\hbar)}{i\langle q,\omega\rangle}\in i\mathbb{R}\,;\qquad \mathcal{B}_k(\xi)=\mathcal{M}_{k,0}\in\mathbb{R}. \tag{4.6.38}$$

In order to prove ($\mathbf{R_1}$), consider the Fourier expansion of $\mathcal{M}_k$ given by (4.6.30)

$$\begin{aligned}\mathcal{M}_k&=\sum_{r=2}^{k}\frac{1}{r!}\sum_{\substack{j_1+\dots+j_r=k\\ j_s\geq 1}}\{\mathcal{W}_{j_1},\{\mathcal{W}_{j_2},\dots,\{\mathcal{W}_{j_r},\mathcal{L}_\omega\}_M\dots\}_M\\ &+\sum_{r=1}^{k-1}\frac{1}{r!}\sum_{\substack{j_1+\dots+j_r=k-1\\ j_s\geq 1}}\{\mathcal{W}_{j_1},\{\mathcal{W}_{j_2},\dots,\{\mathcal{W}_{j_r},\mathcal{M}\}_M\dots\}_M\\ &=\sum_{q\in\mathbb{Z}^l}\mathcal{M}_{k,q}(\xi,\hbar)e^{i\langle q,x\rangle}.\end{aligned}$$

By (4.6.37), the Fourier coefficients $\mathcal{W}_{j_s,q}$ of each term $\mathcal{W}_{j_s}$, $s = 1, \dots, r$, are purely imaginary, and by Theorem A.1.1(3) each Moyal bracket generates another factor i. Therefore,

$$\Big(\sum_{\substack{j_1+\ldots+j_r=k \\ j_s\geq 1}} \{\mathcal{W}_{j_1}, \{\mathcal{W}_{j_2}, \dots, \{\mathcal{W}_{j_r}, \mathcal{L}_\omega\}_M \cdots\}_M \Big)_q (\xi, \hbar) = (i)^{2r} a_{k,q}(\xi, \hbar),$$

$$a_{k,q}(\xi, \hbar) \in \mathbb{R}$$

$$\Big(\sum_{\substack{j_1+\ldots+j_r=k-1 \\ j_s\geq 1}} \{\mathcal{W}_{j_1}, \{\mathcal{W}_{j_2}, \dots, \{\mathcal{W}_{j_r}, \mathcal{M}\}_M \cdots\}_M \Big)_q (\xi, \hbar) = (i)^{2r} b_{k,q}(\xi, \hbar),$$

$$b_{k,q}(\xi, \hbar) \in \mathbb{R}$$

and, as a consequence, $\forall q \in \mathbb{Z}^l$:

$$\mathcal{M}_{k,q}(\xi, \hbar) = (i)^{2r}[a_{k,q}(\xi, \hbar) + b_{k,q}(\xi, \hbar)] = (-1)^r[a_{k,q}(\xi, \hbar) + b_{k,q}(\xi, \hbar)] \in \mathbb{R}.$$

Hence $\mathcal{B}_k(\xi, \hbar) = \mathcal{M}_{k,0} \in \mathbb{R}$. Moreover, the homological equation (4.6.34) does not involve $\mathcal{W}_{k,0}$; therefore, we can always take $\mathcal{W}_{k,0} = 0$. This concludes the proof of the induction and thus of Assertion (1) of Theorem 4.6.1.

4.6.3 Vanishing of the Odd Terms $\mathcal{B}_{2s+1}$

Let us now prove Assertion (2) of Theorem 4.6.1. This will yield

$$\Sigma(\varepsilon) = \mathcal{B}(\xi; \hbar) = \mathcal{L}_\omega(\xi) + \varepsilon^2 \mathcal{B}_2(\xi, \hbar) + \varepsilon^4 \mathcal{B}_4(\xi, \hbar) + \dots .$$

To see this, first recall that $\mathcal{M}_\omega(\xi, x)$ is odd in x: $\mathcal{M}_\omega(\xi, -x) = -\mathcal{M}_\omega(\xi, x)$, and let $\mathcal{Z}$ denote the set of functions $f : \mathbb{T}^l \to \mathbb{C}$ with a definite parity (either even or odd). Moreover, $\forall f \in \mathcal{Z}$ define

$$Jf = \begin{cases} +1, & \text{if } f \text{ is even,} \\ -1, & \text{if } f \text{ is odd.} \end{cases} \tag{4.6.39}$$

Then $Jf = 1$ if and only if $f_q = f_{-q}$ and $Jf = -1$ if and only if $f_q = -f_{-q}$, $\forall q \in \mathbb{Z}^l$. By assumption $\mathcal{M}_{\omega,q}(\xi) = -\mathcal{M}_{\omega,-q}(\xi)$, $\forall q \in \mathbb{Z}^l, \forall \xi \in \mathbb{R}^l$, that is, $J\mathcal{M}_\omega(\xi) = 1$, and by (4.6.36)

$$J\mathcal{W}_1(\xi, \hbar) = 1, \qquad \forall (\xi, \hbar) \in \mathbb{R}^l \times [0, 1].$$

Now we can prove by induction that

$$J\mathcal{M}_k = (-1)^k, \qquad \forall k = 1, 2, \dots \tag{4.6.40}$$

whence $J\mathcal{M}_{2s+1} = 1$, that is, $\mathcal{M}_{2s+1}(\xi, x, \hbar)$ is odd in x, which entails $\mathcal{B}_{2s+1} = \mathcal{M}_{2s+1,0} = 0, \forall s = 0, 1, \dots$. To prove (4.6.40) inductively, first notice that $J\mathcal{M}_1 = J\mathcal{M}_\omega = 1$ and then let us assume that

$$J\mathcal{M}_j = (-1)^j, \qquad \forall j = 1, \dots, k-1.$$

Then by (4.6.35)

$$J\mathcal{W}_j = (-1)^{j+1}, \qquad \forall j = 1, \dots, k-1.$$

Let us examine the parity of the first summand on the right hand side of (4.6.30), making use of Theorem A.1.1(4):

$$\begin{aligned} J(\{\mathcal{W}_{j_1}, \{\mathcal{W}_{j_2}, \dots \{\mathcal{W}_{j_r}, \mathcal{L}_\omega\}_M \dots \}_M\}_M) &= (-1)^r(-1)^{j_1+1} \dots (-1)^{j_r+1} \\ &= (-1)^k \end{aligned}$$

as $J\mathcal{L}_\omega = 1$ and $j_1 + \cdots + j_r = k$. Similarly for the second summand on the right hand side of (4.6.30), we have

$$\begin{aligned} J(\{\mathcal{W}_{j_1}, \{\mathcal{W}_{j_2}, \dots \{\mathcal{W}_{j_r}, \mathcal{M}\}_M \dots \}_M\}_M) &= (-1)^{r+1}(-1)^{j_1+1} \dots (-1)^{j_r+1} \\ &= (-1)^k \end{aligned}$$

as $J\mathcal{M} = -1$ and $j_1 + \cdots + j_r = k - 1$. This completes the proof of Assertion (2) and hence of Theorem 4.6.1.

Proof of Corollary 4.6.2. It is proved in Ref. (43) that the convergence of the S-QNF

$$\Sigma(\varepsilon) = \mathcal{L}_\omega(\xi) + \sum_{k=1}^{\infty} \mathcal{B}_k(\xi, \hbar) g^k$$

takes place in the $\| \cdot \|_{\rho/2}$-norm, where $\| \cdot \|_\rho$ is the norm defined in (4.6.15). Since (Remark A.2.2(b) and formula (A.2.6)) the $\| \cdot \|_{\rho/2}$-norm majorizes the operator norm in $L^2(\mathbb{T}^l)$ of the corresponding Weyl-quantized operators, we can conclude that

$$S(g) = L(\omega, \hbar) + \sum_{k=1}^{\infty} B_k g^k, \qquad B_{2s+1} = 0, \quad \forall s = 0, 1, \dots,$$

where the convergence takes place in the operator norm sense. As $B_k = B_k^*$, $S(g) = S(g)^*$ and the similarity between $H(\omega, g)$ and a self-adjoint operator is therefore proved.

APPENDIX: MOYAL BRACKETS AND THE WEYL QUANTIZATION

A.1 MOYAL BRACKETS

Theorem A.1.1 *Let $\mathcal{F} = \mathcal{F}(\xi, x; \hbar)$ and $\mathcal{G} = \mathcal{G}(\xi, x; \hbar)$ belong to $C^\infty(\mathbb{R}^l \times \mathbb{T}^l \times [0,1]; \mathbb{C})$ and vanish exponentially fast as $|\xi| \to \infty$, uniformly with respect to $(x, \hbar) \in \mathbb{T}^l \times [0,1]$. Consider their Fourier representation*

$$\mathcal{F}(\xi, x; \hbar) = \int_{\mathbb{R}^l} \sum_{q \in \mathbb{Z}^l} \widehat{\mathcal{F}}_q(p; \hbar) e^{i(\langle p, \xi \rangle + \langle q, x \rangle)} \, dp,$$

$$\mathcal{G}(\xi, x; \hbar) = \int_{\mathbb{R}^l} \sum_{q \in \mathbb{Z}^l} \widehat{\mathcal{G}}_q(p; \hbar) e^{i(\langle p, \xi \rangle + \langle q, x \rangle)} \, dp\,,$$

where

$$\mathcal{F}_q(\xi, \hbar) = (2\pi)^{-l/2} \int_{\mathbb{T}^l} \mathcal{F}(\xi, x, \hbar) e^{-i\langle q, x \rangle} \, dx,$$

$$\mathcal{G}_q(\xi, \hbar) = (2\pi)^{-l/2} \int_{\mathbb{T}^l} \mathcal{G}(\xi, x, \hbar) e^{-i\langle q, x \rangle} \, dx$$

and

$$\widehat{\mathcal{F}}_q(p; \hbar) = (2\pi)^{-l/2} \int_{\mathbb{R}^l} \mathcal{F}_q(\xi, \hbar) e^{-i\langle p, \xi \rangle} \, d\xi,$$

$$\widehat{\mathcal{G}}_q(p; \hbar) = (2\pi)^{-l/2} \int_{\mathbb{R}^l} \mathcal{G}_q(\xi, \hbar) e^{-i\langle p, \xi \rangle} \, d\xi.$$

Then, the following assertions hold:

(1) If both $\mathcal{F}$ and $\mathcal{G}$ do not depend on x, that is, $\mathcal{F}(\xi, x; \hbar) = \mathcal{F}(\xi; \hbar)$ and $\mathcal{G}(\xi, x; \hbar) = \mathcal{G}(\xi; \hbar)$, then $\{\mathcal{F}, \mathcal{G}\}_M \equiv 0$.

(2) If $\mathcal{G}(\xi, x; \hbar) = \langle \omega, \xi \rangle$, for a given constant vector $\omega \in \mathbb{R}^l$, that is, $\mathcal{G}$ does not depend on x and is linear in ξ, then

$$\{\mathcal{F}, \mathcal{G}\}_M = \{\mathcal{F}, \mathcal{G}\} = -\langle \nabla_x \mathcal{F}, \omega \rangle\,.$$

(3) Consider the Fourier expansions of $\mathcal{F}$ and $\mathcal{G}$ in the x variable:

$$\mathcal{F}(\xi, x; \hbar) = \sum_{q \in \mathbb{Z}^l} \mathcal{F}_q(\xi; \hbar) e^{i\langle q, x \rangle}, \quad \mathcal{G}(\xi, x; \hbar) = \sum_{q \in \mathbb{Z}^l} \mathcal{G}_q(\xi; \hbar) e^{i\langle q, x \rangle},$$

where, $\forall q \in \mathbb{Z}^l$,

$$\mathcal{F}_q(\xi;\hbar) = (2\pi)^{-l/2}\int_{\mathbb{R}^l} \widehat{\mathcal{F}}_q(p;\hbar)e^{i\langle p,\xi\rangle}\, dp,$$

$$\mathcal{G}_q(\xi;\hbar) = (2\pi)^{-l/2}\int_{\mathbb{R}^l} \widehat{\mathcal{G}}_q(p;\hbar)e^{i\langle p,\xi\rangle}\, dp\,.$$

If $\mathcal{F}_q(\xi;\hbar) \in \mathbb{R}$, *and* $\mathcal{G}_q(\xi;\hbar) \in \mathbb{R}$, $\forall q \in \mathbb{Z}^l$, *then the Fourier expansion of* $\{\mathcal{F},\mathcal{G}\}_M$*has purely imaginary Fourier coefficients, that is,*

$$(\{\mathcal{F},\mathcal{G}\}_M)_q(\xi;\hbar) := \int_{\mathbb{R}^l} (\widehat{\{\mathcal{F},\mathcal{G}\}_M})_q(p;\hbar)e^{i\langle p,\xi\rangle}\, dp \in i\mathbb{R}\,.$$

(4) *Let* $x \in \mathbb{T}^l \to \mathcal{F}(\xi,x;\hbar) \in \mathbb{C}$ *and* $x \in \mathbb{T}^l \to \mathcal{G}(\xi,x;\hbar) \in \mathbb{C}$ *belong to the space* $\mathcal{Z}$ *of the functions with a definite parity (either even or odd) and let* $J : \mathcal{Z} \to \{-1,1\}$ *be defined by* (4.6.39). *Then*

$$J\{\mathcal{F},\mathcal{G}\}_M = -(J\mathcal{F})(J\mathcal{G}).$$

To prove the theorem, we need the following

Lemma A.1.2 *Let* $\mathcal{F} = \mathcal{F}(\xi,x;\hbar) \in C^\infty(\mathbb{R}^l \times \mathbb{T}^l \times [0,1];\mathbb{C})$.*Then*

(i) $\mathcal{F}_q(\xi;\hbar) \in \mathbb{R}$, $\forall q \in \mathbb{Z}^l$, $\forall \xi \in \mathbb{R}^l$ *if and only if*

$$\overline{\widehat{\mathcal{F}}_q(p,\hbar)} = \widehat{\mathcal{F}}_q(-p,\hbar), \quad \forall q \in \mathbb{Z}^l,\ \forall p \in \mathbb{R}^l\,.$$

(ii) $\mathcal{F}_q(\xi;\hbar) \in i\mathbb{R}$, $\forall q \in \mathbb{Z}^l$, $\forall \xi \in \mathbb{R}^l$ *if and only if*

$$\overline{\widehat{\mathcal{F}}_q(p,\hbar)} = -\widehat{\mathcal{F}}_q(-p,\hbar), \quad \forall q \in \mathbb{Z}^l,\ \forall p \in \mathbb{R}^l\,.$$

Proof of Lemma A.1.2. We prove only (i) because the proof of (ii) is analogous. If $\mathcal{F}_q(\xi;\hbar) \in \mathbb{R}$, $\forall q \in \mathbb{Z}^l$, $\forall \xi \in \mathbb{R}^l$, then

$$\begin{aligned}\overline{\widehat{\mathcal{F}}_q(p,\hbar)} &= (2\pi)^{-l/2}\int_{\mathbb{R}^l} \mathcal{F}_q(\xi,\hbar)e^{i\langle p,\xi\rangle}\, d\xi \\ &= (2\pi)^{-l/2}\int_{\mathbb{R}^l} \mathcal{F}_q(\xi,\hbar)e^{-i\langle -p,\xi\rangle}\, d\xi = \widehat{\mathcal{F}}_q(-p,\hbar)\,.\end{aligned}$$

Conversely, let $\overline{\widehat{\mathcal{F}}_q(p,\hbar)} = \widehat{\mathcal{F}}_q(-p,\hbar)$, $\quad \forall q \in \mathbb{Z}^l$, $\forall p \in \mathbb{R}^l$.Then

$$\begin{aligned}\overline{\mathcal{F}_q(\xi;\hbar)} &= (2\pi)^{-l/2}\int_{\mathbb{R}^l} \overline{\widehat{\mathcal{F}}_q(p,\hbar)}e^{-i\langle p,\xi\rangle}\, dp = (2\pi)^{-l/2}\int_{\mathbb{R}^l} \widehat{\mathcal{F}}_q(-p,\hbar)e^{i\langle -p,\xi\rangle}\, dp \\ &= (2\pi)^{-l/2}\int_{\mathbb{R}^l} \widehat{\mathcal{F}}_q(p,\hbar)e^{i\langle p,\xi\rangle}\, dp = \mathcal{F}_q(\xi;\hbar),\end{aligned}$$

where to obtain the second equality, we have performed the change of variables $p \to -p$ in the integral. Hence, $\mathcal{F}_q(\xi;\hbar) \in \mathbb{R}$, $\forall q \in \mathbb{Z}^l$, $\forall \xi \in \mathbb{R}^l$ and this completes the proof of the lemma.

Proof of Theorem A.1.1.

(1) If $\mathcal{F}$ and $\mathcal{G}$ do not depend on x, then $\mathcal{F}_q(\xi,\hbar) = \mathcal{G}_q(\xi,\hbar) = 0$, $\forall q \neq 0, \forall \xi \in \mathbb{R}^l$. Therefore all the terms of the expansion in (4.6.26) with $q' \neq 0$ vanish. Then $\forall q \in \mathbb{Z}^l$

$$(\widehat{\{\mathcal{F},\mathcal{G}\}_M})_q(p,\hbar) = \frac{2}{\hbar}\int_{\mathbb{R}^l} \widehat{\mathcal{F}}_q(p-p',\hbar)\widehat{\mathcal{G}}_0(p',\hbar)\sin(\frac{2}{\hbar}\langle p',q\rangle)\,dp'$$

vanishes both for $q \neq 0$ and for $q = 0$, whence $\{\mathcal{F},\mathcal{G}\}_M \equiv 0$ by (4.6.25).

(2) If $\mathcal{G}(\xi,x;\hbar) = \langle \omega,\xi\rangle$, then by (4.6.26)

$$\begin{aligned}
(\widehat{\{\mathcal{F},\mathcal{G}\}_M})_q(p,\hbar) &= \frac{2}{\hbar}\int_{\mathbb{R}^l} \widehat{\mathcal{F}}_q(p-p',\hbar)\widehat{\mathcal{G}}_0(p',\hbar)\sin(\frac{2}{\hbar}\langle p',q\rangle)\,dp' \\
&= \frac{2}{\hbar}\int_{\mathbb{R}^l} \widehat{\mathcal{F}}_q(p-p',\hbar)\langle\omega, i\delta'(p)\rangle\sin(\frac{2}{\hbar}\langle p',q\rangle)\,dp' \\
&= \frac{2i}{\hbar}\sum_{j=1}^{l} \omega_j \frac{\partial}{\partial p_j}[\widehat{\mathcal{F}}_q(p-p',\hbar)\sin(\frac{2}{\hbar}\langle p',q\rangle)]|_{p'=0} \\
&= -i\sum_{j=1}^{l} \omega_j q_j \widehat{\mathcal{F}}_q(p,\hbar) = -\widehat{\mathcal{F}}_q(p,\hbar)\langle\omega, iq\rangle,
\end{aligned}$$

where the Fourier transform $\widehat{\mathcal{G}}_0(p',\hbar)$ of $\mathcal{G}_0(\xi,\hbar) = \langle\omega,\xi\rangle$ exists in the distributional sense, and is given by $i\delta'(p')$, where

$$\int_{\mathbb{R}^l} \langle\omega,\delta'(p')\rangle f(p')\,dp' = \sum_{j=1}^{l}\omega_j \frac{\partial f}{\partial p'_j}|_{p'=0}, \qquad \forall f \in \mathcal{S}(\mathbb{R}^l).$$

Here $\mathcal{S}(\mathbb{R}^l)$ denotes the Schwartz space. Then by (4.6.25)

$$\begin{aligned}
\{\mathcal{F},\mathcal{G}\}_M(\xi,x;\hbar) &= -\int_{\mathbb{R}^l}\sum_{q\in\mathbb{Z}^l}\langle\omega,iq\rangle\widehat{\mathcal{F}}_q(p,\hbar)e^{i(\langle p,\xi\rangle\langle+\langle q,x\rangle)}\,dp \\
&= -\sum_{q\in\mathbb{Z}^l}\langle\omega,iq\rangle\mathcal{F}_q(\xi,\hbar)e^{i\langle q,x\rangle} = -\langle\omega,\nabla_x\mathcal{F}(\xi,x)\rangle.
\end{aligned}$$

(3) By Lemma A.1.2 (i), we have $\overline{\widehat{\mathcal{F}}_q(p,\hbar)} = \widehat{\mathcal{F}}_q(-p,\hbar)$ and

$$\overline{\widehat{\mathcal{G}}_q(p,\hbar)} = \widehat{\mathcal{G}}_q(-p,\hbar),\ \forall q \in \mathbb{Z}^l,\ \forall p \in \mathbb{R}^l.$$

Then, from (4.6.26) we obtain

$$\overline{(\widehat{\{\mathcal{F},\mathcal{G}\}_M})_q(p,\hbar)} =$$
$$\frac{2}{\hbar}\int_{\mathbb{R}^l}\sum_{q'\in\mathbb{Z}^l}\widehat{\mathcal{F}}_{q-q'}(-p+p',\hbar)\widehat{\mathcal{G}}_{q'}(-p',\hbar)\sin[\frac{2}{\hbar}(\langle p',q\rangle-\langle p,q'\rangle)]\,dp'$$

whence, performing the change of variables $p'\to -p'$ in the integral,

$$\begin{aligned}
&\overline{(\widehat{\{\mathcal{F},\mathcal{G}\}_M})_q(p,\hbar)}\\
&=\frac{2}{\hbar}\int_{\mathbb{R}^l}\sum_{q'\in\mathbb{Z}^l}\widehat{\mathcal{F}}_{q-q'}(-p-p',\hbar)\widehat{\mathcal{G}}_{q'}(p',\hbar)\sin[\frac{2}{\hbar}(-\langle p',q\rangle+\langle -p,q'\rangle)]\,dp'\\
&=-\frac{2}{\hbar}\int_{\mathbb{R}^l}\sum_{q'\in\mathbb{Z}^l}\widehat{\mathcal{F}}_{q-q'}(-p-p',\hbar)\widehat{\mathcal{G}}_{q'}(p',\hbar)\sin[\frac{2}{\hbar}(\langle p',q\rangle-\langle -p,q'\rangle)]\,dp'\\
&=-(\widehat{\{\mathcal{F},\mathcal{G}\}_M})_q(-p,\hbar)
\end{aligned}$$

$\forall q\in\mathbb{Z}^l$, $\forall p\in\mathbb{R}^l$. Then, by Lemma A.1.2 (ii), $(\widehat{\{\mathcal{F},\mathcal{G}\}_M})(\xi,\hbar)\in i\mathbb{R}$, $\forall q\in\mathbb{Z}^l$, $\forall\xi\in\mathbb{R}^l$.

(4) First of all recall that $J\mathcal{F}=\pm 1$ if and only if $\mathcal{F}_q(\xi,\hbar)=\pm\mathcal{F}_{-q}(\xi,\hbar)$, $\forall q\in\mathbb{Z}^l$, $\forall(\xi,\hbar)\in\mathbb{R}^l\times[0,1]$. Then by (4.6.26) we have

$$\begin{aligned}
&(\widehat{\{\mathcal{F},\mathcal{G}\}_M})_{-q}(p,\hbar)\\
&=\frac{2}{\hbar}\int_{\mathbb{R}^l}\sum_{q'\in\mathbb{Z}^l}\widehat{\mathcal{F}}_{-q-q'}(p-p',\hbar)\widehat{\mathcal{G}}_{q'}(p',\hbar)\sin[\frac{2}{\hbar}(-\langle p',q\rangle-\langle p,q'\rangle)]\,dp'\\
&=\frac{2}{\hbar}\int_{\mathbb{R}^l}\sum_{q'\in\mathbb{Z}^l}\widehat{\mathcal{F}}_{-q+q'}(p-p',\hbar)\widehat{\mathcal{G}}_{-q'}(p',\hbar)\sin[\frac{2}{\hbar}(-\langle p',q\rangle+\langle p,q'\rangle)]\,dp'\\
&=-\frac{2}{\hbar}\int_{\mathbb{R}^l}\sum_{q'\in\mathbb{Z}^l}\widehat{\mathcal{F}}_{-q+q'}(p-p',\hbar)\widehat{\mathcal{G}}_{-q'}(p',\hbar)\sin[\frac{2}{\hbar}(\langle p',q\rangle-\langle p,q'\rangle)]\,dp',
\end{aligned}$$

where in the second equality, we have performed the change of variables $q'\to -q'$. Assume first that $J\mathcal{F}=J\mathcal{G}$; then $\mathcal{F}_{-q}\mathcal{G}_{-q'}\equiv\mathcal{F}_q\mathcal{G}_{q'}$ and $\widehat{\mathcal{F}}_{-q}\widehat{\mathcal{G}}_{-q'}\equiv\widehat{\mathcal{F}}_q\widehat{\mathcal{G}}_{q'}$, $\forall q,q'\in\mathbb{Z}^l$. Thus,

$$\begin{aligned}
&(\widehat{\{\mathcal{F},\mathcal{G}\}_M})_{-q}(p,\hbar)\\
&=-\frac{2}{\hbar}\int_{\mathbb{R}^l}\sum_{q'\in\mathbb{Z}^l}\widehat{\mathcal{F}}_{q-q'}(p-p',\hbar)\widehat{\mathcal{G}}_{q'}(p',\hbar)\sin[\frac{2}{\hbar}(\langle p',q\rangle-\langle p,q'\rangle)]\,dp'\\
&=-(\widehat{\{\mathcal{F},\mathcal{G}\}_M})_q(p,\hbar)),
\end{aligned}$$

whence

$$(\{\mathcal{F},\mathcal{G}\}_M)_{-q}(\xi,\hbar) = -(\{\mathcal{F},\mathcal{G}\}_M)_q(\xi,\hbar), \quad \forall q \in \mathbb{Z}^l,\ \forall(\xi,\hbar) \in \mathbb{R}^l \times [0,1]$$

and $J\{\mathcal{F},\mathcal{G}\}_M = -1 = -(J\mathcal{F})(J\mathcal{G})$. In a similar way, we obtain $J\{\mathcal{F},\mathcal{G}\}_M = 1$ if $J\mathcal{F} = -J\mathcal{G}$, and this completes the proof of the theorem.

A.2 THE WEYL QUANTIZATION

Let us sum up the canonical (Weyl) quantization procedure for functions (classical observables) defined on the phase space $\mathbb{R}^l \times \mathbb{T}^l$. For more detail, the reader is referred to Ref. (43).

Let $\mathcal{A}(\xi,x,\hbar) : \mathbb{R}^l \times \mathbb{T}^l \times [0,1] \to \mathbb{C}$ be a family of smooth phase-space functions indexed by $\hbar$ fulfilling the assumptions of Theorem A.1.1, written under its Fourier representation

$$\mathcal{A}(\xi,x,\hbar) = \int_{\mathbb{R}^l} \sum_{q\in\mathbb{Z}^l} \widehat{\mathcal{A}}_q(p;\hbar)e^{i(\langle p,\xi\rangle+\langle q,x\rangle)}\,dp,$$

where, as in Section 1:

$$\mathcal{A}(\xi,x,\hbar) = \sum_{q\in\mathbb{Z}^l} \mathcal{A}_q(\xi,\hbar)e^{i\langle q,x\rangle},$$

$$\mathcal{A}_q(\xi,\hbar) := (2\pi)^{-l/2} \int_{\mathbb{T}^l} \mathcal{A}(\xi,x;\hbar)e^{-i\langle q,x\rangle}\,dx$$

$$\widehat{\mathcal{A}}_q(p;\hbar) = (2\pi)^{-l/2} \int_{\mathbb{R}^l} \mathcal{A}_q(\xi;\hbar)e^{-i\langle p,\xi\rangle}\,dx.$$

Then the (Weyl) quantization of $\mathcal{A}(\xi,x;\hbar)$ is the operator acting on $L^2(\mathbb{T}^l)$, defined by:

$$(A(\hbar)f)(x) := \int_{\mathbb{R}^l} \sum_{q\in\mathbb{Z}^l} \widehat{\mathcal{A}}_q(p;\hbar)e^{i(\langle q,x\rangle+\langle p,q\rangle\hbar/2)}f(x+p\hbar)\,dp, \qquad f \in L^2(\mathbb{T}^l). \tag{A.2.1}$$

Remarks

(a) If $\mathcal{A}$ does not depend on ξ, $\mathcal{A}(\xi,x,\hbar) = \mathcal{A}(x,\hbar)$, (A.1) reduces to the standard *multiplicative* action:

$$(A(\hbar)f)(x) = \int_{\mathbb{R}^l} \sum_{q\in\mathbb{Z}^l} \mathcal{A}_q(\hbar)\delta(p)e^{i(\langle q,x\rangle+\langle p,q\rangle\hbar/2)}f(x+\hbar p)\,dp$$

$$= \sum_{q\in\mathbb{Z}^l} \mathcal{A}_q(\hbar)e^{i\langle q,x\rangle}f(x) = \mathcal{A}(x,\hbar)f(x).$$

(b) If $\mathcal{A}$ does not depend on x, then $\widehat{\mathcal{A}}_q = 0, q \neq 0$; thus $\widehat{\mathcal{A}}_0 = \widehat{\mathcal{A}}(p, \hbar)$ and the standard (pseudo) differential action is recovered:

$$\begin{aligned}(A(\hbar)f)(x) &= \int_{\mathbb{R}^l} \widehat{\mathcal{A}}(p, \hbar) f(x + \hbar p)\, dp = \int_{\mathbb{R}^l} \sum_{q \in \mathbb{Z}^l} \widehat{\mathcal{A}}(p, \hbar) f_q e^{i\langle q, x + \hbar p\rangle}\, dp \\ &= \sum_{q \in \mathbb{Z}^l} f_q \mathcal{A}(q\hbar, \hbar) e^{i\langle q, x\rangle} = (\mathcal{A}(-i\hbar\nabla_x, \hbar) f)(x),\end{aligned}$$

whence the formula yielding all the eigenvalues of A:

$$\lambda_n(\hbar) = \langle e_n, A e_n\rangle = \mathcal{A}(n\hbar, \hbar), \tag{A.2.2}$$

where $\{e_n : n \in \mathbb{N}\}$ is the set of the Hermite functions in $L^2(\mathbb{R}^l)$.
(c) Let $\mathcal{M}(t, x; \hbar)$ be a complex-valued, smooth function defined on $\mathbb{R} \times \mathbb{T}^l \times [0, 1]$ vanishing exponentially fast as $|t| \to \infty$ uniformly with respect to $(x, \hbar) \in \mathbb{T}^l \times [0, 1]$, with Fourier expansion

$$\mathcal{M}(t, x; \hbar) = \int_{\mathbb{R}} \sum_{q \in \mathbb{Z}^l} \widehat{\mathcal{M}}_q(p; \hbar) e^{i(pt + \langle q, x\rangle)}\, dp, \tag{A.2.3}$$

where, as above:

$$\begin{aligned}\mathcal{M}(t, x, \hbar) &= \sum_{q \in \mathbb{Z}^l} \mathcal{M}_q(t, \hbar) e^{i\langle q, x\rangle}, \\ \mathcal{M}_q(t, \hbar) &:= (2\pi)^{-l/2} \int_{\mathbb{T}^l} \mathcal{M}(t, x; \hbar) e^{-i\langle q, x\rangle}\, dx \\ \widehat{\mathcal{M}}_q(p; \hbar) &= (2\pi)^{-l/2} \int_{\mathbb{R}} \mathcal{M}_q(t; \hbar) e^{-ipt}\, dt.\end{aligned}$$

Let the smooth function $\mathcal{M}_\omega(\xi, x; \hbar) : \mathbb{R}^l \times \mathbb{T}^l \times [0, 1] \to \mathbb{C}$ be defined as follows:

$$\mathcal{M}_\omega(\xi, x; \hbar) := \mathcal{M}(t, x, \hbar)|_{t = \mathcal{L}_\omega(\xi)} = \mathcal{M}(\langle \omega, \xi\rangle, x; \hbar).$$

Then we have:

$$\mathcal{M}_\omega(\xi, x; \hbar) = \int_{\mathbb{R}} \sum_{q \in \mathbb{Z}^l} \widehat{\mathcal{M}}_q(p, \hbar) e^{i(\langle q, x\rangle + p\mathcal{L}_\omega(\xi))}\, dp$$

and (A.2.1) clearly becomes:

$$(V_\omega(\hbar)f)(x) = \int_{\mathbb{R}} \sum_{q \in \mathbb{Z}^l} \widehat{\mathcal{M}}_q(p; \hbar) e^{i(\langle q, x\rangle + p\langle \omega, q\rangle \hbar/2)} f(x + p\hbar\omega)\, dp. \tag{A.2.4}$$

(d) Let

$$\|\mathcal{M}_\omega\|_\rho := \sup_{\hbar\in[0,1]} \sum_{q\in\mathbb{Z}^l} e^{\rho|q|} \int_{\mathbb{R}} e^{\rho|p|}\, |\widehat{\mathcal{M}}_q(p,\hbar)|\, dp < +\infty, \quad \rho \ge 0. \tag{A.2.5}$$

and remark that

$$\|\mathcal{M}_\omega\|_{L^1} := \sup_{\hbar\in[0,1]} \sum_{q\in\mathbb{Z}^l} \int_{\mathbb{R}} |\widehat{\mathcal{M}}_q(p,\hbar)|\, dp \le \|\mathcal{M}_\omega\|_\rho.$$

Then $V_\omega(\hbar)$ is a bounded operator in $L^2(\mathbb{T}^l)$, uniformly with respect to $\hbar \in [0,1]$, namely:

$$\sup_{\hbar\in[0,1]} \|V_\omega(\hbar)\|_{L^2\to L^2} \le \|\mathcal{M}_\omega\|_{L^1} \le \|\mathcal{M}_\omega\|_\rho \tag{A.2.6}$$

because

$$\|V_\omega(\hbar)f\|_{L^2} \le \sum_{q\in\mathbb{Z}^l} \int_{\mathbb{R}} |\widehat{\mathcal{M}}_q(p,\hbar)|\, dp\, \|f\|_{L^2} \le \|\mathcal{M}_\omega\|_{L^1}\, \|f\|_{L^2}.$$

(e) If the symbol $\mathcal{M}$ is real valued, then its Weyl quantization $V(\hbar)$ is clearly a symmetric operator in $L^2(\mathbb{T}^l)$; if in addition condition (A.2.5) holds its boundedness entails its self-adjointness.

REFERENCES

1. Bender CM, Boettcher S, Meisinger PN. PT-symmetric quantum mechanics. *J Math Phys* 1999;40:2201–2229.
2. Bender CM, Berry MV, Mandilara A. Generalized PT symmetry and real spectra. *J Phys A Math Gen* 2002;35:L467–L471.
3. Bender CM, Brody DC, Jones HF. Must a Hamiltonian be Hermitian? *Am J Phys* 2003;71:1039–1031.
4. Bender CM. Making sense of non-Hermitian Hamiltoniians. *Rep Prog Phys* 2007;70:947–1018, hep-th/0703096.
5. *J Phys A Math Gen* 2006;39(32) Special Issue: *$\mathcal{PT}$-symmetric quantum mechanics.*
6. *J Phys A Math Theor* 2008;21(24) Special Issue: *Papers dedicated to the 6th International Workshop on Pseudo-Hermitian Hamiltonians in Quantum Physics.*
7. *Pramana J Phys* 2009;73(2):269–277 Special Issue: *Non-Hermitian Hamiltonians in quantum physics - Part I.*
8. Bender CM, Boettcher S. Real spectra in Non-Hermitian Hamiltonians having PT symmetry. *Phys Rev Lett* 1998;80:5243—5246.

9. Caliceti E, Graffi S, Maioli M. Perturbation theory of odd anharmonic oscillators. *Commun Math Phys* 1980;75:51–66.
10. Kato T. *Perturbation Theory for Linear Operators*. New York: Springer-Verlag; 1966 (2nd ed. 1976).
11. Caliceti E. Distributional Borel summability of perturbation theory for the quantum Hénon-Heiles model. *Ann Inst Henri Poincare* 2006;7:561–580.
12. Caliceti E, Graffi S, Sjöstrand J. Spectra of $\mathcal{PT}$-symmetric operators and perturbation theory. *J Phys A Math Gen* 2005;38:185–193.
13. Caliceti E, Graffi S, Sjöstrand J. $\mathcal{PT}$-symmetric non self-adjoint operators, diagonizable and non-diagonalizable, with a real discrete spectrum. *J Phys A Math Theor* 2007;40:10155–10170.
14. Berezin F, Shubin MS. *The Schrödinger Equation*. Dordrecht: Kluwer Academic Publishers; 1991.
15. Shin KC. On the reality of the eigenvalues for a class of PT-symmetric oscillators. *Commun Math Phys* 2002;229:543–564.
16. Dorey P, Dunning C, Tateo R. Spectral equivalences, Bethe ansatz equations and reality properties of $\mathcal{PT}$-symmetric quantum mechanics. *J Phys A* 2001;34:5679.
17. Reed M, Simon B. *Methods of Modern Mathematical Physics*. Volume IV. Academic Press; 1978.
18. Simon B. Anharmonic oscillator. *Ann Phys (NY)* 1970;58:76–130.
19. Caliceti E, Graffi S. On a class of non self-adjoint quantum non-linear oscillators with real spectrum. *J Nonlinear Math Phys* 2005;12:138–145.
20. Ahmed Z. Energy band structure due to a complex, periodic, PT invariant potential. *Phys Lett A* 2001;286:231–235.
21. Bender CM, Dunne GV, Meisinger N. Complex periodic potentials with real band spectra. *Phys Lett A* 1999;252:272–276.
22. Cerverò JM. PT-symmetry in one-dimensional periodic potentials. *Phys Lett A* 2003;317:26–31.
23. Cerverò JM, Rodriguez JM. The band spectrum of periodic potentials with PT symmetry. *J Phys A Math Gen* 2004;37:10167–10177.
24. Jones HF. The energy spectrum of complex periodic potentials of Kronig-Penney type. *Phys Lett A* 1999;262:242–244.
25. Shin KC. On the shape of the spectra of non self-adjoint periodic Schrödinger operators. *J Phys A Math Gen* 2004;37:8287–8291.
26. Albeverio S, Gesztesy F, Höegh-Krohn R, Holden H. *Solvable Models in Quantum Mechanics*. Springer-Verlag; 1988.
27. Eastham MSP. *The Spectral Theory of Periodic Differential Equations*. Edinburgh: Scottish Academic Press; 1973.
28. Caliceti E, Graffi S. A criterion for the reality of the spectrum of $\mathcal{PT}$-symmetric Schrödinger operators with complex-valued periodic potentials. *Rend Lincei Mat Appl* 2008;198:163–173.

29. Herbst IW. Dilation analyticity in constant electric field. *Commun Math Phys* 1979;64:279–298.
30. Caliceti E, Graffi S. An existence criterion for the $\mathcal{PT}$-symmetric phase transition. *Discrete Contin Dyn Syst* 2014;19:1955–1967.
31. Sjöstrand J. Private communication.
32. Dorey P, Dunning C, Tateo R. Supersymmetry and the spontaneous breakdown of $\mathcal{PT}$ symmetry. *J Phys A* 2001;34:L391.
33. Dorey P, Dunning C, Lishman A, Tateo R. $\mathcal{PT}$-symmetry breaking for a class of inhomogeneous complex potentials. *J Phys A* 2009;42:465302.
34. Hunziker W, Pillet CA. Degenerate asymptotic perturbation theory. *Commun Math Phys* 1983;90:219–233.
35. Cannata F, Junker G, Trost J. Schrödinger operators with complex potential but real spectrum. *Phys Lett* 1998;A246:219–226.
36. Levai G, Znojil M. Systematic search for PT symmetric potentials with real energy spectra. *J Phys A Math Gen* 2000;33:7165.
37. Mostafazadeh A. Pseudo-Hermiticity versus P T symmetry: the necessary condition for the reality of the spectrum of a non-Hermitian Hamiltonian. *J Math Phys* 2002;43:205–214.
38. Mostafazadeh A. Pseudo-Hermiticity versus P T Symmetry. II. A complete characterization of non-Hermitian Hamiltonians with a real spectrum. *J Math Phys* 2002;43:2814–2816.
39. Mostafazadeh A. Pseudo-Hermiticity versus P T symmetry. III. Equivalence of pseudo-Hermiticity and the presence of antilinear symmetries. *J Math Phys* 2002;43:3944–3951.
40. Caliceti E, Hitrik M, Graffi S, Sjöstrand J. Quadratic $\mathcal{PT}$-symmetric operators and similarity with self-adjoint operators. *J Phys A Math Theor* 2012;45:444007 (20pp).
41. Sjöstrand J. Semi-excited levels in non-degenerate potential wells. *Asymptotic Anal* 1992;6:29–43.
42. Bambusi D, Graffi S, Paul T. Normal forms and quantization formulae. *Commun Math Phys* 1999;207:173–195.
43. Graffi S, Paul T. Convergence of a quantum normal form and an exact quantization formula. *J Funct Anal* 2012;262:3340–3393.
44. Caliceti E, Graffi S. Convergent quantum normal forms, $\mathcal{PT}$-symmetry and reality of the spectrum. *Rend Lincei Mat Appl* 2013;24:385–407.
45. Folland D. *Harmonic Analysis in Phase Space*. Princeton (NJ): Princeton University Press; 1989.
46. Robert D. *Autour de l'approximation Semi-Classique*. Boston (MA): Birkhauser; 1987.

5

ELEMENTS OF SPECTRAL THEORY WITHOUT THE SPECTRAL THEOREM

DAVID KREJČIŘÍK[1] AND PETR SIEGL[1,2]

[1]*Nuclear Physics Institute, ASCR, Řež, Czech Republic*
[2]*Mathematical Institute, University of Bern, Bern, Switzerland*

5.1 INTRODUCTION

Many physical systems can be described by partial differential equations, and the latter can often be viewed as generating abstract operators between Banach spaces. A typical example is quantum mechanics where the traditional mathematical discipline is the functional analysis of self-adjoint operators in Hilbert spaces. There are also effective models (typically describing open quantum systems, including nonreal fields or complex boundary conditions) or more generally nonconservative processes in Nature on the whole where the underlying operator is non-self-adjoint. More intrinsically, there have been recent attempts to build quantum mechanics with physical observables represented by non-self-adjoint operators.

From the mathematical point of view, the theory of self-adjoint operators is well understood, while the non-self-adjoint theory is still in its infancy. Or maybe more appropriate would be to say that the theory is "underdeveloped," as spectral theory of non-self-adjoint operators is an equally old branch of functional analysis. Indeed, the first pioneering works (1908–1913) of G. D. Birkhoff on non-self-adjoint boundary value problems were written almost at the same time as D. Hilbert's famous papers (1904–1910) that initiated self-adjoint spectral theory (*cf.* [1, p. viii]). But it was not until M. V. Keldyš' work (1951) when first abstract results on non-self-adjoint problems appeared in the literature, while the self-adjoint theory was already enjoying all the pleasures of life due to the needs of quantum mechanics at that time.

Non-Selfadjoint Operators in Quantum Physics: Mathematical Aspects, First Edition.
Edited by Fabio Bagarello, Jean-Pierre Gazeau, Franciszek Hugon Szafraniec and Miloslav Znojil.

It is frustrating that the powerful techniques of the self-adjoint theory, such as the spectral theorem and variational principles, are not available for non-self-adjoint operators. Moreover, recent studies have revealed that this lack of tools is fundamental; the non-self-adjointness may lead to new and unexpected phenomena. Although there exist many interesting observations coming from physics and numerical studies of non-self-adjoint operators, the deep theoretical understanding is still missing. The problem is that the non-self-adjoint theory is much more diverse and it is difficult, if not impossible, to find a common thread. *Indeed it can hardly be called a theory.* This is a quotation from the preface of E. B. Davies' 2007 book (2), where a significant amount of work on spectral theory of non-self-adjoint operators can be found. The author continues:

Studying non-self-adjoint operators is like being a vet rather than a doctor: one has to acquire a much wider range of knowledge, and to accept that one cannot expect to have as high a rate of success when confronted with particular cases.

We fully endorse this opinion and understand that the only way how "to acquire the much wider range of knowledge" is by studying many distinct cases. This chapter is particularly concerned with various cases coming from the rapidly developing field of quantum mechanics with non-self-adjoint operators. But we hope that this material will be useful for anybody interested in the methods of spectral theory when the spectral theorem is not available.

The structure of this chapter is as follows. Section 5.2 is mainly devoted to a collection of basic facts from the spectral theory of operators in Hilbert spaces. In Section 5.3, we summarize some efficient methods how to construct a closed operator with nonempty resolvent set. The theory of compact operators and various definitions of essential spectra are recalled in Section 5.4. Section 5.5 is concerned with operators that are similar to self-adjoint (or more generally normal) operators. Finally, in Section 5.6, we recall the notion of pseudospectra as a more reliable information about non-self-adjoint operators than the spectrum itself.

Our exposition is in many respects based on the classical monographs (3, 4) to which we refer for statements presented here without (or just sketchy) proofs. In addition to these references, we use the new edition (5) about Sobolev spaces, which are denoted here by $H^m(\Omega)$, and the book (6) about partial differential equations. Other references are quoted in the text. The majority of the material is standard, but we illustrate the abstract exposition by some unconventional quantum-mechanically motivated examples.

5.2 CLOSED OPERATORS IN HILBERT SPACES

Having in mind the applications of differential operators in quantum mechanics, we concentrate on closed operators acting in *Hilbert spaces*, although many concepts summarized next are relevant in Banach spaces as well.

5.2.1 Basic Notions

Throughout this chapter, $\mathcal{H}$ stands for a separable Hilbert space over the complex number field $\mathbb{C}$. The norm and inner product (antilinear in the first component) in $\mathcal{H}$ will be denoted by $\|\cdot\|$ and $(\cdot,\cdot)$, respectively. A paradigmatic example is the Lebesgue space $L^2(\Omega)$ of square-integrable functions over an open set $\Omega \subset \mathbb{R}^d$.

We define a *linear operator* H in $\mathcal{H}$ to be a pair consisting of a linear subspace $\mathsf{D}(H) \subset \mathcal{H}$ called the *domain* of H and a linear map $H : \mathsf{D}(H) \to \mathcal{H}$. If $\mathsf{D}(H)$ is dense in $\mathcal{H}$, H is said to be *densely defined*. The image $\mathsf{R}(H) := H\mathsf{D}(H)$ is called the *range* of H. The *null space* or *kernel* $\mathsf{N}(H)$ of H is the set of all $\psi \in \mathsf{D}(H)$ such that $H\psi = 0$.

If H_1 and H_2 are two operators in $\mathcal{H}$ such that $\mathsf{D}(H_1) \subset \mathsf{D}(H_2)$ and $H_1\psi = H_2\psi$ for all $\psi \in \mathsf{D}(H_1)$, we write $H_1 \subset H_2$ and say that H_2 is an *extension* of H_1 and H_1 is a *restriction* of H_2.

The following quantities play an important role in spectral theory:

$$\textit{nullity}, \qquad \mathsf{nul}(H) := \dim \mathsf{N}(H),$$

$$\textit{deficiency}, \qquad \mathsf{def}(H) := \operatorname{codim} \mathsf{R}(H).$$

Recall that the codimension of a subspace $\mathcal{H}' \subset \mathcal{H}$ is defined as the dimension of the quotient space $\mathcal{H}/\mathcal{H}'$. If $\mathcal{H}'$ is closed, then $\operatorname{codim} \mathcal{H}' = \dim \mathcal{H}'^{\perp}$, where $\perp$ denotes the orthogonal complement, but this equality does not extend to nonclosed subspaces, as the following example shows.

EXAMPLE 5.1 Identity operator

The *identity operator* I in $L^2(\mathbb{R})$, that is, $I\psi := \psi$, $\mathsf{D}(I) := L^2(\mathbb{R})$, has a closed range, $\mathsf{R}(I) = L^2(\mathbb{R})$, by definition, so $\mathsf{def}(I) = 0$. The situation is very different for its restriction $I'\psi := I\psi$, $\mathsf{D}(I') := C_0^\infty(\mathbb{R})$, when $\mathsf{R}(I') = C_0^\infty(\mathbb{R})$. As $\mathsf{R}(I')$ is dense in $L^2(\mathbb{R})$, we have $\dim \mathsf{R}(I')^\perp = 0$. However, $\mathsf{R}(I')$ is not closed and $\mathsf{def}(I') = +\infty$; indeed, for instance, Hermite functions are supported everywhere in $\mathbb{R}$ and form an orthonormal basis of $L^2(\mathbb{R})$.

The operator $H : \mathsf{D}(H) \to \mathcal{H}$, understood as a mapping between two normed spaces $(\mathsf{D}(H), \|\cdot\|)$ and $\mathcal{H}$, is said to be *bounded* if there exists a nonnegative number M such that $\|H\psi\| \le M\|\psi\|$ for all $\psi \in \mathsf{D}(H)$. The smallest number M with this property is called the *norm* of H and is denoted by $\|H\|_{\mathsf{D}(H)\to\mathcal{H}}$, that is,

$$\|H\|_{\mathsf{D}(H)\to\mathcal{H}} := \sup_{\psi\in\mathsf{D}(H),\psi\neq 0,} \frac{\|H\psi\|}{\|\psi\|}.$$

If H is bounded and $\mathsf{D}(H) = \mathcal{H}$, that is, H is an operator *on* $\mathcal{H}$ to $\mathcal{H}$, we drop the subscript in the notation of the norm, that is, $\|H\| := \|H\|_{\mathcal{H}\to\mathcal{H}}$. The space of all bounded operators on $\mathcal{H}$ to $\mathcal{H}$ is denoted by $\mathscr{B}(\mathcal{H})$. H is bounded if and only if it is *continuous*, that is,

$$\mathsf{D}(H) \ni \psi_n \xrightarrow[n\to\infty]{} \psi \in \mathsf{D}(H) \quad \Longrightarrow \quad H(\psi_n - \psi) \xrightarrow[n\to\infty]{} 0\,.$$

Most of the physically relevant operators are unbounded, including differential operators in $L^2(\Omega)$.

A suitable substitute for the continuity in the more general situation of unbounded operators is the important notion of closedness. We say that H is *closed* if

$$\left.\begin{aligned} \mathsf{D}(H) \ni \psi_n \xrightarrow[n\to\infty]{} \psi \in \mathcal{H} \\ H\psi_n \xrightarrow[n\to\infty]{} \phi \in \mathcal{H} \end{aligned}\right\} \quad \Longrightarrow \quad \left[\psi \in \mathsf{D}(H) \ \wedge \ H\psi = \phi\right].$$

As the spectrum is defined *only* for closed operators, *cf.* Section 5.2.2, checking that a given operator H is closed should be the first step in any spectral analysis of H. In what follows, H is thus typically assumed to be a closed operator in $\mathcal{H}$.

We also assume that H is densely defined, which is convenient in order to have the unique *adjoint* H^* defined as follows:

$$\begin{aligned} \mathsf{D}(H^*) &:= \left\{\phi \in \mathcal{H} \ : \ \exists \phi^* \in \mathcal{H}, \ \forall \psi \in \mathsf{D}(H), \ (\phi, H\psi) = (\phi^*, \psi)\right\}, \\ H^*\phi &:= \phi^*. \end{aligned} \tag{5.2.1}$$

H^* is always a closed operator, regardless whether H is closed or not, but it may happen that $\mathsf{D}(H^*) = \{0\}$. For any densely defined operator H, we have

$$\mathsf{N}(H^*) = \mathsf{R}(H)^{\perp}. \tag{5.2.2}$$

It turns out that differential operators in $L^2(\Omega)$ are closed and densely defined when their domains are properly chosen. We illustrate the situation on several characteristic examples coming from quantum mechanics.

EXAMPLE 5.2 Multiplication operator

Given an open set $\Omega \subset \mathbb{R}^d$, let M_V be the operator of *multiplication* in $L^2(\Omega)$ by a measurable function $V : \Omega \to \mathbb{C}$. It is defined by $M_V\psi := V\psi$ on its maximal domain $\mathsf{D}(M_V) := \{\psi \in L^2(\Omega) : V\psi \in L^2(\Omega)\}$. M_V is densely defined and closed. M_V is bounded on $L^2(\Omega)$ if and only if V is essentially bounded, in which case we have $\|M_V\| = \|V\|_\infty$. The adjoint of M_V is obtained by simply taking the complex conjugate of V, that is, $M_V^* = M_{\overline{V}}$, in particular, $\mathsf{D}(M_V^*) = \mathsf{D}(M_V)$. A quantum-mechanically distinguished example is the *position operator* q in $L^2(\mathbb{R})$, which is associated with the choice $V(x) := x$.

EXAMPLE 5.3 Momentum operator

Given an open interval $\Omega \subset \mathbb{R}$, we introduce the *momentum* operator p in $L^2(\Omega)$ by $p\psi := -i\psi'$ and $\mathsf{D}(p) := H^1(\Omega)$. The operator p is densely defined, closed, and always unbounded. The adjoint acts in the same way, but it satisfies an extra

Dirichlet boundary condition on $\partial\Omega$, that is, $p^*\psi = -i\psi'$ and $\mathsf{D}(p^*) = H_0^1(\Omega)$. In $L^2(\mathbb{R})$, p and q are unitarily equivalent via the Fourier transform.

EXAMPLE 5.4 Creation and annihilation operators

In $L^2(\mathbb{R})$, we introduce the *creation* and *annihilation* operators as follows. The annihilation operator is introduced as $a := ip + q$; by definition, $\mathsf{D}(a) = \mathsf{D}(p) \cap \mathsf{D}(q)$. The operator a is densely defined and it can be proved that it is closed and that its adjoint, the creation operator, reads $a^* = -ip + q$ (with $\mathsf{D}(a^*) = \mathsf{D}(a)$).

The famous harmonic oscillator Hamiltonian

$$H_{\text{HO}} := p^2 + q^2 , \qquad \mathsf{D}(H_{\text{HO}}) = \{\psi \in H^2(\mathbb{R}) \ : \ x^2\psi \in L^2(\mathbb{R})\}, \tag{5.2.3}$$

is closed and it can be verified that $H_{\text{HO}} = a^*a + 1$, that is, particularly the equality of the domains holds (notice that by definition of the product of two operators, $\mathsf{D}(a^*a) = \{\psi \in \mathsf{D}(a) \ : \ a\psi \in \mathsf{D}(a^*)\}$).

EXAMPLE 5.5 Free Hamiltonian and constraints

Given an open connected set Ω in $\mathbb{R}^d$, let us introduce an auxiliary densely defined operator $-\Delta^\Omega$ in $L^2(\Omega)$, which acts on the Sobolev space $\mathsf{D}(-\Delta^\Omega) := H^2(\Omega)$ as the Laplacian, that is, $-\Delta^\Omega\psi := -\Delta\psi$. The case $\Omega = \mathbb{R}^d$ corresponds to the *free Hamiltonian* describing the motion of a quantum particle in the whole space with the absence of external fields. It is well known that $-\Delta^{\mathbb{R}^d}$ is closed, in fact $-\Delta^{\mathbb{R}^d} = (-\Delta^{\mathbb{R}^d})^*$. If the boundary $\partial\Omega$ is not empty, a physically relevant closed realization of $-\Delta^\Omega$ is typically obtained by imposing suitable boundary conditions. For sufficiently regular Ω, such that the boundary traces $H^2(\Omega) \hookrightarrow H^1(\partial\Omega)$ exist, we consider

$$\textit{Dirichlet boundary conditions,} \qquad \psi = 0 \quad \text{on} \quad \partial\Omega , \tag{5.2.4}$$

$$\textit{Neumann boundary conditions,} \qquad \frac{\partial\psi}{\partial n} = 0 \quad \text{on} \quad \partial\Omega , \tag{5.2.5}$$

$$\textit{Robin boundary conditions,} \qquad \frac{\partial\psi}{\partial n} + \alpha\,\psi = 0 \quad \text{on} \quad \partial\Omega , \tag{5.2.6}$$

where $\psi \in H^2(\Omega)$, n denotes the exterior unit normal vector field of $\partial\Omega$ and $\alpha : \partial\Omega \to \mathbb{C}$. We denote by $-\Delta_D^\Omega$, $-\Delta_N^\Omega$ and $-\Delta_\alpha^\Omega$ the operators in $L^2(\Omega)$ that act as $-\Delta^\Omega$ on smaller domains $\mathsf{D}(-\Delta_\iota^\Omega) := \{\psi \in H^2(\Omega) : (\iota) \text{ holds}\}$, where $\iota \in \{D, N, \alpha\}$ and (ι) stands for (5.2.4), (5.2.5) or (5.2.6), respectively. We call the operators the *Dirichlet*, *Neumann* and *Robin Laplacians*, respectively. All these operators are closed if Ω and α are sufficiently regular (*e.g.*, Ω bounded with boundary of class C^2 and $\alpha \in C^1(\partial\Omega)$). Clearly, $-\Delta_N^\Omega = -\Delta_0^\Omega$, while $-\Delta_D^\Omega$ can be formally considered as corresponding to the extreme situation "$\alpha = \infty$."

5.2.2 Spectra

An *eigenvalue* of H is defined as a complex number λ such that the equation $H\psi = \lambda\psi$ has a nonzero solution $\psi \in \mathsf{D}(H)$ called *eigenvector*. In other words, λ is an eigenvalue of H if the null space $\mathsf{N}(H-\lambda)$ is not $\{0\}$; this null space is the *geometric eigenspace* for λ and the nullity $\mathsf{m}_{\mathrm{g}}(\lambda) := \mathsf{nul}(H-\lambda)$ is called the *geometric multiplicity* of λ. The *algebraic* (or *root*) *eigenspace* for λ is defined by

$$\mathsf{M}_\lambda := \bigcup_{n=1}^{\infty} \mathsf{N}([H-\lambda]^n),$$

nonzero elements of M_λ are called *generalized eigenvectors* (or *root vectors*) corresponding to λ and $\mathsf{m}_{\mathrm{a}}(\lambda) := \dim \mathsf{M}_\lambda$ is called the *algebraic multiplicity* of λ. Obviously, $\mathsf{m}_{\mathrm{a}}(\lambda) \geq \mathsf{m}_{\mathrm{g}}(\lambda)$, where the inequality can be strict in general.

EXAMPLE 5.6 Matrices with degenerate eigenvalues

The nilpotent matrix $H := \left(\begin{smallmatrix} 0 & 1 \\ 0 & 0 \end{smallmatrix}\right)$ on $\mathbb{C}^2$ has only one eigenvalue $\lambda = 0$ with an eigenvector $\left(\begin{smallmatrix} 1 \\ 0 \end{smallmatrix}\right)$ and a generalized eigenvector $\left(\begin{smallmatrix} 0 \\ 1 \end{smallmatrix}\right)$, so $\mathsf{m}_{\mathrm{g}}(0) = 1$ and $\mathsf{m}_{\mathrm{a}}(0) = 2$. On the other hand, the null matrix $H := \left(\begin{smallmatrix} 0 & 0 \\ 0 & 0 \end{smallmatrix}\right)$ has one eigenvalue $\lambda = 0$ with two eigenvectors $\left(\begin{smallmatrix} 1 \\ 0 \end{smallmatrix}\right)$ and $\left(\begin{smallmatrix} 0 \\ 1 \end{smallmatrix}\right)$, so $\mathsf{m}_{\mathrm{g}}(0) = 2 = \mathsf{m}_{\mathrm{a}}(0)$. The null (infinite) matrix on $l^2(\mathbb{N})$ has $\lambda = 0$ as an eigenvalue of infinite geometric and algebraic multiplicities (as in fact it has the null operator on any infinite-dimensional Hilbert space), and it is straightforward to construct examples of matrices with arbitrary values of $\mathsf{m}_{\mathrm{g}}(\lambda)$ and $\mathsf{m}_{\mathrm{a}}(\lambda)$.

More interesting examples of (differential) operators will be presented later.

The set of all eigenvalues of H is called

the *point spectrum*, $\qquad \sigma_{\mathrm{p}}(H) := \{\lambda \in \mathbb{C} \ : \ \mathsf{N}(H-\lambda) \neq \{0\}\}.$

If $\lambda \notin \sigma_{\mathrm{p}}(H)$, then the inverse $(H-\lambda)^{-1}$ exists. The *resolvent set* $\rho(H)$ of H is defined to be the set of all λ's for which $(H-\lambda)^{-1} \in \mathscr{B}(\mathcal{H})$, that is, the inverse exists as a bounded operator on $\mathcal{H}$ (*i.e.*, *on* $\mathcal{H}$ to $\mathcal{H}$). The operator-valued function $\lambda \mapsto (H-\lambda)^{-1}$ from $\rho(H)$ to $\mathscr{B}(\mathcal{H})$ is called the *resolvent* of H. The complement $\sigma(H) := \mathbb{C} \setminus \rho(H)$ is called the *spectrum* of H.

It is customary to introduce the spectrum for closed operators only, the reason being that the notion is trivial otherwise.

Proposition 5.2.1 *If H is not closed, then $\sigma(H) = \mathbb{C}$.*

Proof: We prove it by contraposition: if $\lambda \in \rho(H) \neq \varnothing$, then $\mathsf{N}(H-\lambda) = \{0\}$ and $(H-\lambda)^{-1} \in \mathscr{B}(\mathcal{H})$. The latter implies that $(H-\lambda)^{-1}$ is closed. However, an invertible operator is closed if and only if its inverse is closed. Consequently, $H-\lambda$ and hence H are closed operators. ∎

In what follows, we thus assume that H is a closed operator in $\mathcal{H}$.

The spectrum of operators in finite-dimensional Hilbert spaces is exhausted by eigenvalues. In general, however, there are additional subsets:

$$\textit{continuous spectrum,} \qquad \sigma_c(H) := \left\{ \lambda \in \sigma(H) \setminus \sigma_p(H) \ : \ \overline{\mathsf{R}(H-\lambda)} = \mathcal{H} \right\},$$

$$\textit{residual spectrum,} \qquad \sigma_r(H) := \left\{ \lambda \in \sigma(H) \setminus \sigma_p(H) \ : \ \overline{\mathsf{R}(H-\lambda)} \neq \mathcal{H} \right\}.$$

By the closed-graph theorem [4, Section III.5.4], the pathological situation of $\lambda \in \sigma(H) \setminus \sigma_p(H)$ with $\mathsf{R}(H-\lambda) = \mathcal{H}$ cannot occur, therefore

$$\sigma(H) = \sigma_p(H) \cup \sigma_c(H) \cup \sigma_r(H)$$

and the unions are disjoint. In other words, $\lambda \in \sigma(H)$ if and only if $H - \lambda$ is not bijective as an operator from $\mathsf{D}(H)$ to $\mathcal{H}$.

From the Neumann series for the resolvent, it follows that the resolvent set $\rho(H)$ is an open subset of $\mathbb{C}$, and consequently, the spectrum $\sigma(H)$ is closed (it can be empty or cover the whole complex plane).

The spectra of a densely defined closed operator H and its adjoint H^* are simply related via a mirror symmetry with respect to the real axis,

$$\forall \lambda \in \mathbb{C}, \qquad \lambda \in \sigma(H) \quad \Longleftrightarrow \quad \overline{\lambda} \in \sigma(H^*).$$

However, the individual subsets of the spectrum may not satisfy this symmetry; in general, we have the following implications only.

Proposition 5.2.2 *Let H be a densely defined closed operator and $\lambda \in \mathbb{C}$. Then*

$$\begin{aligned} &\bullet \quad \lambda \in \sigma_p(H) \quad \Longrightarrow \quad \overline{\lambda} \in \sigma_p(H^*) \cup \sigma_r(H^*), \\ &\bullet \quad \lambda \in \sigma_r(H) \quad \Longrightarrow \quad \overline{\lambda} \in \sigma_p(H^*), \\ &\bullet \quad \lambda \in \sigma_c(H) \quad \Longleftrightarrow \quad \overline{\lambda} \in \sigma_c(H^*). \end{aligned} \tag{5.2.7}$$

In particular,

$$\sigma_r(H) = \left\{ \lambda \in \mathbb{C} \setminus \sigma_p(H) \ : \ \overline{\lambda} \in \sigma_p(H^*) \right\}. \tag{5.2.8}$$

Proof: We prove the first two implications from which the rest follows. Let $\lambda \in \sigma_p(H)$ and denote by ϕ the corresponding eigenvector. Then, for every $\psi \in \mathsf{D}(H^*)$, $((H^* - \overline{\lambda})\psi, \phi) = (\psi, (H-\lambda)\phi) = 0$, Therefore, $\mathsf{R}(H^* - \overline{\lambda})^\perp \neq \{0\}$; hence, $\overline{\lambda} \notin \sigma_c(H^*)$. Let $\lambda \in \sigma_r(H)$, then (5.2.2) yields $\mathsf{N}(H^* - \overline{\lambda}) = \mathsf{R}(H-\lambda)^\perp \neq \{0\}$, hence $\overline{\lambda} \in \sigma_p(H^*)$. ∎

EXAMPLE 5.7 Spectrum of the multiplication operator

In the full generality of Example 5.2, we have

$$\sigma(M_V) = \left\{\lambda \in \mathbb{C} \ : \ \left|\{x \in \Omega \ : \ |V(x) - \lambda| < \varepsilon\}\right| > 0 \text{ for all } \varepsilon > 0\right\} ,$$
$$\sigma_{\mathrm{p}}(M_V) = \left\{\lambda \in \mathbb{C} \ : \ \left|\{x \in \Omega \ : \ V(x) = \lambda\}\right| > 0\right\} ,$$
$$\sigma_{\mathrm{r}}(M_V) = \varnothing ,$$

where $|\cdot|$ denotes the Lebesgue measure. Note that the spectrum of M_V equals the essential range of the function V. In particular, if V is continuous, then $\sigma(M_V)$ is the closure of the range of V. In the special case of the position operator q in $L^2(\mathbb{R})$, we have $\sigma(q) = \sigma_{\mathrm{c}}(q) = \mathbb{R}$.

EXAMPLE 5.8 Spectrum of the momentum operator

The spectrum of the momentum operator p from Example 5.3 drastically depends on the choice of the configuration space Ω.

Ω	$\sigma_{\mathrm{p}}(p)$	$\sigma_{\mathrm{c}}(p)$	$\sigma_{\mathrm{r}}(p)$	$\sigma_{\mathrm{p}}(p^*)$	$\sigma_{\mathrm{c}}(p^*)$	$\sigma_{\mathrm{r}}(p^*)$
$\mathbb{R}$	$\varnothing$	$\mathbb{R}$	$\varnothing$	$\varnothing$	$\mathbb{R}$	$\varnothing$
$(0, +\infty)$	$\mathbb{C}^+$	$\mathbb{R}$	$\varnothing$	$\varnothing$	$\mathbb{R}$	$\mathbb{C}^-$
$(0, 1)$	$\mathbb{C}$	$\varnothing$	$\varnothing$	$\varnothing$	$\varnothing$	$\mathbb{C}$

Here the notation $\mathbb{C}^\pm := \{\lambda \in \mathbb{C} \ : \ \Im\lambda \gtrless 0\}$ for the upper and lower half-plane is used.

EXAMPLE 5.9 Spectrum of the creation and annihilation operators

Recall Example 5.4. We have $\sigma_{\mathrm{c}}(a) = \sigma_{\mathrm{c}}(a^*) = \varnothing$, $\sigma_{\mathrm{p}}(a) = \sigma_{\mathrm{r}}(a^*) = \mathbb{C}$ and $\sigma_{\mathrm{r}}(a) = \sigma_{\mathrm{p}}(a^*) = \varnothing$. The spectrum of the harmonic oscillator H_{HO} is given by algebraically simple eigenvalues $2n + 1$ with $n = 0, 1, \ldots$.

EXAMPLE 5.10 Spectrum of the Laplacians

The interesting dependence of the spectrum of the operators from Example 5.5 on the geometry of Ω is out of the scope of this chapter. We only mention the well-known result for the free Hamiltonian, $\sigma(-\Delta^{\mathbb{R}^d}) = \sigma_{\mathrm{c}}(-\Delta^{\mathbb{R}^d}) = [0, \infty)$, and henceforth focus on the one-dimensional situation $\Omega = (-a, a)$, $a > 0$. It is well known that

$$\sigma(-\Delta_N^{(-a,a)}) = \sigma_{\mathrm{p}}(-\Delta_N^{(-a,a)}) = \left\{\left(\frac{n\pi}{2a}\right)^2\right\}_{n=0}^{\infty} ,$$
$$\sigma(-\Delta_D^{(-a,a)}) = \sigma_{\mathrm{p}}(-\Delta_D^{(-a,a)}) = \left\{\left(\frac{n\pi}{2a}\right)^2\right\}_{n=1}^{\infty} ,$$

and all the eigenvalues are algebraically simple. The case of general Robin boundary conditions $\alpha : \partial\Omega \to \mathbb{C}$ is investigated in Refs (7, 8). In this one-dimensional situation, it is natural to identify the function α with the couple $\{\alpha(-a), \alpha(+a)\}$. Here and in the sequel, we consider only the special choice $\alpha(\pm a) = \pm i\alpha_0$ with $\alpha_0 \in \mathbb{R}$ that was originally introduced in Refs (9, 10). This choice admits an explicit solution

$$\sigma(-\Delta^{(-a,a)}_{\{-i\alpha_0, i\alpha_0\}}) = \sigma_{\mathrm{p}}(-\Delta^{(-a,a)}_{\{-i\alpha_0, i\alpha_0\}}) = \{\alpha_0^2\} \cup \left\{\left(\frac{n\pi}{2a}\right)^2\right\}_{n=1}^{\infty} .$$

Furthermore, it is easy to check that all the eigenvalues are algebraically simple provided that $2\alpha_0 a \notin \{\pm\pi, \pm 2\pi, \dots\}$, otherwise the eigenvalue α_0^2 is doubly degenerated with geometric and algebraic multiplicity one and two, respectively, and all the other eigenvalues are algebraically simple.

5.2.3 Numerical Range

Despite a usually direct physical interpretation of the spectrum, it is not an easily accessible quantity. Indeed, there is no hope to get such explicit formulae for the spectra as we did for the examples of the preceding section in the more general situation of differential operators with variable coefficients or defined on geometrically more complicated sets. The objective of this section is to estimate the spectrum in terms of a more accessible quantity:

$$\textit{numerical range,} \qquad \Theta(H) := \{(\psi, H\psi) \ : \ \psi \in \mathsf{D}(H),\ \|\psi\| = 1\}.$$

In general, $\Theta(H)$ is neither open nor closed, even when H is a closed operator. It is, however, always convex. Let

$$\Xi(H) := \complement\overline{\Theta(H)} \equiv \mathbb{C} \setminus \overline{\Theta(H)} \tag{5.2.9}$$

denote the exterior of the numerical range of H. In view of the convexity of the numerical range, $\Xi(H)$ is either an open connected set or a union of two half-planes (for this reason, we like to use the disconnected symbol Ξ to denote the exterior).

If $H \in \mathscr{B}(\mathcal{H})$, then the spectrum of H is a subset of the closure of $\Theta(H)$. More generally, we have

Proposition 5.2.3 *Let H be a closed operator such that each connected component of $\Xi(H)$ has a nonempty intersection with $\rho(H)$. Then*

$$\sigma(H) \subset \overline{\Theta(H)} \qquad \textit{and} \qquad \|(H-\lambda)^{-1}\| \le \frac{1}{\mathrm{dist}\left(\lambda, \overline{\Theta(H)}\right)} \tag{5.2.10}$$

for every $\lambda \in \rho(H)$.

Proof: By Theorem [4, Section V.3.2], $\mathsf{R}(H-\lambda)$ is closed and $\mathsf{nul}(H-\lambda)=0$ for each $\lambda \in \Xi(H)$. Furthermore, $\lambda \mapsto \mathsf{def}(H-\lambda)$ is constant in each of the connected components of $\Xi(H)$. Consequently, if a connected component of $\Xi(H)$ has a nonempty intersection with $\rho(H)$, then it follows from (5.4.3) that this component is actually a subset of $\rho(H)$. This proves the set inclusion in the statement of the proposition. To show the inequality for the resolvent norm, we note that

$$\operatorname{dist}\left(\lambda, \overline{\Theta(H)}\right) \leq |(\psi, H\psi) - \lambda| = |(\psi, (H-\lambda)\psi)| \leq \|(H-\lambda)\psi\|$$

for any $\psi \in \mathsf{D}(H)$ with $\|\psi\| = 1$ and every $\lambda \in \mathbb{C}$. Hence, the desired inequality follows by employing the fact that $\Xi(H)$ is a subset of $\rho(H)$ where $H-\lambda$ is bijective. ■

EXAMPLE 5.11 Numerical range of the momentum operator

The assumption in Proposition 5.2.3 about the intersection of the exterior of the numerical range with the resolvent set is absolutely necessary. We demonstrate it on the example of the momentum operator from Example 5.3. The following table to be compared with that of Example 5.8 shows that the spectrum cannot be controlled by the numerical range in general.

Ω	$\Theta(p)$	$\Theta(p^*)$
$\mathbb{R}$	$\mathbb{R}$	$\mathbb{R}$
$(0,+\infty)$	$\mathbb{C}$	$\mathbb{R}$
$(0,1)$	$\mathbb{C}$	$\mathbb{R}$

Indeed, we see that the spectrum is much larger than the numerical range of p^* on the half-line or bounded interval.

5.2.4 Sectoriality and Accretivity

The extra condition in Proposition 5.2.3 that ensures the useful properties (5.2.10) is of course annoying. The good news of this section is that there exists a distinguished class of operators for which we can do better. These are operators for which one can generically ensure that the exterior of the numerical range cannot have two connected components, by employing the convexity of the numerical range. We have already mentioned that it is the case of bounded operators, but this class of operators is insufficient for applications to differential operators. A fairly wide class is given by *sectorial* operators H defined by the property that their numerical range is a subset of a sector, that is,

$$\Theta(H) \subset S_{\gamma,\vartheta} := \{\lambda \in \mathbb{C} \; : \; |\arg(\lambda-\gamma)| \leq \vartheta\} \tag{5.2.11}$$

with some $\gamma \in \mathbb{R}$ and $0 \leq \vartheta < \pi/2$ called a *vertex* and a *semiangle* of H, respectively. As the inequality for the semiangle is strict, the exterior $\Xi(H)$ is clearly a connected set.

It remains to state an extra property that would ensure that $\Xi(H)$ has a nonempty intersection with $\rho(H)$ provided that H is sectorial. This is done *ad hoc* by introducing the notion of m-sectoriality: H is said to be *m-sectorial* if it is sectorial and

$$\rho(H) \cap \complement\overline{S_{\gamma,\vartheta}} \neq \varnothing . \tag{5.2.12}$$

(The latter is equivalent to $\rho(H) \cap \Xi(H) \neq \varnothing$ due to the sectoriality.) Applying Proposition 5.2.3, we may thus conclude with

Proposition 5.2.4 *Let H be an m-sectorial operator. Then* (5.2.10) *holds.*

For applications, however, it is sometimes needed to allow the extreme situation $\vartheta = \pi/2$ in (5.2.11). H is said to be *quasi-accretive* if $\Theta(H) \subset S_{\gamma,\pi/2}$ with some $\gamma \in \mathbb{R}$ and it is said to be *accretive* if the vertex can be chosen at the origin, that is, $\Theta(H) \subset S_{0,\pi/2}$. For the convenience of the reader, we summarize the various notions at one place here: an operator H is called

sectorial,	if (5.2.11) holds with	$\gamma \in \mathbb{R}$	and	$0 \leq \vartheta < \pi/2$,
accretive,	if (5.2.11) holds with	$\gamma = 0$	and	$0 \leq \vartheta \leq \pi/2$,
quasi-accretive,	if (5.2.11) holds with	$\gamma \in \mathbb{R}$	and	$0 \leq \vartheta \leq \pi/2$.

Again, we add the prefix m- to accretive if in addition (5.2.12) holds (which is now stronger than $\rho(H) \cap \Xi(H) \neq \varnothing$, as $\Xi(H)$ can have two disjoint components if $\vartheta = \pi/2$). Obviously, H is quasi-m-accretive if $H + \gamma$ is m-accretive with some $\gamma \in \mathbb{R}$ and H is m-sectorial if it is sectorial and quasi-m-accretive. Inspecting the proof of Proposition 5.2.3, we easily check that H is m-accretive if and only if the standard requirements (*cf.* [4, Eq. (V.3.38)])

$$\begin{aligned} &\bullet \quad \{\lambda \in \mathbb{C} \,:\, \Re\lambda < 0\} \subset \rho(H)\,, \\ &\bullet \quad \forall \lambda \in \mathbb{C},\ \Re\lambda < 0\,, \quad \|(H-\lambda)^{-1}\| \leq \frac{1}{|\Re\lambda|}\,, \end{aligned} \tag{5.2.13}$$

are satisfied.

The meaning of the m-terminology is that any m-accretive (respectively, m-sectorial) operator is *maximal* in the sense that it has no proper accretive (respectively, sectorial) extension (*cf.* [4, Section V.3.10]). Furthermore, any m-accretive operator is automatically closed (*cf.* Proposition 5.2.1) and densely defined.

While checking the condition (5.2.11) on the numerical range for a given operator may be straightforward, more sophisticated tools are usually needed to verify (5.2.12). We shall be concerned with such methods in Section 5.3.

The Laplacians from Example 5.5 together with operators constructed from them by "small perturbations" (*cf.* Section 5.3.4) are m-sectorial. At the same time, the harmonic oscillator Hamiltonian H_{HO} from Example 5.4 is m-sectorial with vertex 1 and semiangle 0. On the other hand, the momentum operators from Example 5.3 (recall also Examples 5.8 and 5.11) are not even sectorial, although $\pm ip$ and $\pm ip^*$ on $\mathbb{R}$ as well as ip^* on the half-line are m-accretive. As a matter of fact, ip^* on the half-line is a warning example for the fact that no general variant of Proposition 5.2.4 for quasi-m-accretive operators is available. Here we present other examples of quasi-accretive operators that are not sectorial:

EXAMPLE 5.12 Imaginary Airy operator

Consider in $L^2(\mathbb{R})$ the operator:

$$H_{\text{Airy}} := p^2 + iq\,, \qquad \mathsf{D}(H_{\text{Airy}}) = \{\psi \in H^2(\mathbb{R}) \ : \ x\psi \in L^2(\mathbb{R})\},$$

where p and q are introduced in Examples 5.3 and 5.2, respectively. H_{Airy} is m-accretive. The accretivity is simple to verify as, for all $\psi \in \mathsf{D}(H_{\text{Airy}})$, the integration by parts yields $(\psi, H_{\text{Airy}}\psi) = \|\psi'\|^2 + i(\psi, x\psi)$, whence $\Theta(H_{\text{Airy}}) \subset S_{0,\pi/2}$. However, it is much more delicate to check (5.2.12) (*cf.* Example 5.24).

EXAMPLE 5.13 Imaginary cubic oscillator

The operator in $L^2(\mathbb{R})$:

$$H_{\text{cubic}} := p^2 + iq^3\,, \qquad \mathsf{D}(H_{\text{cubic}}) = \{\psi \in H^2(\mathbb{R}) \ : \ x^3\psi \in L^2(\mathbb{R})\},$$

is m-accretive. The reasoning is analogous to the previous example.

EXAMPLE 5.14 Generator of the damped wave equation

Given a bounded open connected set $\Omega \subset \mathbb{R}^d$ with smooth boundary $\partial\Omega$, consider the damped wave equation $u_{tt} + a(x)u_t - \Delta u = 0$, where $(x,t) \in \Omega \times (0,\infty)$ and $a \in L^\infty(\Omega)$ is real-valued, subject to Dirichlet boundary conditions $u(x,t) = 0$ for $(x,t) \in \partial\Omega \times (0,\infty)$ and initial conditions $u(\cdot,0) \in H^1(\Omega)$, $u_t(\cdot,0) \in L^2(\Omega)$. Writing $\psi := \begin{pmatrix} u \\ u_t \end{pmatrix}$, the weak formulation of the differential equation leads to an abstract evolution problem $\psi_t = H_a\psi$ in the Hilbert space $\dot{H}^1_0(\Omega) \times L^2(\Omega)$ with $H_a := \begin{pmatrix} 0 & 1 \\ \Delta^\Omega_D & -a \end{pmatrix}$, $\mathsf{D}(H_a) := \mathsf{D}(-\Delta^\Omega_D) \times H^1_0(\Omega)$. Here $-\Delta^\Omega_D$ is the Dirichlet Laplacian from Example 5.5 and $\dot{H}^1_0(\Omega)$ denotes the closure of $C_0^\infty(\Omega)$ with respect to the norm $\|\nabla \cdot\|$ (it is equivalent to the H^1_0-norm as we assume that Ω is bounded). H_a is m-accretive whenever $a \le 0$. We refer to Refs (11, 12) for an application of spectral analysis of H_a to stability issues related to the damped wave equation.

As the last example suggests, quasi-accretive operators play an important role in evolution processes. In fact, by Hille-Yosida's theorem, *cf.* [13, Theorem 7.4], a closed densely defined operator H in a Hilbert space $\mathcal{H}$ is a generator of a γ-contractive semigroup $T(t)$ (*i.e.*, $\|T(t)\| \le e^{\gamma t}$ for all $t \ge 0$) if and only if $H + \gamma$ is m-accretive.

5.2.5 Symmetries

Proposition 5.2.2 reveals that an additional relationship between H and its adjoint H^* might have important consequences on spectral properties of H. In this section, we recall such "symmetry" relations and the corresponding spectral conclusions.

5.2.5.1 Symmetric Operators A (not necessarily closed) operator H in a Hilbert space $\mathcal{H}$ is said to be *symmetric* if it is densely defined and the adjoint H^* is an extension of H, that is,

$$H^* \supset H .$$

A densely defined operator H is symmetric if and only if it is a *formal adjoint* of itself in the sense that $(\phi, H\psi) = (H\phi, \psi)$ for all $\phi, \psi \in \mathsf{D}(H)$, which is equivalent to $\Theta(H) \subset \mathbb{R}$. We say that a symmetric operator H is *nonnegative* if $\inf \Theta(H) \geq 0$. If H is symmetric, then the point and continuous spectra of H are real, but the residual spectrum can be complex. (For instance, p^* from Example 5.3 considered on a bounded interval or on the half-line is symmetric, but it has complex residual spectra, *cf.* Example 5.8.) However, if the resolvent set $\rho(H)$ contains at least one real number, then $\sigma(H) \subset \mathbb{R}$.

5.2.5.2 Self-adjoint Operators If H is densely defined and

$$H^* = H$$

then H is said to be *self-adjoint*. H is automatically closed and has no proper symmetric extensions. With help of the spectral properties of symmetric operators, Proposition 5.2.2 implies that the residual spectrum of self-adjoint operators is empty. Consequently, any self-adjoint operator H satisfies $\sigma(H) \subset \mathbb{R}$. Moreover, the following important identity holds: for every $\lambda \notin \sigma(H)$,

$$\|(H-\lambda)^{-1}\| = \frac{1}{\operatorname{dist}\big(\lambda, \sigma(H)\big)} . \tag{5.2.14}$$

It is usually a straightforward matter to determine whether or not an operator is symmetric, but self-adjointness is a much more delicate property to establish. Regarding our examples, let us mention that the momentum operator from Example 5.3 considered in $L^2(\mathbb{R})$ is self-adjoint (so in fact $p = p^*$ if $\Omega = \mathbb{R}$); the free Hamiltonian $-\Delta^{\mathbb{R}^d}$ from Example 5.5 is self-adjoint, so are the Dirichlet Laplacian $-\Delta_D^\Omega$, the Neumann Laplacian $-\Delta_N^\Omega$ and the Robin Laplacian $-\Delta_\alpha^\Omega$ if α is real-valued; the operator of multiplication M_V by a function V from Example 5.2 is self-adjoint if and only if V is real-valued; finally, iH_0, where H_0 is the generator of the wave equation without damping ($a = 0$) from Example 5.14, is self-adjoint.

By one of von Neumann's axioms, physical observables are represented by self-adjoint operators in quantum mechanics. Contrary to what one can occasionally read in a physical literature, this is not just a mathematical laziness, to have real spectra for free (and thus real-valued outcomes of measurement), but it is in fact required by the conservative nature of the theory itself. Indeed, by Stone's theorem, there is a one-to-one correspondence between self-adjoint operators H and strongly continuous one-parameter unitary groups e^{itH} (that determine the time evolution in quantum mechanics).

5.2.5.3 Normal Operators An operator H is said to be *normal* if it is closed, densely defined and

$$H^*H = HH^*,$$

that is, H commutes with its adjoint H^*. Self-adjoint operators are special cases of normal operators. By the spectral theorem, functions of self-adjoint operators H are normal (including the unitary group e^{itH} and the resolvent $(H-\lambda)^{-1}$ with $\Im\lambda \neq 0$). Normal operators can have complex spectra, but the residual spectrum is again empty. This follows from Proposition 5.2.2 and the property $\mathsf{N}(H) = \mathsf{N}(H^*)$ for any normal operator H. Identity (5.2.14) holds for normal operators as well.

EXAMPLE 5.15 Laplacians arising from momentum operators

Let p and p^* be the momentum operators of Example 5.3. In $L^2(\mathbb{R})$, p is self-adjoint and thus normal; in fact, $p^*p = -\Delta^{\mathbb{R}} = pp^*$, where $-\Delta^{\mathbb{R}}$ is the free Hamiltonian of Example 5.5. In $L^2(\Omega)$ with an arbitrary interval Ω, we have $pp^* = -\Delta_D^{\Omega}$ and $p^*p = -\Delta_N^{\Omega}$.

5.2.5.4 Complex-Self-adjoint Operators We say that H in $\mathcal{H}$ is *complex-self-adjoint* (*with respect to* $\mathcal{J}$) if it is densely defined and there exists an antiunitary operator $\mathcal{J}$ in $\mathcal{H}$ such that

$$H^* = \mathcal{J}H\mathcal{J}^{-1}. \tag{5.2.15}$$

Recall that the *antiunitarity* means that $\mathcal{J} : \mathcal{H} \to \mathcal{H}$ is a bijective operator satisfying $(\mathcal{J}\phi, \mathcal{J}\psi) = (\psi, \phi)$ for any $\phi, \psi \in \mathcal{H}$. (This notion should be compared with *unitarity* for which the inner product is preserved, that is, $(\mathcal{J}\phi, \mathcal{J}\psi) = (\phi, \psi)$). In particular, an antiunitary $\mathcal{J}$ is antilinear (or conjugate-linear) and $\mathcal{J}, \mathcal{J}^{-1}$ are bounded. Any complex-self-adjoint operator is automatically closed, which follows from (5.2.15) and the closedness of the adjoint. If H is complex-self-adjoint, then λ is an eigenvalue of H (with eigenfunction $\psi \in \mathsf{D}(H)$) if and only if $\overline{\lambda}$ is an eigenvalue of H^* (with eigenfunction $\mathcal{J}^{-1}\psi \in \mathsf{D}(H^*)$); consequently, by Proposition 5.2.2,

$$\sigma_{\mathrm{r}}(H) = \varnothing.$$

EXAMPLE 5.16 Time-reversal operators

A simple example of an antiunitary operator in any Lebesgue space $L^2(\Omega)$ is the complex conjugation $\mathcal{T}\psi := \overline{\psi}$. $\mathcal{T}$ represents a *time-reversal* symmetry operation for a scalar (*i.e.*, spinless) Schrödinger equation in $L^2(\mathbb{R}^d)$. For fermionic systems (*i.e.*, half-integer nonzero spin), the time-evolution is described by a Pauli equation in the spinorial Hilbert space $L^2(\mathbb{R}^d) \otimes \mathbb{C}^2$, where the time-reversal operator can be represented by the antiunitary operator $\mathcal{T}_{1/2} := \left(\begin{smallmatrix} 0 & \mathcal{T} \\ -\mathcal{T} & 0 \end{smallmatrix}\right)$. Note that $\mathcal{T}^2 = 1$, while $\mathcal{T}_{1/2}^2 = -1$, *cf.* (14). The imaginary Airy operator H_{Airy} from Example 5.12 as well as the imaginary cubic oscillator H_{cubic} from Example 5.13 are complex-self-adjoint with respect to $\mathcal{T}$. It is easily seen by formal manipulations, but a rigorous verification requires a somewhat more effort as the description of the domain of the adjoint operator is needed; see Theorem 5.3.2 next.

Complex-self-adjoint operators with respect to $\mathcal{J}$ that is involutive (*i.e.*, $\mathcal{J}^2 = I$) are sometimes called *$\mathcal{J}$-self-adjoint* [15, Section I.4], [3, Section III.5] or (somewhat confusingly) *$\mathcal{J}$-symmetric* or *complex symmetric* (16–18). For a recent review on this special class of complex-self-adjoint operators with many references we refer to (19).

5.2.5.5 Pseudo-Self-adjoint Operators We say that an operator H in $\mathcal{H}$ is *pseudo-self-adjoint* (*with respect to* G) if H is densely defined and there exists a self-adjoint operator $G \in \mathscr{B}(\mathcal{H})$ with $G^{-1} \in \mathscr{B}(\mathcal{H})$ such that

$$H^* = GHG^{-1}. \tag{5.2.16}$$

The crucial difference with respect to the notion of complex-self-adjoint operators is that G is assumed to be *linear*. In general, G is indefinite; the case of positive G is very special and will be discussed in more detail in Section 5.5.2.

Any pseudo-self-adjoint operator is closed, the reasoning is the same as for complex-self-adjoint operators. Relation (5.2.16) and Proposition 5.2.2 imply symmetries of the spectra between H and H^*:

$$\sigma_\iota(H) = \sigma_\iota(H^*), \qquad \iota \in \{\mathrm{p}, \mathrm{c}, \mathrm{r}\}. \tag{5.2.17}$$

Contrary to complex-self-adjoint operators, pseudo-self-adjoint operators may have a nonempty residual spectrum (*cf.* Example 5.18).

EXAMPLE 5.17 Parity operator

A simple example of an indefinite operator G in $L^2(\mathbb{R}^d)$ is the *parity* (or *space-reversal*) operator $(\mathcal{P}\psi)(x) := \psi(-x)$, which represents a space-reversal symmetry operation in quantum mechanics. The imaginary Airy operator H_{Airy} from Example 5.12 as well as the imaginary cubic oscillator H_{cubic} from Example 5.13 are pseudo-self-adjoint with respect to $\mathcal{P}$; similarly to the complex-self-adjointness, the proof of this fact is not immediate.

EXAMPLE 5.18 Shifts on a lattice and perturbations

Let $\mathcal{L}$ be the *left shift* operator in $l^2(\mathbb{Z})$ defined by $\mathcal{L}e_j := e_{j-1}$, where $e_j := (\delta_{kj})_{k\in\mathbb{Z}}$ is the canonical basis in $l^2(\mathbb{Z})$. $\mathcal{L}$ is a unitary operator on $l^2(\mathbb{Z})$ and its adjoint is the *right shift* operator $\mathcal{R}e_j := e_{j+1}$, that is, $\mathcal{L}^* = \mathcal{R}$. The spectrum of $\mathcal{L}$ is discussed in Example 5.32. The discrete parity $\mathcal{P}e_j := e_{-j}$ plays the role of an involutive G in this example. It is easy to verify that $\mathcal{L}$ is pseudo-self-adjoint with respect to $\mathcal{P}$.

As a perturbation, let us consider the operator

$$V := -e_0(e_1,\cdot) - e_{-1}(e_0,\cdot) + i\sum_{j=-1}^{-\infty} e_j(e_j,\cdot) - i\sum_{j=1}^{\infty} e_j(e_j,\cdot)\,.$$

As V is again pseudo-self-adjoint with respect to $\mathcal{P}$, the same holds for the sum $H := \mathcal{L} + V$. Clearly $-i \in \sigma_{\text{p}}(H) = \sigma_{\text{p}}(H^*)$ as e_1 and e_{-1} are the corresponding eigenvectors of H and H^*, respectively. By Proposition 5.2.2, i is either in the point or residual spectrum of H and H^*. We can verify directly that i is not in the point spectrum of H. Indeed,

$$(H-i)\sum_{k\in\mathbb{Z}} \alpha_k e_k = 0 \qquad \Longrightarrow \qquad \begin{cases} \alpha_k = 0, & k \leq 0, \\ \alpha_{k+1} = (2i)^k \alpha_1, & k \geq 1, \end{cases}$$

whence $\sum_{k\in\mathbb{Z}} |\alpha_k|^2 = +\infty$, and thus $\mathsf{N}(H-i) = \{0\}$. In summary, H represents a pseudo-self-adjoint operator with nonempty residual spectrum.

If G is indefinite and involutive, the Hilbert space $\mathcal{H}$ equipped additionally to the inner product $(\cdot,\cdot)$ with the *indefinite inner product* $(\cdot, G\cdot)$ is the so-called *Krein space*, *cf.* (20, 21). Then the pseudo-self-adjoint operator H is in fact a self-adjoint operator in this Krein space, that is, H is self-adjoint with respect to the indefinite inner product.

5.2.5.6 Commutativity Finally, we discuss a notion that is probably closest to the term "symmetry" in physics. In quantum mechanics, a symmetry operation is represented either by a unitary or antiunitary operator $\mathcal{S}$ in a Hilbert space $\mathcal{H}$. We say that a closed densely defined operator H *has a symmetry* $\mathcal{S}$ if

$$[H, \mathcal{S}] = 0\,, \tag{5.2.18}$$

that is, H commutes with $\mathcal{S}$. As usual for the commutativity of an unbounded operator with a bounded operator on $\mathcal{H}$, we understand (5.2.18) by the operator relation $\mathcal{S}H \subset H\mathcal{S}$. It means that whenever $\psi \in \mathsf{D}(H)$, $\mathcal{S}\psi$ also belongs to $\mathsf{D}(H)$ and $\mathcal{S}H\psi = H\mathcal{S}\psi$.

We also say that H is $\mathcal{S}$-*symmetric*, but this notion should not be confused with $\mathcal{J}$-symmetry or G-symmetry used by other authors in the context of complex-self-adjoint or pseudo-self-adjoint operators, respectively. Finally, we simply say that H *has a symmetry* if there exists a unitary or antiunitary operator $\mathcal{S}$ with respect to which H is $\mathcal{S}$-symmetric.

If the symmetry $\mathcal{S}$ is antiunitary, we deduce from (5.2.18) that the spectra of H are symmetric with respect to the real axis,

$$\text{(antiunitary symmetry} \Rightarrow) \quad \lambda \in \sigma_\iota(H) \iff \overline{\lambda} \in \sigma_\iota(H)\,, \; \iota \in \{\mathrm{p}, \mathrm{c}, \mathrm{r}\}\,. \tag{5.2.19}$$

EXAMPLE 5.19 $\mathcal{PT}$-symmetry

The composition operator $\mathcal{PT}$, where $\mathcal{T}$ is the time-reversal operator from Example 5.16 and $\mathcal{P}$ is the space-reversal operator from Example 5.17, is the famous (antiunitary) $\mathcal{PT}$-*symmetry*. Both the imaginary Airy operator H_{Airy} from Example 5.12 and the imaginary cubic oscillator H_{cubic} from Example 5.13 are $\mathcal{PT}$-symmetric. The operator $-\Delta^{(-a,a)}_{\{-i\alpha_0, i\alpha_0\}}$ from Example 5.10 is also $\mathcal{PT}$-symmetric.

5.2.5.7 Nonequivalence of the Three Operator Classes While the previous examples may suggest that the classes of complex-self-adjoint, pseudo-self-adjoint and those having a symmetry are related, it is not the case in general. Examples of operators with nonempty residual spectrum (*cf.* Example 5.18 and the right shift on $l^2(\mathbb{N})$ discussed next in this paragraph) show that complex-self-adjoint operators are different from pseudo-self-adjoint operators and from those having an antiunitary symmetry. Moreover, the operator H from Example 5.18 is pseudo-self-adjoint, but it cannot have any antiunitary symmetry as it does not satisfy (5.2.19). Finally, the right shift on $l^2(\mathbb{N})$, that is, the restriction of $\mathcal{R}$ from Example 5.18 to $l^2(\mathbb{N})$, has the antiunitary symmetry $\mathcal{T}$ and its residual spectrum is the open unit ball, *cf.* [22, Section VI.3]. Thus, in view of the Proposition 5.2.2, it cannot satisfy (5.2.17) and therefore cannot be pseudo-self-adjoint.

5.3 HOW TO WHIP UP A CLOSED OPERATOR

In the previous section, we illustrated the abstract notions of spectral theory on concrete examples of differential operators. As the examples are rather standard, we did not include proofs of closedness. However, our spectral-theoretic approach to non-self-adjoint operators would be incomplete if we did not mention at all how to verify this important property for them. Moreover, in addition to closedness, it is needed that the operator associated with an evolution problem is maximal and quasi-accretive. In this section, we thus collect some abstract methods that can be effectively used to construct a quasi-m-accretive operator from a formal expression. Again, because of applications, we focus on differential operators, but most of the techniques can be applied more generally.

5.3.1 Closed Sectorial Forms

Another advantage of m-sectorial operators is that they naturally arise from quadratic forms. Symmetric forms are familiar in quantum mechanics, where they have a physical interpretation of expectation values. For non-self-adjoint operators, a more general class of sectorial forms is needed. Mathematically, the advantage consists in that the theory of forms is simpler than that of operators in several respects.

A *sesquilinear form* (or just *form*) h in a Hilbert space $\mathcal{H}$ is a pair consisting of a linear subspace $\mathsf{D}(h) \subset \mathcal{H}$ called the *domain* of h and a map $h : \mathcal{H} \times \mathcal{H} \to \mathbb{C}$ such that $h(\phi, \psi)$ is linear in $\psi \in \mathsf{D}(h)$ for each fixed $\phi \in \mathsf{D}(h)$ and antilinear in $\phi \in \mathsf{D}(h)$ for each fixed $\psi \in \mathsf{D}(h)$. $h[\psi] := h(\psi, \psi)$ is called the *quadratic form* (or again just *form*) associated with h. We say that h is *densely defined* if $\mathsf{D}(h)$ is dense in $\mathcal{H}$. *Extensions* and *restrictions* of forms are defined in an obvious way as in the case of operators. A form h is said to be *bounded* on $\mathcal{H}$ if there exists a constant $M > 0$ such that $|h[\psi]| \leq M\|\psi\|^2$ for all $\psi \in \mathcal{H}$ and it is *coercive* on $\mathcal{H}$ if there exists a constant $m > 0$ such that $|h[\psi]| \geq m\|\psi\|^2$ for all $\psi \in \mathcal{H}$. The inner product $(\cdot, \cdot)$ is an example of an everywhere defined bounded and coercive sesquilinear form in $\mathcal{H}$ (in fact, $m, M = 1$ in this case).

The *adjoint form* h^* of h is defined in a much simpler way than the adjoint of an operator,

$$h^*(\phi, \psi) := \overline{h(\psi, \phi)}, \qquad \mathsf{D}(h^*) := \mathsf{D}(h).$$

We say that h is *symmetric* if $h^* = h$ and there is no notion of "self-adjoint form." The *real* and *imaginary parts* of h are, respectively,

$$\Re h := \frac{1}{2}(h + h^*), \qquad \Im h := \frac{1}{2i}(h - h^*).$$

This notation is justified by $\Re h[\psi] = \Re(h[\psi])$ and $\Im h[\psi] = \Im(h[\psi])$, although $\Re h(\phi, \psi)$ and $\Im h(\phi, \psi)$ are not real-valued in general and have nothing to do with $\Re(h(\phi, \psi))$ and $\Im(h(\phi, \psi))$.

The *numerical range* of h is defined by

$$\Theta(h) := \{h[\psi] \ : \ \psi \in \mathsf{D}(h), \ \|\psi\| = 1\}.$$

As in the case of operators, $\Theta(h)$ is a convex set in the complex plane. Contrary to the case of operators, however, we have a simple relation $\lambda \in \Theta(h) \Leftrightarrow \overline{\lambda} \in \Theta(h^*)$. A form h is symmetric if and only if $\Theta(h) \subset \mathbb{R}$. A symmetric form h is said to be *nonnegative* if $\Theta(h) \subset [0, \infty)$.

An important class of forms is given by *sectorial forms* h for which $\Theta(h) \subset S_{\gamma,\vartheta}$, where $S_{\gamma,\vartheta}$ is the sector defined in (5.2.11) with a vertex $\gamma \in \mathbb{R}$ and a semiangle $\vartheta \in$

$[0, \pi/2)$. We say that a sectorial form h is *closed* if

$$\left.\begin{aligned} \mathsf{D}(h) \ni \psi_n &\xrightarrow[n\to\infty]{} \psi \in \mathcal{H} \\ h[\psi_n - \psi_m] &\xrightarrow[n,m\to\infty]{} 0 \end{aligned}\right\} \quad \Longrightarrow \quad \left[\psi \in \mathsf{D}(h) \ \wedge \ h[\psi_n - \psi] \xrightarrow[n\to\infty]{} 0\right].$$

If h is sectorial with a vertex $\gamma > -\infty$ and a semiangle $\vartheta < \pi/2$, then

$$\begin{aligned} |(h - \gamma + 1)(\phi, \psi)| &\leq (1 + \tan\vartheta)\sqrt{(\Re h - \gamma + 1)[\phi]}\sqrt{(\Re h - \gamma + 1)[\psi]}, \\ |(h - \gamma + 1)[\psi]| &\geq (\Re h - \gamma + 1)[\psi], \end{aligned}$$

for all $\phi, \psi \in \mathsf{D}(h)$. Consequently, if h is closed, then it is actually bounded and coercive on the Hilbert space $\mathsf{D}(h)$ equipped with the inner product $\Re h(\cdot,\cdot) + (-\gamma + 1)(\cdot,\cdot)$. Applying the celebrated Lax-Milgram theorem [3, Section IV.1], one can conclude with

Theorem 5.3.1 (First representation theorem) *Let h be a densely defined closed sectorial form in $\mathcal{H}$. Then the operator*

$$\begin{aligned} \mathsf{D}(H) &:= \left\{\psi \in \mathsf{D}(h) \ : \ \exists \eta \in \mathcal{H},\ \forall \phi \in \mathsf{D}(h),\ h(\phi, \psi) = (\phi, \eta)\right\}, \\ H\psi &:= \eta, \end{aligned} \tag{5.3.1}$$

is m-sectorial.

We say that H is *associated with h* and that $\mathsf{D}(h)$ is the *form-domain* of H. The adjoint of H is simply given by the operator determined by the same theorem with the adjoint form h^*. The numerical range $\Theta(H)$ is a dense subset of $\Theta(h)$. Clearly, $\mathsf{D}(H) \subset \mathsf{D}(h)$ and $h(\phi, \psi) = (\phi, H\psi)$ for every $\phi \in \mathsf{D}(h)$ and $\psi \in \mathsf{D}(H)$ and these conditions determine H uniquely.

EXAMPLE 5.20 Multiplication operator defined by a sectorial form

If V is the function from Example 5.2, we define a quadratic form $m_V[\psi] := \int_\Omega V|\psi|^2$, $\mathsf{D}(m_V) := \{\psi \in L^2(\Omega) : |V|^{1/2}\psi \in L^2(\Omega)\}$. If $V(\Omega) \subset S_{\gamma,\vartheta}$ with $\gamma \in \mathbb{R}$ and $0 \leq \vartheta < \pi/2$, then the multiplication operator M_V from Example 5.2 coincides with the m-sectorial operator associated with m_V via Theorem 5.3.1.

5.3.2 Friedrichs' Extension

By Theorem 5.3.1, every densely defined closed sectorial form gives rise to an m-sectorial operator. The converse correspondence is also valid. Indeed, if H is m-sectorial, then the form

$$\dot{h}[\psi] := (\psi, H\psi), \qquad \mathsf{D}(\dot{h}) := \mathsf{D}(H), \tag{5.3.2}$$

is clearly densely defined and sectorial. The form $\dot{h}$ is not necessarily closed, however, it is *closable* in the sense that it admits a closed extension. Then $\dot{h}$ has the *closure* h, that is, the smallest closed extensions, defined by

$$\mathsf{D}(h) := \left\{ \psi \in \mathcal{H} \; : \; \exists \{\psi_n\} \subset \mathsf{D}(\dot{h}), \; \psi_n \xrightarrow[n\to\infty]{} \psi \; \wedge \; \dot{h}[\psi_n - \psi_m] \xrightarrow[n,m\to\infty]{} 0 \right\},$$

$$h[\psi] := \lim_{n\to\infty} \dot{h}[\psi_n],$$

and H coincides with the operator associated with h. Summing up, there is a one-to-one correspondence between the set of all m-sectorial operators and the set of all densely defined closed sectorial forms.

EXAMPLE 5.21 Form associated with a Dirac interaction

The fact that the form $\dot{h}$ defined by (5.3.2) is closable employs the special structure of its action (*cf.* [4, Theorem VI.1.27]). An example of a densely defined sectorial form that is not closable is given by the form associated with the (formal) *Dirac potential* δ in $L^2(\mathbb{R})$: $m_\delta[\psi] := |\psi(0)|^2$, $\mathsf{D}(m_\delta) := H^1(\mathbb{R})$. Note that the form is well defined because of the continuous embedding $H^1(\mathbb{R}) \hookrightarrow C^0(\mathbb{R})$.

The aforementioned procedure of constructing a closed form h from a form $\dot{h}$ defined by a sectorial operator H is not limited to closed operators. Indeed, if $\dot{H}$ is just a densely defined sectorial operator, we construct a densely defined sectorial form $\dot{h}$ from it in the same way as in (5.3.2) (with H being replaced by $\dot{H}$). Then we take the closure h of $\dot{h}$ as earlier and associate to it the m-sectorial operator H via Theorem 5.3.1. Such a constructed H is called the *Friedrichs extension* of $\dot{H}$.

Any densely defined sectorial operator is *closable* (*i.e.*, it admits a closed extension). But there might be many closed extensions and the *closure* (*i.e.*, the smallest closed extension) might not be m-sectorial. The importance of the Friedrichs extension lies in the fact that it assigns a special m-sectorial extension to each densely defined sectorial operator. The Friedrichs extension H of $\dot{H}$ is characterized by the properties that, among all m-sectorial extensions of $\dot{H}$, H has the smallest form-domain (*i.e.*, $\mathsf{D}(h)$ is contained in the domain of the form associated with any other of the extensions) and that H is the only extension of $\dot{H}$ with $\mathsf{D}(H) \subset \mathsf{D}(h)$.

EXAMPLE 5.22 The Neumann Laplacian defined by a sectorial form

On the example of the Neumann Laplacian from Example 5.5, let us show how to employ the Friedrichs extension in order to construct a closed operator from a formal differential expression. Let $\Omega \subset \mathbb{R}^d$ be an open connected (possibly unbounded) set of class $C^{0,1}$, so that the normal vector $n(x)$ is defined for almost every $x \in \partial\Omega$ by Rademacher's theorem. We start with an operator $\dot{H}$ on $L^2(\Omega)$ that acts as the Neumann Laplacian on nice functions, namely $\dot{H}\psi := -\Delta\psi$, $\quad \mathsf{D}(\dot{H}) := \{\psi \in L^2(\Omega) : \exists\tilde{\psi} \in C_0^\infty(\mathbb{R}^d) \quad$ such that $\psi = \tilde{\psi} \restriction$

Ω and ψ satisfies (5.2.5)}. The operator $\dot{H}$ is densely defined and sectorial; in fact, $\dot{H}$ is nonnegative due to $(\psi, \dot{H}\psi) = \|\nabla\psi\|^2 \geq 0$ for all $\psi \in \mathsf{D}(\dot{H})$. We define a densely defined sectorial form $\dot{h}$ as in (5.3.2) (where H is replaced by $\dot{H}$) and construct its closure h. Let H be the m-sectorial operator associated with h via Theorem 5.3.1 (in fact, H is self-adjoint and nonnegative). This procedure enable us to define a closed realization of the Laplacian in Ω, subject to Neumann boundary conditions on $\partial\Omega$, under minimal regularity assumptions on Ω.

Unfortunately, unless we impose some additional restrictions on the boundary $\partial\Omega$, H does not have to coincide with $-\Delta_N^\Omega$ defined in Example 5.5 (as $\mathsf{D}(H)$ is not necessarily a subset of $H^2(\Omega)$). Even worse, the boundary condition (5.2.5) that we understand in the sense of traces of $\psi \in H^2(\Omega)$ might not be well defined. That is, contrary to H, $-\Delta_N^\Omega$ is not well defined under our minimal regularity assumption $C^{0,1}$ on Ω. Let us therefore assume for instance that Ω is bounded and of class C^2; then the boundary traces $H^2(\Omega) \hookrightarrow H^1(\partial\Omega)$ certainly exist [5, Theorem 5.36]. On the other hand, by Davies [23, Theorem 7.2.1], we have $\mathsf{D}(h) = H^1(\Omega)$ (it is remarkable that the boundary condition (5.2.5) disappears as soon as one passes from the operator $\dot{H}$ to the closure of its quadratic form). From (5.3.1) we see that $\psi \in \mathsf{D}(H)$ is a solution of the variational problem $(\nabla\phi, \nabla\psi) = (\phi, \eta)$ for every $\phi \in H^1(\Omega)$, which is nothing else than a weak formulation of the Neumann problem $-\Delta\psi = \eta \in L^2(\Omega)$ in Ω, $\partial\psi/\partial n = 0$ on $\partial\Omega$. In particular, $H\psi = -\Delta\psi \in L^2(\Omega)$, where $\Delta\psi$ means the distributional Laplacian of ψ. Using elliptic regularity theory (see, *e.g.*, [13, Theorem 9.26]), we know that the weak solutions ψ belong to $H^2(\Omega)$, which enables us to eventually conclude with $\mathsf{D}(H) = \mathsf{D}(-\Delta_N^\Omega)$. Hence, $H = -\Delta_N^\Omega$, as we wanted to show. In the other extreme situation $\Omega = \mathbb{R}^d$, we verify in the same (in fact easier) manner that $\mathsf{D}(h) = H^1(\mathbb{R}^d)$ and $H = -\Delta_N^{\mathbb{R}^d} = -\Delta^{\mathbb{R}^d}$.

Finally, let us remark that one can introduce the "Neumann Laplacian" for *any* open set Ω by considering the self-adjoint operator associated with the closed form $\tilde{h}[\psi] := \|\psi\|^2$, $\mathsf{D}(\tilde{h}) := H^1(\Omega)$. Again, this definition coincides with $-\Delta_N^\Omega$ from Example 5.5 for sufficiently regular Ω.

EXAMPLE 5.23 The Dirichlet Laplacian defined by a sectorial form

Dirichlet boundary conditions of Example 5.5 can be treated in the same way. To get a more robust result, we take the Friedrichs extension H of the operator $\dot{H}\psi := -\Delta\psi$, $\mathsf{D}(\dot{H}) := C_0^\infty(\Omega)$. Then, in the full generality of any open set $\Omega \subset \mathbb{R}^d$, $\mathsf{D}(h) = H_0^1(\Omega)$ and $H\psi = -\Delta\psi$ with $\mathsf{D}(H) = \{\psi \in H_0^1(\Omega) : \Delta\psi \in L^2(\Omega)\}$ is a self-adjoint nonnegative operator. If Ω is suficiently regular (*e.g.*, bounded and of class C^2), we obtain $H = -\Delta_D^\Omega$. At the same time, $H = -\Delta_D^{\mathbb{R}^d} = -\Delta^{\mathbb{R}^d}$ if $\Omega = \mathbb{R}^d$.

Robin boundary conditions are best regarded as a perturbation and amenable to the stability methods of Section 5.3.4, *cf.* Example 5.28 next.

5.3.3 M-accretive Realizations of Schrödinger Operators

The method of quadratic forms does not apply to the more general class of quasi-m-accretive operators. For instance, the imaginary Airy operator from Example 5.12 and the imaginary cubic oscillator H_{cubic} from Example 5.13 cannot be defined by means of the elegant techniques of Sections 5.3.1 and 5.3.2. To cover these examples, we now present a specific result obtained by Kato in (24) for Schrödinger operators

$$\mathfrak{H} := -\Delta + V \qquad \text{with} \qquad \Re V \geq 0,$$

where $V : \Omega \to \mathbb{C}$ is a function (possibly with singularities). More specifically, we understand $\mathfrak{H}$ as a formal differential expression in an open set $\Omega \subset \mathbb{R}^d$ and are concerned with an m-accretive realization of $\mathfrak{H}$ that is characterized by Dirichlet boundary conditions. We refer to Ref. [3, Section VII.2] for a nice exposition of Kato's result and proofs of the present statements.

Assuming

$$V \in L^p_{\text{loc}}(\Omega) \qquad \text{with} \qquad p \begin{cases} = 1 & \text{if} \quad d = 1\,, \\ > 1 & \text{if} \quad d = 2\,, \\ = 2d/(d+2) & \text{if} \quad d \geq 3\,, \end{cases} \tag{5.3.3}$$

we have $V\psi \in L^1_{\text{loc}}(\Omega)$ for all $\psi \in H^1_0(\Omega)$ and $\mathfrak{H}\psi$ is well defined as a distribution. Then the operator $\tilde{H}\psi := \mathfrak{H}\psi$ with $\mathsf{D}(\tilde{H}) := \{\psi \in H^1_0(\Omega) : \mathfrak{H}\psi \in H^{-1}(\Omega)\}$ is the maximal realization of $\mathfrak{H}$ as an operator from $H^1_0(\Omega)$ to its dual $H^{-1}(\Omega)$. The message of the following theorem is that the restriction of $\tilde{H}$ to $L^2(\Omega)$ is an m-accretive operator provided that $\Re V \geq 0$ holds.

Theorem 5.3.2 (Kato's theorem) *Let $V : \Omega \to \mathbb{C}$ satisfy* (5.3.3) *and $\Re V \geq 0$. Then the operator H defined by*

$$H\psi := \mathfrak{H}\psi\,, \qquad \mathsf{D}(H) := \left\{\psi \in H^1_0(\Omega) \; : \; \mathfrak{H}\psi \in L^2(\Omega)\right\} \tag{5.3.4}$$

is m-accretive in $L^2(\Omega)$. Moreover, the adjoint H^* of H reads

$$H^*\psi = \overline{\mathfrak{H}\psi}\,, \qquad \mathsf{D}(H^*) = \left\{\psi \in H^1_0(\Omega) \; : \; \overline{\mathfrak{H}\psi} \in L^2(\Omega)\right\}. \tag{5.3.5}$$

Consequently, H is complex-self-adjoint with respect to the time-reversal operator $\mathcal{T}$ (complex conjugation) introduced in Example 5.16. The proof of Theorem 5.3.2 leans heavily on a distributional inequality obtained by Kato in (25), which is an interesting result on its own.

EXAMPLE 5.24 M-accretivity of the imaginary Airy operator

Function $V(x) := ix$ clearly belongs to $L^\infty_{\mathrm{loc}}(\mathbb{R})$ and satisfies $\Re V \geq 0$, so the operator H defined by Theorem 5.3.2 is m-accretive. We intend to show that H coincides with the operator of H_{Airy} introduced in Example 5.12. The inclusion $H_{\mathrm{Airy}} \subset H$ is obvious. To show the opposite one, we employ the fact that H coincides with the closure of $H \upharpoonright C_0^\infty(\mathbb{R})$, *cf.* [3, Corollary 2.7]. Hence, $C_0^\infty(\mathbb{R})$ is dense in $\mathsf{D}(H)$ for the graph norm of H. Integrating by parts, we easily check that

$$\begin{aligned}
\|\psi'\|^2 &= (\psi, -\psi'') \leq \|\psi\|\,\|\psi''\| \leq \epsilon\,\|\psi''\|^2 + \epsilon^{-1}\|\psi\|^2 , \\
\|H\psi\|^2 &= \|\psi''\|^2 + \|x\psi\|^2 + 2\Re(i\psi, \psi') \\
&\geq \|\psi''\|^2 + \|x\psi\|^2 - \epsilon\,\|\psi'\|^2 - \epsilon^{-1}\|\psi\|^2 ,
\end{aligned}$$

for every $\psi \in C_0^\infty(\mathbb{R})$ and any $\epsilon > 0$. Combining these inequalities with sufficiently small ϵ and using the density of $C_0^\infty(\mathbb{R})$ in $\mathsf{D}(H)$, we arrive at the nontrivial fact that if $\psi \in \mathsf{D}(H)$, then $\psi \in H^2(\mathbb{R})$ and $x\psi \in L^2(\mathbb{R})$, so $H \subset H_{\mathrm{Airy}}$. Summing up, the m-accretive realization H obtained by Theorem 5.3.2 coincides with H_{Airy} from Example 5.12.

The m-accretivity of the imaginary cubic oscillator from Example 5.13 or the self-adjointness of the harmonic oscillator from Example 5.4 can be established in the same way.

5.3.4 Small Perturbations

Finally, we present two classical perturbation results. If H_0 is a closed operator in $\mathcal{H}$ and V is any operator that belongs to $\mathscr{B}(\mathcal{H})$, then $H_0 + V$ is also closed. For applications, it is necessary to have an extended version of this stability result for a not necessarily bounded perturbation.

5.3.4.1 Relative Boundedness and Subordination Let H_0 and V be two operators in $\mathcal{H}$. We say that V is *relatively bounded* with respect to H_0 if

$$\begin{aligned}
&\bullet \quad \mathsf{D}(V) \supset \mathsf{D}(H_0) , \\
&\bullet \quad \forall \psi \in \mathsf{D}(H_0) , \qquad \|V\psi\| \leq a\,\|H_0\psi\| + b\,\|\psi\| ,
\end{aligned} \tag{5.3.6}$$

where a, b are nonnegative constants. The infimum of such a is called the *relative bound* of V with respect to H_0.

We say that V is *p-subordinated* to H_0 if

$$\begin{aligned}
&\bullet \quad \mathsf{D}(V) \supset \mathsf{D}(H_0) , \\
&\bullet \quad \forall \psi \in \mathsf{D}(H_0) , \qquad \|V\psi\| \leq c\,\|H_0\psi\|^p\,\|\psi\|^{1-p} ,
\end{aligned} \tag{5.3.7}$$

where c is a nonnegative constant and $p \in [0, 1)$.

Obviously, a bounded V is 0-subordinated to H_0. Moreover, by Young's inequality, any p-subordinated perturbation is relatively bounded with respect to H_0 with the relative bound equal to zero.

Theorem 5.3.3 (Stability of closedness for operators) *If H_0 is closed and V is relatively bounded with respect to H_0 with the relative bound smaller than 1, then $H_0 + V$ is closed.*

The converse is also true: if V is relatively bounded with respect to H_0 with the relative bound smaller than 1 and $H_0 + V$ is not closed, then H_0 cannot be closed.

EXAMPLE 5.25 Closedness of the generator of the damped wave equation

If H_a is the operator from Example 5.14, we write $H_a = H_0 + V$, where $V := \left(\begin{smallmatrix} 0 & 0 \\ 0 & -a \end{smallmatrix}\right)$, $\mathsf{D}(V) := \dot{H}_0^1(\Omega) \times L^2(\Omega)$, is bounded and hence relatively bounded with respect to H_0 with the relative bound equal to zero. At the same time, H_0 is m-accretive because iH_0 is self-adjoint. Therefore, H_a is densely defined and closed for any $a \in L^\infty(\Omega)$. It is m-accretive if $a \leq 0$.

EXAMPLE 5.26 Shifted harmonic oscillator

When checking that the operator $H_\alpha := p^2 + (q + \alpha)^2$, where p is the momentum operator from Example 5.3, q is the position operator from Example 5.2 and $\alpha \in \mathbb{C}$, is closed in $L^2(\mathbb{R})$, it helps to regard it as a perturbation of the (self-adjoint) harmonic oscillator H_{HO} from Example 5.4. Indeed, estimates analogous to those in Example 5.24 yield that the graph norm of H_{HO}, that is, $(\|H_{\mathrm{HO}} \cdot\|^2 + \|\cdot\|^2)^{1/2}$, is equivalent to $(\|\partial_x^2 \cdot\|^2 + \|x^2 \cdot\|^2 + \|\cdot\|^2)^{1/2}$. Then it is easily checked that $H_\alpha := H_{\mathrm{HO}} + V$, where $V := 2\alpha q + \alpha^2$, is $\frac{1}{2}$-subordinated to $H_{\mathrm{HO}} + 1$ and therefore it is also relatively bounded with respect to H_{HO} with the relative bound equal to zero. Applying Theorem 5.3.3, we thus know that H_α is closed on $\mathsf{D}(H_\alpha) = \mathsf{D}(H_{\mathrm{HO}})$.

5.3.4.2 Relative Form-boundedness and Subordination The notion of relative boundedness can be introduced for any forms, but we restrict ourselves to sectorial ones. Let h_0 be a sectorial form in $\mathcal{H}$. A form v in $\mathcal{H}$ (which need not be sectorial) is said to be *relatively bounded* with respect to h_0 if

$$
\begin{aligned}
&\bullet \quad \mathsf{D}(v) \supset \mathsf{D}(h_0), \\
&\bullet \quad \forall \psi \in \mathsf{D}(h_0), \qquad |v[\psi]| \leq a\,|h_0[\psi]| + b\,\|\psi\|^2,
\end{aligned}
\tag{5.3.8}
$$

where a, b are nonnegative constants. Again, the infimum of such a is called the *relative bound* of v with respect to h_0.

A form v in $\mathcal{H}$ is said to be *p-subordinated* to h_0 if

$$\bullet \quad \mathsf{D}(v) \supset \mathsf{D}(h_0), \qquad \bullet \quad \forall \psi \in \mathsf{D}(h_0), \qquad |v[\psi]| \le c\,|h_0[\psi]|^p \|\psi\|^{2-2p}, \tag{5.3.9}$$

where c is a nonnegative constant and $p \in [0, 1)$.

In parallel to the operator case, the p-subordinated form is also relatively bounded with respect to h_0 with the relative bound equal to zero.

Theorem 5.3.4 (Stability of closedness for forms) *If h_0 is sectorial and closed and v is relatively bounded with respect to h_0 with the relative bound smaller than 1, then $h_0 + v$ is sectorial and closed.*

Again, the converse is also true: if h_0 is sectorial, v is relatively bounded with respect to h_0 with the relative bound smaller than 1 and $h_0 + v$ (which is sectorial) is not closed, then h_0 cannot be closed.

When H_0 is an m-sectorial operator, Theorem 5.3.4 enables one to define operators "$H_0 + V$" even if V has no operator sense. Indeed, the densely defined closed sectorial form h_0 obtained by the closure of (5.3.2) is associated to H_0 first. Secondly, by Theorem 5.3.4, the sum $h_0 + v$ with a given form v (possibly not arising from an operator or even not closable) is a densely defined closed sectorial form provided that v is relatively bounded with respect to h_0 with the relative bound less than one. Finally, there is an m-sectorial operator associated with $h_0 + v$ via Theorem 5.3.1 (it is sometimes customary to denote this operator by "$H_0 + V$," although the sum may differ from the operator sum).

EXAMPLE 5.27 Schrödinger operator with a complex Dirac interaction

The Dirac potential (distribution) δ cannot be realized as an operator in $L^2(\mathbb{R})$. However, the Schrödinger operator "$p^2 + \alpha\delta(x)$" with $\alpha \in \mathbb{C}$ can be defined using the strategy of quadratic forms described earlier. Indeed, $h_0[\psi] := \|\psi'\|^2$, $\mathsf{D}(h_0) := H^1(\mathbb{R})$, is the densely defined closed nonnegative form associated with the one-dimensional Laplacian p^2 (*cf.* Example 5.15 and Example 5.22). At the same time, the (nonclosable) form m_δ from Example 5.21 is $\frac{1}{2}$-subordinated to h_0, and hence relatively bounded with respect to h_0 with the relative bound 0. This follows from the elementary bounds

$$\|\psi\|_\infty^2 \le 2\|\psi\|\,\|\psi'\| \le \epsilon\,\|\psi'\|^2 + \epsilon^{-1}\,\|\psi\|^2 \tag{5.3.10}$$

valid for every $\psi \in H^1(\mathbb{R})$ and any $\epsilon > 0$. Hence, by Theorem 5.3.4, the sum $h_0 + \alpha m_\delta$ is a densely defined closed sectorial form to which there exists an m-sectorial operator H_α due to Theorem 5.3.1. H_α can be understood as a form-sum version of "$p^2 + \alpha\delta(x)$." Moreover, with some effort, it is possible to deduce from (5.3.1) that $(H_\alpha\psi)(x) = -\psi''(x)$ for every $x \neq 0$ and $\mathsf{D}(H_\alpha) = \{\psi \in H^1(\mathbb{R}) \cap H^2(\mathbb{R} \setminus \{0\}) : \psi'(0+) - \psi'(0-) = \alpha\psi(0)\}$.

EXAMPLE 5.28 The Robin Laplacian defined by a sectorial form

We show that the Robin Laplacian from Example 5.5 can be introduced as a perturbation of the Neumann Laplacian. (The resemblance with the preceeding Example 5.27 is not accidental.) For simplicity, let us assume that Ω is bounded and of class C^2 and the complex-valued function α belongs to $C^1(\partial\Omega)$. Integrating by parts, we easily check

$$\left(\psi, -\Delta_\alpha^\Omega \psi\right) = \int_\Omega |\nabla\psi|^2 + \int_{\partial\Omega} \alpha\, |\psi|^2 =: h_\alpha[\psi]$$

for every $\psi \in \mathsf{D}(-\Delta_\alpha^\Omega)$. However, the right-hand side is well defined on a larger space $\mathsf{D}(h_\alpha) := H^1(\Omega)$; indeed the boundary values exist in the sense of the trace embedding $H^1(\Omega) \hookrightarrow L^2(\partial\Omega)$. We write $h_\alpha = h_0 + v$, where $h_0[\psi] := \|\nabla\psi\|^2$ with $\mathsf{D}(h_0) := H^1(\Omega)$ is the form associated with the Neumann Laplacian $-\Delta_N^\Omega$ and $v_\alpha[\psi] := \int_{\partial\Omega} \alpha\, |\psi|^2$ with $\mathsf{D}(v_\alpha) := H^1(\Omega)$ is its perturbation. By Example 5.22, h_0 is densely defined, closed and sectorial. To show that v_α is relatively bounded with respect to h_0 (actually $\frac{1}{2}$-subordinated), we estimate the function α by its supremum norm and use the bounds

$$\|\psi\|_{L^2(\partial\Omega)}^2 \leq 2\,C\,\|\psi\|\,\|\nabla\psi\| \leq \epsilon\,\|\nabla\psi\|^2 + C^2\epsilon^{-1}\,\|\psi\|^2$$

for every $\psi \in H^1(\Omega)$ and any $\epsilon > 0$, where the constant C depends on curvatures of $\partial\Omega$. (Here the first inequality is actually behind a proof of the trace embedding $H^1(\Omega) \hookrightarrow L^2(\partial\Omega)$.) Hence, by Theorem 5.3.4, the sum h_α is a densely defined closed sectorial form to which there exists an m-sectorial operator H_α due to Theorem 5.3.1. (As a matter of fact, H_α is well defined as an m-sectorial operator for any $\alpha \in L^\infty(\partial\Omega)$.) Continuing as in Example 5.22 with help of elliptic regularity theory (for which the extra smoothness of α is needed), one can conclude with $H_\alpha = -\Delta_\alpha^\Omega$.

5.4 COMPACTNESS AND A SPECTRAL LIFE WITHOUT IT

The theory of compact operators in Hilbert spaces is reminiscent of the theory of operators in finite-dimensional spaces. In this section, we recall basic properties of this important class of operators and develop a spectral theory for noncompact operators.

5.4.1 Compact Operators and Compact Resolvents

An operator $H \in \mathscr{B}(\mathcal{H})$ is said to be *compact* if, for any bounded sequence $\{\psi_n\} \subset \mathcal{H}$, the sequence $\{H\psi_n\}$ contains a convergent subsequence. As every bounded sequence in a Hilbert space contains a weakly converging subsequence, the compactness of H means that H maps weakly converging sequences to strongly converging sequences. We denote by $\mathscr{B}_\infty(\mathcal{H})$ the set of all compact operators of $\mathscr{B}(\mathcal{H})$.

Compact operators H have spectacularly nice spectral properties:

- $\sigma(H) \setminus \{0\} = \sigma_p(H) \setminus \{0\}$,
- $\sigma_p(H)$ is at most countable and has no accumulation point except possibly 0,
- $\forall \lambda \in \sigma_p(H) \setminus \{0\}, \quad \mathsf{R}(H-\lambda)$ is closed $\wedge$ $\mathsf{m}_a(\lambda) < +\infty$.

That is, every nonzero point λ in the spectrum of a compact operator H is an isolated eigenvalue of finite algebraic multiplicity and the range of $H - \lambda$ is closed. If the Hilbert space $\mathcal{H}$ is infinite-dimensional, zero is always in the spectrum of H, that is, $\sigma(H) = \sigma_p(H) \cup \{0\}$.

Differential operators in $L^2(\Omega)$ are unbounded, so they cannot be compact. However, inverses of differential operators on bounded domains Ω are typically compact. This leads to another important class of operators that have spectra analogous to the spectra of operators in finite-dimensional spaces. We say that a closed operator H in $\mathcal{H}$ has a *compact resolvent* if

- $\rho(H) \neq \varnothing$,
- $(H-\lambda)^{-1} \in \mathscr{B}_\infty(\mathcal{H})$ for some (and hence all) $\lambda \in \rho(H)$.

By virtue of the spectral properties of compact operators and the spectral mapping theorem [3, Theorem IX.2.3], we know that if H has a compact resolvent, then

$$\sigma(H) = \sigma_{\mathrm{disc}}(H),$$

where $\sigma_{\mathrm{disc}}(H)$ is the *discrete spectrum* of H defined (for any closed operator) by

$$\lambda \in \sigma_{\mathrm{disc}}(H) \quad :\Longleftrightarrow \quad \begin{cases} \bullet \quad \lambda \in \sigma_p(H), \\ \bullet \quad \lambda \text{ is isolated (as a point in the spectrum)}, \\ \bullet \quad \mathsf{m}_a(\lambda) < +\infty, \\ \bullet \quad \mathsf{R}(H-\lambda) \text{ is closed}. \end{cases} \tag{5.4.1}$$

EXAMPLE 5.29 Spectrum of the imaginary Airy operator

The operator H_{Airy} from Example 5.12 has a compact resolvent; it follows from Example 5.24 and the compactness of the embedding $\mathsf{D}(H_{\mathrm{Airy}}) \hookrightarrow L^2(\mathbb{R})$, where the former space is assumed to be equipped with the graph norm of H_{Airy}. Consequently, the spectrum of H_{Airy} is purely discrete. If there existed a nonzero $\psi \in \mathsf{D}(H_{\mathrm{Airy}})$ and $\lambda \in \mathbb{C}$ such that $H_{\mathrm{Airy}}\psi = \lambda\psi$, then, by shifting $\tau_c : x \mapsto x + c$, the function $\psi_c := \psi \circ \tau_c$ would solve $H_{\mathrm{Airy}}\psi_c = (\lambda - ic)\psi$ with any $c \in \mathbb{C}$ and we would thus have $\sigma(H_{\mathrm{Airy}}) = \sigma_{\mathrm{p}}(H_{\mathrm{Airy}}) = \mathbb{C}$, which contradicts the discreteness of the spectrum. Hence, $\sigma(H_{\mathrm{Airy}}) = \varnothing$.

The compact embedding argument shows that the imaginary cubic oscillator from Example 5.13 and the shifted harmonic oscillator from Example 5.26 have compact resolvents as well. However, their spectra are not empty. In fact, all eigenvalues of both the operators are real and there are infinitely many of them. While the proof of these facts for the latter operator is rather simple (*e.g.*, it follows by solving the spectral problem in terms of special functions), the proof for the former is nontrivial (26–28).

EXAMPLE 5.30 Green's function of the Laplacian

The Neumann, Dirichlet and Robin Laplacians in $L^2(\Omega)$ from Example 5.5 have compact resolvents provided that Ω is bounded and smooth and α is smooth; it follows from the compactness of the Sobolev embedding $H^1(\Omega) \hookrightarrow L^2(\Omega)$. We have the integral representation

$$\left[(-\Delta_\iota^\Omega - k^2)^{-1}\psi\right](x) = \int_\Omega G_{\iota,k}^\Omega(x,y)\,\psi(y)\,dy\,,$$

where $\iota \in \{D, N, \alpha\}$ and $k^2 \in \rho(-\Delta_\iota^\Omega)$. The integral kernel G_ι^Ω is sometimes referred to as the *Green function.*

The compactness can be checked by hand for the one-dimensional Laplacians considered in Example 5.10, where we have explicit formulae

$$G_{D,k}^{(-a,a)}(x,y) = \frac{-\sin(k(x+a))\,\sin(k(y-a))}{k\,\sin(2ka)}\,,$$

$$G_{N,k}^{(-a,a)}(x,y) = \frac{-\cos(k(x+a))\,\cos(k(y-a))}{k\,\sin(2ka)}\,,$$

$$G_{\alpha,k}^{(-a,a)}(x,y) = \frac{-\left[k\cos(k(x+a)) - i\alpha\sin(k(x+a))\right]}{(k^2-\alpha^2)\,k\,\sin(2ka)} \times \left[k\cos(k(y-a)) - i\alpha\sin(k(y-a))\right],$$

for $x < y$ and the role of x, y should be exchanged for $x > y$.

The compactness of the resolvent of many operators can be proved using stability results, which are parallel to the results on the stability of closedness.

Theorem 5.4.1 (Stability of compact resolvent for operators) *Let H_0 be an m-accretive operator that has a compact resolvent. If V is relatively bounded with respect to H_0 with the relative bound smaller than 1, then $H_0 + V$ has a compact resolvent.*

The assumption on m-accretivity of H_0 is not necessary. However, the condition on the relative bound becomes more complicated otherwise. Namely, the conclusion of the theorem holds if there exists $z \in \rho(H_0)$ such that the inequality

$$a\|H_0(H_0 - z)^{-1}\| + b\|(H_0 - z)^{-1}\| < 1$$

is satisfied, *cf.* [4, Theorem IV.3.17], where a and b are the constants appearing in (5.3.6).

Theorem 5.4.1 provides an alternative proof of the compactness of the resolvent for the shifted oscillator from Example 5.26.

Theorem 5.4.2 (Stability of compact resolvent for forms) *Let h_0 be a densely defined, closed, sectorial form with $\Re h_0 \geq 0$ and let the associated m-sectorial operator H_0 have a compact resolvent. If v is relatively bounded with respect to h_0 with the relative bound smaller than 1, then the operator associated with $h_0 + v$ has a compact resolvent.*

EXAMPLE 5.31 Harmonic oscillator with a Dirac interaction

We find an m-sectorial realization of "$H_{HO} + \alpha\delta(x)$" with $\alpha \in \mathbb{C}$ via the sum of forms. The harmonic oscillator is associated with the form

$$h_{HO}[\psi] := \|\psi'\|^2 + \|x\psi\|^2 , \qquad \mathsf{D}(h_{HO}) := \{\psi \in H^1(\mathbb{R}) \ : \ x\psi \in L^2(\mathbb{R})\} .$$

The inequality (5.3.10) shows that the form αm_δ is $\frac{1}{2}$-subordinated to h_{HO}, thus relatively bounded with the relative bound 0. Therefore, by Theorems 5.3.4 and 5.4.2, the form $h_{HO} + \alpha m_\delta$ determines an m-sectorial operator that has a compact resolvent.

5.4.2 Essential Spectra

We define the *essential spectrum* of any closed operator H as the complement of the discrete spectrum defined in (5.4.1), that is,

$$\sigma_{ess}(H) := \sigma(H) \setminus \sigma_{disc}(H) . \tag{5.4.2}$$

There is considerable divergence in the literature concerning the definition of the essential spectrum for non-self-adjoint operators. Our definition is the largest within these and was originally introduced by Browder (29). It makes the essential spectrum harder to locate, but on the other hand, the remaining discrete eigenvalues have very pleasant properties.

Following [3, Chapter IX] (see also (30)), let us compare our definition of the essential spectrum with the others. We assume that H is closed and recall that the resolvent set can be characterized as

$$\begin{aligned}\rho(H) = \{\, \lambda \in \mathbb{C} :\ & \\ & \mathsf{R}(H-\lambda) \text{ is closed } \wedge \ \mathsf{nul}(H-\lambda) = 0 = \mathsf{def}(H-\lambda)\} \,.\end{aligned} \tag{5.4.3}$$

For $k = 0, 1, \dots, 5$, we set

$$\sigma_{ek}(H) := \mathbb{C} \setminus \rho_{ek}(H) \,,$$

where

$$\begin{aligned}
\textit{Goldberg,}\quad & \rho_{e0}(H) := \{\lambda \in \mathbb{C} : \mathsf{R}(H-\lambda) \text{ is closed}\} \,,\\
\textit{Kato,}\quad & \rho_{e1}(H) := \left\{\lambda \in \rho_{e0}(H) : \mathsf{nul}(H-\lambda) < \infty \vee \mathsf{def}(H-\lambda) < \infty\right\} ,\\
& \rho_{e2}(H) := \left\{\lambda \in \rho_{e0}(H) : \mathsf{nul}(H-\lambda) < \infty\right\} ,\\
\textit{Wolf,}\quad & \rho_{e3}(H) := \left\{\lambda \in \rho_{e0}(H) : \mathsf{nul}(H-\lambda) < \infty \wedge \mathsf{def}(H-\lambda) < \infty\right\} ,\\
\textit{Schechter,}\quad & \rho_{e4}(H) := \left\{\lambda \in \rho_{e0}(H) : \mathsf{nul}(H-\lambda) = \mathsf{def}(H-\lambda) < \infty\right\} ,\\
\textit{Browder,}\quad & \rho_{e5}(H) := \text{union of all the components of } \rho_{e1}(H) \text{ intersecting } \rho(H) \,.
\end{aligned}$$

Clearly,

$$\rho_{e0}(H) \supset \rho_{e1}(H) \supset \rho_{e2}(H) \supset \rho_{e3}(H) \supset \rho_{e4}(H) \supset \rho_{e5}(H) \supset \rho(H) \,,$$

so $\sigma_{e0}(H)$ is the most restrictive and $\sigma_{e5}(H)$ is the widest. The names refer to people who are usually associated with the given definition of the essential spectrum. The set $\sigma_{e2}(H)$ is called the "continuous spectrum" in Ref. (15).

The operator $H - \lambda$ is said to be *normally soluble*, *semi-Fredholm* or *Fredholm* if and only if $\lambda \in \rho_{e0}(H)$, $\lambda \in \rho_{e1}(H)$ or $\lambda \in \rho_{e3}(H)$, respectively. The operator H is normally soluble if and only if $\mathsf{R}(H) = \mathsf{N}(H^*)^{\perp}$, so the condition $\phi \perp \mathsf{N}(H^*)$ is both necessary and sufficient for the equation $H\psi = \phi$ to have a solution ψ (as in finite-dimensional spaces).

Let us show that our definition (5.4.2) indeed coincides with $\sigma_{e5}(H)$.

Proposition 5.4.3 *Let H be closed. One has*

$$\sigma_{e5}(H) = \sigma_{ess}(H) \,.$$

Proof: The statement is equivalent to showing $\rho_{e5}(H) = \rho(H) \cup \sigma_{disc}(H)$. The set $\rho_{e1}(H)$ is an open subset of the complex plane and it can be written as the union of countably many components (*i.e.*, connected open sets) that we denote by $\triangle_n$, $n \in \mathbb{N}$. Kato [4, Section IV.5] shows that $\lambda \mapsto \mathsf{nul}(H-\lambda)$ and $\lambda \mapsto \mathsf{def}(H-\lambda)$ are constant in each $\triangle_n$, save possibly at some isolated values of λ. Denoting by ν_n, μ_n these constant values and by $\lambda_n^j, j \in \mathbb{N}$, these exceptional points in $\triangle_n$, we have

$$\begin{aligned} &\forall \lambda \in \triangle_n \setminus \{\lambda_n^j\}_{j\in\mathbb{N}}, &&\mathsf{nul}(H-\lambda) = \nu_n, &&\mathsf{def}(H-\lambda) = \mu_n, \\ &\forall j \in \mathbb{N}, &&\mathsf{nul}(H-\lambda_n^j) = \nu_n + r_n^j, &&\mathsf{def}(H-\lambda_n^j) = \mu_n + r_n^j, \end{aligned}$$

where $0 < r_n^j < \infty$. If $\triangle_n \cap \rho(H) \neq \emptyset$, then $\nu_n = 0 = \mu_n$ and $\triangle_n$ is a subset of $\rho(H)$ except for the λ_n^j, which are isolated eigenvalues of H with finite algebraic multiplicities (r_n^j are their geometric multiplicities). Hence, $\rho_{\mathrm{e5}}(H) \subset \rho(H) \cup \sigma_{\mathrm{disc}}(H)$. To prove the converse inclusion, we first note that $\rho(H) \subset \rho_{\mathrm{e5}}(H)$ is obvious due to (5.4.3). Finally, if $\lambda \in \sigma_{\mathrm{disc}}(H)$, then of course $\lambda \in \rho_{\mathrm{e1}}(H)$. As λ is isolated, it must belong to a component $\triangle_n$ with $\nu_n = 0 = \mu_n$. But such a component is a subset of $\rho(H)$ except for the exceptional points, λ being one of them; hence, $\triangle_n \cap \rho(H) \neq \emptyset$. ■

If H is self-adjoint, the sets $\sigma_{\mathrm{e}k}(H)$ with $k = 1, \dots, 5$ are identical and $\sigma_{\mathrm{disc}}(H)$ consists of isolated eigenvalues of finite multiplicity ($\mathsf{m_a}(\lambda) = \mathsf{m_g}(\lambda)$ and $\mathsf{R}(H-\lambda)$ is automatically closed). If H is complex-self-adjoint, the sets $\sigma_{\mathrm{e}k}(H)$ with $k = 1, \dots, 4$ are identical. In general, however, the inclusion between the sets may be strict, as the following example shows.

EXAMPLE 5.32 Shift operator and its compact perturbation

Let $\mathcal{L}$ be the *left shift* in $l^2(\mathbb{Z})$ defined in Example 5.18. The operator $\mathcal{L}$ is unitary and

$$\sigma(\mathcal{L}) = \sigma_{\mathrm{e5}}(\mathcal{L}) = \partial B_1,$$

where $B_1 := \{\lambda \in \mathbb{C} : |\lambda| < 1\}$ is an open unit disc. Let V be the compact (in fact of rank 1) operator in $l^2(\mathbb{Z})$ defined by $V := -e_{-1}(e_0, \cdot)$. The sum $H := \mathcal{L} + V$ belongs to $\mathscr{B}(l^2(\mathbb{Z}))$. It is easily shown (*cf.* [3, Ex. IX.2.2]) that

$$\sigma_{\mathrm{e5}}(H) \supset B_1, \qquad \text{while} \qquad \sigma_{\mathrm{e4}}(H) \subset \partial B_1.$$

Hence, $\sigma_{\mathrm{e5}}(H) \neq \sigma_{\mathrm{e4}}(H)$ and $\sigma_{\mathrm{e5}}(\mathcal{L})$ is not preserved by the compact perturbation V.

Fortunately, there is a simple way how to exclude pathological situations of the type we encountered in the precedent example.

Proposition 5.4.4 *Let H be closed. If each component of $\rho_{\mathrm{e1}}(H)$ intersects $\rho(H)$, then the sets $\sigma_{\mathrm{e}k}(H)$ with $k = 1, \dots, 5$ are identical. In particular, the conclusion holds if $\rho_{\mathrm{e1}}(H)$ is connected.*

Proof: The result follows at once from the definition of $\rho_{\mathrm{e5}}(H)$. ■

As far as we know, σ_{e2} is associated with no name, but it is useful because of the following characterization. Although it resembles Weyl's criterion for self-adjoint operators, its proof is quite different (*cf.* (31)).

Theorem 5.4.5 (Weyl's criterion) *Let H be a closed and densely defined operator in $\mathcal{H}$. Then*

$$\lambda \in \sigma_{e2}(H) \qquad \Longleftrightarrow \qquad \exists \{\psi_n\}_{n\in\mathbb{N}} \subset \mathsf{D}(H), \quad \begin{cases} \bullet \quad \forall n \in \mathbb{N}, \quad \|\psi_n\| = 1, \\ \bullet \quad \psi_n \xrightarrow[n\to\infty]{w} 0, \\ \bullet \quad (H-\lambda)\psi_n \xrightarrow[n\to\infty]{s} 0. \end{cases}$$

The sequence from the theorem is called a *singular sequence.*

5.4.3 Stability of the Essential Spectra

In applications, it often happens that the operator H of interest is obtained from a simpler operator H_0 by a "small" perturbation V, say $H = H_0 + V$. If the essential spectrum of H_0 is easy to locate for some reason, it is of great interest to have criteria on the "smallness" of V, which ensure that H and H_0 have the same essential spectrum. By a celebrated result of Weyl, it happens if V is compact. More generally, it is enough to assume that V is "relatively compact" with respect to H_0. Instead of introducing the notion of relative compactness, we state the Weyl's result in the following form (*cf.* [3, Theorem IX.2.4]).

Theorem 5.4.6 (Weyl's theorem) *Let H_1, H_2 be closed operators in $\mathcal{H}$ such that*

$$\exists \lambda \in \rho(H_1) \cap \rho(H_2), \qquad (H_1 - \lambda)^{-1} - (H_2 - \lambda)^{-1} \in \mathscr{B}_\infty(\mathcal{H}). \tag{5.4.4}$$

Then

$$\sigma_{ek}(H_1) = \sigma_{ek}(H_2) \qquad \textit{for} \qquad k = 1, \dots, 4.$$

Unfortunately, the theorem does not apply to our definition of essential spectrum (5.4.2), which coincides with σ_{e5} by Proposition 5.4.3. In fact, the stability result does not hold for our essential spectrum in general, as Example 5.32 clearly demonstrates. Fortunately, Proposition 5.4.4 enables one to use Theorem 5.4.6 also for σ_{e5} in some situations (*e.g.*, Example 5.33 next).

The last result we would like to mention is not related to essential spectra, but it may turn out to be useful when locating the (essential) spectrum of the "unperturbed" operator H_0 from the opening to this section. The following theorem is just [32, Corollary 2 of Theorem XIII.35] translated to the present terminology. Note that the conclusion represents some sort of "separation of variables," which is not at all automatic for non-self-adjoint operators.

Theorem 5.4.7 (Ichinose's lemma) *Let H_1, H_2 be m-sectorial operators in Hilbert spaces $\mathcal{H}_1, \mathcal{H}_2$. Let H denote the closure of $H_1 \otimes I + I \otimes H_2$ on $\mathsf{D}(H_1) \otimes \mathsf{D}(H_2) \subset \mathcal{H}_1 \otimes \mathcal{H}_2$. Then H is m-sectorial and*

$$\sigma(H) = \sigma(H_1) + \sigma(H_2).$$

EXAMPLE 5.33 $\mathcal{PT}$-symmetric waveguide

Let $H_{\alpha_0+\beta} := -\Delta_\alpha^\Omega$ be the Robin Laplacian in $L^2(\Omega)$ from Example 5.5 with $\Omega := \mathbb{R} \times (-a,a)$, $a > 0$, and $\alpha := \pm i(\alpha_0 + \beta)$ on $\mathbb{R} \times \{\pm a\}$, where α_0 is a real number and β can be identified with a function $\beta : \mathbb{R} \to \mathbb{R}$ that we suppose to be smooth and compactly supported. It is shown in Ref. (33) that $H_{\alpha_0+\beta}$ is m-sectorial (the proof is analogous to that given in Example 5.28, but notice that Ω is unbounded now). Moreover, $H_{\alpha_0+\beta}$ is easily seen to be complex-self-adjoint with respect to $\mathcal{T}$ and $\mathcal{PT}$-symmetric, where $\mathcal{T}$ is the complex conjugation (*cf.* Example 5.16) and $(\mathcal{P}\psi)(x_1,x_2) := \psi(x_1,-x_2)$.

The operator $H_{\alpha_0+\beta}$ can be considered as obtained from H_{α_0} by a "small" perturbation. More specifically, the form of $H_{\alpha_0+\beta} - H_{\alpha_0}$ is relatively bounded with respect to the sectorial form of H_{α_0} with the relative bound 0. The "unperturbed" operator H_{α_0} admits the decomposition $-\Delta^{\mathbb{R}} \otimes I + I \otimes -\Delta^{(-a,a)}_{\{-i\alpha_0,i\alpha_0\}}$ in $L^2(\Omega) \simeq L^2(\mathbb{R}) \otimes L^2((-a,a))$, where the one-dimensional Laplacians have been introduced in Examples 5.5 and 5.10. Applying Theorem 5.4.7 (or employing basis properties of the "transverse" Laplacian as in Ref. (33)), we get $\sigma(H_{\alpha_0}) = [\mu_0^2,\infty)$, where $\mu_0^2 := \min\{\alpha_0^2,(\pi/2a)^2\}$ is the lowest eigenvalue of the operator $-\Delta^{(-a,a)}_{\{-i\alpha_0,i\alpha_0\}}$.

The spectrum of H_{α_0} is purely essential, because it has no isolated points, that is, $\sigma(H_{\alpha_0}) = \sigma_{\text{ess}}(H_{\alpha_0})$. It is proved in Ref. (33) that (5.4.4) holds with the operators $H_1 = H_{\alpha_0}$ and $H_2 = H_{\alpha_0+\beta}$. Combining thus Theorem 5.4.6 with Proposition 5.4.4, we conclude with the stability result

$$\sigma_{\text{ess}}(H_{\alpha_0+\beta}) = \sigma_{\text{ess}}(H_{\alpha_0}) = [\mu_0^2,\infty)\,.$$

Note that the essential spectrum is purely real, though the operator $H_{\alpha_0+\beta}$ is not self-adjoint. Sufficient conditions to guarantee the existence of (real) discrete eigenvalues are also established in Ref. (33) (see Refs (34) and (35) for further studies of the model).

5.5 SIMILARITY TO NORMAL OPERATORS

In finite-dimensional Hilbert spaces, every linear operator is similar to a block diagonal Jordan matrix, whose eigenvalues are elementarily computable. Although there is no general replacement of this result in infinite-dimensional Hilbert spaces, the idea of reducing a given operator to a simpler one by a similarity transformation might work in concrete examples. There are certainly many situations of this type in applications, but we focus on conceptually new approach in quantum mechanics that was suggested by physicists in Ref. (36): represent physical observables by (possibly non-self-adjoint!) operators that are merely similar to self-adjoint ones. In the following text and notably in the examples, we argue that the similarity transformations should be necessarily bounded in order to build a consistent quantum mechanics using this unconventional representation.

5.5.1 Similarity Transforms

We say that an operator H_1 is *similar* to another operator H_2 (*via a transformation* A) in the same Hilbert space $\mathcal{H}$ if there exists an injective operator $A \in \mathscr{B}(\mathcal{H})$ with $A^{-1} \in \mathscr{B}(\mathcal{H})$ such that

$$H_2 = AH_1A^{-1} . \tag{5.5.1}$$

This notion is a straightforward generalization of unitary equivalence with which it shares many important properties such as the preservation of the spectrum.

Proposition 5.5.1 *Let H_1 be a closed operator in a Hilbert space $\mathcal{H}$. If H_2 is similar to H_1, then it is closed and*

$$\sigma_\iota(H_2) = \sigma_\iota(H_1), \qquad \textit{where} \qquad \iota \in \{\ , \mathrm{p}, \mathrm{c}, \mathrm{r}, \mathrm{disc}, \mathrm{ess}\} .$$

Moreover, if λ is an eigenvalue of H_1 of geometric multiplicity $\mathsf{m}_\mathrm{g}(\lambda)$ and algebraic multiplicity $\mathsf{m}_\mathrm{a}(\lambda)$, then λ is an eigenvalue of H_2 of the same geometric multiplicity $\mathsf{m}_\mathrm{g}(\lambda)$ and algebraic multiplicity $\mathsf{m}_\mathrm{a}(\lambda)$.

Proof: Let A be a similarity transformation establishing (5.5.1). As $A, A^{-1} \in \mathscr{B}(\mathcal{H})$, A and A^{-1} are bijective operators on $\mathcal{H}$ and the relation (5.5.1) yields

$$\begin{aligned} \mathsf{D}(H_2 - \lambda) &= A\,\mathsf{D}(H_1 - \lambda), \\ \mathsf{R}(H_2 - \lambda) &= A\,\mathsf{R}(H_1 - \lambda), \\ \mathsf{N}([H_2 - \lambda]^n) &= A\,\mathsf{N}([H_1 - \lambda]^n), \end{aligned} \tag{5.5.2}$$

for any $\lambda \in \mathbb{C}$ and $n \in \mathbb{N}$. The closedness of H_2 can be checked by definition by using the first two identities with $\lambda = 0$. The spectral equivalences can be deduced from the last two identities in (5.5.2); in particular,

$$\mathsf{nul}(H_2 - \lambda) = \mathsf{nul}(H_1 - \lambda), \qquad \mathsf{def}(H_2 - \lambda) = \mathsf{def}(H_1 - \lambda),$$

for any $\lambda \in \mathbb{C}$. We leave the details to the reader. ∎

Similarity is sometimes understood in a weaker sense, *e.g.*, as $AH_1 = H_2A$ without boundedness and invertibility assumptions on A or even as $AH_1\psi = H_2A\psi$ valid for all ψ from a subspace of $\mathcal{H}$ only. The differences in the notions are not always reflected in the terminology. If the assumptions on A are relaxed, many pathologies may occur, particularly the spectra may not be preserved, as the following example demonstrates.

EXAMPLE 5.34 Gauged and rotated oscillators

A formal (unbounded) similarity transform of the harmonic oscillator leads to the *gauged* (or *Swanson's* (37, 38)) oscillator. For $\phi \in C_0^\infty(\mathbb{R})$, we first define the action of the latter by

$$H_{\text{gauged}}\phi := (\omega\, a^*a + \alpha\, a^2 + \beta\, (a^*)^2 + \omega)\phi\,, \tag{5.5.3}$$

where α, β and ω are real parameters such that $\omega \neq \alpha + \beta$, and a^* and a are the creation and annihilation operators from Example 5.4. Defining $A := \exp\left(\frac{\beta-\alpha}{\omega-\alpha-\beta}\,\frac{x^2}{2}\right)$, we can easily check that

$$AH_{\text{gauged}}A^{-1}\phi = \left[(\omega-\alpha-\beta)p^2 + \frac{\omega^2 - 4\alpha\beta}{\omega-\alpha-\beta}\,q^2\right]\phi =: \tilde{H}_{\text{HO}}\phi \tag{5.5.4}$$

for every $\phi \in C_0^\infty(\mathbb{R})$, where the right-hand side is just the action of a multiple of the self-adjoint harmonic oscillator with a frequency depending on ω, α and β. In spite of the fact that A or A^{-1} are always unbounded as multiplication operators in $L^2(\mathbb{R})$, equality (5.5.4) can be read as a weak version of the similarity relation (5.5.1).

Setting parameters to $\omega = \beta = 0$ and $\alpha = -1$, we thus get an uninteresting "similarity" relation between $-a^2$ and p^2, which are clearly completely different operators. For instance, recalling Examples 5.8 and 5.9, $\sigma(-a^2) = \sigma_{\text{p}}(-a^2) = \mathbb{C}$ versus $\sigma(p^2) = \sigma_{\text{c}}(p^2) = [0, \infty)$. As shown in Refs (39, 40), similar pathologies appear also in less obvious cases, *e.g.* $\omega > 0$, $-\alpha > \omega$, $\beta = 0$, when H_{gauged} is "highly non-self-adjoint," while $\tilde{H}_{\text{HO}}$ is still related to the usual harmonic oscillator.

To avoid such pathologies, the parameters need to be restricted by the condition $\omega - |\alpha + \beta| > 0$ as pointed out in Ref. (40). Then H_{gauged} can be realized as an m-sectorial operator with compact resolvent, for which the spectral equivalence

$$\sigma(H_{\text{gauged}}) = \sigma(\tilde{H}_{\text{HO}}) = \left\{(2k+1)\sqrt{\omega^2 - 4\alpha\beta}\right\}_{k=0}^{\infty}$$

holds. More specifically, H_{gauged} is defined as the operator in $L^2(\mathbb{R})$ associated with the closed sectorial form

$$\begin{aligned} h_{\text{gauged}}[\psi] &:= (\omega+\alpha+\beta)\|\psi'\|^2 + (\omega-\alpha-\beta)\|x\psi\|^2 \\ &\quad + i(\alpha-\beta)\left[(p\psi, x\psi) + (x\psi, p\psi)\right]\,, \\ \mathsf{D}(h_{\text{gauged}}) &:= \{\psi \in H^1(\mathbb{R}) \;:\; x\psi \in L^2(\mathbb{R})\}\,. \end{aligned}$$

As explained in Ref. (40) (*cf.* Example 5.40 next), the connection between the operators H_{gauged} and $\tilde{H}_{\text{HO}}$ given by (5.5.4) is very weak, although the operators share the same eigenvalues. Indeed, other important characteristics of H_{gauged}, such as the pseudospectrum and basis properties, are not preserved by the unbounded transformation A. On the other hand, H_{gauged} is *unitarily* equivalent (hence similar according to our restrictive definition) to the *rotated* (or *Davies'* (41)) oscillator

$$\mathcal{U}^* H_{\text{gauged}} \mathcal{U} = \zeta \left(p^2 + \frac{\bar{\zeta}}{\zeta} q^2 \right). \tag{5.5.5}$$

Here $\zeta := \sqrt{\omega^2 - (\alpha + \beta)^2} + i(\alpha - \beta)$ and $\mathcal{U}$ is a unitary operator with an explicit action (*cf.* (40)).

Other warning examples where eigenvalues or other spectral characteristics are not preserved by unbounded similarity transformations can be found in (40). We do not claim that there are no physical problems where a weaker (*e.g.*, unbounded) similarity transformation could be useful (in fact, there are!). However, without the assumption $A, A^{-1} \in \mathscr{B}(\mathcal{H})$, operators H_1 and H_2 cannot be viewed as "equivalent." In particular, H_2 cannot be used as a representation of a self-adjoint observable H_1 in quantum mechanics, unless A and A^{-1} are both bounded.

5.5.2 Quasi-Self-Adjoint Operators

We say that an operator H in a Hilbert space $\mathcal{H}$ is *quasi-self-adjoint* (*with respect to* Θ) if it is densely defined and there exists a nonnegative operator $\Theta \in \mathscr{B}(\mathcal{H})$ with $\Theta^{-1} \in \mathscr{B}(\mathcal{H})$ such that

$$H^* = \Theta H \Theta^{-1}. \tag{5.5.6}$$

That is, H^* is similar to H via the transformation Θ. Self-adjoint operators are quasi-self-adjoint with respect to the identity operator I. More generally, quasi-self-adjoint operators represent a special class of pseudo-self-adjoint operators briefly discussed in Section 5.2.5.5.

Any quasi-self-adjoint operator H is automatically closed, which follows from identity (5.5.6) and the closedness of the adjoint. In fact, H is self-adjoint with respect to a modified (but topologically equivalent) inner product $(\cdot, \Theta \cdot)$ in $\mathcal{H}$. For this reason, the operator Θ is sometimes called a *metric* (it is obviously not unique).

Equivalently, given any decomposition $\Theta = A^*A$, where necessarily $A, A^{-1} \in \mathscr{B}(\mathcal{H})$, H is similar to the operator

$$H_{\text{sa}} := AHA^{-1}, \tag{5.5.7}$$

which is self-adjoint with respect to the original inner product $(\cdot, \cdot)$ in $\mathcal{H}$. Applying identity (5.2.14) to H_{sa}, we get

$$\|(H - \lambda)^{-1}\| \leq \frac{\kappa}{\operatorname{dist}\big(\lambda, \sigma(H)\big)} \tag{5.5.8}$$

for every $\lambda \notin \sigma(H)$, where $\kappa := \|A\| \, \|A^{-1}\|$ is called the *condition number*.

Let us summarize the properties of quasi-self-adjoint operators in the following proposition.

Proposition 5.5.2 *Let H be a densely defined operator in a Hilbert space $\mathcal{H}$. The following statements are equivalent:*

(i) *H is quasi-self-adjoint,*
(ii) *H is similar to a self-adjoint operator.*

Any quasi-self-adjoint operator H is closed, $\sigma(H) \subset \mathbb{R}$, $\sigma_{\mathrm{r}}(H) = \emptyset$ and (5.5.8) *holds with a constant $\kappa \geq 1$.*

Quasi-self-adjoint operators can be considered as a nonstandard (possibly non-self-adjoint) representation of physical observables in quantum mechanics. It is possible to introduce a more general class of "quasi-self-adjoint" operators by relaxing the conditions on the boundedness of Θ and/or Θ^{-1}, *cf.* (42). However, this approach usually leads to pathological situations and does not seem to be adequate for applications in quantum mechanics as argued in Ref. (40).

A tool how to prove the quasi-self-adjointness is the following resolvent criterion.

Theorem 5.5.3 ((43–45)) *Let H be a densely defined closed operator in $\mathcal{H}$ with real spectrum. The operator H is similar to a self-adjoint operator if and only if there exists a constant M such that, for every $\psi \in \mathcal{H}$, the two following inequalities*

$$\sup_{\varepsilon>0} \varepsilon \int_{\mathbb{R}} \left\| \left(H - (\xi + i\varepsilon)\right)^{-1} \psi \right\|^2 d\xi \leq M \|\psi\|^2,$$

$$\sup_{\varepsilon>0} \varepsilon \int_{\mathbb{R}} \left\| \left(H^* - (\xi + i\varepsilon)\right)^{-1} \psi \right\|^2 d\xi \leq M \|\psi\|^2,$$

are satisfied.

The conditions on the resolvent may be very difficult to verify, unless the resolvent is known explicitly. As an example, let us quote (46), where the conditions are checked for the Laplacian in $L^2(\mathbb{R})$ with point interactions and an explicit formula for the metric is found too. In general, it is not expectable to have closed formulae for the metric operator and similarity transformations, not mentioning the self-adjoint operator to which a given quasi-self-adjoint operator is similar. As another exceptional situation, let us now summarize the complete story about the quasi-self-adjointness of the one-dimensional $\mathcal{PT}$-symmetric Robin Laplacian from Example 5.10.

EXAMPLE 5.35 Quasi-self-adjointness of the complex Robin Laplacian

For the one-dimensional Robin Laplacian $-\Delta^{(-a,a)}_{\{-i\alpha_0, i\alpha_0\}}$ from Example 5.10, a metric operator together with the corresponding similarity transformation and the similar self-adjoint operator are known explicitly due to Refs

(8–10). If $2a\alpha_0/\pi \notin \mathbb{Z} \setminus \{0\}$, the metric Θ and the similarity transform from a decomposition $\Theta = A^*A$ read

$$\Theta = I + K, \quad A = I + L, \quad A^{-1} = I + M, \tag{5.5.9}$$

where K, L, M are (Hilbert-Schmidt) integral operators with kernels

$$\begin{aligned}
\mathcal{K}(x,y) &:= \alpha_0\, e^{-i\alpha_0(y-x)} \big(\tan(\alpha_0 a) - i\,\mathrm{sgn}(y-x) \big), \\
\mathcal{L}(x,y) &:= \frac{i\alpha_0}{2a} \big(y - a\,\mathrm{sgn}(y-x)\big) + \frac{1}{2a} \left(e^{-i\alpha_0(y+a)} - 1\right), \\
\mathcal{M}(x,y) &:= \frac{\alpha_0\, e^{i\alpha_0(a-x)}}{\sin(2\alpha_0 a)} - \frac{\alpha_0}{2}\, e^{-i\alpha_0(x-y)} \big(\cot(2\alpha_0 a) - i\,\mathrm{sgn}(y-x) \big) \\
&\quad - \frac{\alpha_0 e^{-i\alpha_0(x+y)}}{2\sin(2\alpha_0 a)} .
\end{aligned}$$

Moreover,

$$A\big(-\Delta^{(-a,a)}_{\{-i\alpha_0, i\alpha_0\}} \big) A^{-1} = -\Delta_N^{(-a,a)} + \alpha_0^2\, \chi_0^N (\chi_0^N, \cdot),$$

where $\chi_0^N(x) := (2a)^{-1/2}$ is the first Neumann eigenfunction. The main tool to obtain these formulae is the functional calculus for self-adjoint operators, employing the fact that the eigenfunctions of $-\Delta^{(-a,a)}_{\{-i\alpha_0, i\alpha_0\}}$ can be written down in terms of eigenfunctions of Dirichlet and Neumann Laplacians. We refer Ref. (8) for more details and other explicit formulae, even in a more general setting.

Taking the tensor products $I \otimes \Theta$ and $I \otimes A$, $I \otimes A^{-1}$, where I is the identity in $L^2(\mathbb{R})$, we obtain a metric and similarity transformations for the $\mathcal{PT}$-symmetric waveguide H_{α_0} from Example 5.33.

5.5.3 Basis Properties of Eigensystems

In the case of the operators with compact resolvent, the similarity of H to a normal operator is related to the basis properties of the eigenvectors of H. Let us recall some notions first.

We say that $\{\psi_k\}_{k=1}^\infty$ is *complete* in $\mathcal{H}$ if $(\{\psi_k\}_{k=1}^\infty)^\perp = \{0\}$ or equivalently $\mathrm{span}(\{\psi_k\}_{k=1}^\infty)$ is dense in $\mathcal{H}$. We say that $\{\psi_k\}_{k=1}^\infty$ is a *(Schauder or conditional) basis* if every $\psi \in \mathcal{H}$ has a unique expansion in the vectors $\{\psi_k\}$, that is,

$$\forall \psi \in \mathcal{H}, \quad \exists! \{\alpha_k\}_{k=1}^\infty, \quad \psi = \sum_{k=1}^\infty \alpha_k \psi_k . \tag{5.5.10}$$

Finally, we say that $\{\psi_k\}_{k=1}^{\infty}$, normalized to 1 in $\mathcal{H}$, forms a *Riesz (or unconditional) basis* if it forms a basis and the inequality

$$\forall \psi \in \mathcal{H}, \qquad C^{-1}\|\psi\|^2 \leq \sum_{k=1}^{\infty} |\langle \psi_k, \psi \rangle|^2 \leq C\|\psi\|^2 \tag{5.5.11}$$

holds with a positive constant C independent of ψ. Notice that the Riesz bases are a suitable substitute for orthonormal bases, in which case $C = 1$ due to the Parseval equality. We clearly have

$$\text{complete} \supset \text{basis} \supset \text{Riesz basis} \supset \text{orthonormal basis}.$$

A set $\{\psi_k\}_{k=1}^{\infty}$ is a Riesz basis if there exists an operator $A \in \mathscr{B}(\mathcal{H})$ with $A^{-1} \in \mathscr{B}(\mathcal{H})$ and an orthonormal basis $\{e_k\}_{k=1}^{\infty}$ such that $e_k = A\psi_k$, see Ref. [2, Theorem 3.4.5] for other equivalent formulations. The last property already suggests a relation between the similarity to a normal operator and Riesz basis property, which is expressed more precisely in the following proposition.

Proposition 5.5.4 *Let H have a compact resolvent. Then H is similar to a normal operator if and only if the eigenfunctions of H form a Riesz basis. The latter is equivalent to the similarity to a self-adjoint operator if the spectrum of H is in addition real.*

For non-self-adjoint or nonnormal operators, the geometric and algebraic multiplicity of eigenvalues may differ. In that case, the operator cannot be similar to a normal one, nonetheless, the *generalized eigensystem*, that is, the collection of eigenvectors and root vectors, may still contain a Riesz basis. The latter is a suitable generalization of the similarity to a normal operator and several perturbation results guaranteeing such a property are known. To avoid describing how the root vectors are selected and normalized, the following theorems are expressed with help of spectral projections. We will use the following assumptions and notations in the sequel.

$\langle H_0 \rangle$ Let H_0 be a self-adjoint, nonnegative operator that has a compact resolvent. We denote its eigenvalues (sorted in an increasing order) corresponding orthonormal eigenfunctions and spectral projections as μ_n, ψ_n and P_n, respectively.

Theorem 5.5.5 ([4, Theorem V.4.15a]) *Let H_0 satisfy the aforementioned $\langle H_0 \rangle$. Assume that all eigenvalues μ_n are simple and*

$$\mu_{n+1} - \mu_n \xrightarrow[n\to\infty]{} \infty.$$

Let $V \in \mathscr{B}(\mathcal{H})$ and set $H := H_0 + V$. Then H is closed with compact resolvent, and the eigenvalues and spectral projections of H can be indexed as $\{\lambda_{0k}, \lambda_n\}$ and $\{Q_{0k}, Q_n\}$, respectively, where $k = 1, \ldots, N < \infty$ and $n = N+1, N+2, \ldots$ in such a *way that* $|\mu_n - \lambda_n| = O(1)$ *as* $n \to \infty$ *and there exists* $A \in \mathscr{B}(\mathcal{H})$ *with* $A^{-1} \in \mathscr{B}(\mathcal{H})$ *such that*

$$\sum_{k=1}^{N} Q_{0k} = A^{-1} \sum_{k=1}^{N} P_k A, \qquad Q_n = A^{-1} P_n A, \quad n > N.$$

EXAMPLE 5.36 $\mathcal{PT}$-symmetric square well

We consider a perturbation of the Dirichlet Laplacian $-\Delta_D^{(-a,a)}$ from Example 5.10 studied in Refs (47, 48), namely

$$H_Z := -\Delta_D^{(-a,a)} + iZ\,\mathrm{sgn}\,x, \quad \mathsf{D}(H_Z) := H^2((-a,a)) \cap H_0^1((-a,a)),$$

where Z is a real parameter. Both conditions of Theorem 5.5.5 (*i.e.*, the growing gaps of eigenvalues of H_0 and the boundedness of the perturbation) are satisfied, therefore the eigensystem of H_Z contains a Riesz basis for any $Z \in \mathbb{R}$. As H_Z is $\mathcal{PT}$-symmetric and its eigenvalues depend continuously on Z, all eigenvalues of H_Z are real and simple for all sufficiently small Z and, in this case, H_Z is similar to a self-adjoint operator. When increasing Z, eigenvalues with the lowest real part collide and create complex conjugate pairs. If all eigenvalues of H_Z are simple (some of them possibly complex), then H_Z is similar to a normal operator. For specific values of Z, when two real eigenvalues collide, the geometric multiplicity is one, but the algebraic multiplicity is two, therefore the Riesz basis contains a root vector; in this case, operator H_Z is similar to an "almost diagonal" operator, that is, a two by two Jordan block corresponding to the multiple eigenvalue appears.

Theorem 5.5.5 has many generalizations, using various grow conditions on the eigenvalue gaps and strength of perturbation; we mention particularly classical results in Ref. (49) and works using p-subordination, *cf.* (50–52) and references therein. Suitable theorems for perturbations of the harmonic oscillator or similar ones, that is, with asymptotically constant eigenvalue gaps, were proved only recently. We present jointly the operator and form version of the result; further generalizations and related results can be found in Refs (53, 54).

Theorem 5.5.6 ((55, 56)) *Let* H_0 *satisfy the aforementioned* $\langle H_0 \rangle$. *Assume that all eigenvalues* μ_n *are simple and*

$$\forall n \in \mathbb{N}, \quad \mu_{n+1} - \mu_n \geq \delta > 0.$$

- ***Operator version:*** *Let an operator* V, $\mathsf{D}(V) \supset \mathsf{D}(H)$, *satisfy*

$$\|V\psi_n\| \xrightarrow[n\to\infty]{} 0. \tag{5.5.12}$$

Then the claim of Theorem 5.5.5 holds for $H := H_0 + V$. *Moreover,* $|\mu_n - \lambda_n| = o(1)$ *as* $n \to \infty$.

- ***Form version:*** *Let a sesquilinear form v,* $\mathsf{D}(v) \supset \mathsf{D}(h_0)$*, satisfy*

$$\forall m, n \in \mathbb{N}, \quad |v(\psi_m, \psi_n)| \leq \frac{M}{m^\alpha n^\alpha}, \tag{5.5.13}$$

with some $M \geq 0$ *and* $\alpha > 0$*. Then the claim of Theorem 5.5.5 holds for the operator H associated with the sectorial form* $h := h_0 + v$*, where* h_0 *is the form associated. Moreover,* $|\mu_n - \lambda_n| = o(1)$ *as* $n \to \infty$*.*

The classes of potential perturbations of the harmonic oscillator satisfying the conditions (5.5.12) and (5.5.13) are studied in Refs (55, 56); one simple example is the following.

EXAMPLE 5.37 The eigensystem of the harmonic oscillator with $\alpha\,\delta(x)$

Let H be the operator defined in Example 5.31. As the values of Hermite functions ψ_n, being the eigenfunctions of H_{HO}, at zero are known explicitly, it is not difficult to show that the condition (5.5.13) is satisfied with $\alpha = 1/4$, *cf.* (56). The latter holds also for perturbations consisting of finitely many δ interactions.

5.6 PSEUDOSPECTRA

Highly non-self-adjoint operators have properties very different from self-adjoint or normal operators. The notion of pseudospectra is a possibility how to describe these differences and the new phenomena occurring in non-self-adjoint situations. More information on the subject can be found in by now classical monographs by Trefethen and Embree (57) and Davies (2). Our exposition is in many respects based on Ref. (40).

5.6.1 Definition and Basic Properties

Given a positive number ε, we define the *ε-pseudospectrum* (or simply *pseudospectrum*) of a closed operator H as

$$\sigma_\varepsilon(H) := \left\{ z \in \mathbb{C} \,:\, \|(H-z)^{-1}\| > \varepsilon^{-1} \right\}, \tag{5.6.1}$$

with the convention that $\|(H-z)^{-1}\| = \infty$ for $z \in \sigma(H)$. Some basic and well-known properties of pseudospectra are summarized in the following:

- *Topology.* For every $\varepsilon > 0$, $\sigma_\varepsilon(H)$ is a nonempty open subset of $\mathbb{C}$ and any bounded connected component of $\sigma_\varepsilon(H)$ has a nonempty intersection with $\sigma(H)$. (If the spectrum of H is empty, then $\sigma_\varepsilon(H)$ is unbounded for every $\varepsilon > 0$.)

- *Relation to spectra.* The pseudospectrum always contains an ε-neighborhood of the spectrum, and if $\Xi(H)$ defined in (5.2.9) is connected and has a nonempty intersection with the resolvent set of H, the pseudospectrum is in turn contained in an ε-neighborhood of the numerical range:

$$\begin{aligned} \{z \in \mathbb{C} \, : \, \operatorname{dist}(z, \sigma(H)) < \varepsilon\} \subset \sigma_\varepsilon(H) \\ \sigma_\varepsilon(H) \subset \{z \in \mathbb{C} \, : \, \operatorname{dist}(z, \overline{\Theta(H)}) < \varepsilon\}. \end{aligned} \tag{5.6.2}$$

  The first inclusion follows from the bound $\|(H-z)^{-1}\| \geq \operatorname{dist}(z, \sigma(H))^{-1}$, which is valid for any operator. For normal (and thus self-adjoint) operators this inclusion becomes an equality as a consequence of (5.2.14); hence the notion of pseudospectrum is in fact trivial for such operators. The second inclusion in (5.6.2) follows from (5.2.10). If H is "highly non-self-adjoint," the pseudospectrum $\sigma_\varepsilon(H)$ is typically "much larger" than the ε-neighborhood of the spectrum.
- *Spectral instability.* The following result, sometimes referred to as the Roch-Silberman theorem (58), relates the pseudospectra to the stability of the spectrum under small perturbations:

$$\sigma_\varepsilon(H) = \bigcup_{\|V\|<\varepsilon} \sigma(H+V). \tag{5.6.3}$$

  This property is of particular importance in applications, for instance in numerical analysis, where small errors (*e.g.*, rounding) can easily lead to false identifications of computed eigenvalues of H with possibly very distant eigenvalues of $H+V$ if the pseudospectrum of H is huge.
- *Pseudomodes.* A complex number z belongs to $\sigma_\varepsilon(H)$ if and only if $z \in \sigma(H)$ or z is a *pseudoeigenvalue* (or *approximate eigenvalue*), that is,

$$\|(H-z)\psi\| < \varepsilon\|\psi\| \quad \text{for some} \quad \psi \in \mathsf{D}(H). \tag{5.6.4}$$

  Any ψ satisfying (5.6.4) is called a *pseudoeigenvector* (or *pseudoeigenfunction* or *pseudomode*). Again, for operators H that are far from self-adjoint, pseudoeigenvalues may not be close to the spectrum of H. This is particularly striking if we realize that these pseudoeigenvalues can be turned into true eigenvalues by a very small perturbation, *cf.* (5.6.3). What is more, for differential operators, we can often construct very nice (*e.g.*, smooth and with compact support) pseudoeigenfunctions, see Section 5.6.2.
- *Adjoints.* Using the identity $(H^* - \bar{z})^{-1} = (H-z)^{-1}$ for $z \in \rho(H)$, it is easy to see that

$$\lambda \in \sigma_\varepsilon(H) \iff \bar{\lambda} \in \sigma_\varepsilon(H^*). \tag{5.6.5}$$

- *Antiunitary symmetry.* If H has an antiunitary symmetry, *cf.* Section 5.2.5.6, then

$$\lambda \in \sigma_\varepsilon(H) \iff \bar{\lambda} \in \sigma_\varepsilon(H). \tag{5.6.6}$$

- *Similarity.* If the similarity relation (5.5.1) holds, then H_1 and H_2 have the same spectra, but their pseudospectra may be very different, unless the condition number κ is fairly close to one as, due to (5.5.8),

$$\sigma_{\varepsilon/\kappa}(H_2) \subset \sigma_\varepsilon(H_1) \subset \sigma_{\varepsilon\kappa}(H_2)\,. \tag{5.6.7}$$

As a consequence, if an operator H_2 is similar to a normal operator H_1, the pseudospectrum of H_1 is contained in the $\varepsilon\kappa$-neighborhood of $\sigma(H_2)$.

5.6.2 Main Tool from Microlocal Analysis

Pseudospectra of differential operators can be conveniently studied by semiclassical methods, as firstly realized by Davies (59). His observation was followed by important generalizations, we refer particularly to Refs (60, 61). In this section, we state a simple version of these results adapted to the very special case of differential operators with analytic coefficients in one dimension in a formulation given in Ref. [57, Theorem 11.1].

To state the theorem, we need to recall some notions of semiclassical analysis. Let $\hbar > 0$ be a small parameter (inspired by Planck's constant in quantum mechanics; we deliberately avoid frequently used notation h for the small parameter as we reserve this letter for denoting forms) and $a_j : \mathbb{R} \to \mathbb{C}$, with $j = 0, \dots, n$, are smooth functions. Define

$$f(x,\xi) := \sum_{j=0}^{n} a_j(x)(-i\xi)^j\,, \qquad (x,\xi) \in \mathbb{R}^2\,.$$

We say that $H_\hbar$ is a *semiclassical differential operator* associated with *symbol* f if

$$H_\hbar := \sum_{j=0}^{n} a_j(x)\,\hbar^j\,\frac{d^j}{dx^j}\,, \qquad \mathsf{D}(H_\hbar) := C_0^\infty(\mathbb{R})\,. \tag{5.6.8}$$

The *Poisson bracket* $\{\cdot,\cdot\}$ is defined as

$$\{u,v\} := \frac{\partial u}{\partial \xi}\frac{\partial v}{\partial x} - \frac{\partial u}{\partial x}\frac{\partial v}{\partial \xi} \tag{5.6.9}$$

and the closure of the set

$$\Lambda := \left\{ f(x,\xi) \,:\, (x,\xi) \in \mathbb{R}^2,\ \frac{1}{2i}\{f,\bar{f}\}(x,\xi) > 0 \right\} \tag{5.6.10}$$

is referred to as the *semiclassical pseudospectrum* of $H_\hbar$, *cf.* (61). We remark that in the special case of $H_\hbar$ being a Schrödinger operator with analytic potential $a_0 =: V$ the condition $\frac{1}{2i}\{f,\bar{f}\}(x,\xi) > 0$ reduces to $\Im V'(x) \neq 0$ and $\xi \neq 0$.

Now we are in a position to state the result from Ref. [57, Theorem 11.1]; we refer to Ref. (40) for a proof in the special case of Schrödinger operators.

Theorem 5.6.1 (Semiclassical pseudomodes.) *Let the functions* a_j, $j = 0, \dots, n$, *be analytic and let* $H_\hbar$ *be the semiclassical differential operator* (5.6.8). *Then, for every* $z \in \Lambda$, cf. (5.6.10), *there exist* $C = C(z) > 1$, $\hbar_0 = \hbar_0(z) > 0$ *and an* $\hbar$*-dependent family of* $C_0^\infty(\mathbb{R})$ *functions* $\{\psi_\hbar\}_{0<\hbar\leq\hbar_0}$ *with the property that, for all* $0 < \hbar \leq \hbar_0$,

$$\|(H_\hbar - z)\psi_\hbar\| < C^{-1/\hbar}\, \|\psi_\hbar\|.$$

If the coefficients a_j are not analytic, but only smooth, a slower rate of growth is obtained, *cf.* (59, 61); instead of the upper bound of $C^{-1/\hbar}\|\psi_\hbar\|$ one has that, for each $N \in \mathbb{N}$, there exists a constant $C_N > 0$ such that, for all $0 < \hbar \leq \hbar_0$,

$$\|(H_\hbar - z)\psi_\hbar\| < \frac{\hbar^N}{C_N}\|\psi_\hbar\|.$$

Although Theorem 5.6.1 is stated for semiclassical operators, scaling techniques allow its application to non-semiclassical operators where the spectral parameter tends to infinity. This is based on the principle that the semiclassical limit is equivalent to the high-energy limit after a change of variables; this principle is made concrete in the following examples.

EXAMPLE 5.38 Pseudospectrum of the imaginary Airy operator

We explain how one can apply Theorem 5.6.1 for the (non-semiclassical) operator H_{Airy} from Example 5.12. The scaling argument can be adapted accordingly to the other examples presented next.

We introduce the unitary transform $\mathcal{U}$ on $L^2(\mathbb{R})$ defined by

$$(\mathcal{U}\psi)(x) := \tau^{1/2}\psi(\tau x), \tag{5.6.11}$$

where $\tau \in \mathbb{R}$ is positive (and typically large in the sequel). Then

$$\mathcal{U}H_{\text{Airy}}\mathcal{U}^{-1} = \tau H_\hbar \quad \text{with} \quad H_\hbar := -\hbar^2\frac{d^2}{dx^2} + ix \quad \text{and} \quad \hbar := \tau^{-3/2}.$$

For the symbol $f = \xi^2 + ix$ associated with $H_\hbar$, we have $\{f,\overline{f}\} = -4i\xi$, hence $\Lambda = \{z \in \mathbb{C} \,:\, \Re z > 0\}$. The same translation argument, which shows that the spectrum is empty, *cf.* Example 5.29, proves that

$$\|(H_{\text{Airy}} - z)^{-1}\| = \|(H_{\text{Airy}} - \Re z)^{-1}\|.$$

Applying the unitary scaling and Theorem 5.6.1, we know that, for all $\hbar \leq \hbar_0(1)$,

$$\|(H_{\text{Airy}} - \tau)^{-1}\| = \tau^{-1}\|(H_\hbar - 1)^{-1}\| > \hbar^{2/3}C(1)^{1/\hbar}.$$

From this we deduce

$$\sigma_\varepsilon(H_{\text{Airy}}) \supset \left\{ z \in \mathbb{C} \ : \ \Re z \geq \tau_0 \ \wedge \ (\Re z)^{-1} C(1)^{(\Re z)^{3/2}} \geq \varepsilon^{-1} \right\},$$

where $\tau_0 := \hbar_0(1)^{-2/3}$. Another version of this inclusion is stated in Ref. [40, Section 7.1]. A quite precise study of the norm of $(H_{\text{Airy}} - z)^{-1}$ as $\Re z \to \infty$ can be found in Ref. [62, Corollary 1.4].

EXAMPLE 5.39 Pseudospectrum of the imaginary cubic oscillator

Now we make a pseudospectral analysis of the paradigmatic $\mathcal{PT}$-symmetric model H_{cubic} from Example 5.13. Recall that H_{cubic} has a compact resolvent and that all its eigenvalues are known to be real (26–28). On the other hand, its pseudospectrum turns out to be very different from the pseudospectra of self-adjoint operators.

In view of (5.6.2) and the accretivity of H_{cubic}, we *a priori* know that the pseudospectrum $\sigma_\varepsilon(H_{\text{cubic}})$ is contained in $\{z \in \mathbb{C} \ : \ \Re z \geq -\varepsilon\}$. As a consequence of $\mathcal{PT}$-symmetry, we also know that $\sigma_\varepsilon(H_{\text{cubic}})$ is symmetric with respect to the real axis. The unitary transform (5.6.11) and an application of Theorem 5.6.1 lead to the conclusion that, for every $z \in \Lambda = \{z \in \mathbb{C} \ : \ \Re z > 0 \ \wedge \ \Im z \neq 0\}$, there exists $C(z) > 1$ such that, for all $\hbar \leq \hbar_0(z)$,

$$\|(H_{\text{cubic}} - \tau^3 z)^{-1}\| > \hbar^{6/5} C(z)^{1/\hbar}, \quad \hbar := \tau^{-5/2}. \tag{5.6.12}$$

From this we deduce for instance the inclusion

$$\sigma_\varepsilon(H_{\text{cubic}}) \supset \left\{ \tau^3 + i\tau^3 \in \mathbb{C} \ : \ \tau \geq \tau_0 \ \wedge \ \tau^{-3} C(1+i)^{\tau^{5/2}} \geq \varepsilon^{-1} \right\},$$

where $\tau_0 := \hbar_0(1+i)^{-2/5}$. We see that, for every ε, there are complex points with positive real part, nonzero imaginary part, and large magnitude that lie in the pseudospectrum $\sigma_\varepsilon(H)$. Consequently, the pseudospectrum of H_{cubic} is not contained in a uniform neighborhood of $\sigma(H_{\text{cubic}})$, and therefore H_{cubic} is not similar to a self-adjoint operator. From this we also deduce that H_{cubic} is not quasi-self-adjoint (*cf.* Proposition 5.5.2) and that its eigenfunctions do not form a Riesz basis (*cf.* Proposition 5.5.4).

The asymptotic behavior of the pseudospectral lines of H_{cubic} is studied in Ref. [62, Proposition 4.1], while a result of numerical computations can be found in Refs [57, Fig. 11.4] and [40, Fig. 1], *cf.* Figure 5.1 next. As the most recent result about H_{cubic}, let us mention (63) where it is shown that the norms of the spectral projections of H_{cubic} grow (at exponential rate), therefore the eigenfunctions cannot form even a basis. Nonetheless, it was proved in Ref. (64) that the eigenfunctions are complete.

EXAMPLE 5.40 Pseudospectrum of the gauged oscillator

In view of Example 5.34, the pseudospectrum of the gauged oscillator H_{gauged} (always with $\omega - |\alpha + \beta| > 0$) coincides with the pseudospectrum of a rotated oscillator that appears on the right-hand side of (5.5.5). We could again apply the scaling argument (5.6.11) and Theorem 5.6.1 to the present situation (see Ref. [40, Section 7.4]). However, as the pseudospectrum of the rotated oscillator is well studied (see Ref. (2) and references therein), we restrict ourselves to saying that it is again much larger than a tubular neighborhood of the real eigenvalues. Consequently, the rotated oscillator and H_{gauged} are not quasi-self-adjoint and their eigenfunctions do not form a Riesz basis (unless $\alpha = 0 = \beta$). As it is also known from Ref. (65) that the norms of spectral projections grow exponentially, the eigenfunctions do not form even a basis (although they are complete). We refer to Ref. (40) for more details on both the pseudospectra and eigenfunctions and for further references.

Let us point out that H_{gauged} satisfies (5.5.4), which can be interpreted as some sort of weak similarity to the self-adjoint harmonic oscillator H_{HO}. The essential difference in the pseudospectra and basis properties of eigenfunctions of H_{gauged} and H_{HO} clearly demonstrates that the relation (5.5.4) actually represents only a very weak connection between the two operators.

EXAMPLE 5.41 Pseudospectrum of the shifted oscillator

Let us consider the shifted harmonic oscillator H_α from Example 5.26 with $\alpha = i$. Recall that H_i has a compact resolvent and that all its eigenvalues are real; they actually coincide with the eigenvalues of the self-adjoint oscillator H_{HO} from Example 5.4, *cf.* Example 5.9. Indeed, H_i is *formally* similar to H_{HO} via the *formal* imaginary-shift operator $(A\psi)(x) := \psi(x + i)$.

The pseudospectrum of H_i is symmetric with respect to the real axis as a consequence of $\mathcal{PT}$-symmetry of H_i. Applying the unitary transform (5.6.11) to the shifted harmonic oscillator H_α from Example 5.26 with $\alpha = i$, we end up with an operator of the form

$$\tilde{H}_\hbar := -\hbar^2 p^2 + q^2 + 2i\hbar^{1/2} q - 1 .$$

Because of the presence of a fractional power of $\hbar$, we do not obtain an operator of the semiclassical type (5.6.8) and Theorem 5.6.1 is not applicable. Nevertheless, it is still possible to use Ref. [59, Theorem 1], which is suitable for potentials with fractional powers of $\hbar$, and thereby obtain polynomial lower bounds to the norm of the resolvent. The expected exponential bound has been proved only recently in Ref. [40, Theorem 2] by adapting the proof of Theorem 5.6.1 to the present situation. More specifically, we have

$$\sigma_\varepsilon(H_i) \supset \left\{ z \in \mathbb{C} \ : \ \Re z \geq c^{-1} \ \wedge \ |\Im z| \leq \beta\sqrt{\Re z} \ \wedge \ c e^{c\sqrt{\Re z}} \geq \varepsilon^{-1} \right\},$$

where the number $\beta \in (0, 2)$ can be chosen arbitrarily close to 2 and c is a (small) positive constant. Consequently, H_i possesses large complex pseudoeigenvalues in parabolic regions of the complex plane; H_i is not quasi-self-adjoint (*cf.* Proposition 5.5.2) and its eigenfunctions do not form a Riesz basis (*cf.* Proposition 5.5.4). Summing up, although H_i is formally similar to H_{HO} and their spectra coincide, we see that pseudospectral and basis properties exhibit striking differences.

It has been shown recently in Ref. (66) that the eigenfunctions of the shifted oscillator H_i are complete, but do not form a basis as the norms of the spectral projection grow.

EXAMPLE 5.42 Pseudospectra of the harmonic oscillator with $\alpha\,\delta(x)$

As the eigensystem of the operator H from Example 5.31 form a Riesz basis containing only finitely many root vectors, *cf.* Example 5.37, H is similar to an operator of the form $D + N$, where D is a diagonal operator having the eigenvalues of H as diagonal entries and N is a finite rank operator corresponding to the Jordan block structure. Standard arguments show that, for z in a neighborhood of an eigenvalue λ_0 of H, the resolvent satisfies $\|((D + N) - z)^{-1}\| \sim |\lambda_0 - z|^{-n}$, where $n = 1$ if $\mathsf{m}_a(\lambda_0) = \mathsf{m}_g(\lambda_0)$ and $n > 1$ if $\mathsf{m}_a(\lambda_0) > \mathsf{m}_g(\lambda_0)$. The pseudospectrum of H is therefore contained in a neighborhood of the spectrum, but the possible presence of Jordan blocks results in wider peaks around degenerate eigenvalues with nonequal geometric and algebraic multiplicity.

The same reasoning applies to the $\mathcal{PT}$-symmetric square well from Example 5.36.

We refer to Ref. (40) for further advocacy of the usage of pseudospectra in non-Hermitian quantum mechanics.

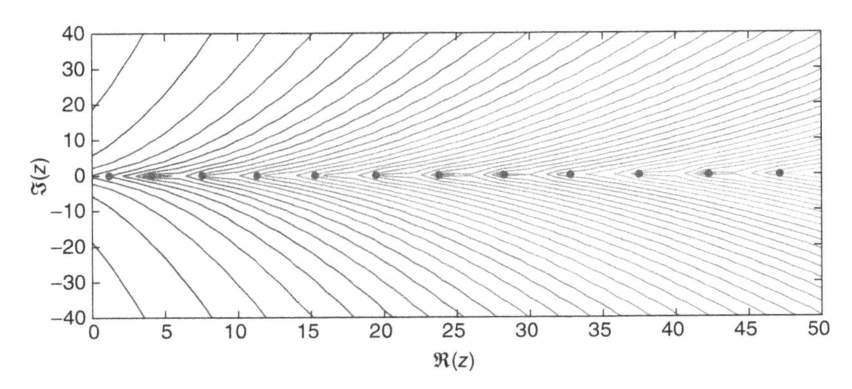

Figure 5.1 Spectrum (dark gray dots) and pseudospectra (enclosed by the light gray contour lines) of the imaginary cubic oscillator. *(Courtesy of Miloš Tater.)*

REFERENCES

1. Tretter C. *Spectral Theory of Block Operator Matrices and Applications*. London: Imperial College Press; 2008.
2. Davies EB. *Linear Operators and their Spectra*. Cambridge: Cambridge University Press; 2007.
3. Edmunds DE, Evans WD. *Spectral Theory and Differential Operators*. New York: Oxford University Press; 1987.
4. Kato T. *Perturbation Theory for Linear Operators*. Berlin: Springer-Verlag; 1966.
5. Adams RA, Fournier JJF. *Sobolev Spaces*. 2nd ed. Amsterdam: Academic Press; 2003.
6. Gilbarg D, Trudinger NS. *Elliptic Partial Differential Equations of Second Order*. Berlin: Springer-Verlag; 1983.
7. Krejčiřík and Siegl D. $\mathcal{PT}$-symmetric models in curved manifolds. *J Phys A Math Theor* 2010;43:485–204.
8. Krejčiřík D, Siegl P, Železný J. On the similarity of Sturm-Liouville operators with non-Hermitian boundary conditions to self-adjoint and normal operators. *Complex Anal Oper Theory* 2014;8:255–281.
9. Krejčiřík D, Bíla H, Znojil M. Closed formula for the metric in the Hilbert space of a $\mathcal{PT}$-symmetric model. *J Phys A* 2006;39:10143–10153.
10. Krejčiřík D. Calculation of the metric in the Hilbert space of a $\mathcal{PT}$-symmetric model via the spectral theorem. *J Phys A Math Theor* 2008;41:244012.
11. Freitas P. On some eigenvalue problems related to the wave equation with indefinite damping. *J Differ Equ* 1996;127(1):320–335.
12. Freitas P, Krejčiřík D. Instability results for the damped wave equation in unbounded domains. *J Differ Equ* 2005;211(1):168–186.
13. Brezis H. *Functional Analysis, Sobolev Spaces and Partial Differential Equations*. New York: Springer; 2011.
14. Kochan D, Krejčiřík D, Novák R, Siegl P. The Pauli equation with complex boundary conditions. *J Phys A Math Theor* 2012;45:444019.
15. Glazman IM. *Direct Methods of Qualitative Spectral Analysis of Singular Differential Operators*. Jerusalem: Israel Program for Scientific Translations; 1965.
16. Garcia SR, Putinar M. Complex symmetric operators and applications. *Trans Am Math Soc* 2006;358:1285–1315.
17. Garcia SR, Putinar M. Complex symmetric operators and applications. *Trans Am Math Soc* 2007;359:3913–3931.
18. Prodan E, Garcia SR, Putinar M. Norm estimates of complex symmetric operators applied to quantum systems. *J Phys A Math Gen* 2006;39:389–400.
19. SR Garcia, Prodan E, Putinar M. Mathematical and physical aspects of complex symmetric operators. *J Phys A Math Theor* 2014;47:353001.
20. Bognar J. *Indefinite Inner Product Spaces*. New York-Heidelberg: Springer-Verlag; 1974.

21. Azizov TYa, Iokhvidov IS. *Linear Operators in Spaces with an Indefinite Metric*. Chichester: John Wiley and Sons; 1989.
22. Reed M, Simon B. *Methods of Modern Mathematical Physics, I. Functional Analysis*. New York: Academic Press; 1972.
23. Davies EB. *Spectral Theory and Differential Operators*. Cambridge: Cambridge University Press; 1995.
24. Kato T. On some Schrödinger operators with a singular complex potential. *Ann Scuola Norm Super Pisa Cl Sci* 1978;5:105–114.
25. Kato T. Schrödinger operators with singular potentials. *Isr J Math* 1972;13:135–148.
26. Dorey P, Dunning C, Tateo R. Spectral equivalences, Bethe ansatz equations, and reality properties in $\mathcal{PT}$-symmetric quantum mechanics. *J Phys A* 2001;34:5679–5704.
27. Shin KC. On the reality of the eigenvalues for a class of $\mathcal{PT}$-symmetric oscillators. *Commun Math Phys* 2002;229:543–564.
28. Giordanelli I, Graf GM. The real spectrum of the imaginary cubic oscillator: an expository proof. *Ann Henri Poincaré*. Annales Henri Poincare, 2015, 16, 99–112.
29. Browder FE. On the spectral theory of elliptic differential operators. I. *Math Ann* 1961;142:22–130.
30. Gustafson K, Weidmann J. On the essential spectrum. *J Math Anal Appl* 1969;25:121–127.
31. Wolf F. On the essential spectrum of partial differential boundary problems. *Commun Pure Appl Math* 1959;12:211–228.
32. Reed M, Simon B. *Methods of Modern Mathematical Physics, IV. Analysis of Operators*. New York: Academic Press; 1978.
33. Borisov D, Krejčiřík D. $\mathcal{PT}$-symmetric waveguides. *Integr Equ Oper Theory* 2008;62(4):489–515.
34. Krejčiřík D, Tater M. Non-Hermitian spectral effects in a $\mathcal{PT}$-symmetric waveguide. *J Phys A Math Theor* 2008;41:244013.
35. Borisov D, Krejčiřík D. The effective Hamiltonian for thin layers with non-Hermitian Robin-type boundary conditions. *Asympt Anal* 2012;76:49–59.
36. Scholtz FG, Geyer HB, Hahne FJW. Quasi-Hermitian operators in quantum mechanics and the variational principle. *Ann Phys* 1992;213:74–101.
37. Ahmed Z. Pseudo-Hermiticity of Hamiltonians under gauge-like transformation: real spectrum of non-Hermitian Hamiltonians. *Phys Lett A* 2002;294:287–291.
38. Swanson MS. Transition elements for a non-Hermitian quadratic Hamiltonian. *J Math Phys* 2004;45:585–601.
39. Pravda-Starov K. On the pseudospectrum of elliptic quadratic differential operators. *Duke Math J* 2008;145:249–279.
40. Krejčiřík D, Siegl P, Tater M, Viola J. Pseudospectra in non-Hermitian quantum mechanics, 2014, preprint on arXiv:1402.1082 [math-SP].
41. Davies EB. Pseudo-spectra, the harmonic oscillator and complex resonances. *Proc R Soc London Ser A* 1999;455:585–599.

42. Dieudonné J. Quasi-Hermitian operators. *Proc Int Sympos Linear Spaces* 1961;115–123.
43. Naboko S. Conditions for similarity to unitary and selfadjoint operators. *Funct Anal Appl* 1984;18:13–22.
44. Malamud M. A criterion for a closed operator to be similar to a selfadjoint operator. *Ukr Math J* 1985;37:41–48.
45. van Casteren JA. Operators similar to unitary or selfadjoint ones. *Pac J Math* 1983;104:241–255.
46. Albeverio S, Kuzhel S. One-dimensional Schrödinger operators with $\mathcal{P}$-symmetric zero-range potentials. *J Phys A Math Theor* 2005;38:4975–4988.
47. Znojil M. $\mathcal{PT}$-symmetric square well. *Phys Lett A* 2001;285:7–10.
48. Siegl P. $\mathcal{PT}$-symmetric square well-perturbations and the existence of metric operator. New York-London-Sydney: *Int J Theor Phys* 2011;50:991–996.
49. Dunford N, Schwartz JT. *Linear Operators, Part 3, Spectral Operators*. Wiley-Interscience; 1971.
50. Wyss C. Riesz bases for p-subordinate perturbations of normal operators. *J Funct Anal* 2010;258:208–240.
51. Markus A. *Introduction to the Spectral Theory of Polynomial Operator Pencils*. Volume 71, *Translations of Mathematical Monographs*. Providence (RI): American Mathematical Society; 1988.
52. Agranovich MS. On series with respect to root vectors of operators associated with forms having symmetric principal part. *Funct Anal Appl* 1994;28:151–167.
53. Adduci J, Mityagin B. Root system of a perturbation of a selfadjoint operator with discrete spectrum. *Integr Equ Oper Theory* 2012;73:153–175.
54. Shkalikov A. On the basis property of root vectors of a perturbed self-adjoint operator. *Proc Steklov Inst Math* 2010;269:284–298.
55. Adduci J, Mityagin B. Eigensystem of an L^2-perturbed harmonic oscillator is an unconditional basis. *Cent Eur J Math* 2012;10:569–589.
56. Mityagin B, Siegl P. Root system of singular perturbations of the harmonic oscillator type operators; 2013, preprint on arXiv:1307.6245.
57. Trefethen LN, Embree M. *Spectra and Pseudospectra*. Princeton (RI): Princeton University Press; 2005.
58. Roch S, Silberman B. C^*-algebra techniques in numerical analysis. *J Oper Theory* 1996;35:241–280.
59. Davies EB. Semi-classical states for non-self-adjoint Schrödinger operators. *Commun Math Phys* 1999;200:35–41.
60. Zworski M. A remark on a paper of E. B. Davies. *Proc Am Math Soc* 2001;129:2955–2957.
61. Dencker N, Sjöstrand J, Zworski M. Pseudospectra of semiclassical (pseudo-) differential operators. *Commun Pure Appl Math* 2004;57:384–415.
62. Bordeaux Montrieux W. Estimation de résolvante et construction de quasimode près du bord du pseudospectre; 2013. arXiv:1301.3102 [math.SP].

63. Henry R. Spectral projections of the complex cubic oscillator. *Ann Henri Poincaré*. Annales Henri Poincare, 2014, 15, 2025–2043.
64. Siegl P, Krejčiřík D. On the metric operator for the imaginary cubic oscillator. *Phys Rev D* 2012;86:121702(R).
65. Davies EB, Kuijlaars ABJ. Spectral asymptotics of the non-self-adjoint harmonic oscillator. *J London Math Soc* 2004;70:420–426.
66. Mityagin B, Siegl P, Viola J. Differential operators admitting various rates of spectral projection growth; 2013, preprint on arXiv:1309.3751.

6

$\mathcal{PT}$-SYMMETRIC OPERATORS IN QUANTUM MECHANICS: KREIN SPACES METHODS

SERGIO ALBEVERIO[1,2,3] AND SERGII KUZHEL[4]

[1] *Institut für Angewandte Mathematik, Universität Bonn, Bonn, Germany*
[2] *CERFIM, Locarno, Switzerland*
[3] *Department of Mathematics and Statistics, King Fahd University of Petroleum and Minerals, Dhahran, Saudi Arabia*
[4] *AGH University of Science and Technology, Kraków, Poland*

6.1 INTRODUCTION

The use of non-self-adjoint operators and indefinite Hilbert space structures in quantum mechanics dates back to the early days of quantum mechanics (1, 2). Nowadays, the steady interest in this subject is growing considerably after it has been discovered numerically (3) and rigorously proved (4) that the spectrum of the Hamiltonian

$$A = -\frac{d^2}{dx^2} + x^2(ix)^{\epsilon}, \qquad 0 \leq \epsilon < 2, \tag{6.1.1}$$

acting in the (complex) $L_2(\mathbb{R})$ space, is real and positive. It was conjectured (3) that the reality of the spectrum of A is a consequence of its $\mathcal{PT}$-symmetry:

$$\mathcal{PT}A = A\mathcal{PT},$$

Non-Selfadjoint Operators in Quantum Physics: Mathematical Aspects, First Edition.
Edited by Fabio Bagarello, Jean-Pierre Gazeau, Franciszek Hugon Szafraniec and Miloslav Znojil.

where the space reflection (parity) operator $\mathcal{P}$ and the complex conjugation operator $\mathcal{T}$ are defined as follows: $(\mathcal{P}f)(x) = f(-x)$ and $(\mathcal{T}f)(x) = \overline{f(x)}$. This gave rise to a consistent complex extension of conventional quantum mechanics into $\mathcal{PT}$ quantum mechanics (PTQM); see, for example, the review papers (5–7) and the references therein.

In general, the Hamiltonians of PTQM are not self-adjoint with respect to the initial Hilbert space's inner product, but they possess a certain "more physical property" of symmetry, which does not depend on the choice of inner product (similarly to the aforementioned $\mathcal{PT}$-symmetry property).

Typically, $\mathcal{PT}$-symmetric Hamiltonians can be interpreted as self-adjoint ones for a suitable choice of Krein space. For instance, the $\mathcal{PT}$-symmetric operator A in (6.1.1) turns out to be self-adjoint with respect to the indefinite metric

$$[f, g] := (\mathcal{P}f, g) = \int_{-\infty}^{\infty} f(-x)\overline{g(x)}dx, \qquad f, g \in L_2(\mathbb{R}).$$

The space $L_2(\mathbb{R})$ with the indefinite metric $[\cdot, \cdot]$ is the Krein space $(L_2(\mathbb{R}), [\cdot, \cdot])$. Hence, A is self-adjoint in the Krein space $(L_2(\mathbb{R}), [\cdot, \cdot])$.

The interpretation of $\mathcal{PT}$-symmetric operators within the Krein spaces framework allows one to apply the Krein spaces' methods for investigation of non-self-adjoint Hamiltonians of PTQM. Recently an essential progress in such a trend has been achieved in Refs (8–10).

However, the self-adjointness of a $\mathcal{PT}$-symmetric Hamiltonian A in a Krein space cannot be completely satisfactory because it does not guarantee the unitarity of the dynamics generated by A. Because the norm of states carries a probabilistic interpretation in the standard quantum theory, the presence of an indefinite metric immediately raises problems of interpretation. This problem can be overcome by finding a new symmetry represented by a linear operator C and such that the sesquilinear form $(\cdot, \cdot)_C := [C\cdot, \cdot]$ is a (positively defined) inner product and A turns out to be self-adjoint with respect to $(\cdot, \cdot)_C$ (11, 12).

The description of a symmetry C for a given $\mathcal{PT}$-symmetric Hamiltonian A is one of the key points in PTQM. Because of the complexity of the problem (as C depends on the particular A one studies), it is not surprising that the majority of the available formulae are still approximative, usually expressed as leading terms of perturbation some series (12–15).

The operator C shows some rough analogy with the charge conjugation operator in quantum field theory that was used in Ref. (16). A generalization from bounded to unbounded C operators was recently discussed in Ref. (17). Another kind of generalized C operator can arise in connection with model classes of interacting relativistic quantum fields with indefinite metrics and satisfying all Morchio-Strocchi axioms; see, for example, (18) (and references therein).

In this chapter, we discuss the general mathematical properties of $\mathcal{PT}$-symmetric operators within the Krein spaces framework, focusing on those aspects of the Krein spaces theory that may be more appealing to mathematical physicists.

We believe that every physically meaningful $\mathcal{PT}$-symmetric operator should be a self-adjoint operator in a suitably chosen Krein space and a proper investigation of a $\mathcal{PT}$-symmetric Hamiltonian A involves the following stages: interpretation of A as a self-adjoint operator in a Krein space $(\mathfrak{H}, [\cdot, \cdot])$; construction of an operator C for A; interpretation of A as a self-adjoint operator in the Hilbert space $(\mathfrak{H}, (\cdot, \cdot)_C)$.

We study the problems mentioned earlier for $\mathcal{PT}$-symmetric operators in some abstract setting. This is a natural problem because various Hamiltonians of PTQM have the property of $\mathcal{PT}$-symmetry realized by different operators $\mathcal{P}$ and $\mathcal{T}$.

In writing this chapter, we try to show how and in what manner the methods of the Krein spaces theory can be applied to the investigation of $\mathcal{PT}$-symmetric operators. One of our goals is to help bridge the gap between the growing community of physicists working in PTQM with the community of mathematicians who study self-adjoint operators in Krein spaces for their own sake.

We adopt the customary notation used in the mathematical literature. For instance, our inner products are linear in the first argument and we use $\overline{x}$ instead of x^* to denote complex conjugation. The adjoint of A is denoted by A^* instead of $A^\dagger$.

In what follows, $\mathcal{D}(A)$ and $\ker A$ denote, respectively, the domain and the kernel space of a linear operator A. The symbol $A \upharpoonright_{\mathcal{D}}$ means the restriction of A to a set $\mathcal{D}$. The resolvent set and the spectrum of A are denoted by $\rho(A)$ and $\sigma(A)$, respectively.

6.2 ELEMENTS OF THE KREIN SPACES THEORY

Here all necessary results of Krein spaces theory are presented in a form convenient for our exposition. Their proofs and detailed analysis can be found, for example, in Refs (19–21).

6.2.1 Definition of the Krein Spaces

Let $\mathfrak{H}$ be a complex Hilbert space with inner product $(\cdot, \cdot)$ linear in the first argument. A linear operator J is called *fundamental symmetry* if

$$(i) \quad J^2 = I, \qquad (ii) \quad (Jf, Jg) = (f, g) \tag{6.2.1}$$

for all $f, g \in \mathfrak{H}$. It follows from (6.2.1) that the operator J is self-adjoint, that is, $J = J^*$, where J^* means the adjoint operator of J with respect to $(\cdot, \cdot)$. An equivalent definition of fundamental symmetry is: an operator J is a fundamental symmetry if J is a bounded self-adjoint operator in $\mathfrak{H}$ and $J^2 = I$.

Let J be a nontrivial fundamental symmetry, that is, $J \neq \pm I$. The Hilbert space $\mathfrak{H}$ equipped with the indefinite inner product

$$[f, g] := (Jf, g), \qquad f, g \in \mathfrak{H} \tag{6.2.2}$$

is called a Krein space $(\mathfrak{H}, [\cdot, \cdot])$.

The Hermitian sesquilinear form (6.2.2) depends on the choice of fundamental symmetry J. For this reason, Krein spaces are called J-spaces.

The principal difference between the initial inner product $(\cdot,\cdot)$ and the indefinite inner product $[\cdot,\cdot]$ is that there exist nonzero elements $f \in \mathfrak{H}$ such that $[f,f]=0$ or $[f,f]<0$. An element $f \neq 0$ is called *neutral*, *positive* or *negative* if $[f,f]=0$, $[f,f]>0$ or $[f,f]<0$, respectively.

A (closed) subspace $\mathfrak{L}$ of the Hilbert space $\mathfrak{H}$ is called *neutral*, *positive* or *negative* if all nonzero elements $f \in \mathfrak{L}$ are, respectively, neutral, positive or negative.

A positive (negative) subspace $\mathfrak{L}$ is called *uniformly positive* (*uniformly negative*) if there exists $\alpha > 0$ such that

$$[f,f] \geq \alpha(f,f) \qquad (-[f,f] \geq \alpha(f,f)) \quad \forall f \in \mathfrak{L}.$$

A positive (closed) subspace $\mathfrak{L}$ of the Hilbert space $\mathfrak{H}$ is called *maximal* if $\mathfrak{L}$ is not a proper subspace of a positive subspace in $\mathfrak{H}$. The maximality of a (negative, neutral, uniformly positive, uniformly negative) closed subspace is defined similarly.

Let $\mathfrak{L}$ be a subset of $\mathfrak{H}$. Then its orthogonal complement with respect to the indefinite inner product[1] (6.2.2)

$$\mathfrak{L}^{[\perp]} = \{f \in \mathfrak{H} \ : \ [f,g]=0, \ \forall g \in \mathfrak{L}\}$$

is a linear closed (in the original topology of $\mathfrak{H}$) subspace of the Hilbert space $\mathfrak{H}$.

A subspace $\mathfrak{L}$ of $\mathfrak{H}$ is called *hypermaximal neutral* if $\mathfrak{L} = \mathfrak{L}^{[\perp]}$.

Let $\mathfrak{L}$ be a hypermaximal neutral subspace. Then $J\mathfrak{L}$ is also hypermaximal neutral and the Hilbert space $\mathfrak{H}$ is decomposed as

$$\mathfrak{H} = \mathfrak{L} \oplus J\mathfrak{L},$$

where $\oplus$ means the orthogonality in the Hilbert space $\mathfrak{H}$.

Let $\mathfrak{L}_+$ be a *maximal positive subspace* of[2] $\mathfrak{H}$. Then its J-orthogonal complement

$$\mathfrak{L}_- = \mathfrak{L}_+^{[\perp]} = \{f \in \mathfrak{H} \ : \ [f,g]=0, \ \forall g \in \mathfrak{L}_+\}$$

is a maximal negative subspace of $\mathfrak{H}$, and the direct sum

$$\mathfrak{D} = \mathfrak{L}_+[\dot{+}]\mathfrak{L}_- \tag{6.2.3}$$

does not coincide with $\mathfrak{H}$. The linear set $\mathfrak{D}$ is *dense* in the Hilbert space $\mathfrak{H}$. The symbol $[\dot{+}]$ in (6.2.3) indicates that elements of the subspaces $\mathfrak{L}_+$ and $\mathfrak{L}_-$ are orthogonal with respect to the indefinite inner product $[\cdot,\cdot]$ (or, J-orthogonal).

[1]or, briefly, J-orthogonal complement

[2]we suppose that $\mathfrak{L}_+$ is not uniformly positive

Let $\mathfrak{L}_+$ be a maximal *uniformly positive* subspace of $\mathfrak{H}$. Then its J-orthogonal complement is a maximal uniformly negative subspace and

$$\mathfrak{H} = \mathfrak{L}_+[\dot{+}]\mathfrak{L}_-. \tag{6.2.4}$$

The decomposition (6.2.4) is called *the fundamental decomposition* of $\mathfrak{H}$.

The simplest example of fundamental decomposition can be constructed directly using the operator of fundamental symmetry J. Namely, let

$$\mathfrak{H}_+ = \frac{1}{2}(I+J)\mathfrak{H}, \qquad \mathfrak{H}_- = \frac{1}{2}(I-J)\mathfrak{H}. \tag{6.2.5}$$

It is easy to see that the subspaces $\mathfrak{H}_\pm$ are maximal uniformly positive/negative with respect to $[\cdot,\cdot]$, they are J-orthogonal, and

$$\mathfrak{H} = \mathfrak{H}_+[\dot{+}]\mathfrak{H}_-. \tag{6.2.6}$$

In this particular case, the subspaces $\mathfrak{H}_\pm$ are also orthogonal with respect to the initial inner product $(\cdot,\cdot)$.

REMARK 1 *The fundamental decomposition (6.2.4) is often used for (an equivalent) definition of Krein spaces. Precisely, let $\mathfrak{H}$ be a complex linear space with a Hermitian sesquilinear form $[\cdot,\cdot]$ (i.e., a mapping $[\cdot,\cdot] : \mathfrak{H}\times\mathfrak{H}\to\mathbb{C}$ such that $[\alpha_1 f_1 + \alpha_2 f_2, g] = \alpha_1[f_1,g]+\alpha_2[f_2,g]$ and $[f,g]=\overline{[g,f]}$ for all $f_1,f_2,f,g\in\mathfrak{H}$, $\alpha_1,\alpha_2\in\mathbb{C}$). Then $(\mathfrak{H},[\cdot,\cdot])$ is called a Krein space if $\mathfrak{H}$ admits a decomposition (6.2.4) such that the linear sets $\mathfrak{L}_\pm$ are orthogonal with respect to $[\cdot,\cdot]$ and these linear sets $\mathfrak{L}_+$ and $\mathfrak{L}_-$ endowed, respectively, with the sesquilinear forms $[\cdot,\cdot]$ and $-[\cdot,\cdot]$ are Hilbert spaces.*

The fundamental decomposition (6.2.4) is not defined uniquely. However, the dimension of the positive (negative) part $\mathfrak{L}_+$ ($\mathfrak{L}_-$) in (6.2.4) does not depend on the choice of $\mathfrak{L}_\pm$. If, in particular, $\kappa = \min\{\dim\mathfrak{L}_-, \ \dim\mathfrak{L}_+\} < \infty$, then $\mathfrak{H}$ is called a π_κ-space or *a Pontryagin space of index κ*.

The indefinite inner product in a Pontryagin space can be interpreted as a "finite-dimensional" perturbation of the initial inner product. In this situation one has a simplification of the general theory of Krein spaces. In particular, a maximal definite (i.e., positive or negative) subspace $\mathfrak{L}$ in a Pontryagin space is always uniformly definite and the direct sum (6.2.3) is impossible.

6.2.2 Bounded Operators C

The existence of various fundamental decompositions (6.2.4) of $\mathfrak{H}$ into J-orthogonal direct sums of positive and negative subspaces is one of the principal features of the Krein spaces theory.

Let a fundamental decomposition (6.2.4) be given. Then an arbitrary element $f \in \mathfrak{H}$ admits the decomposition

$$f = f_{\mathfrak{L}_+} + f_{\mathfrak{L}_-}, \qquad f_{\mathfrak{L}_\pm} \in \mathfrak{L}_\pm \tag{6.2.7}$$

and we can define an operator C via

$$Cf = C(f_{\mathfrak{L}_+} + f_{\mathfrak{L}_-}) = f_{\mathfrak{L}_+} - f_{\mathfrak{L}_-}. \tag{6.2.8}$$

The bounded operator C is defined on the whole space $\mathfrak{H}$ and the subspaces $\mathfrak{L}_\pm$ are recovered by C

$$\mathfrak{L}_+ = \frac{1}{2}(I + C), \qquad \mathfrak{L}_- = \frac{1}{2}(I - C). \tag{6.2.9}$$

Denote $(\cdot, \cdot)_C = [C\cdot, \cdot]$. Then,

$$(f,f)_C = [f_{\mathfrak{L}_+}, f_{\mathfrak{L}_+}] - [f_{\mathfrak{L}_-}, f_{\mathfrak{L}_-}], \qquad f \in \mathfrak{H},$$

where the J-orthogonal components $f_{\mathfrak{L}_\pm}$ are taken from (6.2.7). The sesquilinear form $(\cdot, \cdot)_C$ is an inner product on $\mathfrak{H}$, which is equivalent to the initial inner product $(\cdot, \cdot)$ as

$$(f_{\mathfrak{L}_+}, f_{\mathfrak{L}_+}) \geq [f_{\mathfrak{L}_+}, f_{\mathfrak{L}_+}] \geq \alpha(f_{\mathfrak{L}_+}, f_{\mathfrak{L}_+}),$$

$$(f_{\mathfrak{L}_-}, f_{\mathfrak{L}_-}) \geq -[f_{\mathfrak{L}_-}, f_{\mathfrak{L}_-}] \geq \alpha(f_{\mathfrak{L}_-}, f_{\mathfrak{L}_-}),$$

for some $\alpha > 0$, and for all elements $f_{\mathfrak{L}_\pm}$ of the maximal uniformly positive/negative subspaces $\mathfrak{L}_\pm$.

The operator C is a self-adjoint operator in the Hilbert space $(\mathfrak{H}, (\cdot, \cdot)_C)$. Moreover, $C^2 = I$ due to (6.2.8). Hence, one can view C as a fundamental symmetry of the Krein space $(\mathfrak{H}, [\cdot, \cdot])$ with an underlying Hilbert space $(\mathfrak{H}, (\cdot, \cdot)_C)$.

In view of (6.2.2) and (6.2.8), one has $(\cdot, \cdot)_C = [C\cdot, \cdot] = (JC\cdot, \cdot)$. Hence, JC is a positive bounded self-adjoint operator in the Hilbert space $\mathfrak{H}$ and we can write it as $JC := e^Q$, where Q is a bounded self-adjoint operator in $\mathfrak{H}$. Thus $C = Je^Q$. Then the condition $C^2 = I$ takes the form $Je^QJe^Q = I$ or $Je^Q = e^{-Q}J$. The obtained relation is equivalent to the anticommutation condition $JQ = -QJ$.

Theorem 6.2.1 *There is a one-to-one correspondence between the set of all fundamental decompositions (6.2.4) of the Krein space* $(\mathfrak{H}, [\cdot, \cdot])$ *and the set of all bounded operators of the form* $C = Je^Q$*, where* Q *is a bounded self-adjoint operator in* $\mathfrak{H}$*, which anticommutes with* J*.*

Proof: The implication: fundamental decomposition (6.2.4) $\to$ $C = Je^Q$ was established earlier.

Conversely, let $C = Je^Q$, where the bounded self-adjoint operator Q anticommutes with J. The operator e^Q is positive self-adjoint. This means that the operator

$$T = (I - e^Q)(I + e^Q)^{-1} =$$

$$-\frac{e^{Q/2} - e^{-Q/2}}{2}\left(\frac{e^{Q/2} + e^{-Q/2}}{2}\right)^{-1} = -\frac{\sinh(Q/2)}{\cosh(Q/2)} = -\tanh\frac{Q}{2}$$

is a self-adjoint strong contraction (i.e., $T^* = T$, $\|T\| < 1$) defined on $\mathfrak{H}$. Moreover, T anticommutes with J as $JQ = -QJ$.

Denote

$$\mathfrak{L}_+ = (I + T)\mathfrak{H}_+, \qquad \mathfrak{L}_- = (I + T)\mathfrak{H}_-, \tag{6.2.10}$$

where $\mathfrak{H}_\pm$ are defined in (6.2.5). In some sense, the operator T indicates "a deviation" of subspaces $\mathfrak{L}_\pm$ from $\mathfrak{H}_\pm$. In view of (6.2.10), elements $f_{\mathfrak{L}_\pm} \in \mathfrak{L}_\pm$ have the form

$$f_{\mathfrak{L}_+} = \gamma_+ + T\gamma_+, \qquad f_{\mathfrak{L}_-} = \gamma_- + T\gamma_-, \qquad \gamma_\pm \in \mathfrak{H}_\pm, \tag{6.2.11}$$

where $T\gamma_\pm \in \mathfrak{H}_\mp$ (as T anticommutes with J). We remark also that $J\gamma_\pm = \pm\gamma_\pm$ and $JT\gamma_\pm = \mp\gamma_\pm$ due to the definition of the subspaces $\mathfrak{H}_\pm$ in (6.2.5). Therefore,

$$\begin{aligned}[f_{\mathfrak{L}_+}, f_{\mathfrak{L}_+}] &= (\gamma_+ - T\gamma_+, \gamma_+ + T\gamma_+)\\ &= (\gamma_+, \gamma_+) - (T\gamma_+, T\gamma_+) \geq (1 - \|T\|^2)(\gamma_+, \gamma_+),\end{aligned}$$

On the other hand,

$$\begin{aligned}(f_{\mathfrak{L}_+}, f_{\mathfrak{L}_+}) &= (\gamma_+ + T\gamma_+, \gamma_+ + T\gamma_+)\\ &= (\gamma_+, \gamma_+) + (T\gamma_+, T\gamma_+) \leq (1 + \|T\|^2)(\gamma_+, \gamma_+).\end{aligned}$$

Hence

$$[f_{\mathfrak{L}_+}, f_{\mathfrak{L}_+}] \geq \frac{1 - \|T\|^2}{1 + \|T\|^2}(f_{\mathfrak{L}_+}, f_{\mathfrak{L}_+})$$

and the subspace $\mathfrak{L}_+$ is uniformly positive. Similarly, we establish that $\mathfrak{L}_-$ is uniformly negative. The maximality of $\mathfrak{L}_\pm$ in the classes of uniformly positive/negative subspaces follows from (6.2.10) and Ref. (19).

The subspaces $\mathfrak{L}_\pm$ are J-orthogonal. Indeed, using (6.2.11) and taking into account that the subspaces $\mathfrak{H}_\pm$ are orthogonal with respect to the initial inner product, we obtain

$$[f_{\mathfrak{L}_+}, f_{\mathfrak{L}_-}] = (\gamma_+ - T\gamma_+, \gamma_- + T\gamma_-) = (\gamma_+, T\gamma_-) - (T\gamma_+, \gamma_-) = 0.$$

Summing up, starting with $C = Je^Q$, we determine the maximal uniformly positive/negative subspaces $\mathfrak{L}_\pm$ in the J-orthogonal direct sum (6.2.4). To complete the proof it suffices to show that the obtained decomposition (6.2.4) determines exactly the same operator $C = Je^{Q}$ by formulas (6.2.8). To do that, we show that the projection operators $P_{\mathfrak{L}_{\pm}}$ onto $\mathfrak{L}_{\pm}$ in $\mathfrak{H}$ are determined by the formulas

$$P_{\mathfrak{L}_{+}} = (I-T)^{-1}(P_{+} - TP_{-}), \qquad P_{\mathfrak{L}_{-}} = (I-T)^{-1}(P_{-} - TP_{+}), \tag{6.2.12}$$

where $P_{+} = \frac{1}{2}(I+J)$ and $P_{-} = \frac{1}{2}(I-J)$ are the projections on $\mathfrak{H}_{+}$ and $\mathfrak{H}_{-}$, respectively. First, we note that $P_{+}T = TP_{-}$ as $TJ = -JT$. Then, for any element $\phi = \gamma_{+} + \gamma_{-}$ $(\gamma_{\pm} \in \mathfrak{H}_{\pm})$ from $\mathfrak{H}$,

$$\begin{aligned} P_{\mathfrak{L}_{+}}(I+T)\phi &= (I+T)\gamma_{+} = (I+T)P_{+}\phi = (I-T)^{-1}(I-T^2)P_{+}\phi \\ &= (I-T)^{-1}P_{+}(I-T^2)\phi = (I-T)^{-1}(P_{+} - TP_{-})(I+T)\phi, \end{aligned}$$

which establishes the first formula in (6.2.12) because $(I+T)\mathfrak{H} = \mathfrak{H}$, due to (6.2.4) and (6.2.10). The second formula is proved by similar arguments.

The operator C determined by (6.2.4) is:

$$\begin{aligned} C &= P_{\mathfrak{L}_{+}} - P_{\mathfrak{L}_{-}} \\ &= J(I+T)^{-1}(I-T) = J\left(I - \tanh\frac{Q}{2}\right)^{-1}\left(I + \tanh\frac{Q}{2}\right) = Je^{Q}. \end{aligned}$$

Theorem 6.2.1 is proved. ∎

Theorem 6.2.1 shows that a given indefinite inner product $[\cdot,\cdot]$ generates infinitely many equivalent inner products $(\cdot,\cdot)_C$, which are parameterized by operators C. In fact we do not need to fix an inner product and an operator of fundamental symmetry in order to define a Krein space. It is sufficient to fix some fundamental decomposition (6.2.4) of $\mathfrak{H}$ and to define the corresponding operator C by (6.2.8). Then $[C\cdot,\cdot]$ can be used as an initial inner product $(\cdot,\cdot)_C$ and C turns out to be the operator of fundamental symmetry in the Hilbert space $(\mathfrak{H}, (\cdot,\cdot)_C)$. The indefinite metric $[\cdot,\cdot]$ is then determined by the inner product $(\cdot,\cdot)_C$ and the operator C, that is, $[\cdot,\cdot] := (C\cdot,\cdot)_C = [C^2\cdot,\cdot] = [\cdot,\cdot]$.

6.2.3 Unbounded Operators C

The operator C can also be determined by means of the direct sum (6.2.3). Indeed, let $\mathfrak{L}_{+}$ be a maximal positive subspace (but not uniformly positive). Then (6.2.3) determines a dense linear set $\mathfrak{D}$ in $\mathfrak{H}$. Every element $f \in \mathfrak{D}$ has the decomposition (6.2.7) and we can define C via (6.2.8). The domain of definition of C coincides with $\mathfrak{D}$ and the subspaces $\mathfrak{L}_{\pm}$ in (6.2.3) are recovered by (6.2.9).

Lemma 6.2.2 *The operator C defined on $\mathfrak{D}$ is a closed operator in the Hilbert space $\mathfrak{H}$.*

Proof: Let the sequences of elements $\{f_n\}$ ($f_n \in \mathfrak{D}$) and $\{Cf_n\}$ tend, respectively, to f and g, when $n \to \infty$. Then

$$(I + C)f_n = 2f^n_{\mathfrak{L}_+} \to f + g, \qquad (I - C)f_n = 2f^n_{\mathfrak{L}_-} \to f - g, \tag{6.2.13}$$

where $f_n = f^n_{\mathfrak{L}_+} + f^n_{\mathfrak{L}_-}$ are decomposed with respect to (6.2.3). Hence $f + g \in \mathfrak{L}_+$ and $f - g \in \mathfrak{L}_-$ as $\mathfrak{L}_\pm$ are closed subspaces in $\mathfrak{H}$. This means that the elements f, g belong to $\mathfrak{D}$ and they have the representations $f = f_{\mathfrak{L}_+} + f_{\mathfrak{L}_-}$, $g = g_{\mathfrak{L}_+} + g_{\mathfrak{L}_-}$ with respect to (6.2.3). Using (6.2.13) again, we obtain

$$2f^n_{\mathfrak{L}_+} \to (f_{\mathfrak{L}_+} + g_{\mathfrak{L}_+} + f_{\mathfrak{L}_-} + g_{\mathfrak{L}_-}) \in \mathfrak{L}_+, \qquad 2f^n_{\mathfrak{L}_-} \to (f_{\mathfrak{L}_+} - g_{\mathfrak{L}_+} + f_{\mathfrak{L}_-} - g_{\mathfrak{L}_-}) \in \mathfrak{L}_-$$

that is possible when $f_{\mathfrak{L}_-} + g_{\mathfrak{L}_-} = 0$ and $f_{\mathfrak{L}_+} - g_{\mathfrak{L}_+} = 0$.

Summing up, the element f belongs to $\mathfrak{D}$ (the domain of definition of C) and

$$Cf = C(f_{\mathfrak{L}_+} + f_{\mathfrak{L}_-}) = g = g_{\mathfrak{L}_+} + g_{\mathfrak{L}_-} = f_{\mathfrak{L}_+} - f_{\mathfrak{L}_-}.$$

Therefore, C is a closed operator in $\mathfrak{H}$. ∎

The operator C corresponding to (6.2.3) is an unbounded operator in $\mathfrak{H}$. Indeed, if C was bounded then it could be extended by continuity to the whole space $\mathfrak{H}$. Namely, if $f \in \mathfrak{H}$, then there would exist a sequence of $f_n \in \mathfrak{D}$ such that $f_n \to f$ (as $\mathfrak{D}$ is a densely defined set in $\mathfrak{H}$) and $Cf = \lim_{n\to\infty} Cf_n$. Hence $f \in \mathfrak{D}$ (as C is a closed operator on $\mathfrak{D}$). Therefore, $\mathfrak{D} = \mathfrak{H}$, which is impossible. Thus C is unbounded.

The spectral properties of operators C corresponding to (6.2.3) or (6.2.4) are also different. The operators C determined by fundamental decompositions (6.2.4) have only point spectrum that consists of two eigenvalues $-1, 1$. The corresponding eigensubspaces are $\mathfrak{L}_-$ and $\mathfrak{L}_+$, respectively. If C is determined by the direct sum (6.2.3), then a continuous spectrum arises additionally and it coincides with $\mathbb{C} \setminus \{-1, 1\}$.

Theorem 6.2.3 (cf. Theorem 6.2.1) *There is a one-to-one correspondence between the set of all direct sums (6.2.3) of the Krein space $(\mathfrak{H}, [\cdot, \cdot])$ and the set of all unbounded operators $C = Je^Q$, where Q is an unbounded self-adjoint operator in $\mathfrak{H}$, which anticommutes with J.*

Proof: Let C be determined by the direct J-orthogonal sum (6.2.3). The subspaces $\mathfrak{L}_\pm$ in (6.2.3) are maximal definite and hence, they are determined by (6.2.10) where T is a self-adjoint contraction anticommuting with J and satisfying the inequality (19, 22)

$$\|Tf\| < \|f\|, \qquad f(\neq 0) \in \mathfrak{H}. \tag{6.2.14}$$

The condition (6.2.14) is weaker than the condition $\|T\| < 1$ characterizing J-orthogonal maximal uniformly definite subspaces, however it also guarantees the existence of $(I \pm T)^{-1}$. Repeating the proof of Theorem 6.2.1 we obtain that the projection operators $P_{\mathfrak{L}_\pm}$ onto $\mathfrak{L}_\pm$ in $\mathfrak{D}$ are determined by (6.2.12). Hence,

$$Cf = (P_{\mathfrak{L}_+} - P_{\mathfrak{L}_-})f = J(I+T)^{-1}(I-T)f, \qquad f \in \mathfrak{D}. \tag{6.2.15}$$

The spectrum of T is contained in the segment $\mathrm{I} = [-1, 1]$ and ± 1 cannot be eigenvalues of T due to (6.2.14). In this case

$$(I+T)^{-1}(I-T) = e^{Q}, \quad \text{where} \quad Q = s(T) \text{ and } s(\lambda) = \ln \frac{1-\lambda}{1+\lambda}, \quad \lambda \in \mathrm{I}$$

is an unbounded self-adjoint operator in $\mathfrak{H}$. As $JT = -TJ$, the projection valued measure E_δ associated with T satisfies the relation $JE_\delta = E_{-\delta}J$ for an arbitrary Borel set δ in I (23). Using this relation and taking into account that $s(\lambda) = \ln \frac{1-\lambda}{1+\lambda}$ is an odd function on I we obtain

$$JQ = J \int_{\mathrm{I}} s(\lambda) dE_\lambda = \int_{\mathrm{I}} s(\lambda) dE_{-\lambda} J = -QJ.$$

Conversely, let $C = Je^{Q}$, where the unbounded self-adjoint operator Q anticommutes with J. The operator e^{Q} is positive self-adjoint. This means that the operator $T = (I - e^{Q})(I + e^{Q})^{-1}$ is a self-adjoint contraction, which satisfies (6.2.14) and anticommutes with J.

In that case, the subspaces $\mathfrak{L}_\pm$ defined by (6.2.10) are maximal definite. Similarly to the proof of Theorem 6.2.1, we establish the J-orthogonality of $\mathfrak{L}_\pm$. Then the direct sum (6.2.3) gives rise to an operator C'. Using again the proof of Theorem 6.2.1 we arrive at the conclusion that $C' = Je^{Q}$. Theorem 6.2.3 is proved. ■

Theorem 6.2.3 shows that the J-orthogonal direct sums (6.2.3) give rise to infinitely many sesquilinear forms

$$(\cdot, \cdot)_C = [C\cdot, \cdot] = (e^{Q}\cdot, \cdot)$$

which are inner products on $\mathfrak{D}$. The spaces $(\mathfrak{D}, (\cdot, \cdot)_C)$ are pre-Hilbert spaces because $(\cdot, \cdot)_C$ are not equivalent to the initial inner product $(\cdot, \cdot)$ and the linear sets $\mathfrak{D}$ are not closed with respect to $(\cdot, \cdot)_C$. Denote by $\mathfrak{H}_C$ the completion of $\mathfrak{D}$ with respect to $(\cdot, \cdot)_C$. Then we can say that the J-orthogonal direct sums (6.2.3) generate infinitely many Hilbert spaces $(\mathfrak{H}_C, (\cdot, \cdot)_C)$, which do not coincide[3] with the initial Hilbert space $\mathfrak{H}$.

[3]do not coincide as linear sets

6.2.3.1 Reasons for Nonuniqueness of the Unbounded Operators C Let $\mathfrak{L}'_+$ and $\mathfrak{L}'_-$ be J-orthogonal positive and negative subspaces. It may happen that the direct J-orthogonal sum

$$\mathfrak{D}' = \mathfrak{L}'_+[\dot{+}]\mathfrak{L}'_- \tag{6.2.16}$$

remains a dense set in $\mathfrak{H}$ without the assumption of maximality of $\mathfrak{L}'_\pm$. This phenomenon was first observed by Langer (24). His paper provides a mathematically rigorous explanation based on the fact that the indefinite inner product $[\cdot, \cdot]$ is singular with respect to the initial inner product $(\cdot, \cdot)$.

If at least one of the definite subspaces $\mathfrak{L}'_\pm$ is not maximal but the direct sum $\mathfrak{D}'$ is a dense set, then the formula (6.2.8) allows one to define a densely defined operator C' such that the restriction of C' onto $\mathfrak{L}'_\pm$ coincides with $\pm$ identity operator. The relation $C'^2 = I$ is obvious. Repeating the proof of Lemma 6.2.2 we conclude that the operator C' is a closed operator on $\mathfrak{D}'$.

The positive $\mathfrak{L}'_+$ and negative $\mathfrak{L}'_-$ subspaces in (6.2.16) can be extended to a J-orthogonal direct sum (6.2.3) of positive $\mathfrak{L}_+$ and maximal negative $\mathfrak{L}_-$ subspaces in different ways that gives different extensions of C' to the operators $C = Je^Q$ described in Theorem 6.2.3. From this point of view, the J-orthogonal direct sum (6.2.16) does not determine the operator $C = Je^Q$ uniquely.

It follows from (6.2.16) and definition of C' that

$$(JC'f, g) = [C'f, g] = [f_{\mathfrak{L}'_+}, g_{\mathfrak{L}'_+}] - [f_{\mathfrak{L}'_-}, g_{\mathfrak{L}'_-}] = [f, C'g] = (f, JC'g)$$

for all elements $f = f_{\mathfrak{L}'_+} + f_{\mathfrak{L}'_-}, g = g_{\mathfrak{L}'_+} + g_{\mathfrak{L}'_-}$ from $\mathfrak{D}'$. Therefore, the operator $G' = JC'$ is a positive symmetric operator on $\mathfrak{D}$ (the positivity of G' follows from the fact that $(G'f,f) = [f_{\mathfrak{L}'_+}, f_{\mathfrak{L}'_+}] - [f_{\mathfrak{L}'_-}, f_{\mathfrak{L}'_-}] > 0$ for nonzero $f \in \mathfrak{D}'$). Moreover, $JG'JG'f = f$ for all $f \in \mathfrak{D}'$ as $C'^2 = I$ on $\mathfrak{D}'$.

Let G be a positive self-adjoint extension of G' with the additional condition $JGJG = I$ on the domain of definition of G. The positivity of G yields that $G = e^Q$, where Q is a self-adjoint operator in $\mathfrak{H}$. Then the condition $JGJG = I$ takes the form $Je^Q = e^{-Q}J$.

Denote (cf. the proof of Theorem 6.2.1)

$$T = (I - e^Q)(I + e^Q)^{-1} = -\tanh\frac{Q}{2}.$$

The operator T is a self-adjont contraction satisfying the inequality (6.2.14). Taking the relation $Je^Q = e^{-Q}J$ into account, we obtain

$$JT = (I - e^{-Q})(I + e^{-Q})^{-1}J = (e^{-Q} - I)(e^{-Q} + I)^{-1}J = -TJ.$$

This means that Q also anticommutes with J and the maximal definite subspaces $\mathfrak{L}_\pm$ defined by (6.2.10) constitute the extension of (6.2.16) to the J-orthogonal direct sum (6.2.3), which is determined up the choice of a positive self-adjoint extension G.

Summing up, *the lack of maximality of* $\mathfrak{L}'_{\pm}$ *in (6.2.16) means that the corresponding operator* C' *satisfies the condition* $C'^2 = I$ *but it cannot be presented as* Je^{Q}, *where* e^{Q} *is a positive self-adjoint operator. The desired presentation can be obtained via the extension of* $\mathfrak{L}'_{\pm}$ *to* J*-orthogonal maximal definite subspaces* $\mathfrak{L}_{\pm}$ *or, that is equivalent, via the extension of a positive symmetric operator* $G' = JC'$ *to a positive self-adjoint operator* $G = JC$ *in* $\mathfrak{H}$ *with the "boundary condition"* $JGJG = I$ *on* $\mathcal{D}(G)$. *The different choices of extensions* G *give rise to different operators* $C = JG$, *which are extensions of the initial operator* $C' = JG'$ *determined by (6.2.16).*

In conclusion, we note that the condition of positivity/negativity imposed on $\mathfrak{L}_{\pm}$ in (6.2.16) is essential. If, for instance, $\mathfrak{L}_{+}$ is a nonnegative subspace (i.e., there exists a nonzero $f \in \mathfrak{L}_{+}$ such that $[f,f] = 0$), then the linear set $\mathfrak{D}'$ cannot be densely defined (19).

6.3 SELF-ADJOINT OPERATORS IN KREIN SPACES

6.3.1 Definitions and General Properties

Let A be a linear densely defined operator acting in a Krein space $(\mathfrak{H}, [\cdot, \cdot])$. Repeating the standard definition of the adjoint operator[4] with the use of an indefinite inner product $[\cdot, \cdot]$ instead of an inner product $(\cdot, \cdot)$ we define the adjoint operator A^{+} of A in the Krein space $(\mathfrak{H}, [\cdot, \cdot])$. In that case

$$[Af, g] = [f, A^{+}g], \qquad f \in \mathcal{D}(A), \quad g \in \mathcal{D}(A^{+}). \tag{6.3.1}$$

Let A^{*} be the adjoint operator of A with respect to the inner product $(\cdot, \cdot)$. The relationship between A^{+} and A^{*}:

$$JA^{+} = A^{*}J \tag{6.3.2}$$

follows directly from (6.2.2) and (6.3.1).

REMARK 2 *In what follows we will often use operator identities like*

$$XA = BX, \tag{6.3.3}$$

where A *and* B *are (possible) unbounded operators in a Hilbert space* $\mathfrak{H}$ *and* X *is a bounded operator in* $\mathfrak{H}$. *In that case, we* always assume that (6.3.3) holds on $\mathcal{D}(A)$. *This means that* $X : \mathcal{D}(A) \to \mathcal{D}(B)$ *and the identity* $XAu = BXu$ *holds for all* $u \in \mathcal{D}(A)$. *If* A *is bounded, then (6.3.3) should hold on the whole* $\mathfrak{H}$.

An operator A is called *self-adjoint* in the Krein space $(\mathfrak{H}, [\cdot, \cdot])$ if $A^{+} = A$. Due to (6.3.2), the condition of self-adjointness is equivalent to the relation

$$JA = A^{*}J. \tag{6.3.4}$$

[4] see, for example, [25, Section 39]

An operator A satisfying (6.3.4) is called *J-self-adjoint*. Obviously, J-self-adjointness of A means that A is self-adjoint in the Krein space $(\mathfrak{H}, [\cdot,\cdot])$.

In general, J-self-adjoint operators A are non-self-adjoint in the Hilbert space $(\mathfrak{H}, [J\cdot,\cdot])$ and their spectra $\sigma(A)$ are only symmetric with respect to the real axis:

$$\mu \in \sigma(A) \quad \text{if and only if} \quad \overline{\mu} \in \sigma(A). \tag{6.3.5}$$

Moreover, the situation where $\sigma(A) = \mathbb{C}$ (i.e., A has the empty resolvent set $\rho(A) = \emptyset$) is also possible and it may indicate a special structure of A (26).

It is simple to construct J-self-adjoint operators with empty resolvent set. For instance, let $\mathcal{K}$ be a Hilbert space and let L be a symmetric (non-self-adjoint) operator in $\mathcal{K}$. Consider the operators

$$A := \begin{pmatrix} L & 0 \\ 0 & L^* \end{pmatrix}, \qquad J = \begin{pmatrix} 0 & I \\ I & 0 \end{pmatrix} \tag{6.3.6}$$

in the Hilbert space $\mathfrak{H} = \mathcal{K} \oplus \mathcal{K}$. Then J is a fundamental symmetry in $\mathfrak{H}$ and A is a J-self-adjoint operator. As $\rho(L) = \emptyset$, it is clear that $\rho(A) = \emptyset$.

An operator A acting in the Krein space $(\mathfrak{H}, [\cdot,\cdot])$ is called *J-positive* (*J-nonnegative*) if

$$[Af,f] > 0, \quad \forall f \in \mathcal{D}(A) \setminus \{0\} \qquad ([Af,f] \geq 0, \quad \forall f \in \mathcal{D}(A)).$$

If A is a J-self-adjoint and a J-nonnegative operator then A has no residual spectrum ($\sigma_r(A) = \emptyset$). Moreover, if $\sigma(A) \neq \mathbb{C}$, then A has a real spectrum. The "marginal case" $\sigma(A) = \mathbb{C}$ can be illustrated by an unbounded operator C.

Lemma 6.3.1 *If C is determined by the direct sum (6.2.3), then C is a J-self-adjoint, J-positive operator in the Krein space $(\mathfrak{H}, [\cdot,\cdot])$ and $\sigma(C) = \mathbb{C}$.*

Proof: The sesquilinear form $[Cf,f]$ $(f \in \mathfrak{D})$ coincides with the inner product $(f,f)_C$ defined in Section 6.2.3. Therefore, C is a J-positive operator. By Theorem 6.2.3, $C = Je^Q$. Then $JA = e^Q$ and $A^*J = e^Q JJ = e^Q$. Therefore, C is a J-self-adjoint operator. The relation $\sigma(C) = \mathbb{C}$ was established in Section 6.2.3. Lemma 6.3.1 is proved. ∎

If C' is defined by the J-orthogonal direct sum (6.2.16), where the J-orthogonal definite subspaces $\mathfrak{L}'_\pm$ are not assumed to be maximal, then the operator C' remains J-positive. However, we lose the J-self-adjointness of C'.

A self-adjoint operator A in the Krein space $(\mathfrak{H}, [\cdot,\cdot])$ is called definitizable if $\rho(A) \neq \emptyset$ and there exists a real polynomial $p \neq 0$ such that $[p(A)f,f] \geq 0$ for all $f \in \mathcal{D}(A^k)$, where k is the degree of p (21).

The nice feature of the definitizable operators is that they "are not too far' from the class of self-adjoint operators. In particular, for definitizable operators, there exist analogs of spectral functions and of the functional calculus of self-adjoint operators that ensure additional possibilities for spectral analysis (21, 27, 28).

A physically oriented discussion about the Krein spaces theory can be found in Refs (29, 30).

In conclusion, we consider some classes of operators that admit a natural interpretation as self-adjoint operators in Krein spaces.

6.3.1.1 Pseudo-Hermitian Operators

A linear densely defined operator A acting in a Hilbert space $\mathfrak{H}$ is said to be *pseudo-Hermitian* (6) if there exists a bounded and boundedly invertible self-adjoint operator $\eta : \mathfrak{H} \to \mathfrak{H}$ such that

$$\eta A = A^* \eta. \tag{6.3.7}$$

It follows from (6.3.1) and (6.3.7) that self-adjoint operators in Krein spaces are pseudo-Hermitian. The inverse implication is also true. Indeed, let A be pseudo-Hermitian. Then (6.3.7) holds for some η. Denote

$$|\eta| = \sqrt{\eta^2}, \qquad J = |\eta|^{-1}\eta$$

and consider the Hilbert space $(\mathfrak{H}, (\cdot,\cdot)_{|\eta|})$ endowed with new (equivalent to $(\cdot,\cdot)$) inner product $(\cdot,\cdot)_{|\eta|} = (|\eta|\cdot,\cdot)$. The operator $J = |\eta|^{-1}\eta$ is a fundamental symmetry in the Hilbert space $(\mathfrak{H}, (\cdot,\cdot)_{|\eta|})$ and A is self-adjoint with respect to the indefinite metric

$$[\cdot,\cdot] = (J\cdot,\cdot)_{|\eta|} = (|\eta|J\cdot,\cdot) = (\eta\cdot,\cdot)$$

constructed with the use of fundamental symmetry $J = \eta|\eta|^{-1}$ and the inner product $(\cdot,\cdot)_{|\eta|}$.

6.3.1.2 Self-adjoint Operators in S-spaces

Let a linear densely defined operator A satisfy the relation

$$UA = A^* U, \tag{6.3.8}$$

where U is a *unitary operator* acting in a Hilbert space $\mathfrak{H}$.

Denote $< \cdot,\cdot > = (U\cdot,\cdot)$. The sesquilinear form $< \cdot,\cdot >$ is not Hermitian, in general. The relation (6.3.8) means that

$$< Af, g > = < f, Ag >, \qquad f, g \in \mathcal{D}(A).$$

The operator A is called *a self-adjoint operator* in the S-space $(\mathfrak{H}, < \cdot,\cdot >)$ (see Ref. (31) for detail). In a particular case, where U is additionally a self-adjoint operator, the sesquilinear form $< \cdot,\cdot >$ turns out to be an indefinite metric (as U will be a fundamental symmetry) and the S-space $(\mathfrak{H}, < \cdot,\cdot >)$ is transformed into a Krein space.

Assume that A satisfies (6.3.8) or, that is equivalent, the operator A is self-adjoint in the S-space $(\mathfrak{H}, < \cdot,\cdot >)$. If $\rho(A) \neq \emptyset$, then A is self-adjoint in the Krein space $(\mathfrak{H}, [\cdot,\cdot])$ for a certain choice of indefinite metric $[\cdot,\cdot]$ [31, Theorem 3.13].

It is interesting to note that spectral properties of a unitary operator U in (6.3.8) may guarantee the reality of the spectrum $\sigma(A)$. Namely [31, Theorem 3.12]: let A satisfy (6.3.8), $\rho(A) \neq \emptyset$ and there exists $t \in [0, \pi)$ such that $e^{-it}, -e^{-it} \in \rho(U)$. Let $\mathbb{T} = \mathbb{T}_1 \cup \mathbb{T}_2$ be a decomposition of the unit circle, where

$$\mathbb{T}_1 := \{e^{is} \ : \ -t \leq s < -t + \pi\}, \qquad \mathbb{T}_2 := \{e^{is} \ : \ -t + \pi \leq s < -t + 2\pi\}$$

If $\mathbb{T}_1 \cap \sigma(U) = \emptyset$ or $\mathbb{T}_2 \cap \sigma(U) = \emptyset$, then $\sigma(A) \subset \mathbb{R}$.

6.3.2 Similarity to Self-adjoint Operators

A closed densely defined operator A acting in a Hilbert space $\mathfrak{H}$ is called *similar* to a self-adjoint operator H if there exists a bounded and boundedly invertible operator Z such that

$$ZA = HZ. \tag{6.3.9}$$

If A is similar to a self-adjoint operator, then A is self-adjoint in the Hilbert space $\mathfrak{H}$ endowed with the new (equivalent to $(\cdot, \cdot)$) inner product $(Z^*Z\cdot, \cdot) = (Z\cdot, Z\cdot)$.

The following integral-resolvent criterion of similarity can be useful:

Theorem 6.3.2 ((32)) *An operator A is similar to a self-adjoint one if and only if the spectrum of A is real and there exists a constant M such that*

$$\begin{aligned} &\sup_{\varepsilon>0}\varepsilon \int_{-\infty}^{\infty} \|(A - zI)^{-1}g\|^2 d\xi \leq M\|g\|^2, \\ &\sup_{\varepsilon>0}\varepsilon \int_{-\infty}^{\infty} \|(A^* - zI)^{-1}g\|^2 d\xi \leq M\|g\|^2, \quad \forall g \in \mathfrak{H}, \end{aligned} \tag{6.3.10}$$

where the integrals are taken along the line $z = \xi + i\varepsilon$ ($\varepsilon > 0$ is fixed) of the upper half-plane $\mathbb{C}_+ = \{z \in \mathbb{C} \ : \ \mathrm{Im}\, z > 0\}$.

If A is a self-adjoint operator in a Krein space, then a (possible) similarity of A to a self-adjoint operator in a Hilbert space can be characterized in terms of bounded operators C (see Section 6.2.2).

Definition 6.3.3 *A linear densely defined operator A in a Krein space $(\mathfrak{H}, [\cdot, \cdot])$ possesses the property of C-symmetry if $AC = CA$ for a bounded linear operator $C = Je^{Q}$, where the self-adjoint operator Q anticommutes with J.*

The commutation condition $AC = CA$ is central in the definition of C-symmetry of A. The conditions imposed on C (i.e., $C = Je^{Q}$, $Q = Q^*$, and $JQ = -QJ$) are often reformulated in the following equivalent forms: (i) $C^2 = I$, JC is a metric operator (i.e., JC is a boundedly invertible positive self-adjoint operator); (ii) $C^2 = I$ and $[C\cdot, \cdot]$ is an inner product on the Hilbert space $(\mathfrak{H}, (\cdot, \cdot))$, which is equivalent to the initial one $(\cdot, \cdot)$.

The notion of C-symmetry in Definition 6.3.3 coincides with the notion of fundamentally reducible operator used in the mathematical literature (see, e.g., (28)).

Let A be a J-self-adjoint operator in the Krein space $(\mathfrak{H}, [\cdot,\cdot])$ and let A have the property of C-symmetry. Then $AC = CA$ for a certain $C = Je^{Q}$. Taking (6.3.4) into account, we rewrite the commutation relation

$$AC = AJe^{Q} = JA^{*}e^{Q} = CA = Je^{Q}A.$$

Hence, $e^{Q}A = A^{*}e^{Q}$ or $e^{Q/2}Ae^{-Q/2} = e^{-Q/2}A^{*}e^{Q/2}$. This means that the operator $H = e^{Q/2}Ae^{-Q/2}$ is self-adjoint in $\mathfrak{H}$ and A satisfies the condition of similarity (6.3.9) with $Z = e^{Q/2}$. The operator A is self-adjoint with respect to the new inner product

$$(e^{Q/2}\cdot, e^{Q/2}\cdot) = (e^{Q}\cdot,\cdot) = [C\cdot,\cdot] = (\cdot,\cdot)_{C}.$$

Thus, a J-self-adjoint operator A with the property of C-symmetry turns out to be self-adjoint in the Hilbert space $(\mathfrak{H}, (\cdot,\cdot)_{C})$.

Theorem 6.3.4 *A J-self-adjoint operator A has the property of C-symmetry if and only if A is similar to a self-adjoint operator in $\mathfrak{H}$.*

Proof: The implication: *property of C-symmetry $\to$ similarity to a self-adjoint operator* was already established earlier.

The converse statement: *the similarity $\to$ the existence of C-symmetry* is a direct consequence of the Phillips theorem [19, Chapter 2, Corollary 5.20]. Precisely, if (6.3.9) holds, then the spectrum of A is real. Then the Cayley transform

$$U = (A - iI)(A + iI)^{-1}$$

is a well-defined unitary operator in the Krein space $(\mathfrak{H}, [\cdot,\cdot])$, which satisfies the relation $U = Z^{-1}WZ$, where $W = (H - iI)(H + iI)^{-1}$ is a unitary operator in the Hilbert space $(\mathfrak{H}, (\cdot,\cdot))$. Therefore,

$$\|U^{n}\| = \|Z^{-1}W^{n}Z\| \leq \|Z^{-1}\| \cdot \|Z\|, \qquad \forall n \in \mathbb{N}$$

and U is a *stable operator* (19). Then, due to the Phillips theorem, there exists a fundamental decomposition (6.2.4) with subspaces $\mathfrak{L}_{\pm}$ that are invariant with respect to U. Obviously, the subspaces $\mathfrak{L}_{\pm}$ are also invariant with respect to A. Therefore, A can be decomposed with respect to (6.2.4):

$$A = A_{+}\dot{+}A_{-}, \qquad A_{+} = A\restriction_{\mathfrak{L}_{+}}, \quad A_{-} = A\restriction_{\mathfrak{L}_{-}},$$

where $A_{\pm}$ are operators acting in the subspaces $\mathfrak{L}_{\pm}$. Let C be determined by the formula (6.2.8) with respect to the fundamental decomposition (6.2.4). Then

$$CAf = C(A_{+}f_{\mathfrak{L}_{+}} + A_{-}f_{\mathfrak{L}_{-}}) = A_{+}f_{\mathfrak{L}_{+}} - A_{-}f_{\mathfrak{L}_{-}} = ACf, \quad f \in \mathcal{D}(A).$$

Hence, A has the property of bounded C-symmetry. Theorem 6.3.4 is proved. ■

Summing up, for the case of J-self-adjoint operators, the property of similarity (6.3.9) is equivalent to the existence of C-symmetry. If a J-self-adjoint operator A has the C-symmetry $C = Je^Q$, then A becomes self-adjoint in the Hilbert space $(\mathfrak{H}, (\cdot,\cdot)_C)$, where $(\cdot,\cdot)_C = [C\cdot,\cdot] = (e^Q\cdot,\cdot)$. The finding of C is equivalent to the construction of the metric operator $e^Q = JC$, which ensures the self-adjointness of A.

Let A be a pseudo-Hermitian operator (i.e., A satisfies (6.3.7)). Then A is also a J-self-adjoint operator and we can apply Theorem 6.3.4.

Corollary 6.3.5 *A pseudo-Hermitian operator A is similar to a self-adjoint operator if and only there exists an operator $C = Je^Q$, where $J = |\eta|^{-1}\eta$, the bounded operator Q satisfies the relations*

$$Q^*|\eta| = |\eta|Q, \qquad -Q^*\eta = \eta Q \tag{6.3.11}$$

and $AC = CA$. In that case, the operator A turns out to be self-adjoint in the Hilbert space $\mathfrak{H}$ endowed with inner product $(|\eta|e^Q\cdot,\cdot)$.

Proof: The operator A is self-adjoint in the Krein space where the indefinite metric is determined by the operator of fundamental symmetry $J = |\eta|^{-1}\eta$ acting in the Hilbert space $(\mathfrak{H}, (\cdot,\cdot)_{|\eta|})$, where $(\cdot,\cdot)_{|\eta|} = (|\eta|\cdot,\cdot)$.

According to Theorem 6.3.4, the similarity of A to a self-adjoint operator is equivalent to the commutation relation $CA = AC$, where $C = Je^Q$ and Q is a bounded self-adjoint operator in $(\mathfrak{H}, (\cdot,\cdot)_{|\eta|})$, which anticommutes with $J = |\eta|^{-1}\eta$. The self-adjointness of Q with respect $(|\eta|\cdot,\cdot)$ gives $Q^*|\eta| = |\eta|Q$, where Q^* is the adjoint of Q with respect to the initial inner product $(\cdot,\cdot)$. Taking this relation into account, we rewrite the anticommutation relation $QJ = -JQ$ in the equivalent form $-Q^*\eta = \eta Q$.

The operator A is J-self-adjoint in the Krein space $(\mathfrak{H}, [\cdot,\cdot])$, where $[\cdot,\cdot] = (\eta\cdot,\cdot)$. The existence of C-symmetry guarantees the self-adjointness of A with respect to the inner product $[C\cdot,\cdot] = (\eta Je^Q\cdot,\cdot) = (|\eta|e^Q\cdot,\cdot)$. The proof is complete. ■

6.3.2.1 Reasons for Uniqueness of Bounded C Operators Let $\mathfrak{M}$ be the set of all equivalent inner products on a Hilbert space $(\mathfrak{H}, (\cdot,\cdot))$. The elements of $\mathfrak{M}$ are in one-to-one correspondence with the set of all positive bounded and boundedly invertible self-adjoint operators $\{G\}$ acting in the Hilbert space $(\mathfrak{H}, (\cdot,\cdot))$. Denoting $G = e^Q$, we describe $\mathfrak{M}$ as

$$\mathfrak{M} = \{(e^Q\cdot,\cdot)\},$$

where Q runs over the set of all bounded self-adjoint operators in $(\mathfrak{H}, (\cdot,\cdot))$.

Let J be an operator of fundamental symmetry in $(\mathfrak{H}, (\cdot,\cdot))$ and let $(\mathfrak{H}, [\cdot,\cdot])$ be the corresponding Krein space. The set $\mathfrak{M}_1$ of inner products that are constructed by means of fundamental decompositions of the Krein space is parameterized by bounded operators $C = Je^Q$. Obviously, $\mathfrak{M}_1$ is a subset of $\mathfrak{M}$, which is distinguished by the anticommutation condition $JQ = -QJ$.

If A is a J-self-adjoint operator, then its similarity to a self-adjoint operator means that A turns out to be self-adjoint with respect to an inner product from $\mathfrak{M}_1$.

Is this inner product determined uniquely in $\mathfrak{M}_1$? Precisely, is the operator $C = Je^Q$ determined uniquely for a given J-self-adjoint operator A with the property of C-symmetry?

We analyze this problem in detail for the case where A has a *complete set of eigenvectors* $\{f_n\}$ in $\mathfrak{H}$. In this context, "complete set" means that the linear span of eigenvectors $\{f_n\}$, that is, the set of all possible finite linear combinations

$$\text{span}\{f_n\} \equiv \left\{ \sum_{n=1}^{d} c_n f_n \ : \ \forall d \in \mathbb{N}, \ \forall c_n \in \mathbb{C} \right\},$$

is a dense subset in $\mathfrak{H}$.

We recall that a nonzero vector $f \in \mathcal{D}(A)$ is called a root vector of A corresponding to the eigenvalue λ if $(A - \lambda I)^n f = 0$ for some $n \in \mathbb{N}$. A root vector f is called eigenvector (generalized eigenvector) if $f \in \ker(A - \lambda I)$ $(f \notin \ker(A - \lambda I))$. The set of all roots vectors of A corresponding to a given eigenvalue λ, together with the zero vector, form a linear subspace $\mathcal{L}_\lambda$, which is called the root subspace. The dimension of the root subspace $\mathcal{L}_\lambda$ is called the *algebraic multiplicity* of the eigenvalue λ. The *geometric multiplicity* of λ is defined as the dimension of the kernel subspace $\ker(A - \lambda I)$ (i.e., as the dimension of the linear subspace of eigenfunctions of A corresponding to λ).

Lemma 6.3.6 *Let A be a J-self-adjoint operator with a complete set of eigenvectors corresponding to real eigenvalues $\{\lambda_n\}$ of the discrete spectrum. Then the geometric and the algebraic multiplicities of λ_n coincide.*

Proof: Assume that the geometric and the algebraic multiplicities of $\lambda \in \{\lambda_n\}$ are different. Then there exist a generalized eigenvector f and an eigenvector g of A such that $Af = \lambda f + g$. For any $\gamma \in \ker(A - \lambda I)$,

$$[g, \gamma] = [(A - \lambda I)f, \gamma] = [Af, \gamma] - \lambda[f, \gamma] = [f, A\gamma] - \lambda[f, \gamma] = 0.$$

Thus, g is J-orthogonal to $\ker(A - \lambda I)$.

Let $\lambda' \in \{\lambda_n\}$ and $\lambda' \neq \lambda$. Then, the subspaces $\ker(A - \lambda I)$ and $\ker(A - \lambda' I)$ are J-orthogonal (as A is a J-self-adjoint operator). Therefore, g is J-orthogonal to all eigenvectors corresponding to $\{\lambda_n\}$. The latter means that Jg is orthogonal to the complete set of eigenvectors. Hence, $g = 0$ and f cannot be a generalized eigenvector. Lemma 6.3.6 is proved. ∎

In general, the completeness of a linearly independent sequence of eigenvectors $\{f_n\}$ does not mean that $\{f_n\}$ is a Schauder basis[5] of $\mathfrak{H}$. The difference is that the completeness of $\{f_n\}$ allows us to approximate an arbitrary $f \in \mathfrak{H}$ by finite linear

[5] A sequence $\{f_n\}$ is a Schauder basis for $\mathfrak{H}$ if for each $f \in \mathfrak{H}$, there exist unique scalar coefficients $\{c_n\}$ such that $f = \sum_{n=1}^{\infty} c_n f_n$ (33)

combinations $\sum_{n=1}^{d} c_n^d f_n \to f$ as $d \to \infty$, where the coefficients c_n^d depend on the choice of d, while the definition of a Schauder basis requires that c_n^d does not depend on d.

If a J-self-adjoint operator A has the property of C-symmetry, then the complete linearly independent sequence of eigenvectors $\{f_n\}$ corresponding to the discrete spectrum of A turns out to be a Riesz basis.[6] Indeed, the property of C-symmetry means that A is self-adjoint with respect to the new inner product $(\cdot,\cdot)_C$. Then, without loss of generality, we may assume that the eigenvectors $\{f_n\}$ of A form an orthonormal system in $(\mathfrak{H},(\cdot,\cdot)_C)$. The completeness of $\{f_n\}$ with respect to $(\cdot,\cdot)$ yields the completeness of $\{f_n\}$ in $(\mathfrak{H},(\cdot,\cdot)_C)$. Hence, $\{f_n\}$ is an orthonormal basis of $(\mathfrak{H},(\cdot,\cdot)_C)$. Denote $\psi_n = e^{Q/2}f_n$. The sequence $\{\psi_n\}$ is an orthonormal basis of $(\mathfrak{H},(\cdot,\cdot))$ and $f_n = R\psi_n$, where $R = e^{-Q/2}$. Hence, $\{f_n\}$ is a Riesz basis of the initial Hilbert space $(\mathfrak{H},(\cdot,\cdot))$.

Let a J-self-adjoint operator A have a Riesz basis $\{f_n\}$ of eigenvectors and let A have the property of C-symmetry.

We show that the (possible) nonuniqueness of C operators for A deals with the appearance of neutral elements with respect to the indefinite metric $[\cdot,\cdot]$ in at least one of the kernel subspaces $\ker(A - \lambda I)$.

Let us illustrate this phenomenon by assuming for simplicity that the first k eigenvectors $f_{11},\dots,f_{1k}$ correspond to the eigenvalue λ_1, and the other eigenvalues $\lambda_2,\lambda_3,\dots$ of A are *simple*; that is, $\dim\ker(A-\lambda_m) = 1$, $m \geq 2$ and hence, $\ker(A-\lambda_m I)$ coincides with the span of f_m.

The operator C determines the decomposition (6.2.4) of $\mathfrak{H}$, with subspaces $\mathfrak{L}_\pm$ defined by (6.2.9). Every eigenvector f_m of A is decomposed along (6.2.4) as[7]

$$f_m = f_m^+ + f_m^-, \qquad f_m^\pm \in \mathfrak{L}_\pm = \frac{1}{2}(I \pm C)\mathfrak{H}. \tag{6.3.12}$$

The sequences of elements $\{f_m^\pm\}$ are Riesz bases of $\mathfrak{L}_\pm$; that is, $\mathfrak{L}_+$ and $\mathfrak{L}_-$ coincide with the closures of the linear spans of $\{f_m^+\}$ and $\{f_m^-\}$, respectively.

Recalling that $AC = CA$ we conclude that the elements $f_m^\pm$ in (6.3.12) are also eigenvectors of A corresponding to the same eigenvalue. Therefore, due to the simplicity of the eigenvalues λ_m, $m \geq 2$, the decomposition (6.3.12) of the corresponding element f_m may contain only one nonzero element. This means that

$$f_m = \begin{cases} f_m^+ & \text{if} \quad [f_m,f_m] > 0, \\ f_m^- & \text{if} \quad [f_m,f_m] < 0 \end{cases} \qquad \forall m \geq 2.$$

(The case $[f_m,f_m] = 0$ is impossible because it gives two linearly independent eigenvectors $f_m^\pm$ of A, which contradicts the simplicity of λ_m.) Therefore, the elements $f_m^\pm$ are uniquely determined by f_m when $m \geq 2$.

[6] A sequence $\{f_n\}$ is a Riesz basis if there exists a bounded and boundedly invertible operator R and an orthonormal basis $\{\psi_n\}$ in $\mathfrak{H}$ such that $f_n = R\psi_n$.

[7] one of $f_m^\pm$ may vanish.

The span of the first k eigenvectors $f_{11}, \dots, f_{1k}$ coincides with $\ker(A - \lambda_1 I)$. If this finite-dimensional subspace contains neutral elements with respect to $[\cdot,\cdot]$, then $\ker(A - \lambda_1 I)$ contains positive elements with respect to $[\cdot,\cdot]$ as well as negative ones. This means that the Krein space $(\ker(A - \lambda_1 I), [\cdot,\cdot])$ has different fundamental decompositions

$$\ker(A - \lambda_1 I) = \mathcal{M}_+[\dot{+}]\mathcal{M}_-,$$

where $\mathcal{M}_\pm$ are maximal uniformly definite subspaces.

One of possible decompositions is

$$\mathcal{M}_+ = \operatorname{span}\{f_{1j}^+\}_{j=1}^k \quad \text{and} \quad \mathcal{M}_- = \operatorname{span}\{f_{1j}^-\}_{j=1}^k,$$

where the elements $f_{1j}^\pm$ are determined by f_{1j} by the use of the decomposition (6.3.12). Fixing another fundamental decomposition $\ker(A - \lambda_1 I) = \mathcal{M}'_+[\dot{+}]\mathcal{M}'_-$, we obtain other decompositions of the elements

$$f_{1j} = f'^+_{1j} + f'^-_{1j} \qquad j = 1, \dots, k$$

onto positive and negative parts with respect to $[\cdot,\cdot]$.

Let us define $\mathfrak{L}'_+$ and $\mathfrak{L}'_-$, respectively, as the closure (with respect to $(\cdot,\cdot)$) of the linear spans of the J-orthogonal functions

$$[\{f'^+_{1j}\}_{j=1}^k \cup \{f_m^+\}_{m=2}^\infty] \qquad \text{and} \qquad [\{f'^-_{1j}\}_{j=1}^k \cup \{f_m^-\}_{m=2}^\infty].$$

By the construction, $\mathfrak{L}'_\pm$ are maximal uniformly definite subspaces in the Krein space $(\mathfrak{H}, [\cdot,\cdot])$ and

$$\mathfrak{H} = \mathfrak{L}'_+[\dot{+}]\mathfrak{L}'_- \tag{6.3.13}$$

is a fundamental decomposition of $\mathfrak{H}$.

It is clear that $\mathfrak{L}_+ \neq \mathfrak{L}'_+$ and $\mathfrak{L}_- \neq \mathfrak{L}'_-$. Therefore, the operator C' determined by (6.3.13) (that is, $C' \restriction_{\mathfrak{L}'_+} = I$ and $C' \restriction_{\mathfrak{L}'_-} = -I$) commutes with A but it does not coincide with the initial operator C.

Summing up the aforementioned reasoning we obtain

Theorem 6.3.7 *For a J-self-adjoint operator A with Riesz basis of eigenvectors corresponding to the real eigenvalues $\{\lambda_n\}$ of the discrete spectrum, the bounded C operator is determined uniquely if and only if the kernel subspaces* $\ker(A - \lambda_n I)$ *have no nonzero neutral eigenvectors.*

Assume that the eigenvalues $\{\lambda_n\}$ in Theorem 6.3.7 are simple and denote by $\{f_n\}$ the corresponding Riesz basis. Without loss of generality we may assume that $\{f_n\}$ is an orthonormal system in the Krein space $(\mathfrak{H}, [\cdot,\cdot])$. Then the bounded operator C is uniquely determined by the formula

$$Cf = \sum_{n=1}^{\infty} [f, f_n] f_n, \qquad f \in \mathfrak{H}. \tag{6.3.14}$$

6.3.3 The Property of Unbounded C-symmetry

The operators C in section 6.3.2 correspond to the fundamental decompositions of the Krein space $(\mathfrak{H}, [\cdot,\cdot])$ and they are bounded operators in $\mathfrak{H}$. On the other hand, many operators C appearing in the literature are definitely unbounded (34). For this reason, a natural question arises: is it possible to properly define the concept of *unbounded* C-symmetry for J-self-adjoint operators?

Formal preservation of the commutation relation $AC = CA$ on the *whole domain of definition* of A leads to the confusing conclusion that all nonreal points belong to the continuous spectrum of A. Indeed, an unbounded operator C determines the direct sum (6.2.3), which does not coincide with $\mathfrak{H}$. The commutation $AC = CA$ on the whole domain of A means that A is decomposed along (6.2.3)

$$A = A_+\dot{+}A_-, \qquad A_+ = A\upharpoonright_{\mathfrak{L}_+}, \quad A_- = A\upharpoonright_{\mathfrak{L}_-},$$

where $\mathcal{D}(A) = \mathcal{D}(A_+)\dot{+}\mathcal{D}(A_-)$. Assume that A has a nonreal eigenvalue $z \in \mathbb{C}\setminus\mathbb{R}$. Then at least one of $A_\pm$ should have z as an eigenvalue. However, this is impossible because $A_\pm$ are symmetric in the pre-Hilbert spaces $\mathfrak{L}_\pm$ with scalar products $[\cdot,\cdot]$ and $-[\cdot,\cdot]$, respectively. Let z belong to the residual spectrum of A. Then $\bar{z}$ is an eigenvalue of the adjoint A^*. Owing to (6.3.4), nonreal $\bar{z}$ belongs to the point spectrum of A, that is impossible due to aforementioned reasonings. It follows from the commutation relation $AC = CA$ that $\mathcal{R}(A - zI) \subseteq \mathfrak{L}_+[\dot{+}]\mathfrak{L}_- = \mathfrak{D} \neq \mathfrak{H}$ for any nonreal z. Hence, $\sigma_c(A) \supset \mathbb{C}\setminus\mathbb{R}$.

The obtained result does not relate to an inherent structure of A but rather indicates the wrong choice of underlying Hilbert space $\mathfrak{H}$. Thus, the proper definition of unbounded C-symmetry should involve the commutation between A and C on an appropriately chosen subset S of $\mathcal{D}(A)$.

Let S be a linear subset of $\mathcal{D}(A)$ such that S is a dense set in $\mathfrak{H}$.

Definition 6.3.8 (cf. Definition 6.3.3) *A linear densely defined operator A in a Krein space $(\mathfrak{H}, [\cdot,\cdot])$ possesses the property of unbounded C-symmetry if*

$$ACf = CAf, \qquad \forall f \in S \tag{6.3.15}$$

for an unbounded linear operator $C = Je^{\mathcal{Q}}$, where the self-adjoint operator $\mathcal{Q}$ anticommutes with J.

An unbounded operator C determines the J-orthogonal direct sum (6.2.3): $\mathcal{D}(C) = \mathfrak{D} = \mathfrak{L}_+[\dot{+}]\mathfrak{L}_-$ and the commutation relation (6.3.15) means that

$$\mathcal{D}(C) \supset S \qquad C : S \to \mathcal{D}(A), \qquad A : S \to \mathcal{D}(C).$$

Let us illustrate Definition 6.3.8 for the case of a J-self-adjoint operator A with a complete set of eigenvectors $\{f_n\}$ corresponding to real simple eigenvalues. We do not assume that $\{f_n\}$ is a Riesz basis.

The eigenvectors $\{f_n\}$ of A are mutually J-orthogonal (as A is J-self-adjoint). Without loss of generality, we can suppose that $\{f_n\}$ form an orthonormal system in the Krein space $(\mathfrak{H}, [\cdot,\cdot])$, that is, $[f_n,f_n] \in \{-1, 1\}$. (The case $[f_n,f_n] = 0$ is impossible due to the completeness of $\{f_n\}$.) Separating the sequence $\{f_n\}$ by the signs of the indefinite inner products $[f_n,f_n]$, we obtain two sequences of elements $\{f_n^+\}$ and $\{f_n^-\}$.

Let $\mathfrak{L}'_+$ and $\mathfrak{L}'_-$ be the closure of span$\{f_n^+\}$ and span$\{f_n^-\}$ with respect to the initial inner product $(\cdot,\cdot)$. By the construction, $\mathfrak{L}'_\pm$ are J-orthogonal positive/negative subspaces and the direct sum $\mathfrak{L}'_+[\dot{+}]\mathfrak{L}'_-$ is a dense set in $\mathfrak{H}$ (due to the completeness of $\{f_n\}$). It may happen that $\mathfrak{L}'_\pm$ are proper subspaces[8] of maximal J-orthogonal positive/negative subspaces $\mathfrak{L}_\pm$ (that is, $\mathfrak{L}'_\pm \subset \mathfrak{L}_\pm$ and $\mathfrak{L}'_\pm \neq \mathfrak{L}_\pm$). Let us fix one of such extensions $\mathfrak{L}_\pm$. Then $\mathfrak{D} = \mathfrak{L}_+[\dot{+}]\mathfrak{L}_-$ is the J-orthogonal direct sum (6.2.3), which determines an operator $C = Je^Q$ (Theorem 6.2.3).

Denote by S the linear span of eigenvectors $\{f_n\}$ of A. Obviously, the relation (6.3.15) holds for all $f \in S$ and A has the property of unbounded C-symmetry with the operator $C = Je^Q$ mentioned earlier.

If the eigenvectors $\{f_n\}$ of A form a Schauder basis, then the subspaces $\mathfrak{L}'_\pm$ are maximal positive/negative in $\mathfrak{H}$ [19, Statement 10.12 in Chapter 1] and the decomposition $\mathfrak{L}'_+[\dot{+}]\mathfrak{L}'_-$ defines a *unique* unbounded operator $C = Je^Q$ for A. The operator C is closed and its restriction C' to the elements of S is determined by the formula (cf. (6.3.14))

$$C'f = \sum_{n=1}^{\infty} [f,f_n]f_n, \qquad f \in S. \tag{6.3.16}$$

The operator $G' = JC'$ is essentially self-adjoint on S and its closure coincides with the positive self-adjoint operator $e^Q = JC$ defined on the J-orthogonal direct sum $\mathfrak{L}'_+[\dot{+}]\mathfrak{L}'_-$ of the form (6.2.3).

In the general case where $\{f_n\}$ is a complete set of eigenvectors, the subspaces $\mathfrak{L}'_\pm$ may be only positive/negative (without the property of maximality). In that case the direct sum $\mathfrak{L}'_+[\dot{+}]\mathfrak{L}'_-$ (see (6.2.16)) cannot properly define an operator $C = Je^Q$, where Q is a self-adjoint operator. To this end, due to theorem 6.2.3, we have to extend $\mathfrak{L}'_\pm$ to maximal positive/negative subspaces. The nonuniqueness of such kind of extensions lead to the nonuniqueness of unbounded operators C for A. In other words, the operator $G' = JC'$ where C' is defined by (6.3.16), cannot be an essentially self-adjoint operator and its closure gives us a positive symmetric operator $\overline{G'} = \overline{JC'}$, which is defined on the set $\mathfrak{L}'_+[\dot{+}]\mathfrak{L}'_-$ in $\mathfrak{H}$. The different positive self-adjoint extensions G of $\overline{G'}$ (with the condition $JGJG = I$) give rise to different operators $C = Je^Q = JG$ associated with the initial direct sum $\mathfrak{L}'_+[\dot{+}]\mathfrak{L}'_-$.

Thus, *if a J-self-adjoint operator A has many unbounded operators C, then the complete set of eigenfunctions $\{f_n\}$ cannot be a Schauder basis.*

REMARK 3 *The existence of a bounded operator $C = Je^Q$ for a J-self-adjoint operator A implies that A is self-adjoint with respect to the inner product $(e^Q\cdot,\cdot)$. In*

[8] see Section 6.2.3.1 for details

other words, the finding of bounded C is equivalent to the construction of a suitable (bounded) metric operator e^{Q}. This relationship can be extended to the case of unbounded operators C and unbounded metric operators e^{Q}. Various properties of unbounded metric operators were studied and discussed in Refs (34–37).

6.3.3.1 Imaginary Cubic Oscillator The space reflection operator

$$\mathcal{P}f(x) = f(-x), \qquad f \in L_2(\mathbb{R}) \tag{6.3.17}$$

is an operator of fundamental symmetry in the Hilbert space $\mathfrak{H} = L_2(\mathbb{R})$.

The operator

$$A = -\frac{d^2}{dx^2} + ix^3, \qquad \mathcal{D}(A) = \{f \in L_2(\mathbb{R}) \ : \ -f'' + ix^3 f \in L_2(\mathbb{R})\}$$

is self-adjoint ($\mathcal{P}$-self-adjoint) in the Krein space $(L_2(\mathbb{R}), [\cdot, \cdot])$ with the indefinite inner product

$$[f, g] = (\mathcal{P}f, g) = \int_{\mathbb{R}} f(-x)\overline{g(x)}dx.$$

The operator A has a complete set $\{f_n\}$ of eigenvectors corresponding to real simple eigenvalues. It is known (37) that A is not similar to a self-adjoint operator in $\mathfrak{H}$. Therefore, $\{f_n\}$ cannot be a Riesz basis and there exist unbounded C-symmetries of A. However the question whether the eigenvectors of A form a Schauder basis remains open. Similar conclusions can also be proved for the case where the potential ix^3 is replaced by the potential $x^2 + ix^3$ and in many other cases (37).

It should be noted that for the operators

$$A = -\frac{d^2}{dx^2} + \mu^2 x^2 + i\epsilon x^3, \qquad \mu, \epsilon \in \mathbb{R}$$

infinitely many unbounded operators C were constructed by formal perturbative calculation methods (13, 14). These results allow us to suppose that the corresponding complete sets of eigenvectors do not form Schauder basis.

6.3.3.2 The General Scheme We present next a general scheme (inspired by Langer (24)) for the construction of unbounded C-symmetries.

Let $\{\gamma_n^+\}$ and $\{\gamma_n^-\}$ be orthonormal bases of subspaces $\mathfrak{H}_\pm$ in the fundamental decomposition (6.2.6) of the Krein space $(\mathfrak{H}, [\cdot, \cdot])$. Any element $\phi \in \mathfrak{H}$ has the representation

$$\phi = \gamma^+ + \gamma^- = \sum_{n=1}^{\infty} (c_n^+ \gamma_n^+ + c_n^- \gamma_n^-), \qquad \gamma^\pm = \sum_{n=1}^{\infty} c_n^\pm \gamma_n^\pm \in \mathfrak{H}_\pm,$$

where the sequences $\{c_n^\pm\}$ belong to the Hilbert space l_2; that is, $\sum_{n=1}^{\infty} |c_n^\pm|^2 < \infty$.

The operator

$$T\phi = \sum_{n=1}^{\infty} i\alpha_n(c_n^+\gamma_n^- - c_n^-\gamma_n^+), \qquad \alpha_n = (-1)^n\left(1 - \frac{1}{n}\right), \quad n \in \mathbb{N} \tag{6.3.18}$$

plays a key role in our construction and has many useful properties that can be directly deduced from (6.3.18). In particular, the operator T satisfies (6.2.14) and T is a self-adjoint contraction in $\mathfrak{H}$ that anticommutes with J. The anticommutation with J means that T interchanges the subspaces $\mathfrak{H}_\pm$: $T : \mathfrak{H}_+ \to \mathfrak{H}_-$ and vice versa.

Let $\mathfrak{L}_\pm$ be subspaces of $\mathfrak{H}$ defined by (6.2.10). It follows from (6.2.10) and the properties of T that the subspaces $\mathfrak{L}_\pm$ are J-orthogonal and they are maximal positive/negative. For instance, every $f^+ \in \mathfrak{L}_+$ has the form

$$f^+ = \gamma^+ + T\gamma^+ = \sum_{n=1}^{\infty} c_n^+(\gamma_n^+ + i\alpha_n\gamma_n^-)$$

and

$$[f^+,f^+] = (\gamma^+ - T\gamma^+, \gamma^+ + T\gamma^+) = \|\gamma^+\|^2 - \|T\gamma^+\|^2 = \sum_{n=1}^{\infty}(1 - \alpha_n^2)|c_n^+|^2 > 0,$$

Hence, the subspace $\mathfrak{L}_+$ is positive.

On the other hand,

$$(f^+,f^+) = (\gamma^+ + T\gamma^+, \gamma^+ + T\gamma^+) = \|\gamma^+\|^2 + \|T\gamma^+\|^2 = \sum_{n=1}^{\infty}(1 + \alpha_n^2)|c_n^+|^2.$$

Comparing $[f^+,f^+]$ and (f^+,f^+) we conclude that the indefinite inner product $[\cdot,\cdot]$ is not topologically equivalent to the initial inner product $(\cdot,\cdot)$ on the subspace $\mathfrak{L}_+$ (because $\lim_{n\to\infty}(1-\alpha_n^2) = \lim_{n\to\infty}\frac{1}{n}\left(2 - \frac{1}{n}\right) = 0$). Therefore, the subspace $\mathfrak{L}_+$ *cannot be uniformly positive.* Similar arguments show that the maximal negative subspace $\mathfrak{L}_-$ cannot be uniformly negative.

The J-orthogonal direct sum (6.2.3) of the subspaces $\mathfrak{L}_\pm$ is densely defined in $\mathfrak{H}$ and it determines an operator $C = Je^{Q}$ (Theorem 6.2.3). This operator C can also be expressed via the operator T, see (6.2.15). Elementary transformations of (6.2.15) with the use of the definition (6.3.18) of T lead to the conclusion that

$$\begin{aligned} C\phi = \sum_{n=1}^{\infty} \frac{1}{1-\alpha_n^2}\big(&[(1+\alpha_n^2)c_n^+ + 2i\alpha_n c_n^-]\gamma_n^+ \\ &+[-(1+\alpha_n^2)c_n^- + 2i\alpha_n c_n^+]\gamma_n^-\big). \end{aligned} \tag{6.3.19}$$

Let us fix an element $\chi \in \mathfrak{H}$,

$$\chi = \sum_{n=1}^{\infty} \frac{1}{n^{\delta}}(\gamma_n^+ + \gamma_n^-), \qquad \frac{1}{2} < \delta \leq \frac{3}{2}$$

and define the following subspaces of $\mathfrak{H}_{\pm}$:

$$M_+ = \{\gamma^+ \in \mathfrak{H}_+ : (\gamma^+, \chi) = 0\}, \qquad M_- = \{\gamma^- \in \mathfrak{H}_- : (\gamma^-, \chi) = 0\}.$$

The elements $\gamma^+ = \sum_{n=1}^{\infty} c_n^+ \gamma_n^+ \in M_+$ and $\gamma^- = \sum_{n=1}^{\infty} c_n^- \gamma_n^- \in M_-$ are characterized by the condition

$$\sum_{n=1}^{\infty} \frac{c_n^+}{n^{\delta}} = \sum_{n=1}^{\infty} \frac{c_n^-}{n^{\delta}} = 0. \tag{6.3.20}$$

The subspaces

$$\begin{aligned} \mathfrak{L}'_+ &= \{f^+ = \gamma^+ + T\gamma^+ : \gamma^+ \in M_+\}, \\ \mathfrak{L}'_- &= \{f^- = \gamma^- + T\gamma^- : \gamma^- \in M_-\} \end{aligned} \tag{6.3.21}$$

are proper subspaces of $\mathfrak{L}_+$ and $\mathfrak{L}_-$, respectively (as $M_+ \subset \mathfrak{H}_+$ and $M_- \subset \mathfrak{H}_-$). Therefore, the positive/negative subspaces $\mathfrak{L}'_{\pm}$ lose in this case the property of maximality.

We show that the J-orthogonal direct sum $\mathfrak{L}'_+[\dotplus]\mathfrak{L}'_-$ is densely defined in $\mathfrak{H}$. To this end, we suppose that an element $y = \sum_{n=1}^{\infty}(y_n^+\gamma_n^+ + y_n^-\gamma_n^-)$ is orthogonal to $\mathfrak{L}'_+[\dotplus]\mathfrak{L}'_-$. It follows from (6.3.18) and (6.3.21) that the condition $y \perp \mathfrak{L}'_{\pm}$ is equivalent to the relations

$$\sum_{n=1}^{\infty}(y_n^+ - i\alpha_n y_n^-)c_n^+ = \sum_{n=1}^{\infty}(y_n^- + i\alpha_n y_n^+)c_n^- = 0,$$

where $\{c_n^{\pm}\}$ are all elements of the Hilbert space l_2 that satisfy (6.3.20); that is, $\{c_n^{\pm}\}$ are orthogonal to the element $\{1/n^{\delta}\}$ in l_2. This means that

$$y_n^+ - i\alpha_n y_n^- = \frac{k_1}{n^{\delta}}, \qquad y_n^- + i\alpha_n y_n^+ = \frac{k_2}{n^{\delta}}, \tag{6.3.22}$$

where the constants $k_j, j = 1, 2$, do not depend on n. It follows from (6.3.22) that

$$\left(2 - \frac{1}{n}\right) y_n^+ = n(1 - \alpha_n^2) y_n^+ = (k_1 + i\alpha_n k_2)\frac{1}{n^{\delta-1}}, \qquad \forall n \in \mathbb{N}.$$

This relation holds only for $k_1 = k_2 = 0$, because the sequence $\{y_n^+\}$ belongs to the Hilbert space l_2 but $\{1/n^{\delta-1}\}$ does not belong to it (as $\delta - 1 \leq \frac{1}{2}$). Then (6.3.22) imply that $y_n^{\pm} = 0$. Hence, $y = 0$ and the J-orthogonal direct sum $\mathfrak{L}'_+[\dotplus]\mathfrak{L}'_-$ is densely defined in $\mathfrak{H}$.

Proposition 6.3.9 *If $\frac{1}{2} < \delta \leq 1$, then there exists a unique extension of $\mathfrak{L}'_+[\dot{+}]\mathfrak{L}'_-$ to a J-orthogonal pair $\mathfrak{L}_+[\dot{+}]\mathfrak{L}_-$ of maximal definite subspaces.*

If $1 < \delta \leq \frac{3}{2}$, then the direct sum $\mathfrak{L}'_+[\dot{+}]\mathfrak{L}'_-$ can be extended to different pairs of maximal definite subspaces $\mathfrak{L}_+[\dot{+}]\mathfrak{L}_-$.

Proof: Denote $K\gamma_n^+ = \alpha_n\gamma_n^+$ ($n \in \mathbb{N}$). Obviously, K is a self-adjoint contraction in $\mathfrak{H}_+$ and $\|K\| = 1$. Consider also the operator $U : \mathfrak{H}_+ \to \mathfrak{H}_-$, which is determined by the relations $U\gamma_n^+ = i\gamma_m^-$. The operator U maps isometrically $\mathfrak{H}_+$ onto $\mathfrak{H}_-$ and its adjoint $U^* : \mathfrak{H}_- \to \mathfrak{H}_+$ acts on basis elements γ_n^- as follows: $U^*\gamma_n^- = -i\gamma_n^+$.

The operator T defined by (6.3.18) can be expressed in terms of K and U

$$T = UKP_+ + KU^*P_-, \tag{6.3.23}$$

where $P_+ = \frac{1}{2}(I + J)$ and $P_- = \frac{1}{2}(I - J)$ are projections on $\mathfrak{H}_+$ and $\mathfrak{H}_-$, respectively. It is important that the formula (6.3.23) remains true for the restriction $T' = T\restriction_{M_+[\dot{+}]M_-}$ of T onto $M_+[\dot{+}]M_-$. In that case the self-adjoint operator K should be replaced by the symmetric operator $K' = K\restriction_{M_+}$ acting in $\mathfrak{H}_+$. Moreover, a simple analysis of (6.3.23) shows that every extension of T' to a self-adjoint contraction $\tilde{T}$, which anticommutes with J has the form $\tilde{T} = U\tilde{K}P_+ + \tilde{K}U^*P_-$, where the self-adjoint contraction $\tilde{K}$ is an extension of K' onto $\mathfrak{H}_+$. Conversely, any extension of K' to a self-adjoint contraction $\tilde{K}$ in $\mathfrak{H}_+$ gives rise to a self-adjoint contraction $\tilde{T} = U\tilde{K}P_+ + \tilde{K}U^*P_-$, which anticommutes with J in $\mathfrak{H}$. Therefore, there is a one-to-one correspondence between the sets of self-adjoint contractions $\{\tilde{T}\}$ and $\{\tilde{K}\}$ acting in $\mathfrak{H}$ and $\mathfrak{H}_+$, respectively.

Every operator $\tilde{T}$ determines the J-orthogonal direct sum $\tilde{\mathfrak{L}}_+[\dot{+}]\tilde{\mathfrak{L}}_-$ of maximal definite subspaces

$$\tilde{\mathfrak{L}}_+ = \{f^+ = \gamma^+ + \tilde{T}\gamma^+ : \gamma^+ \in \mathfrak{H}_+\}, \quad \tilde{\mathfrak{L}}_- = \{f^- = \gamma^- + \tilde{T}\gamma^- : \gamma^- \in \mathfrak{H}_-\},$$

which is an extension of the J-orthogonal direct sum $\mathfrak{L}'_+[\dot{+}]\mathfrak{L}'_-$, where the subspaces $\mathfrak{L}'_\pm$ are determined by (6.3.21). Hence, all possible extensions of $\mathfrak{L}'_+[\dot{+}]\mathfrak{L}'_-$ to J-orthogonal direct sums of maximal definite subspaces are described by self-adjoint contractions $\tilde{K}$.

Using Ref. [38, Theorem 4] (where $\phi = \sum_{n=1}^{\infty} \frac{1}{n^\delta}\gamma_n^+$ and $E(\lambda)$ is the spectral function of K), we arrive at the conclusion that the set of self-adjoint contractions $\{\tilde{K}\}$ contains only one element[9] if and only if

$$\sum_{n=1}^{\infty} \frac{1}{n^{2\sigma}(1 + \alpha_n)} = \infty, \qquad \sum_{n=1}^{\infty} \frac{1}{n^{2\sigma}(1 - \alpha_n)} = \infty. \tag{6.3.24}$$

Taking into account that $\alpha_n = (-1)^n(1 - \frac{1}{n})$ we conclude that (6.3.24) holds if and only if $\frac{1}{2} < \delta \leq 1$. Proposition 6.3.9 is proved. ∎

[9] which, of course, coincides with K

The next step involves the interpretation of $\mathfrak{L}'_{\pm}$ as the closure of linear spans of the J-orthogonal elements $\{f_n^{\pm}\}$. These vectors can be determined in different ways. A "constructive" approach uses (6.3.18) to establish that

$$(I - T^2)\phi = \sum_{n=1}^{\infty}(1 - \alpha_n^2)(c_n^+\gamma_n^+ + c_n^-\gamma_n^-), \qquad \phi \in \mathfrak{H}. \tag{6.3.25}$$

The operator $I - T^2$ is self-adjoint by the construction and $I - T^2$ is a compact operator as $\lim_{n\to\infty}(1 - \alpha_n^2) = 0$ [39, problem 132]. Therefore, its restrictions $P_{M_+}(I - T^2)P_{M_+}$ and $P_{M_-}(I - T^2)P_{M_-}$ to[10] M_+ and M_- are compact self-adjoint operators in the Hilbert spaces M_+ and M_-, respectively. These operators have complete sets of orthonormal eigenvectors, which corresponds to simple real eigenvalues. An elementary calculation with the use of (6.3.20) and (6.3.25) shows that the operators $P_{M_{\pm}}(I - T^2)P_{M_{\pm}}$ have the same set of eigenvalues $\{\mu_m\}$, which are determined as the roots of the equation

$$\sum_{n=1}^{\infty}\frac{1}{(1 - \alpha_n^2 - \mu)n^{2\sigma}} = 0, \qquad \mu > 0.$$

The corresponding orthonormal eigenvectors ξ_m^+ (ξ_m^-) in M_+ (M_-) are

$$\xi_m^{\pm} = \beta_m \sum_{n=1}^{\infty}\frac{1}{(1 - \alpha_n^2 - \mu_m)n^{\sigma}}\gamma_n^{\pm},$$

where

$$\beta_m = \frac{1}{\sqrt{\sum_{n=1}^{\infty}(1 - \alpha_n^2 - \mu_m)^2 n^{2\sigma}}}$$

are normalizing coefficients.

Denote

$$f_m^+ = \xi_m^+ + T\xi_m^+, \qquad f_m^- = \xi_m^- + T\xi_m^-.$$

The elements $\{f_m^{\pm}\}$ are J-orthogonal. Indeed, $[f_m^+, f_k^-] = 0$ because $f_m^{\pm} \in \mathfrak{L}'_{\pm}$ by the construction and the subspaces $\mathfrak{L}'_{\pm}$ are J-orthogonal. Furthermore,

$$[f_m^+, f_k^+] = (\xi_m^+ - T\xi_m^+, \xi_k^+ + T\xi_k^+) = ((I - T^2)\xi_m^+, \xi_k^+) = \mu_m(\xi_m^+, \xi_k^+) = \mu_m\delta_{mk}$$

Similarly, $[f_m^-, f_k^-] = -\mu_m\delta_{mk}$. Hence, the elements $\{f_m^{\pm}\}$ are J-orthogonal.

[10] P_{M_+} and P_{M_-} are orthogonal projections onto M_+ and M_- in $\mathfrak{H}$

Assume that an element $f^+ \in \mathfrak{L}'_+$ is orthogonal to span$\{f_m^+\}$. Then $f^+ = \gamma^+ + T\gamma^+$ (due to (6.3.21)) and

$$0 = (f^+, f_m^+) = (\gamma^+ + T\gamma^+, \xi_m^+ + T\xi_m^+)$$
$$= (\gamma^+, \xi_m^+) + (T\gamma^+, T\xi_m^+) = (\gamma^+, (I + T^2)\xi_m^+) = (2 - \mu_m)(\gamma^+, \xi_m^+),$$

where, obviously, $2 - \mu_m \neq 0$. This means that the element $\gamma^+ \in M_+$ is orthogonal to the basis $\{\xi_m^+\}$ of M_+. Hence, $\gamma^+ = 0$ and the closure of span$\{f_m^+\}$ coincides with $\mathfrak{L}'_+$. Similar arguments show that the closure of span$\{f_m^-\}$ coincides with $\mathfrak{L}'_-$.

Summing up, the closures of the linear spans of J-orthogonal vectors $\{f_m^\pm\}$ coincide with the definite subspaces $\mathfrak{L}'_\pm$. Their J-orthogonal direct sum $\mathfrak{L}'_+[\dot{+}]\mathfrak{L}'_-$ is a dense set in $\mathfrak{H}$. The positive/negative subspaces $\mathfrak{L}'_\pm$ are not maximal. The operator C' corresponding to $\mathfrak{L}'_+[\dot{+}]\mathfrak{L}'_-$ cannot be presented as Je^Q, where e^Q is a positive self-adjoint operator.

In general, there are many extensions of the subspaces $\mathfrak{L}'_\pm$ to maximal subspaces $\mathfrak{L}_\pm$ and hence, there are many extensions of C' to unbounded operators $C = Je^Q$. It follows from Proposition 6.3.9 that such an extension is unique if and only if $\frac{1}{2} < \delta \leq 1$. In that case, we can say that the direct sum $\mathfrak{L}'_+[\dot{+}]\mathfrak{L}'_-$ determines a unique unbounded operator C, which is defined by (6.3.19). The corresponding pair of maximal subspaces $\mathfrak{L}_\pm$ is determined by (6.2.10).

On the other hand, if $1 < \delta \leq \frac{3}{2}$, then $\mathfrak{L}'_+[\dot{+}]\mathfrak{L}'_-$ can be extended to *different J-orthogonal direct sums* of maximal positive/negative subspaces $\mathfrak{L}_+[\dot{+}]\mathfrak{L}_-$. This gives infinitely many unbounded operators $C = Je^Q$, which are extensions of C'. Hence, the direct sum $\mathfrak{L}'_+[\dot{+}]\mathfrak{L}'_-$ does not determine the operator C uniquely.

If the parameter δ increases and $\delta > \frac{3}{2}$, then the direct set $\mathfrak{L}'_+[\dot{+}]\mathfrak{L}'_-$ loses the property of being densely defined in $\mathfrak{H}$.

Denote by S the linear span of the J-orthogonal system $\{f_m^\pm\}$ and consider the elements $f_m^\pm$ as eigenvectors of a J-self-adjoint operator A. Then, according to Definition 6.3.8, the operator A has the property of unbounded C-symmetry realized by the operators C mentioned earlier. If $\frac{1}{2} < \delta \leq 1$, then the complete set of eigenvectors $\{f_m^\pm\}$ determines a unique unbounded operator C for A.

6.4 ELEMENTS OF $\mathcal{PT}$-SYMMETRIC OPERATORS THEORY

6.4.1 Definition of $\mathcal{PT}$-symmetric Operators and General Properties

Let $\mathfrak{H}$ be a complex Hilbert space with inner product $(\cdot, \cdot)$ linear in the first argument.

A bounded operator $\mathcal{T}$ defined on $\mathfrak{H}$ is called *conjugation* if

$$(i) \quad \mathcal{T}^2 = I, \qquad (ii) \quad (\mathcal{T}f, \mathcal{T}g) = (g, f)$$

for all $f, g \in \mathfrak{H}$. The conjugation operator is *antilinear* in the sense that

$$\mathcal{T}(\alpha f + \beta g) = \overline{\alpha}\mathcal{T}f + \overline{\beta}\mathcal{T}g, \qquad \alpha, \beta \in \mathbb{C}, \quad f, g \in \mathfrak{H}.$$

Let us fix a fundamental symmetry J in the Hilbert space $\mathfrak{H}$. In what follows, taking into account the terminology of $\mathcal{PT}$-symmetric quantum mechanics (5, 6, 11, 12), we often will use the symbol $\mathcal{P}$ instead of J. Thus, $\mathcal{P}$ is an arbitrary fundamental symmetry in the Hilbert space $\mathfrak{H}$.

Let $\mathcal{T}$ be a conjugation, which *commutes* with $\mathcal{P}$, that is, $\mathcal{PT} = \mathcal{TP}$.

Definition 6.4.1 *A closed densely defined linear operator A in $\mathfrak{H}$ is called $\mathcal{PT}$-symmetric if the relation*

$$\mathcal{PT}A = A\mathcal{PT} \tag{6.4.1}$$

holds on the domain $\mathcal{D}(A)$ of A.

REMARK **4** *The commutation condition between $\mathcal{P}$ and $\mathcal{T}$ means that $G := \mathcal{PT}$ is also a conjugation operator in $\mathfrak{H}$. The definition 6.4.1 of $\mathcal{PT}$-symmetry gives that A commutes with the conjugation operator G. Sometimes in the mathematical literature, an operator A that commutes with a given conjugation G is called* real with respect to *G (25). Real symmetric operators are examples of complex symmetric (G-symmetric) operators, which are characterized by the condition $A \subseteq GA^*G$, where G is a conjugation. Recent advances in the theory of complex symmetric operators with discussion about possible applications in non-Hermitian quantum mechanics can be found in the survey (40).*

Lemma 6.4.2 *If A is $\mathcal{PT}$-symmetric in a Hilbert space $\mathfrak{H}$, then its adjoint operator A^* is also $\mathcal{PT}$-symmetric.*

Proof: Let A be $\mathcal{PT}$-symmetric. It follows from (6.4.1) that for all $f \in \mathcal{D}(A)$ and all $g \in \mathcal{D}(A^*)$

$$(\mathcal{PT}Af, g) = (A\mathcal{PT}f, g) = (\mathcal{PT}f, A^*g) = (\mathcal{PT}A^*g, f).$$

On the other hand,

$$(\mathcal{PT}Af, g) = (\mathcal{T}Af, \mathcal{P}g) = (\mathcal{TP}g, Af) = (\mathcal{PT}g, Af).$$

Comparing the obtained relations, we conclude that $\mathcal{PT}g \in \mathcal{D}(A^*)$ and $\mathcal{PT}A^*g = A^*\mathcal{PT}g$ for all $g \in \mathcal{D}(A^*)$. Hence, the adjoint operator A^* is also $\mathcal{PT}$-symmetric. Lemma 6.4.2 is proved. ■

It follows from Definition 6.4.1 that the point spectrum σ_p, the residual spectrum σ_r, and the continuous spectrum σ_c of the $\mathcal{PT}$-symmetric operator A are all symmetric with respect to the real axis, that is,

$$\lambda \in \sigma_\alpha(A) \iff \overline{\lambda} \in \sigma_\alpha(A), \qquad \alpha \in \{p, r, c\}. \tag{6.4.2}$$

The definition of $\mathcal{PT}$-symmetry is quite general and the set of $\mathcal{PT}$-symmetric operators may contain operators with various curious properties.

6.4.1.1 Example of a $\mathcal{PT}$-symmetric Operator That Cannot Be Self-adjoint in Krein Spaces In the space $\mathfrak{H} = L_2(\mathbb{R})$, we consider the "standard" space reflection operator $\mathcal{P}$ (see (6.3.17) and the conjugation $\mathcal{T}$ operator. The operator

$$A = -\frac{d^2}{dx^2}, \qquad \mathcal{D}(A) = W_2^2(\mathbb{R} \setminus \{0\})$$

is $\mathcal{PT}$-symmetric. Let us suppose the existence of an operator of fundamental symmetry J in $L_2(\mathbb{R})$ such that the operator A turns out to be J-self-adjoint. Hence, the operator identity (6.3.4) holds, where $A^* = -\frac{d^2}{dx^2}$ is defined on

$$\mathcal{D}(A^*) = \{f \in W_2^2(\mathbb{R} \setminus \{0\}) : f(0+) = f(0-) = f'(0+) = f'(0-) = 0\}.$$

The point spectrum of A coincides with $\mathbb{C} \setminus \mathbb{R}_+$. According to our assumption, the operators A^* and A are related by (6.3.4). This means that the point spectrum of A^* also coincides with $\mathbb{C} \setminus \mathbb{R}_+$, that is impossible, because $\mathbb{C} \setminus \mathbb{R}_+$ belongs to the residual spectrum of A^*. The obtained contradiction means that the $\mathcal{PT}$-symmetric operator A cannot be self-adjoint in Krein spaces.

6.4.1.2 Example of a Self-adjoint Operator in a Krein Space That Cannot Be $\mathcal{PT}$-symmetric Let L be a maximal symmetric operator in the Hilbert space $\mathcal{K}$. In the Hilbert space $\mathfrak{H} = \mathcal{K} \oplus \mathcal{K}$ we consider the operator A and the fundamental symmetry J defined by (6.3.6). It is known (Section 6.3.1) that A is a self-adjoint operator in the Krein space $(\mathcal{K} \oplus \mathcal{K}, [\cdot, \cdot])$.

As L is a maximal symmetric operator, without loss of generality, we can assume that the upper half-plane $\mathbb{C}_+ = \{k \in \mathbb{C} \; : \; \mathrm{Im}\, k > 0\}$ belongs to the resolvent set $\rho(L)$, that is, $\rho(L) \supset \mathbb{C}_+$. In that case $\sigma_r(L) \supset \mathbb{C}_-$. For the adjoint operator L^* these inclusions are transformed into $\rho(L^*) \supset \mathbb{C}_-$ and $\sigma_p(L^*) \supset \mathbb{C}_+$. Taking into account the definition of A in (6.3.6), we conclude that

$$\sigma_p(A) \supset \mathbb{C}_+, \quad \text{and} \quad \sigma_r(L) \supset \mathbb{C}_-.$$

In view of (6.4.2)), the obtained inclusions are impossible for $\mathcal{PT}$-symmetric operators. Hence, the J-self-adjoint operator A cannot be interpreted as a $\mathcal{PT}$-symmetric operator.

The difference between $\mathcal{PT}$-symmetric operators and self-adjoint operators in Krein spaces was analyzed in Ref. (41) for operators defined by the differential expression

$$\tau_V(u)(x) = -u''(x) - V(x)u, \qquad V(x) = V(-x),$$

where V is a real valued function such that τ_V is in limit circle case at $\pm\infty$ (42).

6.4.1.3 Self-adjoint Operators as $\mathcal{PT}$-Symmetric Operators In contrast to the case of Krein spaces, any self-adjoint operator in a Hilbert space can be interpreted as a $\mathcal{PT}$-symmetric operator.

Proposition 6.4.3 *Every self-adjoint operator A acting in a separable Hilbert space $\mathfrak{H}$ is a $\mathcal{PT}$-symmetric operator for a suitable choice of $\mathcal{P}$ and $\mathcal{T}$.*

Proof: Let A be a self-adjoint operator in $\mathfrak{H}$. Then the spectral function of A commutes with some conjugation operator $\mathcal{T}'$ [43, Lemma 4]. This means that $\mathcal{T}'A = A\mathcal{T}'$. For the conjugation $\mathcal{T}'$, there exists an orthonormal basis $\{e_k\}$ of $\mathfrak{H}$ such that (see Ref. [25, Section 50])

$$\mathcal{T}'f = \sum_{n=1}^{\infty} \overline{\alpha}_n e_n, \qquad \text{if } f = \sum_{n=1}^{\infty} \alpha_n e_n, \quad \alpha_n \in \mathbb{C}.$$

Set

$$\mathcal{P}f = \sum_{n=1}^{\infty} (-1)^n \alpha_n e_n, \qquad \text{if } f = \sum_{n=1}^{\infty} \alpha_n e_n.$$

It is clear that $\mathcal{P}$ is a fundamental symmetry in $\mathfrak{H}$ and $\mathcal{PT}' = \mathcal{T}'\mathcal{P}$. Then, $\mathcal{T} = \mathcal{PT}'$ is a conjugation operator in $\mathfrak{H}$, which commutes with $\mathcal{P}$. Taking into account that $\mathcal{T}' = \mathcal{PT}$ we obtain the $\mathcal{PT}$-symmetry property of A. ∎

Proposition 6.4.3 is false for maximal dissipative operators and symmetric operators with different defect indices (n_+, n_-). Such operators cannot be $\mathcal{PT}$-symmetric. This simple result follows from the spectral properties of the corresponding classes of operators and (6.4.2).

6.4.1.4 Physically Meaningful $\mathcal{PT}$-symmetric Operators In general, the set of $\mathcal{PT}$-symmetric operators and the set of self-adjoint operators in Krein spaces are different. However, in our opinion, *every "physically meaningful" $\mathcal{PT}$-symmetric operator should be self-adjoint in a suitably chosen Krein space.*

In many cases, the additional "physical" condition of $\mathcal{PT}$-symmetry leads to a simplification of the Krein spaces theory. For example, the spectral analysis of operators that are self-adjoint in Krein spaces is much more difficult than in the case of Hilbert spaces, partly because the residual spectrum is in general not empty for the former. The additional condition of $\mathcal{PT}$-symmetry allows us to guarantee the absence of the residual spectrum.

Proposition 6.4.4 *Let A be a J-self-adjoint and $\mathcal{PT}$-symmetric operator. Then the residual spectrum of A is empty.*

Proof: If A is $\mathcal{PT}$-symmetric, then the adjoint A^* is also $\mathcal{PT}$-symmetric (Lemma 6.4.2). Assume that $\lambda \in \sigma_r(A)$. Then $\overline{\lambda} \in \sigma_p(A^*)$ and $\lambda \in \sigma_p(A^*)$ (as (6.4.2) holds for A^*). The latter gives $\lambda \in \sigma_p(A)$ due to (6.3.4). Hence, $\lambda \in \sigma_p(A) \cap \sigma_r(A)$, which is impossible unless $\sigma_r(A) = \emptyset$. ∎

REMARK 5 *The method used to prove the absence of the residual spectrum in Proposition 6.4.4 is, in some sense, "universal" and it can be repeated for many other classes of operators, see, for example, (44).*

It should be noted that $\mathcal{PT}$-symmetric operators are not necessarily realized as $\mathcal{P}$-self-adjoint operators (45, 46). The Clifford algebra technique is relevant for $\mathcal{PT}$-symmetric studies and it can be useful for the proper choice of fundamental symmetry J (47).

The self-adjointness of a $\mathcal{PT}$-symmetric operator A in a Krein space $(\mathfrak{H}, [\cdot, \cdot])$ is not completely satisfactory because it does not guarantee the unitarity of the dynamics generated by A. To do so one must demonstrate that A is self-adjoint on a *Hilbert space* (not on a Krein space!). This problem can be overcome for a J-self-adjoint operator A by finding a new symmetry represented by a linear operator $C = Je^{Q}$ and such that the semilinear form $(\cdot, \cdot)_C := [C\cdot, \cdot]$ is a (positively defined) inner product and A turns out to be self-adjoint in the Hilbert space $(\cdot, \cdot)_C$ (see Section 6.3.2 for details).

As a rule, the definition of C-symmetry for $\mathcal{PT}$-symmetric operators involves the additional requirement of $\mathcal{PT}$-symmetry of C, that is, $C\mathcal{PT} = \mathcal{PT}C$ (5, 11–14, 17). This condition arises naturally in $\mathcal{PT}$-symmetric quantum mechanics as a direct consequence of unbroken $\mathcal{PT}$-symmetry of a non-self-adjoint Hamiltonian (11). We recall that a $\mathcal{PT}$-symmetry of A is not spontaneously broken if eigenvectors of A are simultaneously eigenvectors of $\mathcal{PT}$. For a $\mathcal{PT}$-symmetric operator A with complete set of eigenvectors $\{f_n\}$ corresponding to simple eigenvalues, the unbroken $\mathcal{PT}$-symmetry means that the corresponding subspaces $\mathfrak{L}_\pm$ are $\mathcal{PT}$-invariant (as $\mathfrak{L}_\pm$ are the closure of the linear spans $\{f_n^\pm\}$ of positive/negative eigenvectors). Hence, the operator C defined by (6.2.8) should be $\mathcal{PT}$-symmetric.

We note that the condition of $\mathcal{PT}$-symmetry imposed on C, in many cases, leads to the simplification of the corresponding formulas (for instance, we can compare the general formula (6.4.11) of C with the expression (6.4.12) for $\mathcal{PT}$-symmetric operators C).

Summing up, we believe that a proper interpretation of $\mathcal{PT}$-symmetric operators as Hamiltonians of $\mathcal{PT}$-symmetric quantum mechanics includes the following stages:

- $\mathcal{PT}$-symmetric operator A in a Hilbert space $(\mathfrak{H}, (\cdot, \cdot))$;
- interpretation of A as a self-adjoint operator in a Krein space $(\mathfrak{H}, [\cdot, \cdot])$, where the corresponding fundamental symmetry J commutes with $\mathcal{PT}$;
- finding of the $\mathcal{PT}$-symmetric operator $C = Je^{Q}$ for A;
- interpretation of A as a self-adjoint operator in the Hilbert space $(\mathfrak{H}, (\cdot, \cdot)_C)$.

In the following section, we illustrate how this scheme works for the case of $\mathcal{PT}$-symmetric operators acting in the Hilbert space $(\mathbb{C}^2, (\cdot, \cdot))$ with the inner product

$$(f, g) = x_1\overline{y_1} + x_2\overline{y_2}, \qquad f = \begin{pmatrix} x_1 \\ x_2 \end{pmatrix}, \quad g = \begin{pmatrix} y_1 \\ y_2 \end{pmatrix}.$$

6.4.2 Two-dimensional Case

6.4.2.1 $\mathcal{PT}$-symmetric Operators We will identify operators acting in $\mathbb{C}^2$ with (2×2)-matrices. An arbitrary (2×2)-matrix A admits the decomposition

$$A = \sum_{j=0}^{3} a_j \sigma_j, \qquad a_j \in \mathbb{C}, \tag{6.4.3}$$

where $\sigma_0 = \begin{pmatrix} 1 & 0 \\ 0 & 1 \end{pmatrix}$ and

$$\sigma_1 = \begin{pmatrix} 0 & 1 \\ 1 & 0 \end{pmatrix}, \quad \sigma_2 = \begin{pmatrix} 0 & -i \\ i & 0 \end{pmatrix}, \quad \sigma_3 = \begin{pmatrix} 1 & 0 \\ 0 & -1 \end{pmatrix}$$

are the Pauli matrices. The Pauli matrices determine the operators of fundamental symmetry in $\mathbb{C}^2$.

Let A be defined by (6.4.3). Then

$$\mathbf{det}\, A = a_0^2 - \sum_{j=1}^{3} a_j^2. \tag{6.4.4}$$

We choose the operator of fundamental symmetry $\mathcal{P}$ and the conjugation operator $\mathcal{T}$ in $\mathbb{C}^2$ as follows:

$$\mathcal{P} = \sigma_1, \qquad \mathcal{T}\begin{pmatrix} x_1 \\ x_2 \end{pmatrix} = \begin{pmatrix} \overline{x}_1 \\ \overline{x}_2 \end{pmatrix}.$$

For this reason, we will use the term $\sigma_1\mathcal{T}$-symmetry instead of $\mathcal{PT}$-symmetry.

Taking into account the standard properties of the Pauli matrices

$$\begin{aligned} &\sigma_j\sigma_k = -\sigma_k\sigma_j,\ j \neq k, \\ &\sigma_3\sigma_1 = i\sigma_2,\ \sigma_1\sigma_2 = i\sigma_3,\ \sigma_2\sigma_3 = i\sigma_1,\ \sigma_j^2 = \sigma_0 \end{aligned} \tag{6.4.5}$$

and the relations

$$\mathcal{T}\sigma_1 = \sigma_1\mathcal{T}, \qquad \mathcal{T}\sigma_2 = -\sigma_2\mathcal{T}, \qquad \mathcal{T}\sigma_3 = \sigma_3\mathcal{T}$$

we conclude that the operator A is $\sigma_1\mathcal{T}$-symmetric if and only if its coefficients a_j in (6.4.3) satisfy the relations

$$a_0,\ a_1, a_2 \in \mathbb{R}, \qquad \text{and} \qquad a_3 \in i\mathbb{R}. \tag{6.4.6}$$

In view of (6.4.4) and (6.4.6), the determinant $\mathbf{det}\, A$ and the trace $\mathbf{tr}\, A = 2a_0$ are real numbers.

Lemma 6.4.5 *The spectrum of a $\sigma_1\mathcal{T}$-symmetric operator A is real if and only if* $(\mathbf{tr}\ A)^2 \geq 4\mathbf{det}\ A$.

Proof: A point $\lambda \in \mathbb{C}$ belongs to $\sigma(A)$ if $\mathbf{det}(A - \lambda\sigma_0) = 0$. Using (6.4.4),

$$\mathbf{det}(A - \lambda\sigma_0) = (a_0 - \lambda)^2 - \sum_{j=1}^{3} a_j^2 = \lambda^2 - (\mathbf{tr}\ A)\lambda + \mathbf{det}\ A = 0.$$

The obtained polynomial has real roots if and only if $(\mathbf{tr}\ A)^2 - 4\mathbf{det}\ A \geq 0$. ∎

Corollary 6.4.6 *The spectrum of a $\sigma_1\mathcal{T}$-symmetric operator A is real if and only if*

$$a_1^2 + a_2^2 - |a_3|^2 \geq 0,$$

where a_j are coefficients of the decomposition (6.4.3) of A.

Proof: This follows directly from (6.4.4), (6.4.6) and Lemma 6.4.5. ∎

6.4.2.2 Fundamental Symmetries in $\mathbb{C}^2$ Denote $B = \sum_{j=0}^{3} b_j\sigma_j$. Taking (6.4.5) into account, it is easy to check

$$A \cdot B = \sum_{j=0}^{3} a_jb_j\sigma_0 + \sum_{j=1}^{3}(a_0b_j + a_jb_0)\sigma_j + i\vec{a} \times \vec{b}, \tag{6.4.7}$$

where $\vec{a} = (a_1, a_2, a_3)$, $\vec{b} = (b_1, b_2, b_3)$ and the "cross product" $\vec{a} \times \vec{b}$ associated with Pauli matrices is determined via the formal determinant

$$\vec{a} \times \vec{b} = \begin{vmatrix} \sigma_1 & \sigma_2 & \sigma_3 \\ a_1 & a_2 & a_3 \\ b_1 & b_2 & b_3 \end{vmatrix},$$

where

$$\begin{vmatrix} \sigma_1 & \sigma_2 & \sigma_3 \\ a_1 & a_2 & a_3 \\ b_1 & b_2 & b_3 \end{vmatrix} = \begin{vmatrix} a_2 & a_3 \\ b_2 & b_3 \end{vmatrix}\sigma_1 - \begin{vmatrix} a_1 & a_3 \\ b_1 & b_3 \end{vmatrix}\sigma_2 + \begin{vmatrix} a_1 & a_2 \\ b_1 & b_2 \end{vmatrix}\sigma_3.$$

Lemma 6.4.7 *The operator A is a nontrivial fundamental symmetry in $\mathbb{C}^2$ if and only if $A = \sum_{j=1}^{3} a_j\sigma_j$, where $\vec{a} = (a_1, a_2, a_3)$ belongs to the unit sphere $\mathbb{S}^2$ in $\mathbb{R}^3$.*

Proof: It suffices to check the self-adjointness of A and the condition $A^2 = \sigma_0$. The self-adjointness is equivalent to the condition $a_j \in \mathbb{R}$ $(j = 0, \ldots 3)$.

Using (6.4.7) for $A = B$, we get

$$A^2 = \sum_{j=0}^{3} a_j^2\sigma_0 + 2\sum_{j=1}^{3} a_0a_j\sigma_j.$$

Hence, $A^2 = \sigma_0$ if and only if either A is a trivial fundamental symmetry $A = \pm\sigma_0$ or $A = \sum_{j=1}^{3} a_j\sigma_j$, where $\sum_{j=1}^{3} a_j^2 = 1$. ∎

Lemma 6.4.7 shows that the nontrivial fundamental symmetries in the Hilbert space $(\mathbb{C}^2, (\cdot,\cdot))$ are parameterized by points of the unit sphere $\mathbb{S}^2$.

Corollary 6.4.8 *If a nontrivial fundamental symmetry A is $\sigma_1\mathcal{T}$-symmetric, then $A = \sigma_1 e^{i\phi\sigma_3}$, where $\phi \in \mathbb{R}$.*

Proof: A nontrivial fundamental symmetry $A = \sum_{j=1}^{3} a_j\sigma_j$ will commute with $\sigma_1\mathcal{T}$ if and only if $a_3 = 0$. Hence, $A = a_1\sigma_1 + a_2\sigma_2$, where $a_1^2 + a_2^2 = 1$. Denote $\cos\phi = \alpha_1$ and $\sin\phi = \alpha_2$, $\phi \in \mathbb{R}$. Then

$$A = (\cos\phi)\sigma_1 + (\sin\phi)\sigma_2 = \sigma_1(\cos\phi\sigma_0 + i\sin\phi\sigma_3) = \sigma_1 e^{i\phi\sigma_3}.$$

∎

Thus, every $\sigma_1\mathcal{T}$-symmetric, nontrivial fundamental symmetry in $\mathbb{C}^2$ has the form $\sigma_{1\phi} := \sigma_1 e^{i\phi\sigma_3}$ and it defines the indefinite inner product

$$[\cdot,\cdot]_\phi = (\sigma_{1\phi}\cdot,\cdot), \qquad \phi \in \mathbb{R},$$

which preserves the conjugation property of $\sigma_1\mathcal{T}$: $[\sigma_1\mathcal{T}f, \sigma_1\mathcal{T}g]_\phi = [g,f]_\phi$.

6.4.2.3 Self-adjointness in the Krein Space $(\mathbb{C}^2, [\cdot,\cdot]_\phi)$

Lemma 6.4.9 *A $\sigma_1\mathcal{T}$-symmetric operator A is self-adjoint in the Krein space $(\mathbb{C}^2, [\cdot,\cdot]_\phi)$, where the parameter ϕ is determined by the relation*

$$a_1\sin\phi = a_2\cos\phi,$$

which involves the coefficients a_1, a_2 of the decomposition (6.4.3) of A.

Proof: In view of (6.4.3) and (6.4.6), the adjoint of A is

$$A^* = a_0\sigma_0 + a_1\sigma_1 + a_2\sigma_2 - a_3\sigma_3. \tag{6.4.8}$$

By virtue of (6.3.4), the operator A is self-adjoint in the Krein space $(\mathbb{C}^2, [\cdot,\cdot]_\phi)$ if and only if

$$\sigma_{1\phi}A = A^*\sigma_{1\phi}. \tag{6.4.9}$$

Taking into account that $\sigma_{1\phi} = (\cos\phi)\sigma_1 + (\sin\phi)\sigma_2$ and the operators A, A^* are defined by (6.4.3) and (6.4.8), respectively, we rewrite both sides of (6.4.9) as follows:

$$\begin{aligned}\sigma_{1\phi}A &= (a_1\cos\phi + a_2\sin\phi)\sigma_0 + (a_0\cos\phi + ia_3\sin\phi)\sigma_1\\ &\quad + (a_0\sin\phi - ia_3\cos\phi)\sigma_2 + (ia_2\cos\phi - ia_1\sin\phi)\sigma_3,\end{aligned}$$

$$A^*\sigma_{1\phi} = (a_1 \cos\phi + a_2 \sin\phi)\sigma_0 + (a_0 \cos\phi + ia_3 \sin\phi)\sigma_1$$
$$+ (a_0 \sin\phi - ia_3 \cos\phi)\sigma_2 + (-ia_2 \cos\phi + ia_1 \sin\phi)\sigma_3.$$

Comparing the obtained decompositions we arrive at the conclusion that the relation (6.4.9) holds if and only if $ia_2 \cos\phi - ia_1 \sin\phi = 0$. The proof is complete. ■

The Krein space $(\mathbb{C}^2, [\cdot,\cdot]_\phi)$ is not determined uniquely in Lemma 6.4.9. For instance, if A is self-adjoint with respect to the indefinite metric $[\cdot,\cdot]_\phi$, then A will also be self-adjoint with respect to $[\cdot,\cdot]_{\phi+\pi} = -[\cdot,\cdot]_\phi$. It looks natural to identify such Krein spaces. For this reason, we will assume that the Krein space $(\mathbb{C}^2, [\cdot,\cdot]_\phi)$ in Lemma 6.4.9 is not determined uniquely if there exists at least two parameters ϕ_1, ϕ_2 such that $|\phi_1 - \phi_2| \neq k\pi$ $(k \in \mathbb{N})$ and operator A is self-adjoint with respect to both indefinite inner products $[\cdot,\cdot]_{\phi_j}$ $(j = 1, 2)$.

Corollary 6.4.10 *If the Krein spaces $(\mathbb{C}^2, [\cdot,\cdot]_\phi)$ in Lemma 6.4.9 is not determined uniquely for a $\sigma_1\mathcal{T}$-symmetric operator A, then either $A = a_0\sigma_0$ $(a_0 \in \mathbb{R})$ or the spectrum of A is nonreal.*

Proof: In view of Lemma 6.4.9, the Krein space $(\mathbb{C}^2, [\cdot,\cdot]_\phi)$ is not determined uniquely for A if and only if $a_1 = a_2 = 0$. Using Corollary 6.4.6, we complete the proof. ■

6.4.2.4 Operators C in the Krein Space $(\mathbb{C}^2, [\cdot,\cdot]_\phi)$ According to Theorem 6.2.1, every operator C in $(\mathbb{C}^2, [\cdot,\cdot]_\phi)$ has the form $C = \sigma_{1\phi}e^Q$, where $\sigma_{1\phi} = \sigma_1 e^{i\phi\sigma_3}$ and Q is a self-adjoint operator in $(\mathbb{C}^2, (\cdot,\cdot))$, which anticommutes with the fundamental symmetry $\sigma_{1\phi}$.

It is convenient to characterize the anticommutation relation between Q and $\sigma_{1\phi}$ in terms of coefficients of the decomposition of Q with respect to a basis involving $\sigma_{1\phi}$. To this end, we consider a fundamental symmetry $\sigma_{1(\phi+\frac{\pi}{2})}$ and prove that $\sigma_{1\phi}\sigma_{1(\phi+\frac{\pi}{2})} = -\sigma_{1(\phi+\frac{\pi}{2})}\sigma_{1\phi}$. As,

$$\sigma_{1(\phi+\frac{\pi}{2})} = \cos(\phi + \frac{\pi}{2})\sigma_1 + \sin(\phi + \frac{\pi}{2})\sigma_2 = -\sin\phi\sigma_1 + \cos\phi\sigma_2,$$

the matrices $\sigma_{1\phi}$ and $\sigma_{1(\phi+\frac{\pi}{2})}$ can be presented as the matrices A and B in (6.4.7) with $\vec{a} = (\cos\phi, \sin\phi, 0)$ and $\vec{b} = (-\sin\phi, \cos\phi, 0)$. Then (6.4.7) gives

$$\sigma_{1\phi}\sigma_{1(\phi+\frac{\pi}{2})} = i\vec{\alpha} \times \vec{\beta} = -i\vec{\beta} \times \vec{\alpha} = -\sigma_{1(\phi+\frac{\pi}{2})}\sigma_{1\phi}.$$

We define the third fundamental symmetry of the basis by $i\sigma_{1(\phi+\frac{\pi}{2})}\sigma_{1\phi}$. Obviously, it anticommutes with $\sigma_{1\phi}$ and with $\sigma_{1(\phi+\frac{\pi}{2})}$. Using (6.4.7) again we obtain

$$i\sigma_{1(\phi+\frac{\pi}{2})}\sigma_{1\phi} = \sigma_3. \tag{6.4.10}$$

Let us decompose Q with respect to the basis $\sigma_0, \sigma_{1\phi}, \sigma_{1(\phi+\frac{\pi}{2})}, \sigma_3$, that is,

$$Q = q_0\sigma_0 + q_1\sigma_{1\phi} + q_2\sigma_{1(\phi+\frac{\pi}{2})} + q_3\sigma_3, \qquad q_j \in \mathbb{C}.$$

The condition of self-adjointness of Q implies that $q_j \in \mathbb{R}$. On the other hand, the anticommutation with $\sigma_{1\phi}$ is possible only for $q_0 = q_1 = 0$. Hence,

$$Q = q_2\sigma_{1(\phi+\frac{\pi}{2})} + q_3\sigma_3 = \rho(\cos\xi\sigma_{1(\phi+\frac{\pi}{2})} + \sin\xi\sigma_3) = \rho Z,$$

where $\rho = \sqrt{q_2^2+q_3^2}$, $\cos\xi = \frac{q_2}{\sqrt{q_2^2+q_3^2}}$, $\sin\xi = \frac{q_3}{\sqrt{q_2^2+q_3^2}}$, and

$$Z = \cos\xi\sigma_{1(\phi+\frac{\pi}{2})} + \sin\xi\sigma_3$$

is a fundamental symmetry in the Hilbert space $(\mathbb{C}^2, (\cdot,\cdot))$ (as $\sigma_{1(\phi+\frac{\pi}{2})}$ anticommutes with σ_3). In that case

$$\begin{aligned} C = \sigma_{1\phi}e^{\rho Z} &= \sigma_{1\phi}(\cosh\rho\sigma_0 + \sinh\rho Z) \\ &= \cosh\rho\sigma_{1\phi} - i\sinh\rho\sin\xi\sigma_{1(\phi+\frac{\pi}{2})} + i\sinh\rho\cos\xi\sigma_3. \end{aligned} \tag{6.4.11}$$

Thus, we proved:

Proposition 6.4.11 *Every operator $C = \sigma_{1\phi}e^Q$ acting in the Krein space $(\mathbb{C}^2, [\cdot,\cdot]_\phi)$ is determined by (6.4.11), where $\xi \in \mathbb{R}$ and $\rho \geq 0$.*

The additional condition of $\sigma_1\mathcal{T}$-symmetry leads to the simplification of (6.4.11).

Corollary 6.4.12 *If an operator C in the Krein space $(\mathbb{C}^2, [\cdot,\cdot]_\phi)$ is $\sigma_1\mathcal{T}$-symmetric then*

$$C = \sigma_{1\phi}e^{\chi\sigma_{1(\phi+\frac{\pi}{2})}}, \qquad \chi \in \mathbb{R}. \tag{6.4.12}$$

Proof: In view of Corollary 6.4.8, the fundamental symmetries $\sigma_{1\phi}$ and $\sigma_{1(\phi+\frac{\pi}{2})}$ are $\sigma_1\mathcal{T}$-symmetric. On the other hand, σ_3 anticommutes with $\sigma_1\mathcal{T}$. Hence, the operator C in (6.4.11) is $\sigma_1\mathcal{T}$-symmetric if and only if $\sinh\rho\sin\xi = 0$. Assume firstly that $\sin\xi = 0$. Then $\cos\xi \in \{-1, 1\}$. Denote $\chi = \rho$ if $\cos\xi = 1$, otherwise $\chi = -\rho$. Then the formula (6.4.11) is rewritten as

$$\begin{aligned} C &= \sigma_{1\phi}(\cosh\chi\sigma_0 + i\sinh\chi\sigma_{1\phi}\sigma_3) \\ &= \sigma_{1\phi}(\cosh\chi\sigma_0 + \sinh\chi\sigma_{1(\phi+\frac{\pi}{2})}) = \sigma_{1\phi}e^{\chi\sigma_{1(\phi+\frac{\pi}{2})}}. \end{aligned}$$

If $\sinh\rho = 0$, then $\cosh\rho = 1$ and (6.4.11) is reduced to $C = \sigma_{1\phi}$. The corollary is proved. ■

6.4.2.5 The Property of C-symmetry for $\sigma_1\mathcal{T}$-symmetric Operators

Lemma 6.4.13 *A $\sigma_1\mathcal{T}$-symmetric operator A has a unique operator of C-symmetry if and only if* $(\mathbf{tr}\ A)^2 > 4\mathbf{det}\ A$. *The corresponding operator C also is $\sigma_1\mathcal{T}$-symmetric and it is determined by (6.4.12).*

Proof: The inequality $(\mathbf{tr}\ A)^2 > 4\mathbf{det}\ A$ is equivalent to the existence of two real eigenvalues of A (Lemma 6.4.5). Let f_1, f_2 be the corresponding eigenvectors. The operator A is self-adjoint in the Krein space $(\mathbb{C}^2, [\cdot,\cdot]_\phi)$ for a special choice of parameter ϕ (Lemma 6.4.9). Hence, the eigenvectors f_1, f_2 are orthogonal with respect to the inner product $[\cdot,\cdot]_\phi$. Moreover $[f_1,f_1]_\phi[f_2,f_2]_\phi < 0$ (as the elements f_1, f_2 form an orthogonal basis in the Krein space $(\mathbb{C}^2, [\cdot,\cdot]_\phi)$). Assume without loss of generality that $[f_1,f_1]_\phi > 0$ and $[f_1,f_1]_\phi < 0$. Then

$$\mathbb{C}^2 = \mathfrak{L}_+[\perp]\mathfrak{L}_-, \qquad \mathfrak{L}_+ = \text{span}\{f_1\}, \quad \mathfrak{L}_- = \text{span}\{f_2\} \tag{6.4.13}$$

and the operator A has the property of C-symmetry, where the corresponding operator C is defined by (6.4.13).

The eigenvectors f_j satisfy the relation $\sigma_1\mathcal{T}f_j = f_j$ as A is $\sigma_1\mathcal{T}$-symmetric. Hence, the subspaces $\mathfrak{L}_\pm$ in (6.4.13) are $\sigma_1\mathcal{T}$-invariant. Recalling (6.2.8) we conclude that C is $\sigma_1\mathcal{T}$-symmetric. This means that C is determined by (6.4.12). The uniqueness of C follows from Theorem 6.3.7. ■

It is easy to check that the equality $(\mathbf{tr}\ A)^2 = 4\mathbf{det}\ A$ corresponds to the case where either A commutes with any operator C and coincides with $a_0\sigma_0$ $(a_0 \in \mathbb{R})$ or A has no C-symmetries.

Lemma 6.4.13 shows that the condition of $\sigma_1\mathcal{T}$-symmetry of C is natural for the case of $\sigma_1\mathcal{T}$-symmetric operators.

It is interesting to note that the property of C-symmetry determines the principal structure of the corresponding $\sigma_1\mathcal{T}$-symmetric operator. It seems that this fact was firstly mentioned in Ref. (48).

Theorem 6.4.14 *The set of all $\sigma_1\mathcal{T}$-symmetric operators in $\mathbb{C}^2$ with the property of C-symmetry is determined by the formula*

$$A = \gamma_1\sigma_0 + \gamma_2\sigma_{1\phi}e^{\chi\sigma_{1(\phi+\frac{\pi}{2})}}, \tag{6.4.14}$$

where $\gamma_1, \gamma_2, \chi, \phi$ are real numbers.

Proof: The operator A determined by (6.4.14) is $\sigma_1\mathcal{T}$-symmetric as $\sigma_{1\phi}$ and $\sigma_{1(\phi+\frac{\pi}{2})}$ are $\sigma_1\mathcal{T}$-symmetric. Furthermore A commutes with $C = \sigma_{1\phi}e^{\chi\sigma_{1(\phi+\frac{\pi}{2})}}$. Therefore, A has the property of C-symmetry.

Conversely, assume that a $\sigma_1\mathcal{T}$-symmetric operator A has the property of C-symmetry. In view of Lemma 6.4.9, there exists $\phi \in \mathbb{R}$ such that A is a self-adjoint operator in the Krein space $(\mathbb{C}^2, [\cdot,\cdot]_\phi)$. Similarly to the proof of Proposition 6.4.11 we decompose A with respect to the basis σ_0, $\sigma_{1\phi}$, $\sigma_{1(\phi+\frac{\pi}{2})}$, σ_3, that is,

$$A = b_0\sigma_0 + b_1\sigma_{1\phi} + b_2\sigma_{1(\phi+\frac{\pi}{2})} + b_3\sigma_3, \qquad b_j \in \mathbb{C}. \tag{6.4.15}$$

The property of $\sigma_1\mathcal{T}$-symmetry of A is equivalent to the following conditions on b_j:

$$b_0 = \overline{b}_0, \quad b_1 = \overline{b}_1, \quad b_2 = \overline{b}_2, \quad b_3 = -\overline{b}_3.$$

The property of C-symmetry of A means the existence of $C = \sigma_{1\phi}e^{\chi\sigma_{1(\phi+\frac{\pi}{2})}}$ such that $AC = CA$. Using (6.4.10), (6.4.15) and taking into account that (see the proof of Corollary 6.4.12)

$$\sigma_{1\phi}e^{\chi\sigma_{1(\phi+\frac{\pi}{2})}} = \cosh\chi\sigma_{1\phi} + i\sinh\chi\sigma_3$$

we calculate

$$\begin{aligned} AC = {} & b_0C + b_1(\cosh\chi\sigma_0 + \sinh\chi\sigma_{1(\phi+\frac{\pi}{2})}) \\ & + b_2(-i\cosh\chi\sigma_3 - \sinh\chi\sigma_{1\phi}) + b_3(i\cosh\chi\sigma_{1(\phi+\frac{\pi}{2})} + i\sinh\chi\sigma_0). \end{aligned}$$

and

$$\begin{aligned} CA = {} & b_0C + b_1(\cosh\chi\sigma_0 - \sinh\chi\sigma_{1(\phi+\frac{\pi}{2})}) \\ & + b_2(i\cosh\chi\sigma_3 + \sinh\chi\sigma_{1\phi}) + b_3(-i\cosh\chi\sigma_{1(\phi+\frac{\pi}{2})} + i\sinh\chi\sigma_0). \end{aligned}$$

By grouping the coefficients with respect to the basic matrices σ_0, $\sigma_{1\phi}$, $\sigma_{1(\phi+\frac{\pi}{2})}$, σ_3 and equating them we arrive at the conclusion that $AC = CA$ if and only if the coefficients b_j in (6.4.15) satisfy the relations

$$b_2 = 0, \qquad b_1\sinh\chi + ib_3\cosh\chi = 0. \tag{6.4.16}$$

Hence, the decomposition (6.4.15) can be modified:

$$A = b_0\sigma_0 + \frac{b_1}{\cosh\chi}(\cosh\chi\sigma_{1\phi} + i\sinh\chi\sigma_3) = b_0\sigma_0 + \frac{b_1}{\cosh\chi}\sigma_{1\phi}e^{\chi\sigma_{1(\phi+\frac{\pi}{2})}}.$$

Denoting $\gamma_1 = b_0$ and $\gamma_2 = \frac{b_1}{\cosh\chi}$ we complete the proof. ∎

Denote

$$\mathrm{Im}\,A = \frac{A - A^*}{2i}, \qquad \mathrm{Re}\,A = \frac{A + A^*}{2}.$$

REMARK 6 *The parameter χ in (6.4.14) can be determined by the relation*

$$\tanh^2 \chi = \frac{(\mathbf{tr}\ Im\ A)^2 - 4\mathbf{det}\ Im\ A}{(\mathbf{tr}\ Re\ A)^2 - 4\mathbf{det}\ Re\ A}. \tag{6.4.17}$$

Indeed, in view of (6.4.15) and (6.4.16),

$$A = b_0\sigma_0 + b_1\sigma_{1\phi} + ib_1 \tanh \chi\sigma_3, \qquad b_0, b_1 \in \mathbb{R}.$$

Hence,

$$\mathrm{Re}\, A = b_0\sigma_0 + b_1\sigma_{1\phi}, \qquad \mathrm{Im}\, A = b_1 \tanh \chi\sigma_3$$

and $\mathbf{tr}\ \mathrm{Re}\, A = 2b_0$, $\mathbf{tr}\ \mathrm{Im}\, A = 0$, $\mathbf{det}\ \mathrm{Re}\, A = b_0^2 - b_1^2$, $\mathbf{det}\ \mathrm{Im}\, A = -b_1^2 \tanh^2 \chi$, which proves (6.4.17).

Corollary 6.4.15 *Let A be a $\sigma_1\mathcal{T}$-symmetric operator in $\mathbb{C}^2$. The following statements are equivalent:*

(i) the inequality $(\mathbf{tr}\ A)^2 > 4\mathbf{det}\ A$ holds;
(ii) the operator A is determined by (6.4.14), where $\gamma_1, \gamma_2, \chi, \phi$ are real numbers and $\gamma_2 \neq 0$.

Proof: This follows easily from Lemma 6.4.13 and Theorem 6.4.14. ■

In conclusion we note that the formula (6.4.14) is highly informative. Indeed, if A is given by (6.4.14), then A is $\sigma_{1\phi}$-self-adjoint (i.e., A is self-adjoint in the Krein space $(\mathbb{C}^2, [\cdot,\cdot]_\phi)$). Its operator of C-symmetry has the form $C = \sigma_{1\phi}e^{\chi\sigma_{1(\phi+\frac{\pi}{2})}}$ and it is determined uniquely when $\gamma_2 \neq 0$. The corresponding metric operator is $e^{\chi\sigma_{1(\phi+\frac{\pi}{2})}}$ and A is self-adjoint in the Hilbert space $(\mathbb{C}^2, (\cdot,\cdot)_C)$, where $(\cdot,\cdot)_C = (e^{\chi\sigma_{1(\phi+\frac{\pi}{2})}}\cdot,\cdot)$.

If $\gamma_2 \neq 0$, then A has two real eigenvalues $\lambda_{1,2}$, which are expressed via $\gamma_{1,2}$:

$$\lambda_{1,2} = \gamma_1 \pm \gamma_2. \tag{6.4.18}$$

To establish (6.4.18) we compare (6.4.15) with (6.4.14) that gives

$$b_0 = \gamma_1, \qquad b_1 = \gamma_2 \cosh \chi, \qquad b_2 = 0.$$

Then, recalling that $b_3 = ib_1 \tanh \chi$ (see (6.4.16)), the determinant of A can be calculated:

$$\mathbf{det}\, A = b_0^2 - \sum_{j=1}^{3} b_j^2 = b_0^2 - b_1^2(1 - \tanh^2 \chi) = \gamma_1^2 - \gamma_2^2(\cosh^2 \chi - \sinh^2 \chi) = \gamma_1^2 - \gamma_2^2$$

.

It follows from (6.4.15) that $\mathbf{tr}\ A = 2b_0 = 2\gamma_1$. Hence, $(\mathbf{tr}\ A)^2 - 4\mathbf{det}\ A = 4\gamma_2^2$. Taking into account that the eigenvalues of A coincide with the roots of

$$\lambda^2 - (\mathbf{tr}\ A)\lambda + \mathbf{det}\ A = 0,$$

we prove (6.4.18).

6.4.3 Schrödinger Operator with $\mathcal{PT}$-symmetric Zero-range Potentials

6.4.3.1 Preliminaries The results of Section 6.4.2 can easily be used for the investigation of Schrödinger operator with $\mathcal{PT}$-symmetric point interactions (47, 49–53). We illustrate this by considering a one-dimensional Schrödinger operator with $\mathcal{PT}$-symmetric zero-range potential at the point $x = 0$, which can be defined by the formal expression

$$-\frac{d^2}{dx^2} + a < \delta, \cdot > \delta(x) + b < \delta', \cdot > \delta(x) + c < \delta, \cdot > \delta'(x) + d < \delta', \cdot > \delta'(x), \tag{6.4.19}$$

where δ and δ' are, respectively, the Dirac δ-function and its derivative (with support at 0) and a, b, c, d are complex numbers. For $\mathcal{P}$ and $\mathcal{T}$, we mean the standard space reflection (6.3.17) and the conjugation operator in $L_2(\mathbb{R})$.

The $\mathcal{PT}$-symmetry property of the singular potential in (6.4.19) is characterized by the condition $a, d \in \mathbb{R}$ and $c, b \in i\mathbb{R}$ (47).

Using the regularization method suggested in Ref. (54), a direct relationship between parameters of the singular potential and operator realizations of (6.4.19) in the Hilbert space $L_2(\mathbb{R})$ can be established (52). Precisely, the formal expression (6.4.19) gives rise to the operator

$$A_{\mathbf{T}} = -\frac{d^2}{dx^2}, \qquad \mathbf{T} = \begin{pmatrix} a & b \\ c & d \end{pmatrix}$$

defined on smooth functions $f \in W_2^2(\mathbb{R}\backslash\{0\})$, which satisfy the boundary condition

$$\mathbf{T}\begin{pmatrix} \dfrac{f(0+) + f(0-)}{2} \\ \dfrac{-f'(0+) - f'(0-)}{2} \end{pmatrix} = \begin{pmatrix} f'(0+) - f'(0-) \\ f(0+) - f(0-) \end{pmatrix}. \tag{6.4.20}$$

Having in mind to apply the results of Section 6.4.2, we rewrite the boundary condition (6.4.20) in the more convenient form

$$\mathfrak{T}\begin{pmatrix} f(0+) + f'(0+) \\ f(0-) - f'(0-) \end{pmatrix} = \frac{1}{2}\begin{pmatrix} f(0+) \\ f(0-) \end{pmatrix}, \tag{6.4.21}$$

where $\mathfrak{T}$ is a (2×2)-matrix.

The set of operators $A_{\mathfrak{T}} = -\frac{d^2}{dx^2}$ with boundary condition (6.4.21) does not coincide with the set of operators $A_{\mathbf{T}}$ defined by (6.4.20). Namely, the domain of definition of $A_{\mathbf{T}}$ can be rewritten as (6.4.21) if and only $-1 \in \rho(A_{\mathbf{T}})$ or, that is equivalent (55), if

$$\Xi = 4 - \mathbf{det}\ \mathbf{T} + 2(a - d) \neq 0,$$

where $\mathbf{det}\ \mathbf{T} = ad - bc$. In that case, the matrix $\mathfrak{T}$ is expressed in terms of $\mathbf{T}$:

$$\mathfrak{T} = \frac{1}{4\Xi}\begin{pmatrix} \Xi + 2(b+c-a-d) & 4 + \mathbf{det}\ \mathbf{T} - 2(b-c) \\ 4 + \mathbf{det}\ \mathbf{T} + 2(b-c) & \Xi - 2(b+c+a+d) \end{pmatrix}.$$

It should be noted that the condition $-1 \in \rho(A_{\mathbf{T}})$ is just technical and the following results can be easily modified for any operator $A_{\mathbf{T}}$ with nonempty resolvent set.

In what follows we will assume that $\mathfrak{T}$ is an arbitrary (2×2)-matrix in (6.4.21).

The operators $A_{\mathfrak{T}}$ are extensions of the symmetric operator

$$A_s = -\frac{d^2}{dx^2}, \qquad \mathcal{D}(A_s) = \left\{ f \in W_2^2(\mathbb{R}\backslash\{0\}) \ : \ \begin{array}{l} f(0+) = f(0-) = 0 \\ f'(0+) = f'(0-) = 0 \end{array} \right\}$$

and restrictions of its adjoint

$$A_s^* = -\frac{d^2}{dx^2}, \qquad \mathcal{D}(A_s^*) = W_2^2(\mathbb{R}\backslash\{0\}).$$

For this reason, the operators $A_{\mathfrak{T}}$ coincide with the restrictions of A_s^* onto the domains

$$\mathfrak{D}(A_{\mathfrak{T}}) = \{f \in W_2^2(\mathbb{R}\backslash\{0\}) \ : \ \mathfrak{T}\Gamma_1 f = \Gamma_0 f\}, \tag{6.4.22}$$

where

$$\Gamma_0 f = \frac{1}{2}\begin{pmatrix} f(0+) \\ f(0-) \end{pmatrix}, \qquad \Gamma_1 f = \begin{pmatrix} f(0+) + f'(0+) \\ f(0-) - f'(0-) \end{pmatrix}. \tag{6.4.23}$$

REMARK 7 *Not every operator $A_{\mathfrak{T}}$ defined by (6.4.22) (or, that is equivalent, by (6.4.21)) can be interpreted as pseudo-Hermitian or $\mathcal{PT}$-symmetric. The operators $A_{\mathfrak{T}}$ are examples of* quasi-self-adjoint operators *(25). It should be noted that the notion of quasi-self-adjoint operators in the mathematical literature completely differs from the Dieudonné's notion of quasi-Hermitian operators (56). Precisely, let A_s be a symmetric operator with deficiency indices (m, m) $(m < \infty)$ acting in a Hilbert space $\mathfrak{H}$. An operator A is a quasi-self-adjoint extension of A_s if $A_s \subset A \subset A_s^*$ (i.e., A is a proper extension of A_s) and* $\dim \mathcal{D}(A) = m$ *(mod $\mathcal{D}(A_s)$).*

Denote

$$\mathcal{R}f(x) = (\text{sgn}\, x)f(x), \qquad f \in L_2(\mathbb{R}). \tag{6.4.24}$$

The operator $\mathcal{R}$ is a fundamental symmetry in $L_2(\mathbb{R})$, which anticommutes with $\mathcal{P}$. It is easy to check that

$$\Gamma_j \mathcal{P} f = \sigma_1 \Gamma_j f, \quad \Gamma_j \mathcal{R} f = \sigma_3 \Gamma_j f, \quad \Gamma_j i\mathcal{P}\mathcal{R} f = i\sigma_1\sigma_3 \Gamma_j f = \sigma_2 \Gamma_j f, \tag{6.4.25}$$

for all $f \in W_2^2(\mathbb{R}\backslash\{0\})$. It is also clear that

$$\Gamma_j \mathcal{T} f = \mathcal{T} \Gamma_j f, \qquad f \in W_2^2(\mathbb{R}\backslash\{0\}). \tag{6.4.26}$$

(We use the same symbol $\mathcal{T}$ for operators of conjugation acting in $L_2(\mathbb{R})$ and in $\mathbb{C}^2$.)

Lemma 6.4.16 *An operator $A_{\mathfrak{T}}$ is $\mathcal{PT}$-symmetric if and only if the matrix $\mathfrak{T}$ is $\sigma_1\mathcal{T}$-symmetric.*

Proof: It is clear that A_s^* is a $\mathcal{PT}$-symmetric operator. As $A_{\mathfrak{T}}$ is the restriction of A_s^* onto $\mathcal{D}(A_{\mathfrak{T}})$, the commutation relation $\mathcal{PT}A_{\mathfrak{T}} = A_{\mathfrak{T}}\mathcal{PT}$ holds if and only if the subspace $\mathcal{D}(A_{\mathfrak{T}})$ is invariant with respect to $\mathcal{PT}$. In view of (6.4.22), the invariance of $\mathcal{D}(A_{\mathfrak{T}})$ with respect to $\mathcal{PT}$ is equivalent to the following implication:

$$\mathfrak{T}\Gamma_1 f = \Gamma_0 f \Longrightarrow \mathfrak{T}\Gamma_1 \mathcal{PT} f = \Gamma_0 \mathcal{PT} f, \qquad \forall f \in \mathcal{D}(A_{\mathfrak{T}}).$$

The latter is equivalent to the commutation relation $\mathfrak{T}\sigma_1\mathcal{T} = \sigma_1\mathcal{T}\mathfrak{T}$ (as $\Gamma_j\mathcal{PT} = \sigma_1\mathcal{T}\Gamma_j$ due to the first equality in (6.4.25) and (6.4.26)). ■

6.4.3.2 The Spectrum of $A_{\mathfrak{T}}$ It is known (see, e.g., (25)) that the continuous spectrum of operators $A_{\mathfrak{T}}$ defined by (6.4.22) coincides with $[0, \infty)$ and there are no eigenvalues of $A_{\mathfrak{T}}$ embedded in the continuous spectrum. The points of the discrete spectrum may appear only on $\mathbb{C} \setminus \mathbb{R}_+$.

Lemma 6.4.17 *Let $A_{\mathfrak{T}}$ be a $\mathcal{PT}$-symmetric operator. Then $z = k^2 \in \mathbb{C} \setminus \mathbb{R}_+$ belongs to the discrete spectrum of $A_{\mathfrak{T}}$ if and only if $k \in \mathbb{C}_+ = \{k \in \mathbb{C} \ : \ \mathrm{Im}\, k > 0\}$ is the root of the equation*

$$\lambda_k^2 - (\mathbf{tr}\ \mathfrak{T})\lambda_k + \mathbf{det}\ \mathfrak{T} = 0, \quad \textit{where} \quad \lambda_k = \frac{1}{2(1+ik)}. \tag{6.4.27}$$

Proof: It is useful to prove this statement with the use of general results of the extension theory of symmetric operators (25, 57). The triplet $(\mathbb{C}^2, \Gamma_0, \Gamma_1)$, where Γ_j are defined by (6.4.23), is the boundary triplet of A_s^* (58, 59). The Weyl function of A_s associated with $(\mathcal{H}, \Gamma_0, \Gamma_1)$ is defined as follows:

$$M(z)\Gamma_0 f_z = \Gamma_1 f_z, \quad \forall f_z \in \ker(A_s^* - zI), \quad \forall z \in \mathbb{C} \setminus \mathbb{R}_+.$$

The Weyl function is a holomorphic matrix-valued function on $\mathbb{C} \setminus \mathbb{R}_+$ and $\mathrm{Im}\, z \cdot \mathrm{Im}\, M(z) > 0$. If an operator $A_{\mathfrak{T}}$ is defined by (6.4.22), then $z \in \mathbb{C} \setminus \mathbb{R}_+$ is an eigenvalue of $A_{\mathfrak{T}}$ if and only if $\mathbf{det}\ [\mathfrak{T} - M^{-1}(z)] = 0$ (60). In our case,

$M(z) = 2(1 + ik)\sigma_0$, where $z = k^2$, $k \in \mathbb{C}_+$ [55, Lemma 3.1]. Hence, $z = k^2$ belongs to the discrete spectrum of $A_{\mathfrak{T}}$ if and only if $M^{-1}(z)$ is an eigenvalue of $\mathfrak{T}$. Repeating the proof of Lemma 6.4.5 (with $A = \mathfrak{T}$ and $\lambda = M^{-1}(z)$) we complete the proof. ■

Corollary 6.4.18 (cf. Lemma 6.4.5) *A $\mathcal{PT}$-symmetric operator $A_{\mathfrak{T}}$ has real spectrum if* $(\mathbf{tr}\ \mathfrak{T})^2 \geq 4\mathbf{det}\ \mathfrak{T}$.

Proof: If $(\mathbf{tr}\ \mathfrak{T})^2 \geq 4\mathbf{det}\ \mathfrak{T}$, then the roots λ_k of the eq. (6.4.27) are real. Then $k \in i\mathbb{R}$ and $z = k^2 < 0$. The operator $A_{\mathfrak{T}}$ has real spectrum. ■

6.4.3.3 ***Self-adjointness of $A_{\mathfrak{T}}$ in Krein Spaces*** Denote $\mathcal{P}_\phi := \mathcal{P}e^{i\phi\mathcal{R}}$, where $\mathcal{R}$ is defined by (6.4.24) and $\phi \in \mathbb{R}$. Obviously, $\mathcal{P}_\phi$ is a fundamental symmetry in $L_2(\mathbb{R})$ and $\mathcal{P}_\phi$ commutes with $\mathcal{PT}$. Define the indefinite inner product $[\cdot,\cdot]_\phi = (\mathcal{P}_\phi\cdot,\cdot)$ in $L_2(\mathbb{R})$ and consider the collection of Krein spaces $(L_2(\mathbb{R}), [\cdot,\cdot]_\phi)$ that depend on the choice of ϕ.

Proposition 6.4.19 (cf. Lemma 6.4.9) *Every $\mathcal{PT}$-symmetric operator $A_{\mathfrak{T}}$ is self-adjoint in the Krein space $(L_2(\mathbb{R}), [\cdot,\cdot]_\phi)$, where ϕ is determined by the relation*

$$\gamma_1 \sin\phi = \gamma_2 \cos\phi, \tag{6.4.28}$$

which involves the coefficients γ_1, γ_2 of the decomposition $\mathfrak{T} = \sum_{j=0}^{3} \gamma_j\sigma_j$.

Proof: Let $A_{\mathfrak{T}}$ be a $\mathcal{PT}$-symmetric operator. In view of Lemma 6.4.16, the matrix $\mathfrak{T}$ is $\sigma_1\mathcal{T}$-symmetric. Then, in accordance with Lemma 6.4.9, $\mathfrak{T}$ is self-adjoint in the Krein space $(\mathbb{C}^2, [\cdot,\cdot]_\phi)$, that is equivalent to the identity $\sigma_{1\phi}\mathfrak{T} = \mathfrak{T}^*\sigma_{1\phi}$, where the parameter ϕ is determined by the relation $\gamma_1 \sin\phi = \gamma_2 \cos\phi$.

Let us show that the identities $\sigma_{1\phi}\mathfrak{T} = \mathfrak{T}^*\sigma_{1\phi}$ and $\mathcal{P}_\phi A_{\mathfrak{T}} = A_{\mathfrak{T}}^*\mathcal{P}_\phi$ are equivalent. Indeed, as $A_{\mathfrak{T}}$ is determined by (6.4.22), the corresponding formula for $A_{\mathfrak{T}}^*$ takes the form (see, e.g., (59)))

$$A_{\mathfrak{T}}^* = -\frac{d^2}{dx^2}, \qquad \mathcal{D}(A_{\mathfrak{T}}^*) = \{f \in W_2^2(\mathbb{R}\backslash\{0\}) \ :\ \mathfrak{T}^*\Gamma_1 f = \Gamma_0 f\}. \tag{6.4.29}$$

It follows from (6.3.17) and (6.4.24) that $\mathcal{P}$ and $\mathcal{R}$ commute with $A_s^* = -\frac{d^2}{dx^2}$. Hence, $\mathcal{P}_\phi = \mathcal{P}e^{i\phi\mathcal{R}}$ commutes with A_s^* too. This means that the identity $\mathcal{P}_\phi A_{\mathfrak{T}} = A_{\mathfrak{T}}^*\mathcal{P}_\phi$ holds if and only if $\mathcal{P}_\phi : \mathcal{D}(A_{\mathfrak{T}}) \to \mathcal{D}(A_{\mathfrak{T}}^*)$. Taking (6.4.22) and (6.4.29) into account we conclude that the latter condition is equivalent to the implication

$$\mathfrak{T}\Gamma_1 f = \Gamma_0 f \Longrightarrow \mathfrak{T}^*\Gamma_1\mathcal{P}_\phi f = \Gamma_0\mathcal{P}_\phi f, \qquad \forall f \in \mathcal{D}(A_{\mathfrak{T}}).$$

Employing (6.4.25) we find that $\Gamma_j\mathcal{P}_\phi = \sigma_{1\phi}\Gamma_j$. Hence, the implication earlier is equivalent to the relation $\sigma_{1\phi}\mathfrak{T} = \mathfrak{T}^*\sigma_{1\phi}$.

Using the equivalence between $\sigma_{1\phi}\mathfrak{T} = \mathfrak{T}^*\sigma_{1\phi}$ and $\mathcal{P}_\phi A_{\mathfrak{T}} = A_{\mathfrak{T}}^*\mathcal{P}_\phi$ and recalling (6.3.4) we establish the self-adjointness of $A_{\mathfrak{T}}$ in the Krein space $(L_2(\mathbb{R}), [\cdot,\cdot]_\phi)$. ■

6.4.3.4 *The Property of $\mathcal{C}$-symmetry for $\mathcal{PT}$-symmetric Operators $A_{\mathfrak{T}}$*

Theorem 6.4.20 (cf. Lemma 6.4.13) *Let $A_{\mathfrak{T}}$ be a $\mathcal{PT}$-symmetric operator. If* $(\mathbf{tr}\ \mathfrak{T})^2 > 4\mathbf{det}\ \mathfrak{T}$, *then $A_{\mathfrak{T}}$ has the property of $\mathcal{C}$-symmetry. The corresponding operator*

$$\mathcal{C} = \mathcal{P}_{\phi}e^{\chi \mathcal{P}_{(\phi+\frac{\pi}{2})}}$$

is $\mathcal{PT}$-symmetric. The parameter ϕ is determined by (6.4.28) and it indicates in which Krein space $(L_2(\mathbb{R}), [\cdot,\cdot]_{\phi})$ the operator $A_{\mathfrak{T}}$ is self-adjoint. The parameter χ is determined by the relation

$$\tanh^2 \chi = \frac{(\mathbf{tr}\ Im\ \mathfrak{T})^2 - 4\mathbf{det}\ Im\ \mathfrak{T}}{(\mathbf{tr}\ Re\ \mathfrak{T})^2 - 4\mathbf{det}\ Re\ \mathfrak{T}}.$$

Proof: It follows from Lemma 6.4.13 and Theorem 6.4.14 that the matrix $\mathfrak{T}$ has the property of $\mathcal{C}$-symmetry with $\mathcal{C} = \sigma_{1\phi}e^{\chi\sigma_{1(\phi+\frac{\pi}{2})}}$. Let us show that this property ensures the property of $\mathcal{C}$-symmetry for $A_{\mathfrak{T}}$ with $\mathcal{C} = \mathcal{P}_{\phi}e^{\chi \mathcal{P}_{(\phi+\frac{\pi}{2})}}$.

Indeed, in view of (6.4.25), we conclude that $\Gamma_j \mathcal{P}_{\phi}e^{\chi \mathcal{P}_{(\phi+\frac{\pi}{2})}} = \sigma_{1\phi}e^{\chi\sigma_{1(\phi+\frac{\pi}{2})}}\Gamma_j$. The operator $\mathcal{P}_{\phi}e^{\chi \mathcal{P}_{(\phi+\frac{\pi}{2})}}$ commutes with the symmetric operator A_s and with its adjoint A_s^*. Hence, $\mathcal{P}_{\phi}e^{\chi \mathcal{P}_{(\phi+\frac{\pi}{2})}}A_{\mathfrak{T}} = A_{\mathfrak{T}}\mathcal{P}_{\phi}e^{\chi \mathcal{P}_{(\phi+\frac{\pi}{2})}}$ holds if and only if

$$\mathcal{P}_{\phi}e^{\chi \mathcal{P}_{(\phi+\frac{\pi}{2})}} : \mathcal{D}(A_{\mathfrak{T}}) \to \mathcal{D}(A_{\mathfrak{T}}).$$

As $\mathcal{D}(A_{\mathfrak{T}})$ is determined by (6.4.22), the latter condition is equivalent to the implication

$$\mathfrak{T}\Gamma_1 f = \Gamma_0 f \Longrightarrow \mathfrak{T}\Gamma_1 \mathcal{P}_{\phi}e^{\chi \mathcal{P}_{(\phi+\frac{\pi}{2})}}f = \Gamma_0 \mathcal{P}_{\phi}e^{\chi \mathcal{P}_{(\phi+\frac{\pi}{2})}}f, \qquad \forall f \in \mathcal{D}(A_{\mathfrak{T}}).$$

which is satisfied due to $\Gamma_j \mathcal{P}_{\phi}e^{\chi \mathcal{P}_{(\phi+\frac{\pi}{2})}} = \sigma_{1\phi}e^{\chi\sigma_{1(\phi+\frac{\pi}{2})}}\Gamma_j$. Taking Remark 6 and Proposition 6.4.19 into account we complete the proof. ■

The $\mathcal{C}$-symmetries of $\mathcal{PT}$-symmetric operators $A_{\mathfrak{T}}$ described in Theorem 6.4.20 are the "image" of the corresponding $\mathcal{C}$-symmetries of $\sigma_1\mathcal{T}$-symmetric matrices $\mathfrak{T}$ studied in Section 6.4.2. In fact, these $\mathcal{C}$-symmetries are determined directly by the matrices of boundary conditions $\mathfrak{T}$ distinguishing the operators $A_{\mathfrak{T}}$ via a suitable chosen boundary triplet $(\mathbb{C}^2, \Gamma_0, \Gamma_1)$; see (6.4.22). Such $\mathcal{C}$-symmetries arise naturally for general quasi-self-adjoint operators (see Remark 7) and they were called "stable $\mathcal{C}$-symmetries" in Ref. (26, 61). Taking Theorem 6.4.14 into account, we can state that the existence of a stable $\mathcal{C}$-symmetry for $A_{\mathfrak{T}}$ directly depends on the form of $\mathfrak{T}$. Precisely, a stable $\mathcal{C}$-symmetry $\mathcal{P}_{\phi}e^{\chi \mathcal{P}_{(\phi+\frac{\pi}{2})}}$ exists for $A_{\mathfrak{T}}$ if the matrix $\mathfrak{T}$ has the form $\mathfrak{T} = \gamma_1\sigma_0 + \gamma_2\sigma_{1\phi}e^{\chi\sigma_{1(\phi+\frac{\pi}{2})}}$. Another useful criterion for the existence of a stable $\mathcal{C}$-symmetry for $A_{\mathfrak{T}}$ is the condition

$$(\mathbf{tr}\ \mathfrak{T})^2 > 4\mathbf{det}\ \mathfrak{T},$$

which guarantees two different real eigenvalues of $\mathfrak{T}$. In view of Lemma 6.4.13 this inequality is equivalent to the property of $\mathcal{C}$-symmetry for a non-self-adjoint $\sigma_1\mathcal{T}$-symmetric matrix $\mathfrak{T}$.

However, in contrast to the matrix case, non-self-adjoint $\mathcal{PT}$-symmetric operators $A_{\mathfrak{T}}$ may have the property of $\mathcal{C}$-symmetry even in the case where $(\mathbf{tr}\ \mathfrak{T})^2 \le 4\mathbf{det}\ \mathfrak{T}$, that is, in the case where $\mathcal{T}$ have nonreal eigenvalues (52).

Next we prove the existence of such kind of $\mathcal{C}$-symmetries. Our method does not provide "a recipe" for the construction of the corresponding operators $\mathcal{C} = \mathcal{P}_\phi e^{\mathcal{Q}}$ in an explicit form. We just state that such operators exist. (More accurately, the parameter ϕ is immediately determined with the use of Proposition 6.4.19, but the construction of the metric operator $e^{\mathcal{Q}}$ remains an open problem.)

Let us set $\mathfrak{T} = 0$ and $\mathfrak{T} = \frac{1}{2}\sigma_0$ in (6.4.22). The obtained self-adjoint operators

$$A_0 = -\frac{d^2}{dx^2}, \qquad \mathcal{D}(A_0) = \{f \in W_2^2(\mathbb{R}\backslash\{0\}) \ :\ f(0+) = f(0-) = 0\},$$

$$A_{\frac{1}{2}\sigma_0} = -\frac{d^2}{dx^2}, \qquad \mathcal{D}(A_{\frac{1}{2}\sigma_0}) = \{f \in W_2^2(\mathbb{R}\backslash\{0\}) \ :\ f'(0+) = f'(0-) = 0\}$$

are, respectively, the Friedrichs and the Krein extensions of the symmetric operator A_s.

Denote $\theta_k = 2(1 + ik)$, $k \in \mathbb{C}_+$. Then the polynomial $\lambda_k^2 - (\mathbf{tr}\ \mathfrak{T})\lambda_k + \mathbf{det}\ \mathfrak{T}$ in (6.4.27) can be rewritten as $\frac{1}{\theta_k^2}p_{\mathfrak{T}}(k)$, where

$$p_{\mathfrak{T}}(k) = \theta_k^2 \mathbf{det}\ \mathfrak{T} - (\mathbf{tr}\ \mathfrak{T})\theta_k + 1.$$

Lemma 6.4.21 *(62) Let $A_{\mathfrak{T}}$ be defined by (6.4.22). Then, for all $g \in L_2(\mathbb{R})$ and for all $z = k^2$ ($k \in \mathbb{C}_+$) from the resolvent set of $A_{\mathfrak{T}}$*

$$\|[(A_{\mathfrak{T}} - zI)^{-1} - (A_0 - zI)^{-1}]g\|^2 = \frac{1}{Im\ k}\left\|\frac{(\sigma_0 + i\sigma_2)[\mathfrak{T} - \theta_k \mathbf{det}\ \mathfrak{T}\sigma_0]}{p_{\mathfrak{T}}(k)}Fg\right\|_{\mathbb{C}^2}^2,$$

where $\theta_k = 2(1 + ik)$, $Fg = \begin{pmatrix} \int_0^\infty e^{iks}g(s)ds \\ \int_{-\infty}^0 e^{-iks}g(s)ds \end{pmatrix}$ *and* $\|\cdot\|_{\mathbb{C}^2}$ *is the norm in the Hilbert space* $\mathbb{C}^2$.

Theorem 6.4.22 *A $\mathcal{PT}$-symmetric operator $A_{\mathfrak{T}}$ has the property of $\mathcal{C}$-symmetry if all roots k of the equation (6.4.27) belong to the lower half-plane $\mathbb{C}_- = \{k \in \mathbb{C} : Im\ k < 0\}$.*

Proof: A $\mathcal{PT}$-symmetric operator $A_{\mathfrak{T}}$ is self-adjoint in the Krein space $(L_2(\mathbb{R}), [\cdot,\cdot]_\phi)$ for a certain choice of ϕ, see Proposition 6.4.19. Hence, by Theorem 6.3.4, the existence of $\mathcal{C}$-symmetry $\mathcal{C} = \mathcal{P}_\phi e^{\mathcal{Q}}$ for $A_{\mathfrak{T}}$ is equivalent to the similarity of $A_{\mathfrak{T}}$ to a self-adjoint operator. To prove the similarity property we will use Theorem 6.3.2. In our case, it suffices to verify the first inequality in (6.3.10) as

$$\|(A_{\mathfrak{T}} - zI)^{-1}g\| = \|(A_{\mathfrak{T}}^{*} - zI)^{-1}\mathcal{P}_{\phi}g\|, \qquad \forall g \in L_2(\mathbb{R}),$$

due to $\mathcal{P}_{\phi}$-self-adjointness of $A_{\mathfrak{T}}$.

The Friedrichs extension A_0 satisfies (6.3.10) as a self-adjoint operator. Therefore, the first inequality in (6.3.10) holds for $(A_{\mathfrak{T}} - zI)^{-1}g$ if and only if for all $g \in L_2(\mathbb{R})$

$$\sup_{\varepsilon>0}\varepsilon \int_{-\infty}^{\infty} \|[(A_{\mathfrak{T}} - zI)^{-1} - (A_0 - zI)^{-1}]g\|^2 d\xi \leq M\|g\|^2. \tag{6.4.30}$$

This means that the similarity of $A_{\mathfrak{T}}$ to a self-adjoint operator is equivalent to the validity of (6.4.30).

Let us first consider the particular case where $A_{\mathfrak{T}}$ coincides with the Krein extension $A_{\frac{1}{2}\sigma_0}$. Then (6.4.30) holds, because $A_{\frac{1}{2}\sigma_0}$ is self-adjoint. Using Lemma 6.4.21, and taking into account that $\mathbf{det}\ \frac{1}{2}\sigma_0 = \frac{1}{4}$ and $p_{\frac{1}{2}\sigma_0}(k) = -k^2$, we get

$$\|[(A_{\frac{1}{2}\sigma_0} - zI)^{-1} - (A_0 - zI)^{-1}]g\|^2 = \frac{1}{\mathrm{Im}\, k}\left\|\frac{(\sigma_0 + i\sigma_2)}{2k}Fg\right\|_{\mathbb{C}^2}^2.$$

Therefore, in view of (6.4.30),

$$\int_{-\infty}^{\infty} \frac{\varepsilon}{\mathrm{Im}\, k}\left\|\frac{(\sigma_0 + i\sigma_2)}{k}Fg\right\|_{\mathbb{C}^2}^2 d\xi \leq M\|g\|^2. \tag{6.4.31}$$

The integral in (6.4.31) is taken along the line $z = k^2 = \xi + i\varepsilon$ ($\varepsilon > 0$ is fixed) of upper half-plane $\mathbb{C}_+$. This means that

$$\varepsilon = 2(\mathrm{Re}\, k)(\mathrm{Im}\, k) > 0, \qquad \xi = (\mathrm{Re}\, k)^2 - (\mathrm{Im}\, k)^2.$$

Therefore, the variable k belongs to $\mathbb{C}_{++} = \{k \in \mathbb{C}_+ : \mathrm{Re}\, k > 0\}$ when $k^2 = \xi + i\varepsilon$.

Let $A_{\mathfrak{T}}$ be a $\mathcal{PT}$-symmetric operator such that all roots of (6.4.27) belong to $\mathbb{C}_-$. Then the spectrum of $A_{\mathfrak{T}}$ is real (Lemma 6.4.17) and the roots of the polynomial $p_{\mathfrak{T}}(k)$ also belong to $\mathbb{C}_-$. In that case, the entries of the matrix

$$\Psi(k) = \frac{k}{p_{\mathfrak{T}}(k)}[\mathfrak{T} - \theta_k \mathbf{det}\ \mathfrak{T}\sigma_0]$$

are uniformly bounded when k runs in $\mathbb{C}_{++}$. Taking this property in mind and using Lemma 6.4.21 and the inequality (6.4.31), we obtain

$$\begin{aligned}
&\varepsilon \int_{-\infty}^{\infty} \|[(A_{\mathfrak{T}} - zI)^{-1} - (A_F - zI)^{-1}]g\|^2 d\xi \\
&= \int_{-\infty}^{\infty} \frac{\varepsilon}{\operatorname{Im} k} \left\| \frac{(\sigma_0 + i\sigma_2)}{k} \Psi(k) F g \right\|_{\mathbb{C}^2}^2 d\xi \\
&\leq M_1 \int_{-\infty}^{\infty} \frac{\varepsilon}{\operatorname{Im} k} \left\| \frac{(\sigma_0 + i\sigma_2)}{k} F g \right\|_{\mathbb{C}^2}^2 d\xi < M M_1 \|g\|^2,
\end{aligned}$$

which justifies (6.4.30). Therefore, $A_{\mathfrak{T}}$ is similar to a self-adjoint operator and $A_{\mathfrak{T}}$ has the property of $\mathcal{C}$-symmetry. Theorem 6.4.22 is proved. ■

The obtained results illustrate the difference between $\mathcal{PT}$-symmetric matrix models and Schrödinger operator models involving $\mathcal{PT}$-symmetric singular potentials. Typically, the Schrödinger-type differential expression A with $\mathcal{PT}$-symmetric singular potential can be interpreted as an extension of a symmetric operator A_s and, simultaneously, as the restriction of the adjoint A_s^* that is, $A_s \subset A \subset A_s^*$. The operator A is distinguished by a matrix $\mathfrak{T}$ in a suitable chosen boundary triplet $(\mathbb{C}^2, \Gamma_0, \Gamma_1)$. However, the spectral properties of A are determined by the pair $\mathfrak{T}, M(\cdot)$, where $M(\cdot)$ is the Weyl function of the symmetric operator A_s.

Thus, in contrast to the matrix case, the existence of $\mathcal{C}$-symmetry for a $\mathcal{PT}$-symmetric operator A depends on the "boundary condition matrix" $\mathfrak{T}$ and a holomorphic parameter (Weyl function or Nevanlinna function or Herglotz function (63)) $M(\cdot)$, which characterizes the impact of the symmetric operator A_s. By changing the Weyl function $M(\cdot)$, we can obtain a variety of exactly solvable models of Schrödinger operator with $\mathcal{PT}$-symmetric singular potentials. In this trend, we mention a convenient approach for the construction of boundary triplets for Schrödinger operators with singular potentials (64). The case of $\mathcal{PT}$-symmetric Coulomb potential on the real axis was discussed in Ref. (65).

REFERENCES

1. Dirac PAM. The physical interpretation of quantum mechanics. *Proc R Soc London Ser A* 1942;180:1.
2. Pauli W. On Dirac's new method of field quantization. *Rev Mod Phys* 1943;15:175.
3. Bender CM, Boettcher S. Real spectra in non-Hermitian Hamiltonians having $\mathcal{PT}$-symmetry. *Phys Rev Lett* 1998;80:5243.
4. Dorey P, Dunning C, Tateo R. Spectral equivalence, Bethe ansatz, and reality properties in $\mathcal{PT}$-symmetric quantum mechanics. *J Phys A* 2001;34:5679.
5. Bender CM. Introduction to $\mathcal{PT}$-symmetric quantum theory. *Contemp Phys* 2005;46:277.

6. Mostafazadeh A. Pseudo-Hermitian representation of quantum mechanics. *Int J Geom Methods Mod Phys* 2010;7:1191.
7. Znojil M. Three-Hilbert-space formulation of quantum mechanics. *SIGMA Symmetry Integrability Geom Methods Appl* 2009;5:1.
8. Langer H, Tretter C. A Krein space approach to $\mathcal{PT}$-symmetry. *Czech J Phys* 2004;54:1113.
9. Langer H, Tretter C. Corrigendum to 'A Krein space approach to $\mathcal{PT}$-symmetry. *Czech J Phys* 2004;54:1113—1120; *Czech J Phys* 2006;56:1063.
10. Nesemann J. *$\mathcal{PT}$-symmetric Schrödinger operators with unbounded potentials*. Heidelberg: Vieweg-Teubner Verlag; 2011.
11. Bender CM, Brody DC, Jones HF. Complex extension of quantum mechanics. *Phys Rev Lett* 2002;89:401.
12. Bender CM. Making sense of non-Hermitian Hamiltonians. *Rep Prog Phys* 2007;70:947.
13. Bender CM, Gianfreda M. Nonuniqueness of the C operator in $\mathcal{PT}$-symmetric quantum mechanics. *J Phys A* 2013;46:275306.
14. Bender CM, Klevansky SP. Nonunique C operator in $\mathcal{PT}$ quantum mechanics. *Phys Lett A* 2009;373:2670.
15. Bender CM, Tan B. Calculation of the hidden symmetry operator for a $\mathcal{PT}$-symmetric square well. *J Phys A* 2006;39:1945.
16. Caliceti E, Francesco F, Znojil M, Ventura A. Construction of $\mathcal{PT}$-asymmetric non-Hermitian Hamiltonians with $C\mathcal{PT}$ symmetry. *Phys Lett A* 2005;335:26.
17. Bender CM, Kuzhel S. Unbounded C-symmetries and their nonuniqueness. *J Phys A* 2012;45:444005.
18. Albeverio S, Gottschalk H. Scattering theory for quantum fields with indefinite metric. *Commun Math Phys* 2001;216:491.
19. Azizov TYa, Iokhvidov IS. *Linear Operators in Spaces with Indefinite Metric*. Chichester: John Wiley and Sons; 1989.
20. Bognar J. *Indefinite Inner Product Spaces*. Berlin: Springer; 1974.
21. Langer H. *Spectral functions of definitizable operators in Krein spaces. Functional Analysis (Dubrovnik, 1981)*. Volume 948, *Lecture Notes in Mathematics*. New York: Springer-Verlag Berlin Heidelberg; 1982. p 1-46.
22. Kuzhel S. On pseudo-Hermitian operators with generalized C-symmetries. *Oper Theory Adv Appl* 2009;190:375.
23. Pedersen S. Anticommuting self-adjoint operators. *J Funct Anal* 1990;89:428.
24. Langer H. Maximal dual pairs of invariant subspaces of J-self-adjoint operators. *Mat Zametki* 1970;7:443, (in Russian).
25. Akhiezer NI, Glazman IM. *Theory of Linear Operators in Hilbert Spaces*. New York: Dover Publication Inc.; 1993.
26. Kuzhel S, Trunk C. On a class of J-self-adjoint operators with empty resolvent set. *J Math Anal Appl* 2011;379:272.

27. Behrndt J, Möws R, Trunk C. On finite rank perturbations of self-adjoint operators in Krein spaces and eigenvalues in spectral gaps. *Complex Anal Oper Theory* 2014;8:925.
28. Jonas P. On a class of selfadjoint operators in Krein spaces and their compact perturbations. *Integr Equ Oper Theory* 1988;11:351.
29. Mostafazadeh A. Krein-space formulation of $\mathcal{PT}$-symmetry, $\mathcal{CPT}$-inner products, and pseudo-Hermiticity. *Czech J Phys* 2006;56:919.
30. Tanaka T. General aspects of $\mathcal{PT}$-symmetric and $\mathcal{P}$-self-adjoint quantum theory in a Krein space. *J Phys A* 2006;39:14175.
31. Phillip F, Szafraniec FH, Trunk C. Self-adjoint operators in S-spaces. *J Funct Anal* 2011;260:1045.
32. Naboko SN. Conditions for similarity to unitary and self-adjoint operators. *Funct Anal Appl* 1984;18(1):13.
33. Gohberg IC, Krein MG. *Introduction to Theory of Linear Non-Self-Adjoint Operators in Hilbert Spaces*. Volume 18, *Translations of mathematical monographs*. American Mathematical Society; Providence, R.I.,1969.
34. Kretschmer R, Szymanowski L. Quasi-Hermiticity in infinite-dimensional Hilbert spaces. *Phys Lett A* 2004;325:112.
35. Bagarello F, Znojil M. Nonlinear pseudo-bosons versus hidden Hermiticity: II. The case of unbounded operators. *J Phys A* 2012;45:115311.
36. Mostafazadeh A. Pseudo-Hermitian quantum mechanics with unbounded metric operators. *Philos Trans R Soc Lond Ser A* 2013;371:20120050.
37. Siegl P, Krejčiřík D. On the metric operator for the imaginary cubic oscillator. *Phys Rev D* 2012;86:121702.
38. Krein MG. Theory of self-adjoint extensions of semibounded operators and its applications I. *Math Trans* 1947;20:431.
39. Halmos PR. *A Hilbert Space Problem Book*. New York: Springer; 1982.
40. Garcia SR, Prodan E, Putinar M. Mathematical and physical aspects of complex symmetric operators. *J Phys A* 2014;47:353001.
41. Azizov TYa, Trunk C. $\mathcal{PT}$-symmetric, Hermitian and $\mathcal{P}$-self-adjoint operators related to potentials in $\mathcal{PT}$ quantum mechanics. *J Math Phys* 2012;53:012109.
42. Zettl A. *Sturm-Liouville Theory*. Providence (RI): American Mathematical Society; 2005.
43. Godič VI, Lucenko IE. On the representation of a unitary operator in the form of a product of two involutions. *Uspehi Mat Nauk* 1965;20(6):64, (in Russian).
44. Borisov D, Krejčiřík D. $\mathcal{PT}$-symmetric waveguides. *Integr Equ Oper Theory* 2008;62:489.
45. Ahmed Z. Pseudo-Hermiticity of Hamiltonians under gauge-like transformation: real spectrum of non-Hermitian Hamiltonians. *Phys Lett A* 2002;294:287.
46. Mostafazadeh A. On the pseudo-Hermiticity of a class of $\mathcal{PT}$-symmetric Hamiltonians in one dimension. *Mod Phys Lett A* 2002;17: 1973.
47. Günther U, Kuzhel S. $\mathcal{PT}$−symmetry, Cartan decompositions and Krein space related Clifford and Pauli algebras. *J Phys A* 2010;43:392002.

48. Günther U, Rotter I, Samsonov B. Projective Hilbert space structures at exceptional points. *J Phys A* 2007;40:8815.
49. Albeverio S, Günther U, Kuzhel S. J-self-adjoint operators with C-symmetries: extension theory approach. *J Phys A* 2009;42:105205.
50. Albeverio S, Fei SM, Kurasov P. Point interactions: $\mathcal{PT}$-Hermiticity and reality of the spectrum. *Lett Math Phys* 2002;59:227.
51. Albeverio S, Kuzhel S. Pseudo-Hermiticity and theory of singular perturbations. *Lett Math Phys* 2004;67:223.
52. Albeverio S, Kuzhel S. One-dimensional Schrödinger operators with $\mathcal{P}$-symmetric zero-range potentials. *J Phys A* 2005;38:4975.
53. Mostafazadeh A. Spectral singularities of a general point interaction. *J Phys A* 2011;44:375302.
54. Albeverio S, Kurasov P. *Singular Perturbations of Differential Operators and Solvable Schrödinger Type Operators, London Mathematical Society Lecture Note Series 271*. Cambridge: Cambridge University Press; 2000.
55. Hrod A, Kuzhel S. Scattering theory for 0-perturbed $\mathcal{PT}$-symmetric operators. *Ukr Math J* 2014;65:1180.
56. Antoine J-P, Trapani C. Some remarks on quasi-Hermitian operators. *J Math Phys* 2014;55:013503.
57. Kuzhel A, Kuzhel S. *Regular Extensions of Hermitian Operators*. Utrecht: VSP; 1998.
58. Gorbachuk ML, Gorbachuk VI. *Boundary-Value Problems for Operator-Differential Equations*. Dordrecht: Kluwer Academic Publishers; 1991.
59. Gorbachuk VI, Gorbachuk ML, Kochubei AN. The theory of extensions of symmetric operators, and boundary value problems for differential equations. *Ukr Math J* 1989;41:1299.
60. Derkach VA, Malamud MM. Generalized resolvents and the boundary value problems for Hermitian operators with gaps. *J Funct Anal* 1991;95:1.
61. Hassi S, Kuzhel S. On J-self-adjoint operators with stable C-symmetries. *Proc R Soc Edinburgh* 2013;143A:141.
62. Cojuhari PA, Grod A, Kuzhel S. On S-matrix of Schrödinger operators with non-symmetric zero-range potentials. *J Phys A* 2014;47:315201.
63. Gesztesy F, Tsekanovskii E. On matrix-valued Herglotz functions. *Math Nachr* 2000;218:61.
64. Kochubei AN. Self-adjoint extensions of the Schrödinger operator with singular potential. *Sib Mat Zh* 1991;32(3):60.
65. Kuzhel S, Patsyuk O. On the theory of $\mathcal{PT}$-symmetric operators. *Ukr Math J* 2012;64:35.

7

METRIC OPERATORS, GENERALIZED HERMITICITY AND LATTICES OF HILBERT SPACES

JEAN-PIERRE ANTOINE[1] AND CAMILLO TRAPANI[2]
[1]*Institut de Recherche en Mathématique et Physique, Université catholique de Louvain, Louvain-la-Neuve, Belgium*
[2]*Dipartimento di Matematica e Informatica, Università di Palermo, Palermo, Italy*

7.1 INTRODUCTION

Pseudo-Hermitian quantum mechanics (QM) is a recent, unconventional, approach to QM, based on the use of non-self-adjoint Hamiltonians, whose self-adjointness can be restored by changing the ambient Hilbert space, via a so-called metric operator.[1] Although not self-adjoint, such Hamiltonians have a real spectrum, usually discrete. Instead they are in general $\mathcal{PT}$-symmetric, that is, invariant under the joint action of space reflection ($\mathcal{P}$) and complex conjugation ($\mathcal{T}$). Typical examples are the $\mathcal{PT}$-symmetric, but non-self-adjoint, Hamiltonians $H = p^2 + ix^3$ and $H = p^2 - x^4$. Surprisingly, both of them have a purely discrete spectrum, real and positive. A full analysis of $\mathcal{PT}$-symmetric Hamiltonians may be found in the review paper of Bender (1). The motivation comes from a number of physical problems, mostly from condensed matter physics, but also from scattering theory (complex scaling), relativistic QM and quantum cosmology, or electromagnetic wave propagation in dielectric media.

[1]Self-adjoint operators are usually called *Hermitian* by physicists.

Non-Selfadjoint Operators in Quantum Physics: Mathematical Aspects, First Edition.
Edited by Fabio Bagarello, Jean-Pierre Gazeau, Franciszek Hugon Szafraniec and Miloslav Znojil.

One may note also that the whole topic of Pseudo-Hermitian QM is covered in a series of annual international workshops, called "Pseudo-Hermitian Hamiltonians in Quantum Physics," starting in 2003, the 12th edition having taken place in Istanbul in August 2013. In addition, several special issues of the Journal of Physics A have been devoted to it, the last one in November 2012 (2). This special issue presents a panorama of theoretical/mathematical problems and also a long list of physical applications such that "studies of classical shock-waves and tunneling, supersymmetric models, spin chain models, models with ring structure, random matrix models, the Pauli equation, the nonlinear Schrödinger equation, quasi-exactly solvable models, integrable models such as the Calogero model, Bose–Einstein condensates, thermodynamics, nonlinear oligomers, quantum catastrophes, the Landau-Zener problem, and pseudo-Fermions." Yet another, more recent, special issue was published in the Philosophical Transactions of the Royal Society (3), with the aim of giving an up-to-date survey of the various attempts to use the techniques of $\mathcal{PT}$ QM for solving some outstanding problems in physics as well as for a number of concrete applications.

The $\mathcal{PT}$-symmetric Hamiltonians are usually pseudo-Hermitian operators, a term introduced a long time ago by Dieudonné (4) (under the name "quasi-Hermitian") for characterizing those bounded operators A that satisfy a relation of the form

$$GA = A^*G, \tag{7.1.1}$$

where G is a *metric operator*, that is, a strictly positive self-adjoint operator. This operator G then defines a new metric (hence the name) and a new Hilbert space (sometimes called physical) in which A is symmetric and may possess a self-adjoint extension. For a systematic analysis of pseudo-Hermitian QM, we may refer to the review of Mostafazadeh (5) and, of course, the special issue (2). We will come back to the terminology in Section 7.2.

Now, in most of the literature, the metric operators are assumed to be bounded with bounded inverse. However, the example of the Hamiltonian of the imaginary cubic oscillator, $H = p^2 + ix^3$, shows that bounded metric operators with unbounded inverse do necessarily occur (6). Moreover, unbounded metric operators have also been introduced in several recent works (7–10) and an effort was made to put the whole machinery on a sound mathematical basis. In particular, the Dieudonné relation implies that the operator A is similar to its adjoint A^*, in some sense, so that the notion of similarity plays a central role in the theory. One aim of this chapter is to explore further the structure of unbounded metric operators, in particular, their incidence on similarity. Many of the results presented here are borrowed from our papers (11, 12).

To start with, we examine, in Section 7.3, the notion of similarity between operators induced by a bounded metric operator with bounded inverse. Although this notion is standard, it is too restrictive for applications, thus we are led to introduce several generalizations. The most useful one, called *quasi-similarity*, applies when the metric operator is bounded and invertible, but has an unbounded inverse. The goal here is to study which spectral properties are preserved under such a quasi-similarity relation.

This applies, for instance, to non-self-adjoint operators with a discrete real spectrum. Later on, in Section 7.7, we introduce an even weaker notion, called *semisimilarity*.

Next we notice, in Section 7.4, that an unbounded metric operator G generates a lattice of seven Hilbert spaces, with lattice operations $\mathcal{H}_1 \wedge \mathcal{H}_2 := \mathcal{H}_1 \cap \mathcal{H}_2, \mathcal{H}_1 \vee \mathcal{H}_2 := \mathcal{H}_1 + \mathcal{H}_2$ (Fig. 7.1). In addition, we consider the infinite Hilbert scale generated by the powers of G. This structure is then extended in Section 7.6 to families of metric operators, bounded or not. Such a family, if it contains unbounded operators, defines a rigged Hilbert space, and the latter in turn generates a canonical lattice of Hilbert spaces. This is a particular case of a partial inner product space (PIP-space), a concept described at length in our monograph (13).

Section 7.5 is the heart of this chapter. It is devoted to the notion of *quasi-Hermitian operators*, defined here in a slightly more general form than the original one of Dieudonné (4), in that both the operator A and the metric operator G in (7.1.1) are allowed to be unbounded. If G is bounded with unbounded inverse, it defines a new Hilbert space $\mathcal{H}(G)$ and the whole game consists in determining how operator properties are transferred from the original Hilbert space $\mathcal{H}$ to $\mathcal{H}(G)$. This is exactly the situation encountered for Hamiltonians in Hermitian QM, as explained earlier. Of particular interest is the case where G is self-adjoint in the (physical) Hilbert space $\mathcal{H}(G)$. We derive a number of conditions to that effect. As an aside we present a class of concrete examples, namely, operators defined from Riesz bases.

In Section 7.7, we apply some of the previous results to operators on the scale of Hilbert spaces generated by the metric operator G. The outcome is that the PIP-space structure indeed improves some of them. Finally, in Section 7.8, we present a construction, inspired from Ref. (10), but significantly more general. Indeed, instead of requiring that the original pseudo-Hermitian Hamiltonian H have a countable family of eigenvectors, we only need to assume that H has a (large) set of analytic vectors.

Finally, we summarize in the Appendix the essential facts about PIP-spaces.

To conclude, let us fix our notations. The framework in a separable Hilbert space $\mathcal{H}$, with inner product $\langle\cdot|\cdot\rangle$, linear in the first entry. Then, for any operator A in $\mathcal{H}$, we denote its domain by $D(A)$, its range by $R(A)$ and, if A is positive, its form domain by $Q(A) := D(A^{1/2})$.

7.2 SOME TERMINOLOGY

We start with the central object of the chapter, namely, *metric operators*.

Definition 7.2.1 By a metric operator in a Hilbert space $\mathcal{H}$, we mean a strictly positive self-adjoint operator G, that is, $G > 0$ or $\langle G\xi|\xi\rangle \geq 0$ for every $\xi \in D(G)$ and $\langle G\xi|\xi\rangle = 0$ if and only if $\xi = 0$.

Of course, G is densely defined and invertible, but need not be bounded; its inverse G^{-1} is also a metric operator, bounded or not (in this case, in fact, 0 belongs to the continuous spectrum of G). For future use, we note that, given metric operators. G, G_1, G_2, one has

(1) If G_1 and G_2 are both bounded, then $G_1 + G_2$ is a bounded metric operator;
(2) λG is a metric operator if $\lambda > 0$;
(3) $G^{1/2}$ and, more generally, $G^{\alpha}(\alpha \in \mathbb{R})$, are metric operators.

As we noticed in the introduction, in most of the literature on Pseudo-Hermitian QM, the metric operators are assumed to be bounded with bounded inverse, although there are exceptions. Following our previous work (11, 12), we will envisage in this chapter all cases: G and G^{-1} both bounded, G bounded with G^{-1} unbounded (such an operator G is sometimes called *quasi-invertible* (14)), G and G^{-1} both unbounded.

Before proceeding, it is necessary to clarify the relationship between our metric operators and similar concepts commonly found in Pseudo-Hermitian QM (12). To start with, Dieudonné (4) calls *quasi-Hermitian* a bounded operator A on a Hilbert space $\mathcal{H}$ for which there exists a bounded, positive, self-adjoint operator $T \neq 0$ such that

$$TA = A^*T. \tag{7.2.1}$$

Thus, T is invertible, but its inverse T^{-1} need not be bounded. Notice that Dieudonné calls "essentially trivial" the case where both T and T^{-1} are bounded. In the sequel of this chapter, we drop the boundedness assumption for both T and T^{-1}.

Quasi-Hermitian operators contain, in particular, spectral operators of scalar type and real spectrum, in the sense of Dunford (15) (i.e., operators similar to a self-adjoint operator) and, *a fortiori*, self-adjoint operators. Note the terminology is not uniform in the literature. Kantorovitz (16), for instance, defines quasi-Hermitian operators exactly as these spectral operators. Now, if the operator T^{-1} is bounded, then (7.2.1) implies that A is similar to a self-adjoint operator, thus it is a spectral operator of scalar type and real spectrum. This is the case treated by Scholtz *et al.* (17) and Geyer *et al.* (18), who introduced the concept in the physics literature.

A slightly more general notion is that of *pseudo-Hermitian* operators, namely operators A satisfying (7.2.1), with T and T^{-1} bounded, but not necessarily positive (this will unavoidably lead to indefinite metrics, see next). This is the definition adopted also by Kretschmer-Szymanowski (19), Mostafazadeh (5), or Albeverio *et al.* (20). Later on, Kretschmer-Szymanowski, Mostafazadeh (10), and Bagarello-Znojil (9) adapted the definition to the case of an *unbounded* operator T, claiming this is required for certain physical applications. Note that the latter authors have also coined the term *cryptohermitian* for bounded quasi-Hermitian operators (9).

Another issue to clarify is the relation between pseudo-Hermitian operators and J-self-adjoint operators in a Krein space. Assume that $\mathcal{H}$ is a Hilbert space with a *fundamental symmetry* J, that is, a self-adjoint involution, $J = J^*, J^2 = I$. Defining the projections $P_\pm = \frac{1}{2}(I \pm J)$, we obtain the fundamental decomposition of $\mathcal{H}$:

$$\mathcal{H} = \mathcal{H}_+ \oplus \mathcal{H}_-, \quad \mathcal{H}_\pm := P_\pm \mathcal{H}. \tag{7.2.2}$$

Then, the space $\mathcal{H}$ endowed with the indefinite inner product

$$[\xi, \eta]_J := (J\xi, \eta) \tag{7.2.3}$$

is a *Krein space*. According to Ref. (21), a Krein space is defined as a decomposable, nondegenerate inner product space $\mathcal{K} = \mathcal{K}_+ \oplus \mathcal{K}_-$, with inner product $[\cdot,\cdot]$, where the subspace $\mathcal{K}_+$, respectively, $\mathcal{K}_-$, consists of vectors of positive, respectively, negative norm $[\cdot,\cdot]^{1/2}$ and both subspaces $\mathcal{K}_\pm$ are complete in the so-called intrinsic norm $|[\cdot,\cdot]|^{1/2}$. In that case, the J-inner product

$$(\xi,\eta)_J := [J\xi,\eta] \tag{7.2.4}$$

is positive definite and $\mathcal{K}$ is a Hilbert space for the J-inner product. Notice that exactly the same operators are bounded for the norms $|[\cdot,\cdot]|^{1/2}$ and $\|\cdot\|_J = (\cdot,\cdot)_J^{1/2}$ (21).

Then, a linear densely defined operator A in the Krein space $(\mathcal{H},[\cdot,\cdot]_J)$ is called *J-self-adjoint* if it satisfies the relation $A^*J = JA$. Thus, J-self-adjoint operators are pseudo-Hermitian and constitute the appropriate class to study rigorously, as claimed in Ref. (20).

Another notion commonly used it that of a C-symmetry (1, 20, 22). One says that a J-self-adjoint operator A in a Krein space has a *C-symmetry* if there exists a linear bounded operator C such that (i) $C^2 = I$; (ii) $JC > 0$; and (iii) $AC = CA$. Thus, the inner product

$$(\xi,\eta)_C := [C\xi,\eta]_J = (JC\xi,\eta)$$

is positive definite, that is, JC is a metric operator. Thus, if a Hamiltonian H has the C-symmetry, $(\mathcal{H},(\cdot,\cdot)_C)$ is a Hilbert space, in which the dynamics generated by H should be described. Actually, there are two cases. If the operator C is bounded, then it is unique, up to equivalence. But if it is not unique, it must be unbounded (the definition may be adapted) (22, 23). So here too, one has to consider unbounded operators, as we shall do in this chapter.

Finally, we note that non-self-adjoint operators (in Banach or Hilbert spaces) and their spectral properties are the object of a systematic analysis by Davies (24).

7.3 SIMILAR AND QUASI-SIMILAR OPERATORS

In this section, we collect some basic definitions and facts about similarity of linear operators in Hilbert spaces and discuss several generalizations of this notion. Throughout most of the section, G will denote a *bounded* metric operator. From now on, we will always suppose the domains of the given operators to be dense in $\mathcal{H}$.

7.3.1 Similarity

In order to state precisely what we mean by similarity, we first define intertwining operators (11).

Definition 7.3.1 Let $\mathcal{H},\mathcal{K}$ be Hilbert spaces, $D(A)$ and $D(B)$ dense subspaces of $\mathcal{H}$ and $\mathcal{K}$, respectively, $A : D(A) \to \mathcal{H}$, $B : D(B) \to \mathcal{K}$ two linear operators. A bounded operator $T : \mathcal{H} \to \mathcal{K}$ is called a *bounded intertwining operator* for A and B if

(io$_1$) $T : D(A) \to D(B)$;
(io$_2$) $BT\xi = TA\xi, \ \forall \xi \in D(A)$.

Remark 7.3.2 If T is a bounded intertwining operator for A and B, then $T^* : \mathcal{K} \to \mathcal{H}$ is a bounded intertwining operator for B^* and A^*.

Definition 7.3.3 Let A, B be two linear operators in the Hilbert spaces $\mathcal{H}$ and $\mathcal{K}$, respectively. Then, we say that

(i) A and B are *similar*, and write $A \sim B$, if there exists a bounded intertwining operator T for A and B with bounded inverse $T^{-1} : \mathcal{K} \to \mathcal{H}$, which is intertwining for B and A .

(ii) A and B are *unitarily equivalent* if $A \sim B$ and $T : \mathcal{H} \to \mathcal{K}$ is unitary, in which case we write $A \approx B$.

We notice that $\sim$ and $\approx$ are equivalence relations. Also, in both cases, one has $TD(A) = D(B)$.

The following properties of similar operators are easy (see Ref. (11) for a proof).

Proposition 7.3.4 *Let A and B be linear operators in $\mathcal{H}$ and $\mathcal{K}$, respectively, and T a bounded intertwining operator for A and B. Then the following statements hold.*

(i) $A \sim B$ if, and only if, $B^* \sim A^*$.
(ii) A is closed if, and only if, B is closed.
(iii) A^{-1} exists if, and only if, B^{-1} exists. Moreover, $B^{-1} \sim A^{-1}$.

Similarity of A and B is symmetric, preserves both the closedness of the operators and their spectra. However, in general, it does not preserve self-adjointness, as will result from Corollary 7.3.7 and Proposition 7.3.8 as follows.

As we will see in Proposition 7.3.6 next, similarity preserves also the resolvent set $\rho(\cdot)$ of operators and the parts in which the spectrum is traditionally decomposed: the point spectrum $\sigma_p(\cdot)$, the continuous spectrum $\sigma_c(\cdot)$ and the residual spectrum $\sigma_r(\cdot)$. As we are dealing with closed, non-self-adjoint operators, it is worth recalling the definitions of these various sets (25–27).

Given a closed operator A in $\mathcal{H}$, consider $A - \lambda I : D(A) \to \mathcal{H}$ and the resolvent $R_A(\lambda) := (A - \lambda I)^{-1}$. Then one defines:

- The resolvent set $\rho(A) := \{\lambda \in \mathbb{C} : A - \lambda I$ is one-to-one and $(A - \lambda I)^{-1}$ is bounded$\}$.
- The spectrum $\sigma(A) := \mathbb{C} \setminus \rho(A)$.
- The point spectrum $\sigma_p(A) := \{\lambda \in \mathbb{C} : A - \lambda I$ is not one-to-one$\}$, that is, the set of eigenvalues of A.
- The continuous spectrum $\sigma_c(A) := \{\lambda \in \mathbb{C} : A - \lambda I$ is one-to-one and has dense range, different from $\mathcal{H}\}$, hence $(A - \lambda I)^{-1}$ is densely defined, but unbounded.

- The residual spectrum $\sigma_r(A) := \{\lambda \in \mathbb{C} : A - \lambda I$ is one-to-one, but its range is not dense$\}$, hence $(A - \lambda I)^{-1}$ is not densely defined.

With these definitions, the three sets $\sigma_p(A), \sigma_c(A), \sigma_r(A)$ are disjoint and

$$\sigma(A) = \sigma_p(A) \cup \sigma_c(A) \cup \sigma_r(A). \tag{7.3.1}$$

We note also that $\sigma_r(A) = \overline{\sigma_p(A^*)} = \{\overline{\lambda} : \lambda \in \sigma_p(A^*)\}$. Indeed, for any $\lambda \in \sigma_r(A)$, there exists $\eta \neq 0$ such that

$$0 = \langle (A - \lambda I)\xi | \eta \rangle = \langle \xi | (A^* - \overline{\lambda} I)\eta \rangle, \ \forall \xi \in D(A),$$

which implies $\overline{\lambda} \in \sigma_p(A^*)$. Also $\sigma_r(A) = \emptyset$ if A is self-adjoint.

Remark 7.3.5 Here we follow Dunford-Schwartz (25), but other authors give a different definition of the continuous spectrum, implying that it is no longer disjoint from the point spectrum, for instance, Reed-Simon (26) or Schmüdgen (27). This alternative definition allows for eigenvalues embedded in the continuous spectrum, a situation common in many physical situations, such as the Helium atom, and a typical source of resonance effects in scattering theory (see Ref. [28, Section XII.6]).

We proceed now to show the stability of the different parts of the spectrum under the similarity relation $\sim$, as announced earlier [11, Propositions 3.7 and 3.9].

Proposition 7.3.6 *Let A, B be closed operators such that $A \sim B$ with the bounded intertwining operator T. Then,*

(i) $\rho(A) = \rho(B)$.

(ii) $\sigma_p(A) = \sigma_p(B)$. *Moreover if $\xi \in D(A)$ is an eigenvector of A corresponding to the eigenvalue λ, then $T\xi$ is an eigenvector of B corresponding to the same eigenvalue. Conversely, if $\eta \in D(B)$ is an eigenvector of B corresponding to the eigenvalue λ, then $T^{-1}\eta$ is an eigenvector of A corresponding to the same eigenvalue. Moreover, the multiplicity of λ as eigenvalue of A is the same as its multiplicity as eigenvalue of B.*

(iii) $\sigma_c(A) = \sigma_c(B)$.

(iv) $\sigma_r(A) = \sigma_r(B)$.

Proof. (i) Let $\lambda \in \rho(A)$, so that $(A - \lambda I)^{-1}$ exists and it is bounded. The operator $X_\lambda := T(A - \lambda I)^{-1}T^{-1}$ is bounded. As $(A - \lambda I)^{-1}T^{-1}\eta \in D(A)$, for every $\eta \in \mathcal{K}$, we have

$$\begin{aligned}(B - \lambda I)X_\lambda \eta &= (B - \lambda I)T(A - \lambda I)^{-1}T^{-1}\eta \\ &= T(A - \lambda I)(A - \lambda I)^{-1}T^{-1}\eta = \eta, \quad \forall \eta \in \mathcal{K}.\end{aligned}$$

On the other hand, as $(B - \lambda I)T\xi = T(A - \lambda I)\xi$, for all $\xi \in D(A)$, taking $\xi = T^{-1}\eta$, $\eta \in D(B)$, we obtain $(B - \lambda I)\eta = T(A - \lambda I)T^{-1}\eta$ and then $T^{-1}(B - \lambda I)\eta = (A - \lambda I)T^{-1}\eta$. Then, for every $\eta \in D(B)$, we get

$$\begin{aligned} X_\lambda(B - \lambda I)\eta &= T(A - \lambda I)^{-1}T^{-1}(B - \lambda I)\eta \\ &= T(A - \lambda I)^{-1}(A - \lambda I)T^{-1}\eta = \eta, \quad \forall \eta \in D(B). \end{aligned}$$

Hence $X_\lambda = (B - \lambda I)^{-1}$ and $\lambda \in \rho(B)$. The statement follows by replacing A with B and T with T^{-1}.

(ii) is easy.

(iii) Let $\lambda \in \sigma_c(B)$. If $\eta \in \mathcal{H}$, then $\eta = T^{-1}\eta'$ for some $\eta' \in \mathcal{K}$. As $R(B - \lambda I)$ is dense in $\mathcal{K}$, there exists a sequence $\{\eta'_k\} \subset R(B - \lambda I)$ such that $\eta'_k \to \eta'$. Put $\eta'_k = (B - \lambda I)\xi'_k$, with $\xi'_k \in D(B)$. As $TD(A) = D(B)$, for every $k \in \mathbb{N}$, there exists $\xi_k \in D(A)$ such that $\xi'_k = T\xi_k$. Hence

$$\eta' = \lim_{k\to\infty} (B - \lambda I)T\xi_k = \lim_{k\to\infty} T(A - \lambda I)\xi_k.$$

This implies that

$$\eta = T^{-1}\eta' = \lim_{k\to\infty} (A - \lambda I)\xi_k.$$

Thus, $R(A - \lambda I)$ is dense in $\mathcal{H}$. As $(B - \lambda I)^{-1}$ is unbounded, so is also $(A - \lambda I)^{-1}$. Hence $\sigma_c(A) \subseteq \sigma_c(B)$. Interchanging the roles of A and B, one gets the reverse inclusion.

(iv) follows from (7.3.1) and (i)–(iii). ∎

In the proof of (i), the assumption "T^{-1} bounded" guarantees that X_λ (which is in any case a left inverse) is bounded, whereas the assumption "$TD(A) = D(B)$" allows one to prove that X_λ is also a right inverse. They both seem to be unavoidable.

Taking into account that, if A is self-adjoint, its residual spectrum is empty, we obtain

Corollary 7.3.7 *Let A, B be closed operators with A self-adjoint. Assume that $A \sim B$. Then B has real spectrum and $\sigma_r(B) = \emptyset$.*

In other words, B is then a spectral operator of scalar type with real spectrum, as discussed in Section 7.2. This corollary can be used to show the existence of non-symmetric operators having real spectrum and empty residual spectrum.

Actually, it is contained in the following result of Williams (29).

Proposition 7.3.8 *If the operator A satisfies the conditions $A = T^{-1}A^*T$, where T^{-1} is bounded and $0 \notin \overline{W(T)}$, then it is similar to a self-adjoint operator, hence has real spectrum.*

In this proposition, $\overline{W(T)}$ denotes the closure of the numerical range of T, that is, $W(T) := \{\langle T\xi|\xi\rangle : \xi \in D(T), \|\xi\| = 1\}$ [24, Section 9.3]. The set $\overline{W(T)}$ is convex and contains $\sigma(T)$. The argument runs as follows. By the condition $0 \notin \overline{W(T)}$, one can separate $\overline{W(T)}$ from 0 in such a way that $\overline{W(T)}$ belongs to a half-plane and 0 does not. Possibly replacing T by $e^{i\theta}T$, this half-plane can be taken as $\operatorname{Re} z \geq \epsilon$, for some $\epsilon > 0$. Defining $B = \frac{1}{2}(T + T^*)$, one sees that $W(B) = \operatorname{Re} W(T)$ lies on the real axis to the right of ϵ, thus B is positive and boundedly invertible. As $A = T^{-1}A^*T$, it follows that $L := B^{-1/2}AB^{1/2}$ is self-adjoint and $A \sim L$. Note that one can assume, in particular, $T > 0$ or $T < 0$.

In other words, for a quasi-Hermitian operator satisfying $TA = A^*T$ to be similar to a self-adjoint operator, one needs both $T > 0$ (or $T < 0$) and T^{-1} bounded. Indeed, Williams gives an example of an operator A (the bilateral shift in ℓ^2) where T^{-1} is bounded but not positive and $\sigma(A)$ is not real. On the other hand, Dieudonné (4) gives an example of a quasi-Hermitian operator A with T^{-1} unbounded and $\sigma(A)$ not real.

7.3.2 Quasi-Similitarity and Spectra

The notion of similarity discussed in the previous section is too strong in many situations, thus we seek a weaker one. A natural step is to drop the boundedness of T^{-1}.

Definition 7.3.9 We say that A is *quasi-similar* to B, and write $A \dashv B$, if there exists a bounded intertwining operator T for A and B, which is invertible, with inverse T^{-1} densely defined (but not necessarily bounded).

Note that, even if T^{-1} is bounded, A and B need not be similar, unless T^{-1} is also an intertwining operator. Indeed, T^{-1} does not necessarily map $D(B)$ into $D(A)$, unless of course if $TD(A) = D(B)$.

As already remarked in Ref. (12), there is a considerable confusion in the literature concerning the notion of quasi-similarity.

(1) First, essentially all authors consider only (quasi-)similarity between two *bounded* operators. Next, a bounded invertible operator T with (possibly) unbounded inverse T^{-1} is called a *quasi-affinity* by Sz.-Nagy and Foiaş [30, Chapter II, Section 3] and a *quasi-invertible* operator by other authors (14). Then, if A, B are two bounded operators such that $TA = BT$, that is, $A \dashv B$, A is called a *quasi-affine transform* of B. In this context, A and B are called *quasi-similar* if $A \dashv B$ *and* $B \dashv A$ (so that quasi-similarity becomes also an equivalence relation).

(2) Tzafriri (31) considers only bounded spectral operators, in Dunford's sense (15, 25). For these, he introduces a different notion of quasi-similarity (but under the same name) based on the resolution of the identity.

(3) Hoover (14) shows that if two bounded spectral operators A and B are quasi-similar (i.e., $A \dashv B$ and $B \dashv A$), then B is quasi-similar to A in the sense of Tzafriri (which he calls weakly similar). On the other hand, Feldzamen (32) considers yet another notion of generalized similarity, called *semisimilarity*, but then Hoover shows that two semisimilar bounded spectral operators A and B are in fact quasi-similar.

(4) Quasi-similarity of unbounded operators is considered by Ôta and Schmüdgen (33). Namely, given two unbounded operators A and B in Hilbert spaces $\mathcal{H}, \mathcal{K}$, respectively, A is said to be *quasi-similar* to B if there exist two (quasi-invertible) intertwining operators (in the sense of Definition 7.3.1) $T_{AB} : D(A) \to D(B)$ and $T_{BA} : D(B) \to D(A)$. In other words, this notion is the straightforward generalization of the quasi-similarity of bounded operators defined by the previous authors.

In the sequel, we will stick to the asymmetrical notion of quasi-similarity given in Definition 7.3.9, namely, $A \dashv B$, because it appears naturally in the presence of a bounded metric operator with unbounded inverse, as shown in Theorem 7.5.15. A concrete example is the Hamiltonian of the imaginary cubic oscillator (6).[2]

Actually, there is a whole class of similar concrete examples, namely, $H = p^2 + x^2 + ix^3$ or, more generally, Schrödinger Hamiltonians of the form

$$H = p^2 + \sum_{m=1}^{2n} c_m x^m,$$

where the constant c_m has positive real and imaginary parts. These operators were already mentioned by Davies (34). Two further examples will be given next, in Examples 7.3.22 and 7.3.23. Another explicit example, that we will describe in detail in Section 7.8.1 next, is the operator of second derivative on the positive half-line analyzed by Samsonov (35).

Accordingly, we will say that two closed operators A and B are *mutually quasi-similar* if they are quasi-similar in the sense of Ôta and Schmüdgen, that is, if we have both $A \dashv B$ and $B \dashv A$, which we denote by $A \dashv\vdash B$. Clearly, $\dashv\vdash$ is an equivalence relation. Moreover, $A \dashv\vdash B$ implies $A^* \dashv\vdash B^*$.

Definition 7.3.10 If $A \sim B$ (respectively, $A \dashv B$) and the intertwining operator is a metric operator G, we say that A and B are *metrically* similar (respectively, quasi-similar).

If $A \sim B$ and T is the corresponding intertwining operator, then $T = UG$, where U is unitary and $G := (T^*T)^{1/2}$ is a metric operator. If we put $B' = U^{-1}BU$, then B' and A are metrically similar. Thus, up to unitary equivalence, one can always consider metric similarity instead of similarity.

Proposition 7.3.11 *If $A \dashv B$, with the bounded intertwining operator T, then $B^* \dashv A^*$ with the bounded intertwining operator T^*.*

This follows from Remark 7.3.2 and from the fact that, as T^{-1} exists, then $(T^*)^{-1}$ exists too and $(T^*)^{-1} = (T^{-1})^*$.

[2]There is a misprint in that paper, on page 2, l.-2. The correct statement is $\Theta[\mathrm{Dom}(H)] \subset \mathrm{Dom}(H^\dagger)$, which indeed satisfies the relation $H \dashv H^\dagger$.

In the case of nonclosed operators, however, we must still weaken the notion of quasi-similarity, replacing conditions (io_1) and (io_2) by the following weak-type condition.

Definition 7.3.12 The operator A is called *weakly quasi-similar* to B, in which case we write $A \dashv_w B$, if B is closable, T is invertible with densely defined inverse T^{-1} and the following condition holds

(ws) $\langle T\xi | B^*\eta \rangle = \langle TA\xi | \eta \rangle, \ \forall \xi \in D(A), \ \eta \in D(B^*)$.

Of course, if the operator B is closed, we recover the original Definition 7.3.9.

Proposition 7.3.13 *$A \dashv_w B$ if and only if $T : D(A) \to D(B^{**})$ and $B^{**}T\xi = TA\xi$, for every $\xi \in D(A)$. In particular, if B is closed, $A \dashv B$ if and only if $A \dashv_w B$.*

Proposition 7.3.14 *If B is closable and $A \dashv_w B$, then A is closable.*

Proof. Assume that $\{\xi_n\}$ is a sequence in $D(A)$ and $\xi_n \to 0$, $A\xi_n \to \eta$. Then, $T\xi_n \to 0$ and $TA\xi_n \to T\eta$. But $TA\xi_n = B^{**}T\xi_n \to T\eta$. The closedness of B^{**} then implies that $T\eta = 0$ and, therefore, $\eta = 0$. ■

The converse of the previous statement does not hold, in general, as shown by the following counterexample (11).

Example 7.3.15 In the Hilbert space $L^2(\mathbb{R})$, consider the operator Q of multiplication by x, defined on the dense domain

$$D(Q) = \left\{ f \in L^2(\mathbb{R}) : \int_{\mathbb{R}} x^2 |f(x)|^2 \, dx < \infty \right\}.$$

Given $\varphi \in L^2(\mathbb{R})$, with $\|\varphi\| = 1$, let $P_\varphi := \varphi \otimes \overline{\varphi}$ denote the projection operator[3] onto the one-dimensional subspace generated by φ and A_φ the operator with domain $D(A_\varphi) = D(Q^2)$ defined by

$$A_\varphi f = \langle (I + Q^2) f | \varphi \rangle (I + Q^2)^{-1} \varphi, \quad f \in D(A_\varphi).$$

Then, it is easily seen that $P_\varphi \dashv A_\varphi$ with the bounded intertwining operator $T := (I + Q^2)^{-1}$. Clearly, P_φ is everywhere defined and bounded, but the operator A_φ is closable if, and only if, $\varphi \in D(Q^2)$. This is seen as follows. Being densely defined, A_φ is closable if and only if it has a densely defined adjoint. If $\varphi \in D(Q^2)$ we have, for every $g \in L^2(\mathbb{R})$,

$$\langle A_\varphi f | g \rangle = \langle (I + Q^2) f | \varphi \rangle \langle (I + Q^2)^{-1} \varphi | g \rangle = \langle f | (I + Q^2) \varphi \rangle \langle (I + Q^2)^{-1} \varphi | g \rangle.$$

[3] In physicists' Dirac notation, $P_\varphi = |\varphi\rangle\langle\varphi|$.

Hence, $A^*_\varphi g = \langle g|(I+Q^2)^{-1}\varphi\rangle(I+Q^2)\varphi = \langle (I+Q^2)^{-1}g|\varphi\rangle(I+Q^2)\varphi$. This proves also that, in this case, A_φ is bounded, as g can be arbitrarily chosen in $\mathcal{H}$. On the other hand, if $g \in D(A^*_\varphi)$,

$$\langle A_\varphi f|g\rangle = \langle (I+Q^2)f|\varphi\rangle\langle (I+Q^2)^{-1}\varphi|g\rangle = \langle f|A^*_\varphi g\rangle.$$

Then the last equality shows that $\varphi \in D(Q^2)$ and

$$A^*_\varphi g = \langle (I+Q^2)^{-1}g|\varphi\rangle(I+Q^2)\varphi.$$

Now we consider the relationship between the spectra of quasi-similar operators, following mostly (11).

Proposition 7.3.16 *Let A and B be closed operators and assume that $A \dashv B$, with the bounded intertwining operator T. Then the following statements hold.*

(i) $\sigma_p(A) \subseteq \sigma_p(B)$ *and for every* $\lambda \in \sigma_p(A)$ *one has* $m_A(\lambda) \leq m_B(\lambda)$, *where* $m_A(\lambda)$, *respectively,* $m_B(\lambda)$, *denotes the multiplicity of* λ *as eigenvalue of the operator A, respectively, B.*

(ii) $\sigma_r(B) \subseteq \sigma_r(A)$.

(iii) If $TD(A) = D(B)$, *then* $\sigma_p(B) = \sigma_p(A)$.

(iv) If T^{-1} *is bounded and* $TD(A)$ *is a core for B, then* $\sigma_p(B) \subseteq \sigma(A)$.

Proof.

The statements (i) and (iii) can be proved as in Proposition 7.3.6. We prove only the statements (ii) and (iv).

(ii) By Proposition 7.3.11, $B^* \dashv A^*$, with the bounded intertwining operator T^*. Then, by (i), $\sigma_p(B^*) \subseteq \sigma_p(A^*)$. The statement follows from the relation $\sigma_r(C) = \overline{\sigma_p(C^*)}$, shown earlier.

(iv): Let $\lambda \in \sigma_p(B)$. Then there exists $\eta \in D(B) \setminus \{0\}$ such that $B\eta = \lambda\eta$. We may suppose that $\|\eta\| = 1$. As $TD(A)$ is a core for B, there exists a sequence $\{\xi_n\} \subset D(A)$ such that $T\xi_n \to \eta$ and $BT\xi_n \to B\eta$. Then,

$$\begin{aligned}\lim_{n\to\infty} T(A\xi_n - \lambda\xi_n) &= \lim_{n\to\infty} TA\xi_n - \lambda \lim_{n\to\infty} T\xi_n = \lim_{n\to\infty} BT\xi_n - \lambda\eta \\ &= B\eta - \lambda\eta = 0.\end{aligned}$$

As T^{-1} is bounded, we get

$$\lim_{n\to\infty} (A\xi_n - \lambda\xi_n) = 0, \tag{7.3.2}$$

Assume that $\lambda \in \rho(A)$. Then $(A-\lambda I)^{-1} \in \mathcal{B}(\mathcal{H})$. We put, $\eta_n = (A-\lambda I)\xi_n$. Then, by (7.3.2), $\eta_n \to 0$. Hence, $\xi_n = (A-\lambda I)^{-1}\eta_n \to 0$. This in turn implies that $T\xi_n \to 0$, which is impossible as $\|\eta\| = 1$. ∎

Proposition 7.3.17 *Let A and B be closed operators. Assume that $A \dashv B$, with the bounded intertwining operator T. Then the following statements hold.*

(a) *Let $\lambda \in \rho(A)$ and define*

$$D(X_\lambda) = D(T^{-1}),$$
$$X_\lambda \eta = T(A - \lambda I)^{-1} T^{-1} \eta, \ \eta \in D(X_\lambda).$$

Then,

(a.1) $(B - \lambda I) X_\lambda \eta = \eta, \ \forall \eta \in D(X_\lambda).$

(a.2) *If $(B - \lambda I)\eta \in D(T^{-1})$, $\forall \eta \in D(B)$, and $\lambda \notin \sigma_p(B)$, then $X_\lambda (B - \lambda I)\eta = \eta$, $\forall \eta \in D(B)$.*

(b) *Let $\lambda \in \rho(B)$ and define*

$$D(Y_\lambda) = \{\xi \in \mathcal{H} : (B - \lambda I)^{-1} T\xi \in D(T^{-1})\},$$
$$Y_\lambda \xi = T^{-1}(B - \lambda I)^{-1} T\xi, \ \xi \in D(Y_\lambda).$$

Then,

(b.1) $Y_\lambda (A - \lambda I)\xi = \xi, \ \forall \xi \in D(A).$

(b.2) *For every $\eta \in \mathcal{H}$ such that $Y_\lambda \eta \in D(A)$, $(A - \lambda I) Y_\lambda \eta = \eta$.*

We skip the easy proof, which is given in Ref. [11, Proposition 3.25].

Corollary 7.3.18 *Let A, B be as in Proposition 7.3.17 and assume that T^{-1} is everywhere defined and bounded. Then $\rho(A) \setminus \sigma_p(B) \subseteq \rho(B)$ and $\rho(B) \setminus \sigma_r(A) \subseteq \rho(A)$.*

Proof. The first inclusion is an immediate application of (a) of the previous proposition and the second one is obtained by taking the adjoints. ■

Actually, we may drop the assumption that T^{-1} is bounded and still get a useful result.

Corollary 7.3.19 *Let A, B be as in Proposition 7.3.17. Assume that $D(B)$ and $R(B)$ are subspaces of $D(T^{-1})$. Then $\rho(A) \setminus \sigma_p(B) \subseteq \rho(B) \cup \sigma_c(B)$.*

Proof. Let $\lambda \in \rho(A) \setminus \sigma_p(B)$. By Proposition 7.3.17(a), the operator $(B - \lambda I)^{-1}$ has a densely defined inverse. If $(B - \lambda I)^{-1}$ is bounded, then it has an everywhere defined bounded closure, which coincides with $(B - \lambda I)^{-1}$, as the latter is closed, being the inverse of a closed operator. In this case, $\lambda \in \rho(B)$. On the other hand, if $(B - \lambda I)^{-1}$ is unbounded, then $\lambda \in \sigma_c(B)$. Therefore, $\rho(A) \setminus \sigma_p(B) \subseteq \rho(B) \cup \sigma_c(B)$. ■

Let us consider again the special case where T^{-1} is also everywhere defined and bounded (but does not necessarily satisfy $TD(A) = D(B)$).

Proposition 7.3.20 *Let A and B be closed operators. Assume that $A \dashv B$, with the bounded intertwining operator T. Assume that T^{-1} is everywhere defined and bounded and $TD(A)$ is a core for B. Then*

$$\sigma_p(A) \subseteq \sigma_p(B) \subseteq \sigma(B) \subseteq \sigma(A).$$

Proof. We simply notice that, in this case, by (iv) of Proposition 7.3.16, $\sigma_p(B) \subset \sigma(A)$. Hence, $\rho(A) \setminus \sigma_p(B) = \rho(A) \subseteq \rho(B)$, by Corollary 7.3.18. ■

Remark 7.3.21 The situation described in Proposition 7.3.20 is quite important for possible applications. Even if the spectra of A and B may be different, it gives a certain number of informations on $\sigma(B)$ once $\sigma(A)$ is known. For instance, if A has a pure point spectrum, then B is isospectral to A. More generally, if A is self-adjoint, then any operator B, which is quasi-similar to A by means of a bounded intertwining operator T whose inverse is bounded too, has real spectrum.

We will illustrate the previous propositions by two examples, both taken from Ref. (11). In the first one, $A \dashv B, A, B$ and T are all bounded, and the two spectra, which are pure point, coincide.

Example 7.3.22 Let us consider the operators P_φ and A_φ of Example 7.3.15 with $\varphi \in D(Q^2)$. In this case A_φ is bounded and everywhere defined and, as noticed before, $P_\varphi \dashv A_\varphi$ with the intertwining operator $T := (I + Q^2)^{-1}$ The spectrum of A_φ is easily computed to be $\sigma(A_\varphi) = \{0, 1\}$. Thus, it coincides with $\sigma(P_\varphi)$, in accordance with Proposition 7.3.16 (iii). To see this, we begin by looking for eigenvalues. The equation

$$\langle (I + Q^2)f | \varphi \rangle (I + Q^2)^{-1}\varphi - \lambda f = 0 \tag{7.3.3}$$

has nonzero solutions in two cases: if $\lambda = 0$, then any element of $\{(I + Q^2)\varphi\}^\perp$ is an eigenvector. If $\lambda \neq 0$, then a solution must be a multiple of $(I + Q^2)^{-1}\varphi$, that is, $f = \kappa(I + Q^2)^{-1}\varphi$. Substituting in (7.3.3) one obtains $\lambda = 1$ and the set of eigenvectors is the one-dimensional subspace generated by $(I + Q^2)^{-1}\varphi$. On the other hand, if $\lambda \notin \{0, 1\}$, then, for every $g \in L^2(\mathbb{R})$, the equation $(A_\varphi - \lambda I)f = g$ has the unique solution

$$f = -\frac{1}{\lambda} g + \frac{\langle g | (I + Q^2)\varphi \rangle}{\lambda(1 - \lambda)} (I + Q^2)^{-1}\varphi.$$

Thus, $(A_\varphi - \lambda I)^{-1}$ is an everywhere defined operator. Next, being the inverse of a closed (in fact, bounded) operator, it is closed. Therefore, by the closed graph theorem, it is bounded, $(A_\varphi - \lambda I)^{-1} \in \mathcal{B}(\mathcal{H})$. We conclude that $\sigma_p(A_\varphi) = \sigma_p(P_\varphi) = \{0, 1\}$.

In the second example, T is bounded, but A and B are both unbounded. In that case, the two spectra coincide as a whole, but not their individual parts. In particular, A has a nonempty residual spectrum, whereas B does not.

Example 7.3.23 Let A be the operator in $L^2(\mathbb{R})$ defined as follows:

$$(Af)(x) = f'(x) - \frac{2x}{1+x^2} f(x), \quad f \in D(A) = W^{1,2}(\mathbb{R}).$$

Then A is a closed operator in $L^2(\mathbb{R})$, being the sum of a closed operator and a bounded one. Let B be the closed operator defined by

$$(Bf)(x) = f'(x), \quad f \in D(B) = W^{1,2}(\mathbb{R}).$$

Then $A \dashv B$ with the intertwining operator $T = (I + Q^2)^{-1}$. Indeed, it is easily seen that $T : W^{1,2}(\mathbb{R}) \to W^{1,2}(\mathbb{R})$. Moreover, for every $f \in W^{1,2}(\mathbb{R})$, one has

$$(TAf)(x) = (1+x^2)^{-1}\left(f'(x) - \frac{2x}{1+x^2} f(x)\right) = \frac{f'(x)}{1+x^2} - \frac{2xf(x)}{(1+x^2)^2}$$

and

$$(BTf)(x) = \frac{d}{dx}\left(\frac{f(x)}{1+x^2}\right) = \frac{f'(x)}{1+x^2} - \frac{2xf(x)}{(1+x^2)^2}.$$

Thus, indeed, $TD(A) \subseteq D(B)$ and $TAf = BTf$, $\forall f \in D(A)$. It is easily seen that $\sigma_p(A) = \emptyset$. As for B, one has, as it is well known, $\sigma(B) = \sigma_c(B) = i\mathbb{R}$. On the other hand, $0 \in \sigma_r(A)$, as, if $h(x) = (1+x^2)^{-1}$, then $\langle Af|h\rangle = 0$, for every $f \in W^{1,2}(\mathbb{R})$, so that the range $R(A)$ is not dense. Actually one has $\sigma_r(A) = \{0\}$, as one can easily check by computing $\sigma_p(A^*)$. Thus, by Corollary 7.3.18, $\sigma(A) = \sigma(B)$, but the quasi-similarity does not preserve the relevant parts of the spectra.

7.3.3 Quasi-Similarity with an Unbounded Intertwining Operator

As shown in Ref. (12), it is easy to generalize the preceding analysis to the case of an unbounded intertwining operator. First we adapt the definition.

Definition 7.3.24 Let A, B be two densely defined linear operators on the Hilbert spaces $\mathcal{H}, \mathcal{K}$, respectively. A closed (densely defined) operator $T : D(T) \subseteq \mathcal{H} \to \mathcal{K}$ is called an *intertwining operator* for A and B if

(io_0) $D(TA) = D(A) \subset D(T)$;
(io_1) $T : D(A) \to D(B)$;
(io_2) $BT\xi = TA\xi$, $\forall \xi \in D(A)$.

The first part of condition (io_0) means that $\xi \in D(A)$ implies $A\xi \in D(T)$. Of course, this definition reduces to the usual one, Definition 7.3.1, if T is bounded, as then condition (io_0) is satisfied automatically.

In terms of this definition, we say again that A is *quasi-similar* to B , and write $A \dashv B$, if there exists a (possibly unbounded) intertwining operator T for A and B, which is invertible, with inverse T^{-1} densely defined (that is, T is quasi-invertible, in the terminology of Ref. (14)). This definition implies easily that A is quasi-similar to B if, and only if, $A \subseteq T^{-1}BT$, where T is a closed densely defined operator, which is injective and has dense range. Indeed, T^{-1} exists by assumption and by (io$_2$), if $\xi \in D(A)$, $BT\xi$ is an element of the range of T; thus, we can apply T^{-1} to both sides and get $T^{-1}BT\xi = A\xi$. This is equivalent to say that $A \subseteq T^{-1}BT$.

Notice that, contrary to Remark 7.3.2, if T is an unbounded intertwining operator for A and B, its adjoint $T^* : D(T^*) \subseteq \mathcal{K} \to \mathcal{H}$ need not be an intertwining operator for B^* and A^*, because Condition (io$_0$) may fail for B^*, unless T is bounded. As a matter of fact, quasi-similarity with an unbounded intertwining operator may occur only under *singular*, even pathological, circumstances, as shown by the next proposition..

Proposition 7.3.25 *Let $A \dashv B$ with intertwining operator T. If the resolvent set $\rho(A)$ is not empty, then T is necessarily bounded.*

Proof. From $A \subseteq T^{-1}BT$ it follows that $A - \lambda I \subseteq T^{-1}(B - \lambda I)T$, for every $\lambda \in \mathbb{C}$. If $\lambda \in \rho(A)$, then, for every $\eta \in \mathcal{H}$, there exists $\xi \in D(A)$ such that $(A - \lambda I)\xi = \eta$. Thus, $\xi \in D(T^{-1}(B - \lambda I)T)$ and $T^{-1}(B - \lambda I)T\xi = \eta$. This clearly implies that $\eta \in D(T)$. Hence $D(T) = \mathcal{H}$ and thus T is bounded, as it is closed by definition. ∎

What remains to do now is to see how much of Propositions 7.3.16–7.3.20 remains true when the intertwining operator T is unbounded. The first result parallels part of Proposition 7.3.16.

Proposition 7.3.26 *Let A and B be closed operators and assume that $A \dashv B$, with the (possibly) unbounded intertwining operator T. Then the following statements hold.*

(i) $\sigma_p(A) \subseteq \sigma_p(B)$. *If $\xi \in D(A)$ is an eigenvector of A corresponding to the eigenvalue λ, then $T\xi$ is an eigenvector of B corresponding to the same eigenvalue. Thus, for every $\lambda \in \sigma_p(A)$, one has for the multiplicities $m_A(\lambda) \leq m_B(\lambda)$.*

(ii) If $TD(A) = D(B)$ and T^{-1} is bounded, then $\sigma_p(A) = \sigma_p(B)$.

(iii) If T^{-1} is bounded and $TD(A)$ is a core for B, then $\sigma_p(B) \subseteq \sigma(A)$.

Proof. (i) Let $\lambda \in \sigma_p(A)$, that is, $\psi \in D(A)$ and $A\psi = \lambda\psi$. Then, by (io$_0$), $A\psi \in D(T)$ and $TA\psi = \lambda T\psi$. The rest is obvious.

(ii) If $\eta \in D(B)$, there exists $\xi \in D(A)$ such that $\eta = T\xi$ and $T^{-1}\eta = \xi$. Then $B\eta = \lambda\eta$ implies $A\xi = \lambda\xi$.

(iii) The proof of Proposition 7.3.16(iv) remains valid. The last argument states that $\xi_n \to 0$ and $T\xi_n \to \eta$. As T is closed, this again implies that $T\xi_n \to 0$, which is impossible as $\|\eta\| = 1$. ∎

Next, Proposition 7.3.17 goes through, with the domain $D(Y_\lambda)$ replaced by

$$D(Y_\lambda) = \{\xi \in D(T) : (B - \lambda I)^{-1}T\xi \in D(T^{-1})\}.$$

As a consequence, the first half of Corollary 7.3.18 goes through.

Corollary 7.3.27 *Let A, B be as in Proposition 7.3.17 and assume that T^{-1} is everywhere defined and bounded. Then*

$$\rho(A) \setminus \sigma_p(B) \subseteq \rho(B).$$

Once again, the second statement, about $\sigma_r(A)$, does not hold in general, as it relies on the adjoints, for which we have no information. In the same way, Corollary 7.3.19 still holds, but Proposition 7.3.20 does not.

Things improve if we assume the operators A and B to be mutually quasi-similar, $A \dashv\vdash B$. First, we can improve Proposition 7.3.26(i).

Proposition 7.3.28 *Let A and B be closed operators and assume that $A \dashv\vdash B$, with possibly unbounded intertwining operators $T_{AB} : D(A) \to D(B)$ and $T_{BA} : D(B) \to D(A)$. Then: $\sigma_p(A) = \sigma_p(B)$. If $\xi \in D(A)$ is an eigenvector of A corresponding to the eigenvalue λ, then $T_{AB}\xi$ is an eigenvector of B corresponding to the same eigenvalue. If $\eta \in D(B)$ is an eigenvector of B corresponding to the eigenvalue μ, then $T_{BA}\xi$ is an eigenvector of A corresponding to the same eigenvalue. In both cases, the multiplicities are the same.*

In order to obtain identity of the spectra, we have to assume that both intertwining operators have a bounded inverse. Indeed, applying Corollary 7.3.27, we get immediately:

Proposition 7.3.29 *Let A and B be closed operators such that $A \dashv\vdash B$. Assume that both intertwining operators T_{AB}, T_{BA} have a bounded inverse. Then one has, in addition to the statements of Proposition 7.3.16, $\rho(A) = \rho(B)$, hence $\sigma(A) = \sigma(B)$.*

Proof. Let T_{AB}^{-1} be everywhere defined and bounded. Then, by Corollary 7.3.27 and Proposition 7.3.28, we have

$$\rho(A) \setminus \sigma_p(B) = \rho(A) \setminus \sigma_p(A) = \rho(A) \subseteq \rho(B).$$

Exchanging A and B, we get $\rho(B) \subseteq \rho(A)$, which proves (i). Then the statement follows immediately. ∎

Under these conditions, it follows that $\sigma_c(A) \cup \sigma_r(A) = \sigma_c(B) \cup \sigma_r(B)$, but we cannot compare separately the two remaining parts of the spectra of A and B, for the same reason as before.

These results show that mutual quasi-similarity is a strong property. As another testimony of that fact, it is worth quoting a result from Ôta and Schmüdgen (33).

Proposition 7.3.30 *Let A and B be closed operators. Then:*

(i) Let A and B be normal (in particular, self-adjoint) and $A \dashv\vdash B$. Then they are unitarily equivalent, $A \approx B$.

(ii) Let A be symmetric and B self-adjoint, with $A \dashv\vdash B$. Then A is self-adjoint and $A \approx B$.

(iii) Let A be symmetric and $A \dashv\vdash A^*$. Then A is self-adjoint.

7.4 THE LATTICE GENERATED BY A SINGLE METRIC OPERATOR

Let us consider first the general case, where both G and G^{-1} may be unbounded. We consider the domain $D(G^{1/2})$ and we equip it with the following norm

$$\|\xi\|_{R_G}^2 := \left\|(I+G)^{1/2}\xi\right\|^2, \ \xi \in D(G^{1/2}). \tag{7.4.1}$$

As this norm is equivalent to the graph norm,

$$\|\xi\|_{\rm gr}^2 := \|\xi\|^2 + \left\|G^{1/2}\xi\right\|^2, \tag{7.4.2}$$

this yields a Hilbert space, denoted $\mathcal{H}(R_G)$, dense in $\mathcal{H}$. Next, we equip that space with the norm $\|\xi\|_G^2 := \|G^{1/2}\xi\|^2$ and denote by $\mathcal{H}(G)$ the completion of $\mathcal{H}(R_G)$ in that norm and corresponding inner product $\langle\cdot|\cdot\rangle_G := \langle G^{1/2}\cdot|G^{1/2}\cdot\rangle$. Hence, we have $\mathcal{H}(R_G) = \mathcal{H} \cap \mathcal{H}(G)$, with the so-called projective norm [13, Section I.2.1], which here is simply the graph norm (7.4.2). Then we define $R_G := I + G$, which justifies the notation $\mathcal{H}(R_G)$, by comparison of (7.4.1) with the norm $\|\cdot\|_G^2$ of $\mathcal{H}(G)$.

Now we perform the construction described in Ref. [13, Section 5.5], and largely inspired by interpolation theory (36). First we notice that the conjugate dual $\mathcal{H}(R_G)^\times$ of $\mathcal{H}(R_G)$ is $\mathcal{H}(R_G^{-1})$ and one gets the triplet

$$\mathcal{H}(R_G) \subset \mathcal{H} \subset \mathcal{H}(R_G^{-1}). \tag{7.4.3}$$

Proceeding in the same way with the inverse operator G^{-1}, we obtain another Hilbert space, $\mathcal{H}(G^{-1})$, and another triplet

$$\mathcal{H}(R_{G^{-1}}) \subset \mathcal{H} \subset \mathcal{H}(R_{G^{-1}}^{-1}). \tag{7.4.4}$$

Then, taking conjugate duals, it is easy to see that one has

$$\mathcal{H}(R_G)^\times = \mathcal{H}(R_G^{-1}) = \mathcal{H} + \mathcal{H}(G^{-1}), \tag{7.4.5}$$

$$\mathcal{H}(R_{G^{-1}})^\times = \mathcal{H}(R_{G^{-1}}^{-1}) = \mathcal{H} + \mathcal{H}(G). \tag{7.4.6}$$

In these relations, the right-hand side is meant to carry the inductive norm (and topology) [13, Section I.2.1], so that both sides are in fact unitary equivalent, hence identified.

By the definition of the spaces $\mathcal{H}(R_{G^{\pm 1}})$ and the relations (7.4.5)-(7.4.6), it is clear that all the seven spaces involved constitute a lattice with respect to the lattice operations

$$\mathcal{H}_1 \wedge \mathcal{H}_2 := \mathcal{H}_1 \cap \mathcal{H}_2,$$

$$\mathcal{H}_1 \vee \mathcal{H}_2 := \mathcal{H}_1 + \mathcal{H}_2.$$

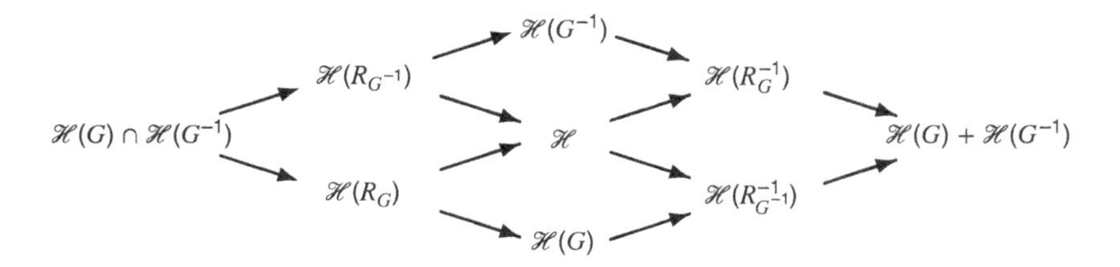

Figure 7.1 The lattice of Hilbert spaces generated by a metric operator.

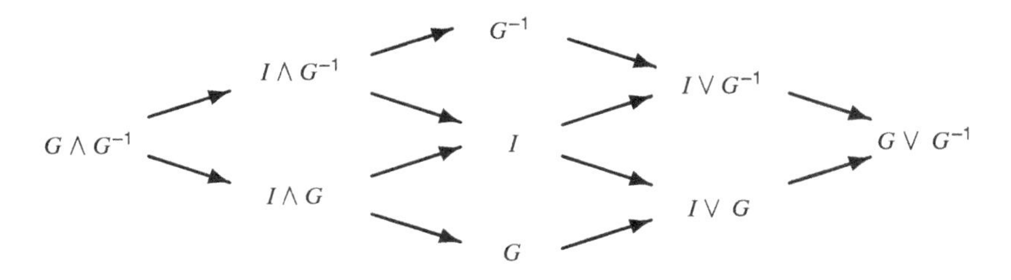

Figure 7.2 The lattice generated by a metric operator.

Completing that lattice by the extreme spaces $\mathcal{H}(R_G) \cap \mathcal{H}(R_{G^{-1}})) = \mathcal{H}(G) \cap \mathcal{H}(G^{-1})$ and $\mathcal{H}(R_G^{-1}) + \mathcal{H}(R_{G^{-1}}^{-1}) = \mathcal{H}(G) + \mathcal{H}(G^{-1})$ (these equalities follow from interpolation), we obtain the diagram shown in Fig. 7.1, which completes the corresponding one from Ref. (11). Here also every embedding is continuous and has dense range.

Next, on the space $\mathcal{H}(R_G)$, equipped with the norm $\|\cdot\|_G^2$, the operator $G^{1/2}$ is isometric onto $\mathcal{H}$, hence it extends to a unitary operator from $\mathcal{H}(G)$ onto $\mathcal{H}$. Analogously, $G^{-1/2}$ is a unitary operator from $\mathcal{H}(G^{-1})$ onto $\mathcal{H}$. In the same way, the operator $R_G^{1/2}$ is unitary from $\mathcal{H}(R_G)$ onto $\mathcal{H}$ and from $\mathcal{H}$ onto $\mathcal{H}(R_G^{-1})$.[4] Hence R_G is the Riesz unitary operator mapping $\mathcal{H}(R_G)$ onto its conjugate dual $\mathcal{H}(R_G^{-1})$, and similarly R_G^{-1} from $\mathcal{H}(R_G^{-1})$ onto $\mathcal{H}(R_G)$, that is, in the triplet (7.4.3). Analogous relations hold for G^{-1}, that is, in the triplet (7.4.4).

As all spaces $\mathcal{H}(A)$ are indexed by the corresponding operator A, we can as well apply the lattice operations on the operators themselves. This would give the diagram shown in Fig. 7.2.

The link between the lattices of Figs 7.1 and 7.2 is given in terms of an order relation: $G_1 \preceq G_2$ if and only if $\mathcal{H}(G_1) \subset \mathcal{H}(G_2)$, where the embedding is continuous and has dense range. In particular, if G is bounded and G^{-1} unbounded, the relation (7.4.8) becomes

$$G^{-1} \preceq I \preceq G.$$

In Section 7.6, we extend these considerations to families of metric operators.

[4]The space $\mathcal{H}(R_G^{-1})$ is (three times) erroneously denoted $\mathcal{H}(R_{G^{-1}})$ in Ref. [11, p. 4]; see Corrigendum.

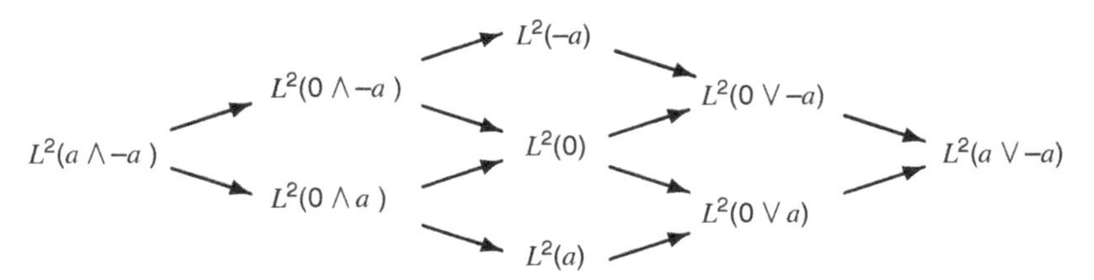

Figure 7.3 The lattice generated by a metric operator. Note that $L^2(0) \equiv L^2$.

Before proceeding, let us give two (easy) examples, in which G and G^{-1} are multiplication operators in $\mathcal{H} = L^2(\mathbb{R}, \mathrm{d}x)$, both unbounded, so that the three middle spaces are mutually noncomparable.

1. The first example comes from Ref. [13, Section 5.5.1], namely, $G = x^2$, so that $R_G = 1 + x^2$. Then all spaces appearing in Fig. 7.1 are weighted L^2 spaces, for instance $\mathcal{H}(G) = L^2(\mathbb{R}, x^2\, \mathrm{d}x)$, $\mathcal{H}(R_G) = L^2(\mathbb{R}, (1 + x^2)\, \mathrm{d}x)$, and so on. The complete lattice is given in Ref. [13, Fig. 2]. As expected, all the norms are equivalent to the corresponding projective norms, respectively inductive norms (see the proof of a similar statement for sequences in Ref. [13, Section 4.3.1]).
2. For the second example, inspired by Ali *et al.* (37) and Kretschmer and Szymanowski (19), one chooses $G = e^{ax}, G^{-1} = e^{-ax}$ and proceeds in the same way. The resulting lattice plays a significant role in quantum scattering theory, as discussed in Ref. [13, Section 4.6.3], so that it is worthwhile to go into some details. Keeping the notation of that reference, we define the Hilbert space

$$L^2(a) := \{f : \int_{-\infty}^{+\infty} e^{ax}\, |f(x)|^2\, \mathrm{d}x < \infty\} = L^2(r_a) \quad \text{with } r_a(t) = e^{-ax}. \tag{7.4.7}$$

Then consider the lattice generated by the family $L^2(a), L^2(0) = L^2$ and $L^2(-a)$. The infimum is $L^2(a) \wedge L^2(b) = L^2(a) \cap L^2(b) = L^2(a \wedge b)$, with $r_{a \wedge b}(x) = \min(r_a(x), r_b(x))$, and the supremum $L^2(a) \vee L^2(b) = L^2(a) + L^2(b) = L^2(a \vee b)$, $r_{a \vee b}(x) = \max(r_a(x), r_b(x))$. As usual, these norms are equivalent to the projective, respectively, inductive, norms. For instance, the following two norms are equivalent

$$\|f\|^2_{L^2(r_{a \wedge -a})} = \int_{-\infty}^{+\infty} e^{a|x|}\, |f(x)|^2\, \mathrm{d}x \asymp \int_{-\infty}^{+\infty} (e^{ax} + e^{-ax})|f(x)|^2\, \mathrm{d}x.$$

The resulting lattice is shown in Fig. 7.3.

As a matter of fact, in both examples, the discrete lattice of nine spaces may be converted into a continuous one by interpolation. In the first case, the spaces of the central column are $\{L^2(\mathbb{R}, x^\alpha \, dx), -2 \le \alpha \le 2\}$. Similarly, the second example yields $\{L^2(a), -1 \le a \le 1\}$. And actually the same holds true in the general, abstract case, where one gets instead the family $\{\mathcal{H}(G^\alpha), -1 \le \alpha \le 1\}$.

7.4.1 Bounded Metric Operators

Now, if G is bounded, the triplet (7.4.3) collapses, in the sense that all three spaces coincide as vector spaces, with equivalent norms. Similarly, one gets $\mathcal{H}(R_{G^{-1}}) = \mathcal{H}(G^{-1})$ and $\mathcal{H}(R_{G^{-1}}^{-1}) = \mathcal{H}(G)$. So we are left with the triplet

$$\mathcal{H}(G^{-1}) \subset \mathcal{H} \subset \mathcal{H}(G). \tag{7.4.8}$$

Then $G^{1/2}$ is a unitary operator from $\mathcal{H}(G)$ onto $\mathcal{H}$ and from $\mathcal{H}$ onto $\mathcal{H}(G^{-1})$, whereas $G^{-1/2}$ is a unitary operator $\mathcal{H}(G^{-1})$ onto $\mathcal{H}$ and from $\mathcal{H}$ onto $\mathcal{H}(G)$.

If G^{-1} is also bounded, then the spaces $\mathcal{H}(G^{-1})$ and $\mathcal{H}(G)$ coincide with $\mathcal{H}$ as vector spaces and their norms are equivalent to (but different from) the norm of $\mathcal{H}$.

Remark 7.4.1 If B is a bounded operator from $\mathcal{H}(G)$ into itself, for every $\eta \in \mathcal{H} \subseteq \mathcal{H}(G)$, we have

$$\|B\xi\|_G \le \gamma \, \|\xi\|_G \le \gamma' \, \|\xi\| \, ,$$

for some $\gamma, \gamma' > 0$.

This means that $B_0 := B\restriction_{\mathcal{H}}$ is bounded from $(\mathcal{H}, \|\cdot\|)$ into $(\mathcal{H}(G), \|\cdot\|_G)$. Then there exists a bounded operator $B_0^\dagger : \mathcal{H}(G) \to \mathcal{H}$ such that

$$\langle B_0\xi | \eta \rangle_G = \langle \xi | B_0^\dagger \eta \rangle, \quad \forall \xi \in \mathcal{H}, \, \eta \in \mathcal{H}(G).$$

This in particular applies to G and $G^{1/2}$, which can be viewed as restrictions to $\mathcal{H}$ of their natural extensions $\widetilde{G}$ and $\widetilde{G^{1/2}}$ to $\mathcal{H}(G)$. In this case, one can easily prove that $G^\dagger = G^{3/2}\widetilde{G^{1/2}}$, $\widetilde{G^{-1/2}} = (G^{1/2})^\dagger$. In particular, if G^{-1} is also bounded, $G^\dagger = G^2$ and $(G^{1/2})^\dagger = G^{-1/2}$.

7.4.2 Unbounded Metric Operators

Actually one can go further, following a construction made in Ref. (38). Let G be *unbounded*, with $G > 1$. Then the norm $\|\cdot\|_G$ is equivalent to the norm $\|\cdot\|_{R_G}$ on $D(G^{1/2})$, so that $\mathcal{H}(G) = \mathcal{H}(R_G)$ as vector spaces and thus also $\mathcal{H}(G^{-1}) = \mathcal{H}(R_G^{-1})$. On the other hand, G^{-1} is bounded. Hence, we get the triplet

$$\mathcal{H}(G) \subset \mathcal{H} \subset \mathcal{H}(G^{-1}). \tag{7.4.9}$$

In the general case, we have $R_G = 1 + G > 1$ and it is also a metric operator. Thus, we have now

$$\mathcal{H}(R_G) \subset \mathcal{H} \subset \mathcal{H}(R_G^{-1}). \tag{7.4.10}$$

In both cases, one recognizes that the triplet (7.4.9), respectively, (7.4.10), is the central part of the discrete scale of Hilbert spaces built on the powers of $G^{1/2}$, respectively, $R_G^{1/2}$. This means, in the first case, $V_{\mathcal{G}} := \{\mathcal{H}_n, n \in \mathbb{Z}\}$, where $\mathcal{H}_n = D(G^{n/2}), n \in \mathbb{N}$, with a norm equivalent to the graph norm, and $\mathcal{H}_{-n} = \mathcal{H}_n^{\times}$:

$$\ldots \subset \mathcal{H}_2 \subset \mathcal{H}_1 \subset \mathcal{H} \subset \mathcal{H}_{-1} \subset \mathcal{H}_{-2} \subset \ldots \tag{7.4.11}$$

Thus $\mathcal{H}_1 = \mathcal{H}(G)$ and $\mathcal{H}_{-1} = \mathcal{H}(G^{-1})$. In the second case, one simply replaces $G^{1/2}$ by $R_G^{1/2}$ and performs the same construction.

As in the original construction, this raises the question of identifying the end spaces of the scale, namely,

$$\mathcal{H}_{\infty}(G) := \bigcap_{n\in\mathbb{Z}} \mathcal{H}_n, \qquad \mathcal{H}_{-\infty}(G) := \bigcup_{n\in\mathbb{Z}} \mathcal{H}_n. \tag{7.4.12}$$

In fact, one can go one more step. Namely, following Ref. [13, Section 5.1.2], we can use quadratic interpolation theory and build a continuous scale of Hilbert spaces $\mathcal{H}_\alpha$, $0 \leq \alpha \leq 1$, between $\mathcal{H}_1$ and $\mathcal{H}$, where $\mathcal{H}_\alpha = D(G^{\alpha/2})$, with the graph norm $\|\xi\|_\alpha^2 = \|\xi\|^2 + \|G^{\alpha/2}\xi\|^2$ or, equivalently, the norm $\|(I+G)^{\alpha/2}\xi\|^2$. Indeed every $G^\alpha, \alpha \geq 0$, is an unbounded metric operator.

Next, we define $\mathcal{H}_{-\alpha} = \mathcal{H}_\alpha^{\times}$ and iterate the construction to the full continuous scale $V_{\widetilde{\mathcal{G}}} := \{\mathcal{H}_\alpha, \alpha \in \mathbb{R}\}$. Then, of course, one can replace $\mathbb{Z}$ by $\mathbb{R}$ in the definition (7.4.12) of the end spaces of the scale.

Let us give three (trivial) examples. Take first $\mathcal{H} = L^2(\mathbb{R}, \mathrm{d}x)$ and define G_x as the operator of multiplication by $(1+x^2)^{1/2}$, which is an unbounded metric operator (and the square root of the operator $\mathcal{H}(R_G)$ of the first example. In the same way, define $G_p := (1 - \mathrm{d}^2/\mathrm{d}x^2)^{1/2} = \mathcal{F}G_x\mathcal{F}^{-1}$, where $\mathcal{F}$ is the Fourier transform. Similarly, in $L^2(\mathbb{R}^3)$, we can take $G_p := (1-\Delta)^{1/2}$. For these examples, the end spaces of the scale (7.4.11) are easy to identify: $\mathcal{H}_\infty(G_x)$ consists of square integrable, fast decreasing functions, whereas the scale built on G_p is precisely the scale of Sobolev spaces, $\mathcal{H}_n = W^{n/2,2}$ [24, Section 3.2]. The notation, of course, refers to the operators of position x and momentum p in QM.

For the third example, take $\mathcal{H} = L^2(\mathbb{R}, \mathrm{d}x)$ and $G_{\rm osc} := (1 - \mathrm{d}^2/\mathrm{d}x^2 + x^2)^{1/2}$, where one recognizes the Hamiltonian of a one-dimensional quantum harmonic oscillator. Then $\mathcal{H}_\infty(G_{\rm osc})$ is simply the Schwartz space $\mathcal{S}$ of smooth fast decreasing functions and $\mathcal{H}_{-\infty}(G_{\rm osc}) = \mathcal{S}^{\times}$, the space of tempered distributions. In addition, $\mathcal{H}_\infty(G_x) \cap \mathcal{H}_\infty(G_p) = \mathcal{H}_\infty(G_{\rm osc})$.

More generally, given any unbounded self-adjoint operator A in $\mathcal{H}$, $G_A := (1+A^2)^{1/2}$ is an unbounded metric operator, larger than 1, and the construction of the corresponding scale is straightforward.

7.5 QUASI-HERMITIAN OPERATORS

Intuitively, a quasi-Hermitian operator A is an operator that is Hermitian when the space is endowed with a new inner product. We will make this precise in the sequel, generalizing the original definition of Dieudonné (4).

Definition 7.5.1 A closed operator A, with dense domain $D(A)$, is called *quasi-Hermitian* if there exists a metric operator G, with dense domain $D(G)$, such that $D(A) \subset D(G)$ and

$$\langle A\xi | G\eta \rangle = \langle G\xi | A\eta \rangle, \quad \xi, \eta \in D(A). \tag{7.5.1}$$

Of course, if the condition $D(A) \subset D(G)$ is not satisfied, the relation (7.5.1) may hold for every $\xi, \eta \in D(G) \cap D(A)$, but the definition is not reliable, as it may happen that $D(G) \cap D(A) = \{0\}$. Thus, to make sense, this more general definition would require additional conditions on G, that will depend whether G and G^{-1} are bounded or not. For that reason, we will keep Definition 7.5.1 in the sequel.

7.5.1 Changing the Norm: Two-Hilbert Space Formalism

Take first G bounded and G^{-1} possibly unbounded. According to the analysis of Section 7.4, we are facing the triplet (7.4.8), namely,

$$\mathcal{H}(G^{-1}) \subset \mathcal{H} \subset \mathcal{H}(G),$$

where $\mathcal{H}(G)$ is a Hilbert space, the completion of $\mathcal{H}$ in the norm $\|\cdot\|_G$. Thus, we have now two different Hilbert spaces and the question is how operator properties are transferred from $\mathcal{H}$ to $\mathcal{H}(G)$. In particular, two different adjoints may be defined and we have to compare them before analyzing quasi-Hermitian operators. Notice that we are recovering here the standard situation in pseudo-Hermitian QM (1, 2), even the three–Hilbert space formulation developed in Ref. (39). Here we follow again (12).

We begin by the following easy result (11).

Proposition 7.5.2 *(i) Given the bounded metric operator G, let A be a linear operator in $\mathcal{H}$ with $D(A)$ dense in $\mathcal{H}$. Then $D(A)$ is dense in $\mathcal{H}(G)$.*

(ii) Let A be a linear operator in $\mathcal{H}$. If A is closed in $\mathcal{H}(G)$, then it closed in $\mathcal{H}$.

Notice that, in both statements of Proposition 7.5.2, one can replace the space $\mathcal{H}(G)$ by $\mathcal{H}(G^\alpha)$, for any $\alpha > 0$. We emphasize that the converse of (ii) does not hold in general: if A closed in $\mathcal{H}$, it need not be closed, not even closable, in $\mathcal{H}(G)$. See Corollary 7.5.8 next.

Next, for the reader's convenience, we prove the following well-known result [40, Theorem 4.19].

Lemma 7.5.3 *Let us consider a closed operator S in $\mathcal{H}$ with dense domain $D(S)$. If B is bounded, one has $(BS)^* = S^*B^*$.*

Proof. It is standard that $(BS)^* \supseteq S^*B^*$. Let $\eta \in D((BS)^*)$. Then there exists $\eta^* \in \mathcal{H}$ such that

$$\langle BS\xi|\eta\rangle = \langle \xi|\eta^*\rangle, \quad \forall\, \xi \in D(S) = D(BS).$$

Thus,

$$\langle S\xi|B^*\eta\rangle = \langle \xi|\eta^*\rangle, \quad \forall\, \xi \in D(S).$$

Hence, $B^*\eta \in D(S^*)$ and $S^*B^*\eta = (BS)^*\eta$. Thus, $(BS)^* \subseteq S^*B^*$. ∎

Let again S be a closed densely defined operator in $\mathcal{H}$. By Proposition 7.5.2, $D(S)$ is dense in $\mathcal{H}(G)$. So S has a well-defined adjoint in $\mathcal{H}(G)$. We denote it by $S^\#$, while we denote by S^* the usual adjoint in $\mathcal{H}$. We compute $S^\#$, recalling that $\widetilde{G}$ is the natural extension of G to $\mathcal{H}(G)$.

Proposition 7.5.4 *Let G be a bounded metric operator in $\mathcal{H}$ and S a closed, densely defined operator in $\mathcal{H}$. Then:*

(i) $\widetilde{G}D(S^\#) \subseteq D(S^*)$ and $S^*\widetilde{G}\eta = \widetilde{G}S^\#\eta$, for every $\eta \in D(S^\#)$, where $\widetilde{G}$ denotes the natural extension of G to $\mathcal{H}(G)$.

(ii) If G^{-1} is also bounded, then $D(S^\#) = G^{-1}D(S^*)$ and $S^\#\eta = G^{-1}S^*G\eta$, for every $\eta \in D(S^\#)$.

Proof. (i) Let $\eta \in D(S^\#)$. Then there exists $\eta^\# \in \mathcal{H}(G)$ such that

$$\langle S\xi|\eta\rangle_G = \langle \xi|\eta^\#\rangle_G, \quad \forall \xi \in D(S)$$

and $S^\#\eta = \eta^\#$. As

$$|\langle S\xi|\eta\rangle_G| = |\langle \xi|\eta^\#\rangle_G| \le \|\xi\|_G\, \|\eta^\#\|_G \le \gamma\|\xi\|\, \|\eta^\#\|_G, \quad \forall\, \xi \in D(S),$$

there exists $\eta^* \in \mathcal{H}$ such that

$$\langle S\xi|\widetilde{G}\eta\rangle = \langle \xi|\eta^*\rangle.$$

This implies that $\widetilde{G}\eta \in D(S^*)$. We recall that $\widetilde{G}\mathcal{H}(G) \subseteq \mathcal{H}$ (see Remark 7.4.1). It is easily seen that $\eta^* = \widetilde{G}\eta^\#$. Hence

$$\langle S\xi|\widetilde{G}\eta\rangle = \langle \xi|S^*\widetilde{G}\eta\rangle = \langle \xi|\widetilde{G}S^\#\eta\rangle, \quad \forall\, \xi \in D(S).$$

This, in turn, implies that $S^*\widetilde{G}\eta = \widetilde{G}S^\#\eta$, for every $\eta \in D(S^\#)$.

(ii) If G and G^{-1} are both bounded, $\mathcal{H}(G^{-1}) = \mathcal{H} = \mathcal{H}(G)$ as vector spaces, but with different norms. From (i), it follows that $GD(S^\#) \subseteq D(S^*)$ and $S^*G\eta = GS^\#\eta$, for every $\eta \in D(S^\#)$.

Let now $\zeta \in D(S^*)$. Then $\zeta = G\eta$ and $S^*G\eta = G\eta^*$, for some $\eta, \eta^* \in \mathcal{H}$. Then we have, for any $\xi \in D(S)$,

$$\langle S\xi|\eta\rangle_G = \langle S\xi|G\eta\rangle = \langle S\xi|\zeta\rangle = \langle \xi|S^*\zeta\rangle = \langle \xi|G\eta^*\rangle = \langle \xi|\eta^*\rangle_G.$$

Hence $\eta \in D(S^\#)$ and $S^\#\eta = \eta^* = G^{-1}S^*\zeta = G^{-1}S^*G\eta$, that is, $GS^\#\eta = S^*G\eta$. ∎

The second part of the proof of (ii) does not hold if G^{-1} is unbounded. In this case, we have only the following partial result.

Proposition 7.5.5 *Let G be a bounded metric operator, with G^{-1} (possibly) unbounded. Then, for every $\zeta \in D(S^*) \cap D(G^{-1/2})$ such that $S^*\zeta \in D(G^{-1/2})$, there exists $\eta \in D(S^\#)$ such that $\widetilde{G}\eta = \zeta$ and $GS^\#\eta = S^*\widetilde{G}\eta$.*

Proof. The proof is similar to the previous one, noting that $D(G^{-1}) = \mathcal{H}(G^{-1})$ and $\zeta, S^*\zeta \in \mathcal{H}(G^{-1})$. Hence there exist $\eta, \eta^* \in \mathcal{H}$ such that

$$\widetilde{G^{-1/2}}\zeta = \widetilde{G^{1/2}}\eta \quad \text{and} \quad \widetilde{G^{-1/2}}\zeta^* = \widetilde{G^{1/2}}\eta^*.$$

The rest is as before. ∎

Lemma 7.5.6 *Let G be bounded in $\mathcal{H}$ and let S be closed and densely defined. Define $D(K) = G^{1/2}D(S)$ and $K\xi = G^{1/2}SG^{-1/2}\xi$, $\xi \in D(K)$. Then K is densely defined, with adjoint $K^* = G^{-1/2}S^*G^{1/2}$.*

Proof. It is easy to see that $D(K)$ is dense in $\mathcal{H}$ and that $G^{-1/2}S^*G^{1/2} \subset K^*$. We prove the converse inclusion. Let $\eta \in D(K^*)$. Then there exists $\eta^* \in \mathcal{H}$ such that

$$\langle G^{1/2}SG^{-1/2}\xi|\eta\rangle = \langle \xi|\eta^*\rangle, \quad \forall \xi \in G^{1/2}D(S).$$

This implies that

$$\langle S\zeta|G^{1/2}\eta\rangle = \langle G^{1/2}\zeta|\eta^*\rangle = \langle \zeta|G^{1/2}\eta^*\rangle, \quad \forall \zeta \in D(S).$$

Hence, $G^{1/2}\eta \in D(S^*)$ and

$$\langle G^{-1/2}\xi|S^*G^{1/2}\eta\rangle = \langle \xi|\eta^*\rangle, \quad \forall \xi \in G^{1/2}D(S).$$

This in turn implies that $S^*G^{1/2}\eta \in D(G^{-1/2})$ and $K^*\eta = G^{-1/2}S^*G^{1/2}\eta$. ∎

In Proposition 7.5.4 (ii), we have obtained the expression of $S^\#$ in the case where G and G^{-1} are both bounded in $\mathcal{H}$, namely, $S^\# = G^{-1}S^*G$. If G^{-1} is unbounded, we can only determine the restriction of $S^\#$ to $\mathcal{H}$.

Proposition 7.5.7 *Given the closed operator S, put*

$$D(S^\star) := \{\eta \in \mathcal{H} : G\eta \in D(S^*),\ S^*G\eta \in D(G^{-1})\} \tag{7.5.2}$$

$$S^\star\eta := G^{-1}S^*G\eta, \quad \forall \eta \in D(S^\star).$$

Then $S^\star$ is the restriction to $\mathcal{H}$ of the adjoint $S^\#$ of S in $\mathcal{H}(G)$.

Proof. Let $\xi \in D(S)$ and $\eta \in D(S^\star)$. Then,

$$\begin{aligned}\langle S\xi|\eta\rangle_G &= \langle GS\xi|\eta\rangle = \langle S\xi|G\eta\rangle \\ &= \langle \xi|S^*G\eta\rangle = \langle G^{-1}G\xi|S^*G\eta\rangle \\ &= \langle G\xi|G^{-1}S^*G\eta\rangle = \langle \xi|G^{-1}S^*G\eta\rangle_G.\end{aligned}$$

■

Corollary 7.5.8 *If the domain $D(S^\star)$ is dense, then $S^\#$ is densely defined and S is closable in $\mathcal{H}(G)$.*

However, we still don't know whether S is closed in $\mathcal{H}(G)$, that is, whether one has $(S^\#)^\# = S$.

Let now A, B be closed operators in $\mathcal{H}$. Assume that they are metrically quasi-similar and let G be the bounded metric intertwining operator for A, B. Then, by (**ws**), we have

$$\langle G\xi|B^*\eta\rangle = \langle GA\xi|\eta\rangle, \quad \forall\, \xi \in D(A),\ \eta \in D(B^*).$$

This equality can be rewritten as

$$\langle \xi|B^*\eta\rangle_G = \langle A\xi|\eta\rangle_G, \quad \forall\, \xi \in D(A),\ \eta \in D(B^*).$$

This means that B^* is a restriction of $A^\#$, the adjoint of A in $\mathcal{H}(G)$.

As $D(B^*)$ is dense in $\mathcal{H}$, then, by Proposition 7.5.2 (i), $A^\#$ is densely defined in $\mathcal{H}(G)$ and $(A^\#)^\# \supseteq A$. Then we can consider the operator $A^\star$ defined in Proposition 7.5.7. This operator is an extension of B^* in $\mathcal{H}$.

Clearly $A^\star$ satisfies the equality

$$\langle \xi|A^\star\eta\rangle_G = \langle A\xi|\eta\rangle_G, \quad \forall\, \xi \in D(A),\ \eta \in D(A^\star)$$

and by Proposition 7.5.7, $A^\star\eta = G^{-1}A^*G\eta, \quad \forall\, \eta \in D(A^\star)$. As $A^\star \supset B^*$, then $(A^\star)^* \subset B$. Put $B_0 := (A^\star)^*$. The previous discussion means that $A \dashv B_0$, and that B_0 is minimal among the closed operators B such that $A \dashv B$, for fixed A and G. From these facts, it follows easily that $GD(A)$ is a core for B_0. Indeed, it is easily checked that $GD(A)$ is dense in $\mathcal{H}$. Let B_1 denote closure of the restriction of B_0 to $GD(A)$. Then $B_1 \subseteq B_0$ and it is easily seen that $A \dashv B_1$. Hence $B_1 = B_0$.

Thus, we have proved [11, Lemma 3.23]:

Lemma 7.5.9 *Let A, B be closed and $A \dashv B$ with a bounded metric intertwining operator G. Then $A^\star$ is densely defined, $B_0 := (A^\star)^*$ is minimal among the closed operators B satisfying, for fixed A and G, the conditions*

$$G : D(A) \to D(B),$$

$$BG\xi = GA\xi,\ \forall\, \xi \in D(A),$$

that is, B_0 is minimal among the closed operators B satisfying $A \dashv B$. Moreover, $GD(A)$ is a core for B_0.

7.5.2 Bounded Quasi-Hermitian Operators

Let A be a bounded operator in $\mathcal{H}$. Assume that A is quasi-Hermitian and that the metric operator G in (7.5.1) is bounded with bounded inverse. Then

$$\langle GA\xi|\eta\rangle = \langle A\xi|G\eta\rangle = \langle G\xi|A\eta\rangle = \langle \xi|GA\eta\rangle, \quad \forall\, \xi, \eta \in \mathcal{H}. \tag{7.5.3}$$

Thus, GA is self-adjoint in $\mathcal{H}$.

Proposition 7.5.10 *Let A be bounded. The following statements are equivalent.*

(i) A is quasi-Hermitian.
(ii) There exists a bounded metric operator G, with bounded inverse, such that GA $(= A^*G)$ is self-adjoint.
(iii) A is metrically similar to a self-adjoint operator K.

Proof. (i)⇒(ii) is easy.
(ii)⇒(iii): We put $K = G^{1/2}AG^{-1/2}$. As A^*G is self-adjoint, we get

$$\begin{aligned} K^* &= G^{-1/2}A^*G^{1/2} = G^{-1/2}(A^*G)G^{-1/2} = G^{-1/2}(GA)G^{-1/2} \\ &= G^{1/2}AG^{-1/2}. \end{aligned}$$

Hence, K is self-adjoint and $A = G^{-1/2}KG^{1/2}$, that is, $A \sim K$.
(iii)⇒(i): Assume $A = G^{-1/2}KG^{1/2}$ with $K = K^*$. Then, for every $\xi, \eta \in \mathcal{H}$

$$\begin{aligned} \langle GA\xi|\eta\rangle &= \langle A\xi|G\eta\rangle = \langle G^{-1/2}KG^{1/2}\xi|G\eta\rangle \\ &= \langle G^{1/2}\xi|KG^{1/2}\eta\rangle = \langle G^{-1/2}G\xi|KG^{-1/2}G\eta\rangle \\ &= \langle G\xi|G^{-1/2}KG^{1/2}\eta\rangle = \langle G\xi|A\eta\rangle. \end{aligned}$$

■

As a consequence of this proposition, bounded quasi-Hermitian operators coincide with bounded spectral operators of scalar type and real spectrum, mentioned in Section 7.2 (15).

7.5.3 Unbounded Quasi-Hermitian Operators

Let again G be bounded, but now we take A unbounded and quasi-Hermitian in the sense of Definition 7.5.1. The first result is immediate.

Proposition 7.5.11 *If G is bounded, then A is quasi-Hermitian if, and only if, GA is symmetric in $\mathcal{H}$.*

Proof. If A is quasi-Hermitian, (7.5.1) implies immediately

$$\langle GA\xi|\eta\rangle = \langle A\xi|G\eta\rangle = \langle G\xi|A\eta\rangle = \langle \xi|GA\eta\rangle, \quad \forall\, \xi,\eta \in D(A). \tag{7.5.4}$$

Hence GA is symmetric.

On the other hand, if GA is symmetric,

$$\langle A\xi|G\eta\rangle = \langle GA\xi|\eta\rangle = \langle \xi|GA\eta\rangle = \langle G\xi|A\eta\rangle, \quad \forall\, \xi,\eta \in D(A). \tag{7.5.5}$$

Thus, A is quasi-Hermitian. ■

Next we investigate the self-adjointness of A as an operator in $\mathcal{H}(G)$. This is equally easy.

Proposition 7.5.12 *Let G be bounded. If A is self-adjoint in $\mathcal{H}(G)$, then GA is symmetric in $\mathcal{H}$ and A is quasi-Hermitian. If G^{-1} is also bounded, then A is self-adjoint in $\mathcal{H}(G)$ if, and only if, GA is self-adjoint in $\mathcal{H}$.*

Proof. Let $A = A^{\#}$. Then, by Lemma 7.5.3, $(GA)^* = A^*G$, $\forall\, \xi \in D(A)$, $GD(A) \subseteq D(A^*)$ and $A^*G\xi = GA\xi$. Hence GA is symmetric and, thus, A is quasi-Hermitian by Proposition 7.5.11.

If G^{-1} is bounded, one has

$$A = A^{\#} = G^{-1}A^*G \iff GA = A^*G = (GA)^* \iff GA \text{ self-adjoint.} \quad ■$$

Now we turn the problem around. Namely, given the closed densely defined operator A, possibly unbounded, we seek whether there is a metric operator G that makes A quasi-Hermitian and self-adjoint in $\mathcal{H}(G)$. The first result is rather strong.

Proposition 7.5.13 *Let A be closed and densely defined. Then the following statements are equivalent:*

(i) There exists a bounded metric operator G, with bounded inverse, such that A is self-adjoint in $\mathcal{H}(G)$.
(ii) There exists a bounded metric operator G, with bounded inverse, such that $GA = A^*G$, that is, A is similar to its adjoint A^*, with intertwining operator G.
(iii) There exists a bounded metric operator G, with bounded inverse, such that $G^{1/2}AG^{-1/2}$ is self-adjoint.
(iv) A is a spectral operator of scalar type with real spectrum.

Proof. The implication (i) ⇒(ii) is clear. Next we prove that (ii) ⇒ (iii) and (iii) ⇒(i), which implies the equivalence of the first three statements.

(ii) ⇒ (iii): Let $K := G^{1/2}AG^{-1/2}$. This operator is self-adjoint. Indeed, by Lemma 7.5.6, we have

$$K^* = G^{-1/2}A^*G^{1/2} = G^{-1/2}A^*GG^{-1/2} = G^{-1/2}GAG^{-1/2} = G^{1/2}AG^{-1/2}$$
$$= K.$$

(iii) ⇒ (i): Let $G^{1/2}AG^{-1/2}$ be self-adjoint. Then by Lemma 7.5.6, we get

$$GA = G^{1/2}(G^{1/2}AG^{-1/2})G^{1/2} = G^{1/2}(G^{-1/2}A^*G^{1/2})G^{1/2} = A^*G = (GA)^*.$$

The statement follows from Proposition 7.5.12.

Now we turn to the fourth statement.

(iii) ⇒ (iv): Put again $K = G^{1/2}AG^{-1/2}$, which is self-adjoint by assumption. Hence $K = \int_{\mathbb{R}} \lambda \, \mathrm{d}E(\lambda)$, where $\{E(\lambda)\}$ is a self-adjoint spectral family. From this, it follows that

$$A = \int_{\mathbb{R}} \lambda \, \mathrm{d}X(\lambda), \quad \text{where} \quad X(\lambda) = G^{-1/2}E(\lambda)G^{1/2}.$$

That is, A is a spectral operator of scalar type with real spectrum, as $\sigma(A) = \sigma(K) \subseteq \mathbb{R}$.

(iv) ⇒ (iii): If A is a spectral operator of scalar type with $\sigma(A) \subseteq \mathbb{R}$, then $A = \int_{\mathbb{R}} \lambda \, \mathrm{d}X(\lambda)$, where $\{X(\lambda)\}$ is a countably additive resolution of the identity (not necessarily self-adjoint) (41). By a result of Mackey [42, Theorem 55], there exists a bounded operator T with bounded inverse and a self-adjoint resolution of the identity $\{E(\lambda)\}$ such that $X(\lambda) = T^{-1}E(\lambda)T$. Put $G = |T|^2$. By the polar decomposition, $T = UG^{1/2}$ with U unitary. Hence, $X(\lambda) = G^{-1/2}U^{-1}E(\lambda)UG^{1/2}$. Put $F(\lambda) = U^{-1}E(\lambda)U$. Then $\{F(\lambda)\}$ is a self-adjoint resolution of the identity. Thus, $K := \int_{\mathbb{R}} \lambda \, \mathrm{d}F(\lambda)$ is self-adjoint. Clearly, $K = G^{1/2}AG^{-1/2}$, as announced.[5] ■

Condition (i) of Proposition 7.5.13 suggests the following definition.

Definition 7.5.14 Let A be closed and densely defined. We say that A is *quasi-self-adjoint* if there exists a bounded metric operator G, such that A is self-adjoint in $\mathcal{H}(G)$.

In particular, if any of the conditions of Proposition 7.5.13 is satisfied, then A is quasi-self-adjoint. Notice, however, that Definition 7.5.14 is slightly more general than the set-up of the proposition, as we do not require G^{-1} to be bounded.

Proposition 7.5.13 characterizes quasi-self-adjointness in terms of similarity of A and A^*, if the intertwining metric operator is bounded with bounded inverse. Instead of requiring that A be similar to A^*, we may ask that they be only quasi-similar. The price to pay is that now G^{-1} is no longer bounded and, therefore, the equivalences stated in Proposition 7.5.13 are no longer true. Instead we have the following weaker result (12).

[5] Using Dunford's result [25, Section XV.6], we may conclude directly that $A = T^{-1}ST$, with S self-adjoint. The rest follows by putting again $T = UG^{1/2}$.

Proposition 7.5.15 *Let A be closed and densely defined. Consider the statements*

(i) There exists a bounded metric operator G such that $GD(A) = D(A^*)$, $A^*G\xi = GA\xi$, for every $\xi \in D(A)$, in particular, A is quasi-similar to its adjoint A^*, with intertwining operator G.
(ii) There exists a bounded metric operator G, such that $G^{1/2}AG^{-1/2}$ is self-adjoint.
(iii) There exists a bounded metric operator G such that A is self-adjoint in $\mathcal{H}(G)$; that is, A is quasi-self-adjoint.
(iv) There exists a bounded metric operator G such that $GD(A) = D(G^{-1}A^*)$, $A^*G\xi = GA\xi$, for every $\xi \in D(A)$, in particular, A is quasi-similar to its adjoint A^*, with intertwining operator G.

Then, the following implications hold :

$$(i) \Rightarrow (ii) \Rightarrow (iii) \Rightarrow (iv).$$

If the range $R(A^*)$ of A^* is contained in $D(G^{-1})$, then the four conditions (i)-(iv) are equivalent.

Proof. (i) $\Rightarrow$ (ii): We put $K := G^{1/2}AG^{-1/2}$ and show it is self-adjoint. As in Lemma 7.5.6, take $\xi \in D(K)$, $\eta \in D(K^*)$. Then, taking into account that $\xi \in D(G^{-1/2})$ and that, as $G^{1/2}\eta \in D(A^*)$ and $D(A^*) = GD(A)$, $G^{1/2}\eta = G\zeta$ for some $\zeta \in D(A)$, we have

$$\begin{aligned}\langle K\xi|\eta\rangle &= \langle \xi|G^{-1/2}A^*G^{1/2}\eta\rangle = \langle G^{-1/2}\xi|A^*G^{1/2}\eta\rangle \\ &= \langle G^{-1/2}\xi|A^*G\zeta\rangle = \langle G^{-1/2}\xi|GA\zeta\rangle \\ &= \langle G^{-1/2}\xi|GAG^{-1/2}\eta\rangle = \langle \xi|G^{1/2}AG^{-1/2}\eta\rangle = \langle \xi|K\eta\rangle.\end{aligned}$$

Hence $K = K^*$ is self-adjoint.
(ii)$\Rightarrow$ (iii): First, we prove that A is symmetric in $\mathcal{H}(G)$, that is, $A \subseteq A^\#$. Indeed, if $\xi, \eta \in D(A)$, we have, by putting $\zeta = G^{1/2}\xi$ and $\varsigma = G^{1/2}\eta$,

$$\langle A\xi|\eta\rangle_G = \langle GA\xi|\eta\rangle = \langle G^{1/2}AG^{-1/2}\zeta|\varsigma\rangle = \langle \zeta|G^{1/2}AG^{-1/2}\varsigma\rangle = \langle \xi|A\eta\rangle_G.$$

Let now $\eta \in D(A^\#) \subseteq \mathcal{H}(G)$. Then, there exists an element $\eta^* \in \mathcal{H}(G)$, such that

$$\langle A\xi|\eta\rangle_G = \langle \xi|\eta^*\rangle, \quad \forall \xi \in D(A);$$

or, equivalently,

$$\langle A\xi|\widetilde{G}\eta\rangle_G = \langle \xi|\widetilde{G}\eta^*\rangle, \quad \forall \xi \in D(A).$$

As noticed before, $\widetilde{G}\mathcal{H}(G) = D(G^{-1/2} = G^{1/2}(\mathcal{H})$ and $\widetilde{G}^{1/2}\mathcal{H}(G) = \mathcal{H}$. Hence we get the equality $\widetilde{G} = G^{1/2}\widetilde{G}^{1/2}$. Then,

$$\langle G^{1/2}A\xi|\widetilde{G}^{1/2}\eta\rangle = \langle G^{1/2}\xi|\widetilde{G}^{1/2}\eta^*\rangle, \quad \forall \xi \in D(A).$$

Let $\zeta := G^{1/2}\xi$, we get

$$\langle G^{1/2}AG^{-1/2}\zeta|\widetilde{G}^{1/2}\eta\rangle = \langle \zeta|\widetilde{G}^{1/2}\eta^*\rangle, \quad \forall \zeta \in G^{1/2}D(A).$$

This implies that $\widetilde{G}^{1/2}\eta \in D((G^{1/2}AG^{-1/2})^*) = D(G^{1/2}AG^{-1/2}) = G^{1/2}D(A)$. Thus, in particular, $\widetilde{G}^{1/2}\eta \in D(G^{-1/2}) = \widetilde{G}\mathcal{H}(G)$. Hence $\widetilde{G}^{1/2}\eta = G^{1/2}\widetilde{G}^{1/2}\varphi$ for some $\varphi \in \mathcal{H}(G)$. The injectivity of $\widetilde{G}^{1/2}$ then implies that $\eta = \widetilde{G}^{1/2}\varphi \in \mathcal{H}$. Therefore, $G^{1/2}\eta = \widetilde{G}^{1/2}\eta \in D(AG^{-1/2})$. This, in turn, implies that $\eta \in D(A)$. In conclusion, A is self-adjoint in $\mathcal{H}(G)$.

(iii)⇒ (iv): Assume that A is self-adjoint in $\mathcal{H}(G)$, that is, $A = A^{\#}$. Then, by Proposition 7.5.7, it follows that

$$D(A) = \{\eta \in \mathcal{H} : G\eta \in D(A^*), A^*G\eta \in D(G^{-1})\}. \tag{7.5.6}$$

Now, $\zeta \in GD(A)$ if and only if $G^{-1}\zeta \in D(A)$. By (7.5.6), this is equivalent to say that $\zeta \in D(A^*)$ and $A^*\zeta \in D(G^{-1})$. The latter two conditions define the domain of $D(G^{-1}A^*)$. Hence, $GD(A) = D(G^{-1}A^*)$. Furthermore, if $\xi \in D(A)$, Proposition 7.5.7 implies also that $A\xi = G^{-1}A^*G\xi$. Then, as $A^*G\xi \in D(G^{-1})$, by applying G to both sides we conclude that $GA\xi = A^*G\xi$.

Finally, if $R(A^*) \subset D(G^{-1})$, then $D(G^{-1}A^*) = D(A^*)$ and (iv)⇒(i) is obvious. ■

Remark 7.5.16 Condition (ii) of Proposition 7.5.13 is equivalent to the self-adjointness of the operator GA (G is there bounded, with bounded inverse). So one could expect that the self-adjointness of GA plays a role also when studying, as in Proposition 7.5.15, the quasi-self-adjointness of A in a more general context. However, it seems not to be so. One can easily prove that the condition (i) in Proposition 7.5.15 implies the self-adjointness of GA. But the self-adjointness of GA seems not to be sufficient for the quasi-self-adjointness of A.

Let us now assume that one of the equivalent conditions (ii), (iii), or (iv) of Proposition 7.5.15 holds for a certain bounded metric operator G and define $H := G^{1/2}AG^{-1/2}$ on $D(H) = G^{1/2}D(A)$. Then, H is self-adjoint and $HG^{1/2}\xi = G^{1/2}A\xi$ for every $\xi \in D(A)$. Clearly, $G^{1/2}$ intertwines A and H and $A \dashv H$. We notice, on the other hand, that $G^{-1/2}$ intertwines H and A in the sense of Definition 7.3.1.

Let now $\{E(\lambda)\}$ denote the spectral family of H. Let $\xi \in \mathcal{H}$ and consider the conjugate linear functional $\Omega_{\lambda,\xi}$ defined on $D(G^{-1/2})$ by

$$\Omega_{\lambda,\xi}(\eta) = \langle E(\lambda)G^{1/2}\xi|G^{-1/2}\eta\rangle, \quad \eta \in D(G^{-1/2}).$$

We consider here again $D(G^{-1/2})$ as a Hilbert space, denoted by $\mathcal{H}(G^{-1})$, with norm $\|\cdot\|_{G^{-1}} = \|G^{-1/2}\cdot\|$ (see Section 7.2). Then,

$$|\Omega_{\lambda,\xi}(\eta)| = |\langle E(\lambda)G^{1/2}\xi|G^{-1/2}\eta\rangle| \le \|G^{1/2}\xi\|\|G^{-1/2}\eta\| = \|G^{1/2}\xi\|\|\eta\|_{G^{-1}}.$$

Hence, $\Omega_{\lambda,\xi}$ can be represented as follows:

$$\Omega_{\lambda,\xi}(\eta) = \langle E(\lambda)G^{1/2}\xi|G^{-1/2}\eta\rangle = \langle\Phi|\eta\rangle, \quad \eta \in D(G^{-1/2}),$$

for a unique $\Phi \in \mathcal{H}(G^{-1})^\times$, the conjugate dual of $\mathcal{H}(G^{-1})$. It is a standard fact that $\mathcal{H}(G^{-1})^\times$ can be identified with $\mathcal{H}(G)$. We define $X(\lambda)\xi = \Phi$. Then $X(\lambda)$ is linear and maps $\mathcal{H}$ into $\mathcal{H}(G)$ continuously. One can easily prove that

$$X(\lambda)\xi = \widetilde{G^{-1/2}E(\lambda)G^{1/2}}\xi, \quad \xi \in \mathcal{H},$$

and it obviously satisfies

$$\langle X(\lambda)\xi|\eta\rangle = \langle E(\lambda)G^{1/2}\xi|G^{-1/2}\eta\rangle, \quad \forall\, \xi \in \mathcal{H}, \eta \in D(G^{-1/2}).$$

Proposition 7.5.17 *The family $\{X(\lambda)\}$ enjoys the following properties.*

(i) $\lim_{\lambda\to-\infty}\langle X(\lambda)\xi|\eta\rangle = 0;\ \lim_{\lambda\to\infty}\langle X(\lambda)\xi|\eta\rangle = \langle\xi|\eta\rangle, \quad \forall\, \xi \in \mathcal{H}, \eta \in D(G^{-1/2}).$

(ii) $\lim_{\lambda\downarrow\mu}\langle X(\lambda)\xi|\eta\rangle = \langle X(\mu)\xi|\eta\rangle, \quad \forall\, \xi \in \mathcal{H}, \eta \in D(G^{-1/2}).$

(iii) The function $f_{\xi,\eta} : \lambda \mapsto \langle X(\lambda)\xi|\eta\rangle$ is of bounded variation, for every $\xi \in \mathcal{H}, \eta \in D(G^{-1/2})$, and its total variation $V(f_{\xi,\eta})$ does not exceed $\|G^{1/2}\xi\|\|G^{-1/2}\eta\|$.

(iv) The following equality holds:

$$\langle A\xi|\eta\rangle = \int_{\mathbb{R}} \lambda\, \mathrm{d}\langle X(\lambda)\xi|\eta\rangle, \quad \forall\, \xi \in D(A), \eta \in D(G^{-1/2}).$$

Proof. The proof of these statements reduces to simple applications of the spectral theorem for a self-adjoint operator, similar to those given in Ref. (43) in an analogous situation. We simply check (iv). One has, in fact, for $\xi \in D(A)$ and $\eta \in D(G^{-1/2})$,

$$\begin{aligned}\int_{\mathbb{R}} \lambda\, \mathrm{d}\langle X(\lambda)\xi|\eta\rangle &= \int_{\mathbb{R}} \lambda\, \mathrm{d}\langle E(\lambda)G^{1/2}\xi|G^{-1/2}\eta\rangle \\ &= \langle HG^{1/2}\xi|G^{-1/2}\eta\rangle \\ &= \langle (G^{1/2}AG^{-1/2})G^{1/2}\xi|G^{-1/2}\eta\rangle \\ &= \langle A\xi|\eta\rangle.\end{aligned}$$

■

Hence, A is a spectral operator of scalar type in a generalized sense. We notice that the representation in (iv) does not imply that $\sigma(A) = \sigma(H)$.

7.5.4 Quasi-Hermitian Operators with Unbounded Metric Operators

Assume now that G is also unbounded. Given a closed and densely defined operator A, we still say that A is *quasi-Hermitian* if it verifies Definition 7.5.1. We say that A is *strictly quasi-Hermitian* if, in addition, $AD(A) \subset D(G)$ or, equivalently, $D(GA) = D(A)$.

In that case, $\eta \in D(A), A\eta \in D(G)$ implies $G\eta \in D(A^*)$, so that we may write

$$\langle \xi | A^* G\eta \rangle = \langle A\xi | G\eta \rangle = \langle G\xi | A\eta \rangle = \langle \xi | GA\eta \rangle, \quad \forall\, \xi, \eta \in D(A).$$

Therefore,

$$A^* G\eta = GA\eta, \quad \forall\, \eta \in D(A). \tag{7.5.7}$$

The relation (7.5.7) means A is quasi-Hermitian in the sense of Dieudonné, that is, it satisfies the relation $A^*G = GA$ on the dense domain $D(A)$.

Now, as a consequence of (7.5.1), the condition $D(GA) = D(A)$ is in fact equivalent to $G : D(A) \to D(A^*)$. Thus, comparing the discussion earlier with the definition of (generalized) quasi-similarity given in Section 7.3.3, we see that A is strictly quasi-Hermitian if, and only if, A is quasi-similar to A^*, $A \dashv A^*$. We will come back to this point in Section 7.8.

Although these results have some interest, they do not solve the main problem, namely, given the quasi-Hermitian operator A, how does one construct an appropriate metric operator G? We suspect that there is no general answer to the question: it has to be analyzed for each specific operator A.

A partial answer may be given if one uses the formalism of PIP-spaces, as we will see in Section 7.7.2.

Another open question is the following. Given two closed operators A, B, under which conditions are they (quasi-)similar to each other? According to the discussion so far, these conditions will be of a spectral nature, such as equality of the spectra or of some parts of the spectra.

7.5.5 Example: Operators Defined from Riesz Bases

For the case where G and G^{-1} are both bounded, an interesting class of examples has been given recently in Ref. (44), namely, quasi-Hermitian operators defined in terms of a Riesz basis (45). We recall that $\mathcal{F}_\phi = \{\phi_n,\ n \geq 0\}$ is a Riesz basis if there exist an orthonormal basis $\{e_n,\ n \geq 0\}$ and a bounded operator T, invertible and with bounded inverse T^{-1} such that $\phi_n = Te_n$ for all $n \geq 0$. Actually, a Riesz basis is the same thing as an exact frame, that is, a frame $\{\phi_n\}$ that ceases to be a frame if one removes any vector from it (45, 46). Moreover, a family $\{\phi_n,\ n \geq 0\}$ is a Riesz basis if and only if it is a bounded unconditional basis, that is, $0 < \inf \|\phi_n\| \leq \sup \|\phi_n\| < \infty$ and the series $f = \sum_n c_n \phi_n$ converges unconditionally for any $f \in \mathcal{H}$(46).

Let A be a closed operator with a purely discrete simple spectrum, that is, the spectrum $\sigma(A)$ consists only of isolated eigenvalues with multiplicity one, but not necessarily real. Assume that the corresponding eigenvectors form a Riesz basis $\mathcal{F}_\phi = \{\phi_n, n \geq 0\}$ for $\mathcal{H}$. This is a rather strong assumption, interesting from a mathematical point of view, but not so frequent in concrete physical models (47).

Writing again $\phi_n = Te_n$, define $\psi_n := (T^{-1})^* e_n$ and $\mathcal{F}_\psi = \{\psi_n, n \geq 0\}$. It is clear that $\mathcal{F}_\psi$ is a Riesz basis, too. Moreover, it is biorthogonal to $\mathcal{F}_\phi$: $\langle \phi_n | \psi_m \rangle = \delta_{n,m}$, and

$$\xi = \sum_{n=0}^{\infty} \langle \xi | \phi_n \rangle \psi_n = \sum_{n=0}^{\infty} \langle \xi | \psi_n \rangle \phi_n, \quad \forall \xi \in \mathcal{H}.$$

Furthermore, the operators S_ϕ and S_ψ defined by

$$S_\phi \xi = \sum_{n=0}^{\infty} \langle \xi | \phi_n \rangle \phi_n, \qquad S_\psi \xi = \sum_{n=0}^{\infty} \langle \xi | \psi_n \rangle \psi_n, \tag{7.5.8}$$

are bounded, everywhere defined in $\mathcal{H}$, positive and self-adjoint. Hence, they are both metric operators, inverse of each other: $S_\phi = (S_\psi)^{-1}$ and $S_\phi = TT^*$. In addition, they are both intertwining operators between A and A^*:

$$S_\phi A^* = AS_\phi, \text{ and } S_\psi A = A^* S_\psi,$$

so that indeed A is quasi-Hermitian. Hence, by Proposition 7.5.11, $S_\psi A$ is a symmetric operator. Moreover, by Proposition 7.5.13, $S_\psi A$ is self-adjoint if and only if A is similar to a self-adjoint operator, $S_\psi^{1/2} A S_\psi^{-1/2}$.

According to Ref. (44), where all the proofs may be found, the Riesz basis $\mathcal{F}_\phi$ generates a whole class of quasi-Hermitian operators, that we describe now. Given a sequence $\boldsymbol{\alpha} = (\alpha_n)$ of complex numbers, we define two operators

$$A^{\boldsymbol{\alpha}}_{\phi,\psi} = \sum_{n=0}^{\infty} \alpha_n \phi_n \otimes \overline{\psi}_n, \text{ and } A^{\boldsymbol{\alpha}}_{\psi,\phi} = \sum_{n=0}^{\infty} \alpha_n \psi_n \otimes \overline{\phi}_n$$

as follows:

$$\begin{cases} D(A^{\boldsymbol{\alpha}}_{\phi,\psi}) = \left\{ \xi \in \mathcal{H} : \sum_{n=0}^{\infty} \alpha_n \langle \xi | \psi_n \rangle \phi_n \text{ exists in } \mathcal{H} \right\} \\ A^{\boldsymbol{\alpha}}_{\phi,\psi} \xi = \sum_{n=0}^{\infty} \alpha_n \langle \xi | \psi_n \rangle \phi_n, \ \xi \in D(A^{\boldsymbol{\alpha}}_{\phi,\psi}), \end{cases}$$

$$\begin{cases} D(A^{\boldsymbol{\alpha}}_{\psi,\phi}) = \left\{ \xi \in \mathcal{H} : \sum_{n=0}^{\infty} \alpha_n \langle \xi | \phi_n \rangle \psi_n \text{ exists in } \mathcal{H} \right\} \\ A^{\boldsymbol{\alpha}}_{\psi,\phi} \xi = \sum_{n=0}^{\infty} \alpha_n \langle \xi | \phi_n \rangle \psi_n, \ \xi \in D(A^{\boldsymbol{\alpha}}_{\phi,\psi}). \end{cases}$$

Then we have immediately

$$\mathcal{D}_\phi := \text{span}\{\phi_n\} \subset D(A^{\boldsymbol{\alpha}}_{\phi,\psi}) \text{ and } \mathcal{D}_\psi := \text{span}\{\psi_n\} \subset D(A^{\boldsymbol{\alpha}}_{\psi,\phi}),$$

$$A^{\boldsymbol{\alpha}}_{\phi,\psi} \phi_k = \alpha_k \phi_k, \ k = 0, 1, \ldots, \text{ and } A^{\boldsymbol{\alpha}}_{\psi,\phi} \psi_k = \alpha_k \psi_k, \ k = 0, 1, \ldots .$$

Hence, $A^{\boldsymbol{\alpha}}_{\phi,\psi}$ and $A^{\boldsymbol{\alpha}}_{\psi,\phi}$ are densely defined.

Proposition 7.5.18 *The following statements hold.*

(i) $D(A^{\alpha}_{\phi,\psi}) = \{\xi \in \mathcal{H} : \sum_{n=0}^{\infty} |\alpha_n|^2 |\langle \xi|\psi_n\rangle|^2 < \infty\}$,
$D(A^{\alpha}_{\psi,\phi}) = \{\xi \in \mathcal{H} : \sum_{n=0}^{\infty} |\alpha_n|^2 |\langle \xi|\phi_n\rangle|^2 < \infty\}$.
(ii) $A^{\alpha}_{\phi,\psi}$ *and* $A^{\alpha}_{\psi,\phi}$ *are closed.*
(iii) $(A^{\alpha}_{\phi,\psi})^* = A^{\overline{\alpha}}_{\psi,\phi}$, *where* $\overline{\boldsymbol{\alpha}} = (\overline{\alpha}_n)$.
(iv) $A^{\alpha}_{\phi,\psi}$ *is bounded if and only if* $A^{\alpha}_{\psi,\phi}$ *is bounded if and only if* $\boldsymbol{\alpha}$ *is a bounded sequence. In particular* $A^{\mathbf{1}}_{\phi,\psi} = A^{\mathbf{1}}_{\psi,\phi} = I$, *where* $\mathbf{1}$ *is the sequence constantly equal to* 1.

It turns out that S_ϕ, S_ψ are still intertwining operators for the new operators. Moreover, one has

Proposition 7.5.19 *The following equalities hold:*

$$S_\psi A^{\alpha}_{\phi,\psi} = A^{\alpha}_{\psi,\phi} S_\psi = S^{\alpha}_{\psi} := \sum_{n=0}^{\infty} \alpha_n \phi_n \otimes \overline{\phi}_n, \tag{7.5.9}$$

$$S_\phi A^{\alpha}_{\psi,\phi} = A^{\alpha}_{\phi,\psi} S_\phi = S^{\alpha}_{\phi} := \sum_{n=0}^{\infty} \alpha_n \psi_n \otimes \overline{\psi}_n. \tag{7.5.10}$$

It follows again that both $A^{\alpha}_{\phi,\psi}$ and $A^{\alpha}_{\psi,\phi}$ are quasi-Hermitian operators. Thus we may proceed as before, in Section 7.5.1, and consider the triplet (7.4.8), that reads now as follows, as $(S_\psi)^{-1} = S_\phi$:

$$\mathcal{H}(S_\phi) \subset \mathcal{H} \subset \mathcal{H}(S_\psi). \tag{7.5.11}$$

The three spaces coincide as vector spaces, with equivalent, but different norms. Thus, on $\mathcal{H}(S_\psi)$, we consider the inner product $\langle\cdot|\cdot\rangle_{S_\psi}$ given by $\langle \xi|\eta\rangle_{S_\psi} = \langle S_\psi \xi|\eta\rangle$, $\xi, \eta \in \mathcal{H}$. Of course, $A^{\alpha}_{\phi,\psi}$ is symmetric with respect to this new inner product, that is,

$$\langle A^{\alpha}_{\phi,\psi}\xi|\eta\rangle_{S_\psi} = \langle \xi|A^{\alpha}_{\phi,\psi}\eta\rangle_{S_\psi}, \text{ for } \xi, \eta \in D(A^{\alpha}_{\phi,\psi}).$$

This is, in a certain sense, not surprising, as the set of eigenvectors of $A^{\alpha}_{\phi,\psi}$, $\mathcal{F}_\phi$, is an orthonormal basis in $\mathcal{H}$, when endowed with the inner product $\langle\cdot|\cdot\rangle_{S_\psi}$; indeed, $\langle \phi_n, \phi_m\rangle_{S_\psi} = \delta_{n,m}$.

However, we are looking for self-adjoint operators. We define the operators $\mathsf{a}^{\alpha}_{\phi,\psi}$ and $\mathsf{a}^{\alpha}_{\psi,\phi}$ as follows:

$$\mathsf{a}^{\alpha}_{\phi,\psi} = S_\psi^{1/2} A^{\alpha}_{\phi,\psi} S_\phi^{1/2}, \quad \mathsf{a}^{\alpha}_{\psi,\phi} = S_\phi^{1/2} A^{\alpha}_{\psi,\phi} S_\psi^{1/2}. \tag{7.5.12}$$

Then, by Proposition 7.5.11, $A^{\alpha}_{\phi,\psi}$ is self-adjoint in $\mathcal{H}(S_\psi)$ whenever the operator $\mathsf{a}^{\alpha}_{\phi,\psi}$ is self-adjoint in $\mathcal{H}$. A criterion to that effect is given by the following result.

Proposition 7.5.20 *The following statements hold:*

(i) $D(\mathsf{a}^{\alpha}_{\phi,\psi}) = \{S_{\psi}^{1/2}\xi : \xi \in D(A^{\alpha}_{\phi,\psi})\}$, *and* $D(\mathsf{a}^{\alpha}_{\psi,\phi}) = \{S_{\phi}^{1/2}\xi : \xi \in D(A^{\alpha}_{\psi,\phi})\}$ *and they are dense in* $\mathcal{H}$.
(ii) $(\mathsf{a}^{\alpha}_{\phi,\psi})^* = \mathsf{a}^{\overline{\alpha}}_{\psi,\phi}$.
(iii) If $\{\alpha_n\} \subset \mathbb{R}$, *then* $\mathsf{a}^{\alpha}_{\phi,\psi}$ *is self-adjoint.*

This is the kind of result one hopes for Hamiltonians in Pseudo-Hermitian QM.
Two remarks in conclusion.

1. Instead of the operators $S^{\alpha}_{\psi}, S^{\alpha}_{\phi}$ defined in (7.5.9) and (7.5.10), one can consider operators $S^{\beta}_{\psi}, S^{\beta}_{\phi}$, with another sequence $\beta = (\beta_n)$ of complex numbers and study their interplay with $A^{\alpha}_{\phi,\psi}$ and $A^{\alpha}_{\psi,\phi}$. The results are very similar to those given in Proposition 7.5.18.
2. As the whole machinery is symmetric in ϕ, ψ, one can interchange S_{ϕ} and S_{ψ}, thus considering, instead of (7.5.11), the triplet

$$\mathcal{H}(S_{\psi}) \subset \mathcal{H} \subset \mathcal{H}(S_{\phi}). \tag{7.5.13}$$

Then on $\mathcal{H}(S_{\phi})$ we consider the inner product $\langle\cdot|\cdot\rangle_{S_{\phi}}$ given by $\langle\xi|\eta\rangle_{S_{\phi}} = \langle S_{\phi}\xi|\eta\rangle, \xi, \eta \in \mathcal{H}$. Of course, $A^{\alpha}_{\psi,\phi}$ is symmetric with respect to this new inner product, and the whole development may be repeated.

7.6 THE LHS GENERATED BY METRIC OPERATORS

Let $\mathcal{M}(\mathcal{H})$ denote the family of all metric operators and $\mathcal{M}_b(\mathcal{H})$ that of all the bounded metric operators. As said in Section 7.2, there is a natural order in $\mathcal{M}(\mathcal{H}) : G_1 \preceq G_2$ if and only if $\mathcal{H}(G_1) \subset \mathcal{H}(G_2)$, where the embedding is continuous and has dense range. If G_1 and G_2 are both bounded, a sufficient condition for $G_1 \preceq G_2$ is that there exists $\gamma > 0$ such that $G_2 \leq \gamma G_1$. Then one has

$$G_2^{-1} \preceq G_1^{-1} \iff G_1 \preceq G_2 \text{ if } G_1, G_2 \in \mathcal{M}_b(\mathcal{H})$$

and

$$G^{-1} \preceq I \preceq G, \quad \forall G \in \mathcal{M}_b(\mathcal{H}).$$

Remark 7.6.1 The family $\mathcal{M}(\mathcal{H})$ is not necessarily directed upward with respect to $\preceq$. For instance, if $X, Y \in \mathcal{M}_b(\mathcal{H})$, then X, Y have the null operator 0 as a lower bound; 0 is not the greatest lower bound [48, Example 2.8.18], but we cannot say that a positive lower bound exists. If a positive lower bound Z exists, then $Z \in \mathcal{M}_b(\mathcal{H})$ and by definition $Z \preceq X, Y$.

As we will see now, the spaces $\{\mathcal{H}(X) : X \in \mathcal{M}(\mathcal{H})\}$ constitute a lattice of Hilbert spaces (LHS) $V_{\mathcal{J}}$ in the sense of Ref. [13, Definition 2.4.8]. For the convenience of the reader, we have summarized in the Appendix the necessary notions about LHSs and, more generally, partial inner product spaces (PIP-spaces), and operators on them.

Let first $\mathcal{O} \subset \mathcal{M}(\mathcal{H})$ be a family of metric operators and assume that

$$\mathcal{D} := \bigcap_{G \in \mathcal{O}} D(G^{1/2})$$

is a dense subspace of $\mathcal{H}$. Of course, the condition is nontrivial only if $\mathcal{O}$ contains unbounded elements, for instance, unbounded inverses of bounded operators. We may always suppose that $I \in \mathcal{O}$.

As shown in Refs (11, 49) and in Ref. [13, Section 5.5.2], the family $\mathcal{O}$ generates a canonical LHS. The lattice operations are defined by means of the operators

$$X \wedge Y := X \dotplus Y,$$
$$X \vee Y := (X^{-1} \dotplus Y^{-1})^{-1},$$

where $\dotplus$ stands for the form sum and $X, Y \in \mathcal{O}$. We recall that the form sum $T_1 \dotplus T_2$ of two positive operators is the positive self-adjoint operator associated to the quadratic form $\mathfrak{t} = \mathfrak{t}_1 + \mathfrak{t}_2$, where $\mathfrak{t}_1, \mathfrak{t}_2$ are the quadratic forms of T_1, T_2, respectively [50, §VI.2.5].

We notice that $X \wedge Y$ is a metric operator, but it need not belong to $\mathcal{O}$. First, it is self-adjoint and bounded from below by a positive quantity. In addition, $(X \wedge Y)\xi = 0$ implies $\xi = 0$, $\forall \xi \in Q(X \dotplus Y) = Q(X) \cap Q(Y)$, which is dense. Indeed, $\langle (X + Y)\xi | \xi \rangle = \langle X\xi | \xi \rangle + \langle Y\xi | \xi \rangle = 0$ implies $\langle X\xi | \xi \rangle = \langle Y\xi | \xi \rangle = 0$, as both X and Y are positive. This in turn implies $\xi = 0$. Thus $X \wedge Y$ is a metric operator, but it need not belong to $\mathcal{O}$. The same argument applies to the operator $X \vee Y$.

In particular, if we take for $\mathcal{O}$ the set $\mathcal{M}(\mathcal{H})$ of all metric operators, we see that it is stable under the lattice operations, that is, it is a lattice by itself (but the corresponding domain $\mathcal{D}$ may fail to be dense). This is not true in the general case envisaged in Ref. [13, Section 5.5.2].

For the corresponding Hilbert spaces, one has

$$\begin{aligned} \mathcal{H}(X \wedge Y) &:= \mathcal{H}(X) \cap \mathcal{H}(Y), \\ \mathcal{H}(X \vee Y) &:= \mathcal{H}(X) + \mathcal{H}(Y), \end{aligned} \tag{7.6.1}$$

equipped, respectively, with the projective and the inductive norm, namely,

$$\begin{aligned} \|\xi\|^2_{X \wedge Y} &= \|\xi\|^2_X + \|\xi\|^2_Y, \\ \|\xi\|^2_{X \vee Y} &= \inf_{\xi=\eta+\zeta} \left(\|\eta\|^2_X + \|\zeta\|^2_Y \right), \ \eta \in \mathcal{H}(X), \zeta \in \mathcal{H}(Y). \end{aligned} \tag{7.6.2}$$

Define the set $\mathcal{R} = \mathcal{R}(\mathcal{O}) := \{G^{\pm 1/2} : G \in \mathcal{O}\}$ and the corresponding domain $\mathcal{D}_{\mathcal{R}} := \bigcap_{X \in \mathcal{R}} D(X)$. Let now Σ denote the minimal set of self-adjoint operators containing $\mathcal{O}$, stable under inversion and form sums, with the property that $\mathcal{D}_{\mathcal{R}}$ is dense in every $H_Z, Z \in \Sigma$ (i.e., Σ is an admissible cone of self-adjoint operators, in the sense of Ref. [13, Definition 5.5.4]). Then, by Ref. [13, Theorem 5.5.6], $\mathcal{O}$ generates a lattice of Hilbert spaces $\mathcal{J} := \mathcal{J}_\Sigma = \{\mathcal{H}(X), X \in \Sigma\}$ and a PIP-space V_Σ with central Hilbert space $\mathcal{H} = \mathcal{H}(I)$ and total space $V = \sum_{G\in\Sigma} \mathcal{H}(G)$. The "smallest" space is $V^\# = \mathcal{D}_\mathcal{R}$. The compatibility and the partial inner product read, respectively, as

$$\xi \# \eta \iff \exists\, G \in \Sigma \text{ such that } \xi \in \mathcal{H}(G),\ \eta \in \mathcal{H}(G^{-1}),$$
$$\langle \xi | \eta \rangle_\mathcal{J} = \langle G^{1/2}\xi | G^{-1/2}\eta \rangle_\mathcal{H}.$$

From now on, we shall denote the partial inner product simply as $\langle \xi|\eta\rangle := \langle \xi|\eta\rangle_\mathcal{J}$, as it coincides with the inner product of $\mathcal{H}$ whenever $\xi, \eta \in \mathcal{H}$.

For instance, if $\mathcal{O} = \{I, G\}$, the set Σ consists of the nine operators of Fig. 7.2. On the other hand, every power of G is a metric operator. Thus, if we take $\mathcal{O} = \{G^\alpha, \alpha \in \mathbb{Z} \text{ or } \mathbb{R}\}$ and some G^{α_0} is bounded with $G^{-\alpha_0}$ unbounded, then the set $\mathcal{O}$ is totally ordered and we obtain the scales $V_\mathcal{G}$ and $V_{\widetilde{\mathcal{G}}}$, which are the PIP-spaces generated by the aforementioned construction.

We denote by $\mathrm{Op}(V_\Sigma)$ the space of operators in V_Σ, described at length in Section A.2. Given $(X, Y) \in \mathsf{j}(A)$, we denote by $A_{YX} : \mathcal{H}(X) \to \mathcal{H}(Y)$ the (X, Y)-representative of A, that is, the restriction of A to $\mathcal{H}(X)$. Then A is identified with the collection of its representatives:

$$A \simeq \{A_{YX} : (X, Y) \in \mathsf{j}(A)\}.$$

In particular, $E_{YX} : \mathcal{H}(X) \to \mathcal{H}(Y)$ is the representative of the identity operator (embedding) when $\mathcal{H}(X) \subset \mathcal{H}(Y)$.

In addition, we define $\mathsf{s}(A) = \{X \in \Sigma : (X, X) \in \mathsf{j}(A)\}$, so that $\mathsf{s}(A^\times) = \{X^{-1} : X \in \mathsf{s}(A)\}$. From the definition (7.6.1), it is clear that the set $\mathsf{s}(A)$ is invariant under the lattice operations $\cap$ and $+$. Coming back to the scale (7.4.11) or its continuous extension $V_{\widetilde{\mathcal{G}}} := \{\mathcal{H}_\alpha, \alpha \in \mathbb{R}\}$, associated to the fixed metric operator G, we may identify $\alpha \in \mathbb{R}$ with $\mathcal{H}_\alpha = \mathcal{H}(G^\alpha)$ and consider the subset $\mathsf{s}_G(A) = \{\alpha \in \mathbb{R} : (\alpha, \alpha) \in \mathsf{j}(A)\} \subset \mathbb{R}$. Then $\mathsf{s}_G(A^\times) = \{-\alpha : \alpha \in \mathsf{s}_G(A)\}$.

7.7 SIMILARITY FOR PIP-SPACE OPERATORS

7.7.1 General PIP-space Operators

Let us assume now that $G \in \mathsf{s}(A)$, that is, $(G, G) \in \mathsf{j}(A)$, for some $G \in \mathcal{M}(\mathcal{H})$, bounded or not. Then A_{GG} is a bounded operator from $\mathcal{H}(G)$ into itself, that is, there exists $c > 0$ such that

$$\|G^{1/2} A_{GG} \xi\| \le c \|G^{1/2}\xi\|, \quad \forall \xi \in \mathcal{H}(G).$$

This means that

$$\|G^{1/2} A_{GG} G^{-1/2}\eta\| \le c\|\eta\|, \quad \forall \eta \in \mathcal{H}.$$

Hence, $\mathsf{B} := G^{1/2} A_{GG} G^{-1/2}$ is a bounded operator on $\mathcal{H}$. Then the operator $A_{GG} \in \mathcal{B}(\mathcal{H}(G))$ is *quasi-similar* to $\mathsf{B} \in \mathcal{B}(\mathcal{H})$, that is, $A_{GG} \dashv \mathsf{B}$, with respect to the (possibly unbounded) intertwining operator $G^{1/2}$, in the sense of Definition 7.3.24.

More generally, by an argument similar to that used earlier for the couple (G, G), one can prove the following

Proposition 7.7.1 *Consider $A \in \mathrm{Op}(V_\Sigma)$. Then, $(X, Y) \in \mathsf{j}(A)$ if, and only if, $Y^{1/2} A X^{-1/2}$ is a bounded operator in $\mathcal{H}$.*

Remark 7.7.2 As $\mathsf{B} = G^{1/2} A_{GG} G^{-1/2}$ is a bounded operator on $\mathcal{H}$, its restriction B_0 to $V^\#$ is continuous from $V^\#$ into V, hence B_0 determines a unique operator $B \in \mathrm{Op}(V_\Sigma)$ [13, Proposition 3.1.2] such that $B\xi = \mathsf{B}_0\xi$, for every $\xi \in V^\#$ and $B_{II} = \mathsf{B}$. The previous statement then reads as follows: $G \in \mathsf{s}(A)$, for some $G \in \mathcal{M}(\mathcal{H})$, implies $I \in \mathsf{s}(B)$ (here I is the identity operator, corresponding to $\mathcal{H}(I) = \mathcal{H}$).

Take first $G \in \mathcal{M}_b(\mathcal{H})$, with unbounded inverse, so that $\mathcal{H}(G^{-1}) \subset \mathcal{H} \subset \mathcal{H}(G)$. Let $A \in \mathrm{Op}(V_\Sigma)$ and assume that $G \in \mathsf{s}(A)$. Then, following the standard approach (see Section A.2.1), we can consider the restriction A of A_{GG} to $\mathcal{H}$, on the domain $D(\mathsf{A}) = \{\xi \in \mathcal{H} : A_{GG}\xi = A\xi \in \mathcal{H}\}$. However, in general, $D(\mathsf{A})$ need not be dense in $\mathcal{H}$. A sufficient condition for the density of $D(\mathsf{A})$ can be given in terms of the adjoint operator $A^\times$, defined in (A.2.3). Indeed, by the assumption, $G^{-1} \in \mathsf{s}(A^\times)$. Then the space $D(\mathsf{A}^\sharp) := \{\xi \in \mathcal{H} : A^\times\xi \in \mathcal{H}\}$ is dense in $\mathcal{H}$ as it contains $\mathcal{H}(G^{-1})$. We define $\mathsf{A}^\sharp\xi = A^\times\xi$ for $\xi \in D(\mathsf{A}^\sharp)$. Then we have

Proposition 7.7.3 *Let $G \in \mathsf{s}(A)$, with $G \in \mathcal{M}_b(\mathcal{H})$. Then the domain $D(\mathsf{A})$ is dense in $\mathcal{H}$ if and only if the operator $\mathsf{A}^{\sharp *}$ is a restriction of A.*

Proof. Let us assume first that $\mathsf{A}^{\sharp *}$ is a restriction of A. Then, as the domain $D(\mathsf{A})$ is maximal in $\mathcal{H}$, we get that $\mathsf{A}^{\sharp *} = \mathsf{A}$. Hence A is densely defined in $\mathcal{H}$. Conversely, for $\xi \in D(\mathsf{A}^{\sharp *}) \subset \mathcal{H} \subset \mathcal{H}(G)$ and $\eta \in \mathcal{H}(G^{-1})$, we have

$$
\begin{aligned}
\langle \mathsf{A}^{\sharp *}\xi | \eta \rangle &= \langle \xi | \mathsf{A}^\sharp \eta \rangle = \langle \xi | A^\times \eta \rangle = \langle \xi | A^\times_{G^{-1}G^{-1}} \eta \rangle \\
&= \langle A_{GG}\xi | \eta \rangle = \langle A\xi | \eta \rangle,
\end{aligned}
$$

the last equality being valid on the dense domain $D(\mathsf{A})$. ■

Assume that the domain $D(\mathsf{A}) = \{\xi \in \mathcal{H} : A\xi \in \mathcal{H}\}$ is dense in $\mathcal{H}$. As $G^{1/2}$ is bounded, we have $G^{1/2} : D(\mathsf{A}) \to D(\mathsf{B}) = \mathcal{H}$, where, as before, $\mathsf{B} = G^{1/2} A_{GG} G^{-1/2}$ and

$$
\mathsf{B}\, G^{1/2}\eta = G^{1/2}\mathsf{A}\,\eta, \quad \forall \eta \in D(\mathsf{A}).
$$

This means that $\mathsf{A} \dashv \mathsf{B}$, with $G^{1/2}$ as intertwining operator.

On the other hand,

$$
\mathsf{B}\, G^{1/2}\eta = G^{1/2} A\,\eta, \quad \forall \eta \in \mathcal{H}(G).
$$

So, if $G \in \mathsf{s}(A)$, then A is similar to a bounded operator in $\mathcal{H}$. But $G^{1/2}$ is a unitary operator from $\mathcal{H}(G)$ onto $\mathcal{H}$; hence A and B are unitarily equivalent, while for the restriction A of A to $\mathcal{H}$ only quasi-similarity may hold.

Next, take G unbounded, with G^{-1} bounded, so that $\mathcal{H}(G) \subset \mathcal{H} \subset \mathcal{H}(G^{-1})$. Then $A : \mathcal{H}(G) \to \mathcal{H}(G)$ is a densely defined operator in $\mathcal{H}$. As before, $\mathsf{B} := G^{1/2} A_{GG} G^{-1/2}$ is bounded and everywhere defined on $\mathcal{H}$. Hence $G^{-1/2} : D(\mathsf{B}) = \mathcal{H} \to D(A_{GG}) = \mathcal{H}(G)$ and $G^{-1/2}\mathsf{B}\xi = A_{GG} G^{-1/2}\xi, \ \forall \xi \in \mathcal{H}$, that is, $\mathsf{B} \dashv A_{GG}$ with respect to the bounded intertwining operator $G^{-1/2}$. On the other hand, we had already $A_{GG} \dashv \mathsf{B}$, with respect to $G^{1/2}$, so that, in this case, $A_{GG} \dashv\vdash \mathsf{B}$. In addition, as $G^{\pm 1/2}$ are unitary between $\mathcal{H}$ and $\mathcal{H}(G)$, it follows that A_{GG} and B are unitarily equivalent.

If G and G^{-1} are both unbounded, then $A : \mathcal{H}(R_G) = \mathcal{H} \cap \mathcal{H}(G) \to \mathcal{H}(G)$ is a densely defined operator in $\mathcal{H}$, and the argument goes through.

7.7.2 The Case of Symmetric PIP-space Operators

In many applications, it is essential to show that a given symmetric operator A in a Hilbert space $\mathcal{H}$ is self-adjoint. This is, for instance, the crucial question in QM, for the Hamiltonian of the system under consideration. The same question can be asked for other observables of the system. More generally, we may ask whether A is similar in a some sense to a self-adjoint operator. In that case, we might start from a pseudo-Hermitian operator or a quasi-Hermitian operator A on $\mathcal{H}$, for instance, a $\mathcal{PT}$-symmetric Hamiltonian. If A is a symmetric, densely defined, operator in the Hilbert space $\mathcal{H}$, it makes sense to ask for the existence of self-adjoint *extensions* of A (if A itself is not self-adjoint). The standard technique is to use quadratic forms (the Friedrichs extension) or von Neumann's theory of self-adjoint extensions (27, 51). This approach is taken, for instance, by Albeverio *et al.* (20), in the language of J-self-adjoint operators in a Krein space (see Section 7.2).

However, there is another possibility. Namely, given a operator A in a space $\mathcal{K} \supset \mathcal{H}$, symmetric in some sense, it is natural to ask directly whether A has *restrictions* that are self-adjoint in $\mathcal{H}$. The answer is given essentially by the KLMN theorem that we recall here.[6] The framework is a scale of three Hilbert spaces, $\mathcal{H}_1 \subset \mathcal{H}_0 \subset \mathcal{H}_{\overline{1}}$, where the embeddings are continuous and have dense range, and $\mathcal{H}_{\overline{1}}$ is the conjugate dual of $\mathcal{H}_1$. Such are, for instance, the triplet (7.4.8) or (7.4.9) or the central triplet of the scale (7.4.11). Then the theorem reads as follows.

Theorem 7.7.4 (KLMN theorem) *Let A_o be a continuous map from $\mathcal{H}_1$ into $\mathcal{H}_{\overline{1}}$, such that $A_o - \lambda I$ is a bijection for some $\lambda \in \mathbb{R}$ and that $\langle f|A_o g\rangle = \langle A_o f|g\rangle, \forall f, g \in \mathcal{H}_1$. Then there is a unique self-adjoint operator A in $\mathcal{H}_0$ with domain $D(A) = \{g \in \mathcal{H}_1 : Ag \in \mathcal{H}_0\} \subset \mathcal{H}_1$ and $\langle f|Ag\rangle = \langle f|A_o g\rangle$, $\forall g \in D(A)$ and $f \in \mathcal{H}_1$.*

This celebrated theorem (which has already a PIP-space flavor in its Hilbert space formulation) can be extended to a PIP-space context [13, Theorems 3.3.27 and 3.3.28].

[6] KLMN stands for Kato, Lax, Lions, Milgram, Nelson.

Thus, we formulate the question in a PIP-space context and we assume that V is a nondegenerate, positive definite indexed PIP-space with central Hilbert space $V_o = V_{\overline{o}} = \mathcal{H}$, for instance, the LHS V_J. However, as every operator $A \in \mathrm{Op}(V_J)$ satisfies the condition $A^{\times\times} = A$, there is no room for extensions, only the KLMN approach may be used. Thus, in order to obtain a self-adjoint representative in $\mathcal{H}$, by restriction from a larger space, we have to start from a *symmetric operator* in the PIP-space sense, that is, an operator $A \in \mathrm{Op}(V_J)$ that satisfies $A = A^{\times}$ (see Section A.2.1). This is the class that can give rise to self-adjoint *restrictions* to $\mathcal{H}$, thanks to the generalized KLMN Theorem A.2.1.

Let $A = A^{\times} \in \mathrm{Op}(V_J)$ be a symmetric operator on the LHS V_J. Then $X \in \mathsf{s}(A)$ if, and only if $X^{-1} \in \mathsf{s}(A)$, and this implies $I \in \mathsf{s}(A)$, by Ref. [13, Corollary 3.3.24] (see Remark 7.7.2 and Section A.2.1). In addition, as $\mathsf{s}(A)$ is invariant under the lattice operations $\cap$ and $+$, it becomes a genuine (involutive) sublattice of $\mathcal{J}$. Thus, if there is a metric operator $G \in \mathcal{M}(\mathcal{H})$, with $G \in \mathsf{s}(A)$, it follows that A has a bounded representative A_{II} in $\mathcal{H}$. Moreover, A fixes all three middle spaces in Fig. 7.1 and, therefore, all nine spaces of the lattice.

This applies, in particular, to all three spaces in the triplet (7.4.8) if G is bounded, or in the triplet (7.4.9) or (7.4.10) if G is unbounded. Moreover, by the interpolation property (iii) of Ref. [13, Section 5.1.2], A leaves invariant every space $\mathcal{H}_\alpha, \alpha \in \mathbb{Z}$ or $\mathbb{R}$, in the scales $V_{\mathcal{G}}$ and $V_{\tilde{\mathcal{G}}}$. In other words, A is a totally regular operator in these PIP-spaces (see Section A.2.2), hence, $\mathsf{s}_G(A) = \mathbb{Z}$ or $\mathbb{R}$, respectively.

Thus we may state:

Proposition 7.7.5 *Every symmetric operator $A \in \mathrm{Op}(V_\Sigma)$ such that $G \in \mathsf{s}(A)$, with $G \in \mathcal{M}(\mathcal{H})$, has a bounded representative A_{II} in $\mathcal{H}$.*

There is a sort of converse to the previous statement. Given a closed unbounded operator B in $\mathcal{H}$, one may consider the self-adjoint operator $G := 1 + (B^*B)^{1/2}$ and the scale $V_{\mathcal{G}}$ built on the powers of $G^{1/2}$ (this is essentially the only PIP-space one can build intrinsically from B alone). Then $T = G^{-1/2}$ is a bounded metric operator. Hence, according to Proposition 7.5.11, B is quasi-Hermitian with respect to T if and only if $A_0 = TB$ is symmetric in $\mathcal{H}$. Next, as $D(A_0)$ is dense in $\mathcal{H}$, A_0 defines a unique symmetric operator $A = A^{\times}$ in the scale $V_{\mathcal{G}}$.

Clearly, the assumption that $G \in \mathsf{s}(A)$ is too strong for applications, as it implies that A has a *bounded* restriction to $\mathcal{H}$. Assume instead that $(G^{-1}, G) \in \mathsf{j}(A)$ for some $G \in \mathcal{M}_b(\mathcal{H})$ with an unbounded inverse. Then, according to (7.4.8), $\mathcal{H}(G^{-1}) \subset \mathcal{H} \subset \mathcal{H}(G)$ and we can apply the KLMN theorem in its PIP-space version [13, Theorem 3.3.27].

Proposition 7.7.6 *Given a symmetric operator $A = A^{\times}$, assume there is a metric operator $G \in \mathcal{M}_b(\mathcal{H})$ with an unbounded inverse, for which there exists a $\lambda \in \mathbb{R}$ such that $A - \lambda I$ has a boundedly invertible representative $(A - \lambda I)_{GG^{-1}} : \mathcal{H}(G^{-1}) \to \mathcal{H}(G)$. Then $A_{GG^{-1}}$ has a unique restriction to a self-adjoint operator A in the Hilbert space $\mathcal{H}$, with dense domain $D(\mathsf{A}) = \{\xi \in \mathcal{H} : A\xi \in \mathcal{H}\}$. In addition, $\lambda \in \rho(\mathsf{A})$.*

In a nutshell, the argument runs as follows. Let $R_{G^{-1}G} = ((A - \lambda I)_{GG^{-1}})^{-1} : \mathcal{H}(G) \to \mathcal{H}(G^{-1})$ be the bounded inverse of the invertible representative

$(A - \lambda I)_{GG^{-1}}$. Define $R_{II} = E_{IG^{-1}} R_{G^{-1}G} E_{GI}$, which is a restriction of $R_{G^{-1}G}$. (we recall that $\mathcal{H}(I) := \mathcal{H}$). Then, by the assumption, R_{II} is bounded and, by Ref. as before [13, Lemma 3.3.26], it has a self-adjoint inverse $\mathsf{A} - \lambda I$, which is a restriction of $(A - \lambda I)_{GG^{-1}}$. The rest is obvious.

Proposition 7.7.7 *Given a symmetric operator $A = A^{\times}$, assume there is a metric operator G such that $A_{GG^{-1}}$ has a restriction to a self-adjoint operator A in the Hilbert space $\mathcal{H}$. Then, if the natural embedding $\mathcal{H}(G^{-1}) \to \mathcal{H}(G)$ is compact, the operator A has a purely point spectrum of finite multiplicity, thus $\sigma(\mathsf{A}) = \sigma_p(\mathsf{A})$, $m_{\mathsf{A}}(\lambda_j) < \infty$ for every $\lambda_j \in \sigma_p(\mathsf{A})$ and $\sigma_c(\mathsf{A}) = \emptyset$.*

Proof. According to the KLMN Theorem A.2.1, the resolvent $(\mathsf{A} - \lambda I)^{-1}$ is compact if and only if the natural embedding $\mathcal{H}(G^{-1}) \to \mathcal{H}(G)$ is compact. In that case, A is a self-adjoint operator with compact resolvent, which implies the statements. ∎

Even if we don't have $(G^{-1}, G) \in \mathsf{j}(A)$ for some $G \in \mathcal{M}_b(\mathcal{H})$ with an unbounded inverse, we can still obtain a self-adjoint restriction to $\mathcal{H}$, by exploiting Theorem 3.3.28 in Ref. (13), that is, Theorem A.2.1, by restricting the PIP-space to one of the scales $V_{\mathcal{G}}$ and $V_{\tilde{\mathcal{G}}}$ built on the powers of $G^{-1/2}$.

Proposition 7.7.8 *Let $V_{\mathcal{G}} = \{\mathcal{H}_n, n \in \mathbb{Z}\}$ be the Hilbert scale built on the powers of the operator $G^{-1/2}$, where $G \in \mathcal{M}_b(\mathcal{H})$ with G^{-1} unbounded. Given $A = A^{\times} \in$ $\mathrm{Op}(V_{\mathcal{G}})$, assume there is a $\lambda \in \mathbb{R}$ such that $A - \lambda I$ has a boundedly invertible representative $(A - \lambda I)_{nm} : \mathcal{H}_m \to \mathcal{H}_n$, with $\mathcal{H}_m \subset \mathcal{H}_n$. Then the conclusions of Proposition 7.7.6 hold true.*

According to the proof of Ref. [13, Theorem 3.3.28], the assumption implies that the operator $R_{mn} = (A_{mn} - \lambda I_{mn})^{-1} : \mathcal{H}_n \to \mathcal{H}_m$ has a self-adjoint representative R_{oo} in $\mathcal{H}$, which is injective and has dense range. Therefore, its inverse $\mathsf{A} - \lambda I = R_{oo}^{-1}$, thus also A itself, is defined on a dense domain and is self-adjoint.

In the discussion earlier, we assumed that G is bounded and G^{-1} unbounded, so that the natural environment is the scale built on the powers of $G^{-1/2}$. If G is unbounded and G^{-1} bounded, one can perform the same construction using the powers of $G^{1/2}$. If G is and G^{-1} are both unbounded, we can proceed by taking the powers of $(R_G)^{1/2}$ or $(R_{G^{-1}})^{1/2}$, thus getting the scales around the triplet (7.4.3) or (7.4.4). Thus globally, we may state

Theorem 7.7.9 *Let $V_{\mathcal{G}} = \{\mathcal{H}_n, n \in \mathbb{Z}\}$ be the Hilbert scale built on the powers of the operator $G^{\pm 1/2}$ or $(R_{G^{\pm 1}})^{-1/2}$, depending on the (un)boundedness of $G^{\pm 1} \in \mathcal{M}(\mathcal{H})$ and let $A = A^{\times}$ be a symmetric operator in $V_{\mathcal{G}}$.*

(i) Assume that there is a $\lambda \in \mathbb{R}$ such that $A - \lambda I$ has a boundedly invertible representative $(A - \lambda I)_{nm} : \mathcal{H}_m \to \mathcal{H}_n$, with $\mathcal{H}_m \subset \mathcal{H}_n$. Then A_{nm} has a unique restriction to a self-adjoint operator A in the Hilbert space $\mathcal{H}$, with dense domain $D(\mathsf{A}) = \{\xi \in \mathcal{H} : A\xi \in \mathcal{H}\}$. In addition, $\lambda \in \rho(\mathsf{A})$.

(ii) If the natural embedding $\mathcal{H}_m \to \mathcal{H}_n$ is compact, the operator A has a purely point spectrum of finite multiplicity, thus $\sigma(\mathsf{A}) = \sigma_p(\mathsf{A})$, $m_{\mathsf{A}}(\lambda_j) < \infty$ for every $\lambda_j \in \sigma_p(\mathsf{A})$ and $\sigma_c(\mathsf{A}) = \emptyset$.

This condition on the natural embeddings is familiar in the theory of topological vector spaces. For instance, the end space of the scale $V_{\mathcal{G}}$, namely, $\mathcal{H}_\infty(G^{1/2}) = \bigcap_{n\in\mathbb{Z}} \mathcal{H}_n$ (see (7.4.12)), is nuclear if, for every $m \in \mathbb{Z}$, there exists $n \in \mathbb{Z}$ such that the natural embedding $\mathcal{H}_m \to \mathcal{H}_n$ is a Hilbert-Schmidt operator (analogous results hold when the embedding is compact).

At this stage, we do have a self-adjoint restriction A of A in $\mathcal{H}$, but we don't know if there is any (quasi-)similarity relation between $A_{GG^{-1}}$ or A and another operator.

On the contrary, assume that G is bounded and G^{-1} unbounded, and that A maps $\mathcal{H}(G^{-1})$ continuously into $\mathcal{H}(G)$. Then, following the discussion preceding Proposition 7.7.1, there exists $c > 0$ such that

$$\|G^{1/2}A_{GG^{-1}}\xi\| \le c\|G^{-1/2}\xi\|, \quad \forall\, \xi \in \mathcal{H}(G^{-1}).$$

This means that

$$\|G^{1/2}A_{GG^{-1}}G^{1/2}\eta\| \le c\|\eta\|, \quad \forall\, \eta \in \mathcal{H}.$$

Hence, $\mathsf{B} := G^{1/2}A_{GG^{-1}}G^{1/2}$ is a bounded operator on $\mathcal{H}$. However, the operator $A_{GG^{-1}}$ is *not* quasi-similar to $\mathsf{B} \in \mathcal{B}(\mathcal{H})$. Indeed, condition ($\mathsf{io}_1$) imposes that $T = G^{-1/2}$, hence unbounded, but the conditions (io_0) and (io_2) cannot be satisfied.

7.7.3 Semisimilarity

So far we have considered only the case of one metric operator G in relation to A. Assume now we take two different metric operators $G_1, G_2 \in \mathcal{M}(\mathcal{H})$. What can be said concerning A if it maps $\mathcal{H}(G_1)$ into $\mathcal{H}(G_2)$?

One possibility is to introduce, following (11), a notion slightly more general than quasi-similarity, called *semisimilarity*.

Definition 7.7.10 Let $\mathcal{H}, \mathcal{K}_1$ and $\mathcal{K}_2$ be three Hilbert spaces, A a closed, densely defined operator from $\mathcal{K}_1$ to $\mathcal{K}_2$, B a closed, densely defined operator on $\mathcal{H}$. Then A is said to be *semisimilar* to B, which we denote by $A \dashv\vdash B$, if there exist two bounded operators $T : \mathcal{K}_1 \to \mathcal{H}$ and $S : \mathcal{K}_2 \to \mathcal{H}$ such that (see Fig. 7.4):

(i) $T : D(A) \to D(B)$;
(ii) $BT\xi = SA\xi, \ \forall\, \xi \in D(A)$.

The pair (T, S) is called an *intertwining couple*.

Of course, if $\mathcal{K}_1 = \mathcal{K}_2$ and $S = T$, we recover the notion of quasi-similarity and $A \dashv B$ (with a bounded intertwining operator).

Now we come back to the case envisaged earlier: $A : \mathcal{H}(G_1) \to \mathcal{H}(G_2)$ continuously, for the two metric operators $G_1, G_2 \in \mathcal{M}_b(\mathcal{H})$, but A is *not* supposed to be

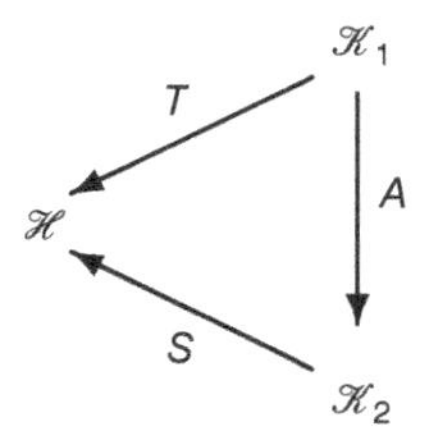

Figure 7.4 The semisimilarity scheme (from Ref. 11).

symmetric. Under this assumption, we essentially recover the previous situation. As $A_{G_2G_1}$ is a bounded operator from $\mathcal{H}(G_1)$ into $\mathcal{H}(G_2)$, there exists $c > 0$ such that

$$\|G_2^{1/2} A_{G_2G_1} \xi\| \le c \|G_1^{1/2} \xi\|, \quad \forall \xi \in \mathcal{H}(G_1).$$

This means that

$$\|G_2^{1/2} A_{G_2G_1} G_1^{-1/2} \eta\| \le c \|\eta\|, \quad \forall \eta \in \mathcal{H}.$$

Hence, $\mathsf{B} := G_2^{1/2} A_{G_2G_1} G_1^{-1/2}$ is bounded in $\mathcal{H}$. Then the operator $A_{G_2G_1}$ is *semisimilar* to $\mathsf{B} \in \mathcal{B}(\mathcal{H})$, that is, $A_{G_2G_1} \dashv\vdash \mathsf{B}$, with respect to the intertwining couple $T = G_1^{1/2}, S = G_2^{1/2}$.

Next we take a symmetric operator $A = A^\times \in \mathrm{Op}(V_\mathcal{I})$, where $V_\mathcal{I}$ is any LHS (or PIP-space) containing $\mathcal{H}(G_1)$ and $\mathcal{H}(G_2)$. Then $A : \mathcal{H}(G_1) \to \mathcal{H}(G_2)$ continuously implies $A : \mathcal{H}(G_2^{-1}) \to \mathcal{H}(G_1^{-1})$ continuously as well. Assume that $G_1 \preceq G_2$, that is, $\mathcal{H}(G_1) \subset \mathcal{H}(G_2)$. This yields the following situation:

$$\mathcal{H}(G_2^{-1}) \subset \mathcal{H}(G_1^{-1}) \subset \mathcal{H} \subset \mathcal{H}(G_1) \subset \mathcal{H}(G_2). \qquad (7.7.1)$$

Therefore, as $\mathcal{H}(G_1^{-1}) \hookrightarrow \mathcal{H}(G_2)$, the operator A maps $\mathcal{H}(G_2^{-1})$ continuously into $\mathcal{H}(G_2)$, that is, we are back to the situation of Proposition 7.7.6 and we can state:

Proposition 7.7.11 *Given a symmetric operator $A = A^\times \in \mathrm{Op}(V_\mathcal{I})$, assume that there exists two metric operators $G_1, G_2 \in \mathcal{M}_b(\mathcal{H})$ such that $G_1 \preceq G_2$ and $(G_1, G_2) \in \mathsf{j}(A)$. Assume there exists a $\lambda \in \mathbb{R}$ such that $A - \lambda I$ has an invertible representative $(A - \lambda I)_{G_2G_2^{-1}} : \mathcal{H}(G_2^{-1}) \to \mathcal{H}(G_2)$. Then there exists a unique restriction of $A_{G_2G_2^{-1}}$ to a self-adjoint operator A in the Hilbert space $\mathcal{H}$. The number λ does not belong to the spectrum of A. The dense domain of A is given by $D(\mathsf{A}) = \{\xi \in \mathcal{H} : A\xi \in \mathcal{H}\}$. The resolvent $(\mathsf{A} - \lambda I)^{-1}$ is compact (trace class, etc.) if and only if the natural embedding $\mathcal{H}(G_1) \to \mathcal{H}(G_2)$ is compact (trace class, etc.).*

The analysis may be extended to the three other cases, assuming again that $A : \mathcal{H}(G_1) \to \mathcal{H}(G_2)$:

1. G_1 unbounded, G_2 bounded: then

$$\mathcal{H}(G_1) \subset \mathcal{H} \subset \mathcal{H}(G_2)$$

and A maps the small space into the large one. Then the KLMN theorem applies.
2. G_1 and G_2 both unbounded, with $\mathcal{H}(G_1) \subset \mathcal{H}(G_2)$; then we are back to the first situation, with a bounded representative $(A - \lambda I)_{G_1^{-1}G_1} : \mathcal{H}(G_1) \to \mathcal{H}(G_1^{-1})$ and the KLMN theorem applies.
3. G_1 bounded, G_2 unbounded: then

$$\mathcal{H}(G_2) \subset \mathcal{H} \subset \mathcal{H}(G_1) \quad \text{and} \quad \mathcal{H}(G_1^{-1}) \subset \mathcal{H} \subset \mathcal{H}(G_2^{-1}),$$

so that, in both cases, A maps the large space into the small one. Then the KLMN theorem does *not* apply.[7]

In conclusion, if $A = A^{\times}$ is symmetric and $(G_1, G_2) \in \mathfrak{j}(A)$, with $G_1 \preceq G_2$, the KLMN theorem applies and yields a self-adjoint restriction in $\mathcal{H}$, in two cases: either G_1 is unbounded, or G_1 and G_2 are both bounded.

7.8 THE CASE OF PSEUDO-HERMITIAN HAMILTONIANS

Metric operators appear routinely in the so-called pseudo-Hermitian QM (1), but in general only bounded ones are considered. In some recent work (9, 10, 19), however, unbounded metric operators have been discussed. The question is, how do these operators fit in the present formalism?

Following the argument of Ref. (10), the starting point is a reference Hilbert space $\mathcal{H}$ and a quasi-Hermitian operator [8] H on $\mathcal{H}$, which means that there exists an unbounded metric operator G satisfying the relation

$$H^* G = GH. \tag{7.8.1}$$

This operator H is the putative non-self-adjoint (but $\mathcal{PT}$-symmetric) Hamiltonian of a quantum system.

In the relation (7.8.1), the two operators are assumed to have the same domain, $D(H^*G) = D(GH)$, which is supposed to be dense. This condition is not necessary, however, if we assume that H is quasi-Hermitian in the sense of Definition 7.5.1. This means that $D(H) \subset D(G)$ and

$$\langle H\xi | G\eta \rangle = \langle G\xi | H\eta \rangle, \quad \forall\, \xi, \eta \in D(H). \tag{7.8.2}$$

[7] The statement of Case 3. given in Ref. [11, Section 5.6] is wrong. See Corrigendum.

[8] The author of Ref. (10) calls this a G-pseudo-Hermitian operator, but in fact it is simply a quasi-Hermitian operator, in the original sense of Dieudonné (4), but unbounded.

If G is bounded, we get $H \dashv H^*$ and then Proposition 7.5.15 implies that $G^{1/2}HG^{-1/2}$ is self-adjoint, that is, H is quasi-self-adjoint.

On the other hand, If G is unbounded and if we assume that H is strictly quasi-Hermitian, we still have $H \dashv H^*$, but we cannot conclude. However, if in addition G^{-1} is bounded, we get $G^{-1}H^*G\eta = H\eta$, $\forall \eta \in D(H)$, which is a more restrictive form of similarity.

Finally, let us assume that the quasi-Hermitian operator H possesses a (large) set of vectors, $\mathcal{D}^{\omega}_{G}(H)$, which are analytic in the norm $\|\cdot\|_G$ and are contained in $D(G)$ (52, 53). This means that every vector $\phi \in \mathcal{D}^{\omega}_{G}(H)$ satisfies the relation

$$\sum_{n=0}^{\infty} \frac{\|H^n \phi\|_G}{n!} t^n < \infty, \text{ for some } t \in \mathbb{R}.$$

This implies that every $\phi \in \mathcal{D}^{\omega}_{G}(H)$ satisfies the relation $H^n\phi \in D(G^{1/2})$, $\forall n = 0, 1, \ldots$. Thus, one has

$$\mathcal{D}^{\omega}_{G}(H) \subset D(H) \subset D(G) \subset D(G^{1/2}) \subset \mathcal{H}. \tag{7.8.3}$$

Then the construction given in Ref. [11, Section 6] can be performed.

Endow $\mathcal{D}^{\omega}_{G}(H)$ with the norm $\|\cdot\|_G$ and take the completion $\mathcal{H}_G$, which is a closed subspace of $\mathcal{H}(G)$, as defined in Section 7.2. An immediate calculation then yields

$$\langle \phi | H\psi \rangle_G = \langle H\phi | \psi \rangle_G, \ \forall \phi, \psi \in \mathcal{D}^{\omega}_{G}(H),$$

that is, H is a densely defined symmetric operator in $\mathcal{H}_G$. As it has a dense set of analytic vectors, it is essentially self-adjoint, by Nelson's theorem (52, 53), [27, Theorem 7.16], hence its closure $\overline{H}$ is a self-adjoint operator in $\mathcal{H}_G$. The pair $(\mathcal{H}_G, \overline{H})$ is then interpreted as the physical quantum system.

Next, by definition, $W_{\mathcal{D}} := G^{1/2} \upharpoonright \mathcal{D}^{\omega}_{G}(H)$ is isometric from $\mathcal{D}^{\omega}_{G}(H)$ into $\mathcal{H}$, hence it extends to an isometry $W = \overline{W_{\mathcal{D}}} : \mathcal{H}_G \to \mathcal{H}$. The range of W is a closed subspace of $\mathcal{H}$, denoted $\mathcal{H}_{\rm phys}$, and the operator W is unitary from $\mathcal{H}_G$ onto $\mathcal{H}_{\rm phys}$. Therefore, the operator $h = W\overline{H}W^{-1}$ is self-adjoint in $\mathcal{H}_{\rm phys}$. This operator h is interpreted as the genuine Hamiltonian of the system, acting in the physical Hilbert space $\mathcal{H}_{\rm phys}$.

Things simplify if $\mathcal{D}^{\omega}_{G}(H)$ is dense in $\mathcal{H}$. Then $W(\mathcal{D}^{\omega}_{G}(H))$ is also dense, $\mathcal{H}_G = \mathcal{H}(G)$, $\mathcal{H}_{\rm phys} = \mathcal{H}$ and $W = G^{1/2}$ is unitary from $\mathcal{H}(G)$ onto $\mathcal{H}$. Also, if G is bounded, it is sufficient to assume that the vectors in $\mathcal{D}^{\omega}_{G}(H)$ are analytic with respect to the original norm of $\mathcal{H}$. But then H has a dense set of analytic vectors, so that it is essentially self-adjoint in $\mathcal{H}$, by Nelson's theorem again. Note that we are back to the situation of Proposition 7.5.13, as then $\mathcal{H}(G) = \mathcal{H}$, with equivalent, but different norms.

Now, every eigenvector of an operator is automatically analytic, hence this construction generalizes that of Ref. (10). This applies, for instance, to the example given there, namely, the $\mathcal{PT}$-symmetric operator $H = \frac{1}{2}(p - i\alpha)^2 + \frac{1}{2}\omega^2 x^2$ in $\mathcal{H} = L^2(\mathbb{R})$, for any $\alpha \in \mathbb{R}$, which has an orthonormal basis of eigenvectors.

7.8.1 An Example

A beautiful example of the situation just analyzed has been given recently by Samsonov (35), namely, the second derivative on the positive half-line (this example stems from Schwartz (54)):

$$H = -\frac{\mathrm{d}^2}{\mathrm{d}x^2}, \quad x \geq 0, \tag{7.8.4}$$

with domain

$$D(H) = \{\xi \in L^2(0,\infty) : \xi'' \in L^2(0,\infty), \xi'(0) + (d+ib)\xi(0) = 0\}. \tag{7.8.5}$$

For $d < 0$, this operator has a purely continuous spectrum. Its adjoint H^* is given again by (7.8.4), on the domain $D(H^*)$ defined as in (7.8.5), with b replaced by $-b$.

Next introduce the unbounded operator

$$G = -\frac{\mathrm{d}^2}{\mathrm{d}x^2} - 2ib\frac{\mathrm{d}}{\mathrm{d}x} + d^2 + b^2, \tag{7.8.6}$$

on the domain $D(G) = D(H)$. Then a direct calculation shows that G is self-adjoint, strictly positive and invertible, that is, it is a metric operator. As its spectrum is $\sigma(G) = \sigma_c(G) = [d^2, \infty)$, it follows that G^{-1} is bounded.

As both H and G are second-order differential operators, an element of the domain $D(GH)$ should have a square integrable fourth derivative. Hence one defines

$$\widetilde{D}(H) = \{\xi \in D(H) : \xi^{(\mathrm{iv})} \in L^2(0,\infty)\} \subset D(H) \tag{7.8.7}$$

and $\widetilde{H} = H \upharpoonright \widetilde{D}(H)$. Then the analysis of Ref. (35) yields the following results:

- (i) H is quasi-Hermitian in the sense of Definition 7.5.1, that is, it satisfies the relation (7.8.2) on $D(G) = D(H)$.
- (ii) G maps $\widetilde{D}(H)$ into $D(H^*)$.
- (iii) H is quasi-Hermitian in the sense of Dieudonné, that is, $GH = H^*G$ on the dense domain $\widetilde{D}(H)$ (we have to restrict ourselves to $\widetilde{D}(H)$ because of the requirement on the fourth derivative).
- (iv) The operator $h = G^{1/2}HG^{-1/2} = G^{-1/2}H^*G^{1/2}$ is self-adjoint on the domain $D(h) = \{\eta = G^{1/2}\xi, \ \xi \in \widetilde{D}(H)\}$.

In conclusion, by (i) and (ii), we get $\widetilde{H} \dashv H^*$.

In fact, one can use as metric operator $T := G^{-1}$, which is bounded, with unbounded inverse $T^{-1} = G$, a more standard situation. Writing $H = H_\lambda$, with $\lambda = (d+ib)$, we have that $H^* = H_{\overline{\lambda}}$ and the relation $GH = H^*G$ becomes

$$GH_\lambda = H_{\overline{\lambda}}G \tag{7.8.8}$$

which yields a symmetry between H_λ and $H_{\overline{\lambda}}$. Multiplying both sides of (7.8.8) from the left and from the right by $T^{-1} = G$, we get

$$H_\lambda T\eta = TH_{\overline{\lambda}}\eta, \ \forall \eta \in \widetilde{D}(H_{\overline{\lambda}}),$$

where $\widetilde{D}(H_{\overline{\lambda}})$ is defined as in (7.8.7). Noting that $T : \widetilde{D}(H_{\overline{\lambda}}) \to D(H_\lambda)$, we can conclude as before that $\widetilde{H_{\overline{\lambda}}} \dashv H_\lambda$, that is, $\widetilde{H^*} \dashv H$, with the bounded intertwining operator $T = G^{-1}$. The problem, of course, is that we don't know the operator T explicitly. Being the inverse of the differential operator G, it is presumably an integral operator. Therefore, this second solution, albeit more standard, is of little use.

7.9 CONCLUSION

In this chapter, we have introduced several generalizations of similarity between operators and we have obtained some results on the preservation of spectral properties under quasi-similarity, but only with a bounded metric operator with unbounded inverse. However, we have seen that the consideration of unbounded metric operators leads naturally to the formalism of PIP-spaces. And indeed it turns out that exploiting the connection between metric operators and PIP-spaces does in certain cases improve the quasi-similarity of operators. More precisely, given a symmetric operator $A = A^\times$ in a PIP-space with central Hilbert space $\mathcal{H}$, one can apply the KLMN theorem, which may yield a self-adjoint restriction of A in $\mathcal{H}$. Then additional quasi-similarity relations follow.

Of course, these results are only a first step, many open problems subsist. In view of the applications, notably in pseudo-Hermitian QM, the most crucial ones concern the behavior of spectral properties under some generalized similarity with an unbounded metric operator. In the same vein, there are few results about the spectral properties of self-adjoint operators derived in a PIP-space context from a symmetric operator via the KLMN theorem. Then, of course, one should investigate the connection between these two types of problems. In particular, one needs to investigate in more details the spectral properties of symmetric operators in a PIP-space, and in an LHS in the first place. Research in this direction is in progress. Preliminary results may be found in Ref. (55), following the approach developed for the case of a rigged Hilbert space (RHS) by Bellomonte, di Bella and one of us (56).

APPENDIX: PARTIAL INNER PRODUCT SPACES

A.1 PIP-SPACES AND INDEXED PIP-SPACES

For the convenience of the reader, we have collected here the main features of partial inner product spaces and operators on them, keeping only what is needed for reading the chapter. Further information may be found in our monograph (13) or our review paper (57).

The general framework is that of a PIP-space V, corresponding to the linear compatibility #, that is, a symmetric binary relation $f\#g$ which preserves linearity. We call *assaying subspace* of V a subspace S such that $S^{\#\#} = S$ and we denote by $\mathcal{F}(V, \#)$ the family of all assaying subspaces of V, ordered by inclusion. The assaying subspaces are denoted by $V_r, V_q, \ldots$ and the index set is F. By definition, $q \leq r$ if and only if $V_q \subseteq V_r$. Thus we may write

$$f\#g \;\Leftrightarrow\; \exists\, r \in F \text{ such that } f \in V_r, g \in V_{\bar{r}}. \tag{A.1.1}$$

General considerations (58) imply that the family $\mathcal{F}(V, \#) := \{V_r, r \in F\}$, ordered by inclusion, is a complete involutive lattice, that is, it is stable under the following operations, arbitrarily iterated:

$$\begin{array}{llll} \text{. involution:} & V_r & \leftrightarrow & V_{\bar{r}} = (V_r)^{\#}, \\ \text{. infimum:} & V_{p\wedge q} & := & V_p \wedge V_q = V_p \cap V_q, \qquad (p, q, r \in F), \\ \text{. supremum:} & V_{p\vee q} & := & V_p \vee V_q = (V_p + V_q)^{\#\#}. \end{array}$$

The smallest element of $\mathcal{F}(V, \#)$ is $V^{\#} = \bigcap_r V_r$ and the greatest element is $V = \bigcup_r V_r$.

By definition, the index set F is also a complete involutive lattice; for instance,

$$(V_{p\wedge q})^{\#} = V_{\overline{p\wedge q}} = V_{\bar{p}\vee\bar{q}} = V_{\bar{p}} \vee V_{\bar{q}}.$$

Given a vector space V equipped with a linear compatibility #, a *partial inner product* on $(V, \#)$ is a Hermitian form $\langle\cdot|\cdot\rangle$ defined exactly on compatible pairs of vectors. A *partial inner product space* (PIP-space) is a vector space V equipped with a linear compatibility and a partial inner product.

From now on, we will assume that our PIP-space $(V, \#, \langle\cdot|\cdot\rangle)$ is *nondegenerate*, that is, $\langle f|g\rangle = 0$ for all $f \in V^{\#}$ implies $g = 0$. As a consequence, $(V^{\#}, V)$ and every couple $(V_r, V_{\bar{r}})$, $r \in F$, are a dual pair in the sense of topological vector spaces (59). Next we assume that every V_r carries its Mackey topology $\tau(V_r, V_{\bar{r}})$, so that its conjugate dual is $(V_r)^{\times} = V_{\bar{r}}$, $\forall\, r \in F$. Then, $r < s$ implies $V_r \subset V_s$, and the embedding operator $E_{sr} : V_r \to V_s$ is continuous and has dense range. In particular, $V^{\#}$ is dense in every V_r. In the sequel, we also assume the partial inner product to be positive definite, $\langle f|f\rangle > 0$ whenever $f \neq 0$.

As a matter of fact, the whole structure can be reconstructed from a fairly small subset of $\mathcal{F}$, namely, a *generating* involutive sublattice $\mathcal{J}$ of $\mathcal{F}(V, \#)$, indexed by J, which means that

$$f\#g \;\Leftrightarrow\; \exists\, r \in J \text{ such that } f \in V_r, g \in V_{\bar{r}}. \tag{A.1.2}$$

The resulting structure is called an *indexed* PIP-*space* and denoted simply by $V_J := (V, \mathcal{J}, \langle\cdot|\cdot\rangle)$.

Then an indexed PIP-space V_J is said to be:

(i) *additive*, if $V_{p \vee q} = V_p + V_q$, $\forall p, q \in J$.
(ii) *projective* if $V_{p \wedge q}|_\tau \simeq (V_p \cap V_q)_{\text{proj}}$, $\forall p, q \in J$; here $V_{p \wedge q}|_\tau$ denotes $V_{p \wedge q}$ equipped with the Mackey topology $\tau(V_{p \wedge q}, V_{\overline{p} \vee \overline{q}})$, the right-hand side denotes $V_p \cap V_q$ with the topology of the projective limit from V_p and V_q and $\simeq$ denotes an isomorphism of locally convex topological spaces.

For practical applications, it is essentially sufficient to restrict oneself to the case of an indexed PIP-space satisfying the following conditions:

(i) every V_r, $r \in J$, is a Hilbert space or a reflexive Banach space, so that the Mackey topology $\tau(V_r, V_{\overline{r}})$ coincides with the norm topology;
(ii) there is a unique self-dual, Hilbert, assaying subspace $V_o = V_{\overline{o}}$.

In that case, the *indexed* PIP-*space* $V_J := (V, \mathcal{J}, \langle \cdot | \cdot \rangle)$ is called, respectively, a *LHS* or a *lattice of Banach spaces* (LBS). We refer to Ref. (13) for more precise definitions, including explicit requirements on norms. In particular, the partial inner product $\langle \cdot | \cdot \rangle$ coincides with the inner product of V_o on the latter. The important facts here are that

(i) Every projective indexed PIP-space is additive.
(ii) A LBS or a LHS is projective if and only if it is additive.

Note that $V^\#$, V themselves usually do *not* belong to the family $\{V_r, r \in J\}$, but they can be recovered as

$$V^\# = \bigcap_{r \in J} V_r, \quad V = \sum_{r \in J} V_r.$$

A standard, albeit trivial, example is that of a RHS $\Phi \subset \mathcal{H} \subset \Phi^\#$ (it is trivial because the lattice $\mathcal{F}$ contains only three elements). One should note that the construction of a RHS from a directed family of Hilbert spaces, via projective and inductive limits, has been investigated recently by Bellomonte and Trapani (60). Similar constructions, in the language of categories, may be found in the work of Mityagin and Shvarts (61) and that of Semadeni and Zidenberg (62).

Let us give some concrete examples.

(i) Sequence spaces

Let V be the space ω of *all* complex sequences $x = (x_n)$ and define on it (i) a compatibility relation by $x \# y \Leftrightarrow \sum_{n=1}^{\infty} |x_n y_n| < \infty$; (ii) a partial inner product $\langle x|y \rangle = \sum_{n=1}^{\infty} \overline{x_n} y_n$. Then $\omega^\# = \varphi$, the space of finite sequences, and the complete lattice $\mathcal{F}(\omega, \#)$ consists of Köthe's perfect sequence spaces [59, § 30]. Among these, a nice example is the lattice of the so-called ℓ_ϕ spaces associated to symmetric norming functions or, more generally, Banach sequence ideals discussed in Ref. [13, Section 4.3.2] and previously in Ref. [61, § 6] (in this example, the extreme spaces are, respectively, ℓ^1 and ℓ^∞).

(ii) Spaces of locally integrable functions

Let V be $L^1_{\text{loc}}(\mathbb{R}, \mathrm{d}x)$, the space of Lebesgue measurable functions, integrable over compact subsets, and define a compatibility relation on it by

$$f\#g \Leftrightarrow \int_{\mathbb{R}} |f(x)g(x)|\, \mathrm{d}x < \infty$$

and a partial inner product given by $\langle f|g\rangle = \int_{\mathbb{R}} \overline{f(x)} g(x)\, \mathrm{d}x$. Then one gets $V^{\#} = L^\infty_{\text{c}}(\mathbb{R}, \mathrm{d}x)$, the space of bounded measurable functions of compact support. The complete lattice $\mathcal{F}(L^1_{\text{loc}}, \#)$ consists of the so-called Köthe function spaces (63, 64). Here again, normed ideals of measurable functions in $L^1([0,1], \mathrm{d}x)$ are described in Ref. [61, § 8].

A.2 OPERATORS ON INDEXED PIP-SPACE S

Let V_J and Y_K be two nondegenerate indexed PIP-spaces (in particular, two LHSs or LBSs). Then an *operator* from V_J to Y_K is a map from a subset $\mathcal{D}(A) \subset V$ into Y, such that

(i) $\mathcal{D}(A) = \bigcup_{q \in \mathsf{d}(A)} V_q$, where $\mathsf{d}(A)$ is a nonempty subset of J;

(ii) For every $r \in \mathsf{d}(A)$, there exists $u \in K$ such that the restriction of A to V_r is a continuous linear map into Y_u (we denote this restriction by A_{ur});

(iii) A has no proper extension satisfying (i) and (ii).

We denote by $\mathrm{Op}(V_J, Y_K)$ the set of all operators from V_J to Y_K and, in particular, $\mathrm{Op}(V_J) := \mathrm{Op}(V_J, V_J)$. The continuous linear operator $A_{ur} : V_r \to Y_u$ is called a *representative* of A. The properties of A are conveniently described by the set $\mathsf{j}(A)$ of all pairs $(r, u) \in J \times K$ such that A maps V_r continuously into Y_u. Thus, the operator A may be identified with the collection of its representatives,

$$A \simeq \{A_{ur} : V_r \to Y_u : (r,u) \in \mathsf{j}(A)\}. \tag{A.2.1}$$

We will also need the following sets:

$$\mathsf{d}(A) = \{r \in J : \text{there is a } u \text{ such that } A_{ur} \text{ exists}\},$$

$$\mathsf{i}(A) = \{u \in K : \text{there is a } r \text{ such that } A_{ur} \text{ exists}\}.$$

The following properties are immediate:

- $\mathsf{d}(A)$ is an initial subset of J: if $r \in \mathsf{d}(A)$ and $r' < r$, then $r \in \mathsf{d}(A)$, and $A_{ur'} = A_{ur}E_{rr'}$, where $E_{rr'}$ is a representative of the unit operator.

- $\mathsf{i}(A)$ is a final subset of K: if $u \in \mathsf{i}(A)$ and $u' > u$, then $u' \in \mathsf{i}(A)$ and $A_{u'r} = E_{u'u}A_{ur}$.

In the case of an operator $A \in \text{Op}(V_J)$, the diagonal of $J \times J$ plays a special role. Hence, the following set is useful

$$\mathsf{s}(A) = \{r \in J : (r, r) \in \mathsf{j}(A)\}. \tag{A.2.2}$$

This set can be conveniently used to describe operators similar or quasi-similar to some representative of A. From the definitions (7.6.1), it is clear that the set $\mathsf{s}(A)$ is invariant under the lattice operations $\cap$ and $+$.

Although an operator may be identified with a separately continuous sesquilinear form on $V^{\#} \times V^{\#}$, it is more useful to keep also the *algebraic operations* on operators, namely:

(i) *Adjoint:* every $A \in \text{Op}(V_J, Y_K)$ has a unique adjoint $A^{\times} \in \text{Op}(Y_K, V_J)$, defined by

$$\langle A^{\times}y|x\rangle = \langle y|Ax\rangle, \;\; \text{for}\, x \in V_r,\; r \in \mathsf{d}(A) \;\text{and}\; y \in Y_{\overline{u}},\; u \in \mathsf{i}(A), \tag{A.2.3}$$

that is, $(A^{\times})_{\overline{r}\overline{u}} = (A_{ur})'$, where $(A_{ur})' : Y_{\overline{u}} \to V_{\overline{r}}$ is the adjoint map of A_{ur}. Furthermore, one has $A^{\times\times} = A$, for every $A \in \text{Op}(V_J, Y_K)$: no extension is allowed, by the maximality condition (iii) of the definition.

(ii) *Partial multiplication:* Let V_J, W_L, and Y_K be nondegenerate indexed PIP-spaces (some, or all, may coincide). Given two operators $A \in \text{Op}(V_J, W_L)$ and $B \in \text{Op}(W_L, Y_K)$, we say that the product BA is defined if and only if there is a $t \in \mathsf{i}(A) \cap \mathsf{d}(B)$, that is, if and only if there is continuous factorization through some W_t:

$$V_r \xrightarrow{A} W_t \xrightarrow{B} Y_u, \;\text{i.e.,}\;\; (BA)_{ur} = B_{ut}A_{tr}, \;\; \text{for some}\;\; r \in \mathsf{d}(A), u \in \mathsf{i}(B). \tag{A.2.4}$$

Concerning the adjoint, we note that $\mathsf{j}(A^{\times}) = \mathsf{j}^{\times}(A) := \{(\overline{u}, \overline{r}) : (r, u) \in \mathsf{j}(A)\} \subset J \times K$. If $V = Y$, $\mathsf{j}(A^{\times})$ is obtained by reflecting $\mathsf{j}(A)$ with respect to the anti-diagonal $\{(r, \overline{r}), r \in J\}$. In particular, if $(r, \overline{r}) \in \mathsf{j}(A)$, then $(r, \overline{r}) \in \mathsf{j}(A^{\times})$ as well. Notice also that $\mathsf{s}(A^{\times}) = \{\overline{r} : r \in \mathsf{s}(A)\}$.

A.2.1 Symmetric Operators

In Section 7.7.2, we have discussed the problem of showing that a given symmetric operator H in a Hilbert space $\mathcal{H}$ is self-adjoint. The standard technique is to use quadratic forms (the Friedrichs extension) or von Neumann's theory of self-adjoint extensions (27, 51).

However, as every operator $A \in \mathrm{Op}(V_J)$ satisfies the condition $A^{\times\times} = A$, there is no room for extensions. Thus the only notion we have at our disposal is that of *symmetric operator*, in the PIP-space sense, namely, an operator $A \in \mathrm{Op}(V_J)$ such that $A = A^{\times}$. If A is symmetric, the set $\mathsf{j}(A)$ is symmetric with respect to the anti-diagonal $\{(r, \overline{r}), r \in J\}$. Now, if $r \in \mathsf{s}(A)$, then $\overline{r} \in \mathsf{s}(A)$ as well, hence by interpolation, $o \in \mathsf{s}(A)$, that is, A has a bounded representative $A_{oo} : \mathcal{H} \to \mathcal{H}$ ([13, Corollary 3.3.24], reformulated as Proposition 7.7.5).

However, one may also ask whether A has *restrictions* that are self-adjoint in $\mathcal{H}$. The answer is given essentially by the KLMN theorem, which can be extended to a PIP-space context [13, Theorems 3.3.27 and 3.3.28]. The most general version, adapted to an LHS, reads as follows.

Theorem A.2.1 (Generalized KLMN theorem) *Let V_J be a LHS with a central Hilbert space $\mathcal{H}$ and $A = A^{\times} \in \mathrm{Op}(V_J)$. Assume there exists a $\lambda \in \mathbb{R}$ such that $A - \lambda$ has a (boundedly) invertible representative $A_{sr} - \lambda E_{sr} : V_r \to V_s$, where $V_r \subseteq V_s$. Then there exists a unique restriction of A_{sr} to a self-adjoint operator A in the Hilbert space $\mathcal{H}$. The number λ does not belong to the spectrum of A. The domain of A is obtained by eliminating from V_r exactly the vectors f that are mapped by A_{sr} beyond $\mathcal{H}$ (i.e., satisfy $A_{sr}f \notin \mathcal{H}$). The resolvent $(\mathsf{A} - \lambda)^{-1}$ is compact (trace class, etc.) if and only if the natural embedding $E_{sr} : V_r \to V_s$ is compact (trace class, etc.).*

A.2.2 Regular Operators, Morphisms, and Projections

Besides symmetric operators, other classes of operators on PIP-spaces may be defined. First, an operator $A \in \mathrm{Op}(V_J, Y_K)$ is called *regular* if $\mathsf{d}(A) = J$ and $\mathsf{i}(A) = K$ or, equivalently, if $A : V^{\#} \to Y^{\#}$ and $A : V \to Y$ continuously for the respective Mackey topologies. This notion depends only on the pairs $(V^{\#}, V)$ and $(Y^{\#}, Y)$, *not* on the particular compatibilities on them. In the case $V_J = Y_K$, an operator $A \in \mathrm{Op}(V_J)$ is regular if and only if $A^{\times}$ is.

Next, an operator $A \in \mathrm{Op}(V_J)$ is called *totally regular* if $\mathsf{j}(A)$ contains the diagonal of $J \times J$, that is, A_{rr} exists for every $r \in J$ or A maps every V_r into itself continuously, in other words, $\mathsf{s}(A) = J$. This class leads to the identification of *-algebras of operators in $\mathrm{Op}(V_J)$ [13, Section 3.3.3].

Among operators on indexed PIP-spaces, a special role is played by morphisms. An operator $A \in \mathrm{Op}(V_J, Y_K)$ is called a *homomorphism* if

(i) for every $r \in J$, there exists $u \in K$ such that both A_{ur} and $A_{\overline{u}\overline{r}}$ exist;
(ii) for every $u \in K$, there exists $r \in J$ such that both A_{ur} and $A_{\overline{u}\overline{r}}$ exist.

We denote by $\mathrm{Hom}(V_J, Y_K)$ the set of all homomorphisms from V_J into Y_K and by $\mathrm{Hom}(V_J)$ those from V_J into itself. The following properties are immediate.

Proposition A.2.2 *Let $V_J, Y_K, \ldots$ be indexed* PIP*-spaces. Then:*

(i) $A \in \mathrm{Hom}(V_J, Y_K)$ if and only if $A^{\times} \in \mathrm{Hom}(Y_K, V_J)$.

(ii) *The product of any number of homomorphisms (between successive* PIP*-spaces) is defined and is a homomorphism.*
(iii) *If* $A \in \mathrm{Hom}(V_J, Y_K)$, *then* $f\#g$ *implies* $Af\#Ag$.
(iv) *If* $A \in \mathrm{Hom}(V_J, Y_K)$, *then* $\mathsf{j}(A^{\times}A)$ *contains the diagonal of* $J \times J$ *and* $\mathsf{j}(AA^{\times})$ *contains the diagonal of* $K \times K$.

Note that an arbitrary homomorphism $A \in \mathrm{Hom}(V_J)$ need not be totally regular, but both $A^{\times}A$ and $AA^{\times}$ are.

The definition of homomorphisms just given is tailored in such a way that one may consider the category of all indexed PIP-spaces, with the homomorphisms as morphisms (arrows), as we have done in Ref. (65). In the same language, we may define particular classes of morphisms, such as monomorphisms, epimorphisms and isomorphisms. In particular, unitary isomorphisms are the proper tool for defining representations of Lie groups and Lie algebras in PIP-spaces. Examples and further properties of morphisms may be found in Ref. [13, Section 3.3] and in Ref. (66).

Finally, an *orthogonal projection* on a nondegenerate indexed PIP-space V_J, in particular, a LBS or a LHS, is a homomorphism $P \in \mathrm{Hom}(V_J)$ such that $P^2 = P^{\times} = P$.

A PIP-subspace W of a PIP-space V is defined in Ref. [13, Section 3.4.2] as an *orthocomplemented* subspace of V, that is, a vector subspace W for which there exists a vector subspace $Z \subseteq V$ such that $V = W \oplus Z$ and

(i) $\{f\}^{\#} = \{f_W\}^{\#} \cap \{f_Z\}^{\#}$ for every $f \in V$, where $f = f_W + f_Z$, $f_W \in W, f_Z \in Z$;
(ii) if $f \in W, g \in Z$ and $f\#g$, then $\langle f|g\rangle = 0$.

Condition (i) means that the compatibility # can be recovered from its restriction to W and Z. In the same Section 3.4.2 of Ref. (13), it is shown that a vector subspace W of a nondegenerate PIP-space is orthocomplemented if and only if it is *topologically regular*, which means that it satisfies the following two conditions:

(i) for every assaying subset $V_r \subseteq V$, the intersections $W_r = W \cap V_r$ and $W_{\bar{r}} = W \cap V_{\bar{r}}$ are a dual pair in V;
(ii) the intrinsic Mackey topology $\tau(W_r, W_{\bar{r}})$ coincides with the Mackey topology $\tau(V_r, V_{\bar{r}})|_{W_r}$ induced by V_r.

Then the fundamental result, which is the analogue to the similar statement for a Hilbert space, says that a vector subspace W of the nondegenerate PIP-space V is orthocomplemented if and only if it is the range of an orthogonal projection :

$$W = PV \text{ and } V = W \oplus W^{\perp} = PV \oplus (1 - P)V.$$

Clearly, this raises the question of identifying the subobjects of any category consisting of PIP-spaces. For instance, in a category consisting of LHS/LBSs only, a subspace is a LHS/LBS if and only if it is topologically regular, thus orthocomplemented. In that case, the subobjects are precisely the orthocomplemented subspaces. However, for more general indexed PIP-spaces, this need not be true. Orthocomplemented subspaces are subobjects, but there might be other ones. A simple example is that of a noncomplete prehilbert space (i.e., $V = V^{\times}$): then every subspace is a subobject, but need not be orthocomplemented. Further details may be found in Ref. (66).

REFERENCES

1. Bender CM. Making sense of non-Hermitian Hamiltonians. *Rep Prog Phys* 2007;70:947–1018.
2. Bender CM, Fring A, Günther U, Jones H. Quantum physics with non-Hermitian operators. *J Phys A Math Theor* 2012;45:440301.
3. Bender CM, DeKieviet M, Klevansky SP. $\mathcal{PT}$ quantum mechanics. *Philos Trans R Soc London Ser A* 2013;371:20120523.
4. Dieudonné J. Quasi-Hermitian operators. *Proceedings of the International Symposium on Linear Spaces, Jerusalem 1960*. Oxford: Pergamon Press; 1961. p 115–122.
5. Mostafazadeh A. Pseudo-Hermitian representation of quantum mechanics. *Int J Geom Methods Mod Phys* 2010;7:1191–1306.
6. Siegl P, Krejčiřík D. On the metric operator for the imaginary cubic oscillator. *Phys Rev D* 2012;86:121702(R).
7. Bagarello F. From self-adjoint to non-self-adjoint harmonic oscillators: physical consequences and mathematical pitfalls. *Phys Rev A* 2013;88:032120.
8. Bagarello F, Fring A. Non-self-adjoint model of a two-dimensional noncommutative space with an unbounded metric. *Phys Rev A* 2013;88:042119.
9. Bagarello F, Znojil M. Nonlinear pseudo-bosons versus hidden Hermiticity, II. The case of unbounded operators. *J Phys A Math Theor* 2012;45:115311.
10. Mostafazadeh A. Pseudo–Hermitian quantum mechanics with unbounded metric operators. *Philos Trans R Soc London Ser A* 2013;371:20120050.
11. Antoine J-P, Trapani C. Partial inner product spaces, metric operators and generalized hermiticity. *J Phys A Math Theor* 2013;46:025204. Corrigendum, *Ibid.* 2013;46:329501.
12. Antoine J-P, Trapani C. Some remarks on quasi-Hermitian operators. *J Math Phys* 2014;55:013503.
13. Antoine J-P, Trapani C. *Partial Inner Product Spaces: Theory and Applications*. Volume 1986, *Lecture Notes in Mathematics*. New York: Springer-Verlag Berlin Heidelberg; 2009.
14. Hoover TB. Quasi-similarity of operators. *Illinois J Math* 1972;16:678–686.
15. Dunford N. A survey of the theory of spectral operators. *Bull Amer Math Soc* 1958;64:217–274.
16. Kantorovitz S. On the characterization of spectral operators. *Trans Amer Math Soc* 1964;111:152–181.
17. Scholtz FG, Geyer HB, Hahne FJW. Quasi-Hermitian operators in Quantum Mechanics and the variational principle. *Ann Phys NY* 1992;213:74–101.

18. Geyer HB, Heiss WD, Scholtz FG. Non-Hermitian Hamiltonians, metric, other observables and physical implications. 2007, preprint, arXiv:0710.5593v1.
19. Kretschmer R, Szymanowski L. Quasi-Hermiticity in infinite-dimensional Hilbert spaces. *Phys Lett A* 2004;325:112–117.
20. Albeverio S, Günther U, Kuzhzel S. J-self-adjoint operators with C-symmetries: an extension theory approach. *J Phys A Math Theor* 2009;42:105205.
21. Bognar J. *Indefinite Inner Product Spaces*. Berlin, New York: Springer-Verlag; 1974.
22. Kuzhel S. On pseudo-Hermitian operators with generalized C-symmetries. In: V. Adamyan et al., editors. *Modern Analysis and Applications. The Mark Krein Centenary Conference*. Vol. 1: *Operator Theory and Related Topics, Operator Theory: Advances and Applications*, Volumes 190. Basel: Birkhäuser; 2009. p 375–385.
23. Bender CM, Kuzhel S. Unbounded C-symmetries and their nonuniqueness. *J Phys A Math Theor* 2012;45:444005.
24. Davies EB. *Linear Operators and their Spectra*. Cambridge: Cambridge University Press; 2007.
25. Dunford N, Schwartz JT. *Linear Operators. Part I: General Theory; Part II: Spectral Theory*. New York: Interscience; 1957, 1963.
26. Reed M, Simon B. *Methods of Modern Mathematical Physics. I. Functional Analysis*. New York and London: Academic Press; 1972, 1980.
27. Schmüdgen K. *Unbounded self-adjoint operators on Hilbert space*. Dordrecht, Heidelberg: Springer-Verlag; 2012.
28. Reed M, Simon B. *Methods of Modern Mathematical Physics. IV. Analysis of Operators*. New York and London: Academic Press; 1978.
29. Williams JP. Operators similar to their adjoints. *Proc Am Math Soc* 1969;20:121–123.
30. Sz-Nagy B, Foiaş C. *Harmonic Analysis of Operators in Hilbert Space*. Budapest, Amsterdam; Akadémiai Kiadó/North-Holland Publishing Company; 1970.
31. Tzafriri L. Quasi-similarity for spectral operators on Banach spaces. *Pacific J Math* 1968;25:197–217.
32. Feldzamen AN. Semi-similarity invariants for spectral operators on Hilbert space. *Trans Amer Math Soc* 1961;100:277–323.
33. Ôta S, Schmüdgen K. On some classes of unbounded operators. *Integr Equ Oper Theory* 1989;12:211–226.
34. Davies EB. Semi-classical states for non-self-adjoint Schrödinger operators. *Commun Math Phys* 1999;200:35–41.
35. Samsonov BF. Hermitian Hamiltonian equivalent to a given non-Hermitian one: manifestation of spectral singularity. *Philos Trans R Soc London Ser A* 2013;371:20120044.
36. Bergh J, Löfström J. *Interpolation Spaces*. Berlin: Springer-Verlag; 1976.
37. Ali ST, Bagarello F, Gazeau J-P. Modified Landau levels, damped harmonic oscillator, and two-dimensional pseudo-bosons. *J Math Phys* 2010;51:123502.
38. Antoine J-P, Balazs P. Frames, semi-frames, and Hilbert scales. *Numer Funct Anal Optim* 2012;33:1–34.

39. Znojil M. Three-Hilbert space formulation of quantum mechanics. *Symmetry Integrability and Geom. Methods Appl. (SIGMA)* 2009;5:001.
40. Weidmann J. *Linear Operators in Hilbert Spaces*. New York: Springer-Verlag Berlin Heidelberg; 1980.
41. Inoue A, Trapani C. Non-self-adjoint resolutions of the identity and associated operators. *Complex Anal Oper Theory* 2014. 2014;8:1531–1546, arXiv:1312.7090v2.
42. Mackey GW. *Commutative Banach Algebras*. Notas de Matematica, n° 17. Rio de Janeiro, RJ: Instituto de Matemática Pura e Aplicada; 1959.
43. Burnap C, Zweifel PF. A note on the spectral theorem. *Integr Equ Oper Theory* 1986;9:305–324.
44. Bagarello F, Inoue A, Trapani C. Non-self-adjoint hamiltonians defined by Riesz bases. *J Math Phys* 2014;55:033501.
45. Christensen O. *An Introduction to Frames and Riesz Bases*. Boston (MA): Birkhäuser; 2003.
46. Young RM. *An Introduction to Nonharmonic Fourier Series*. New York: Academic Press; 1980.
47. Bagarello F. From self-adjoint to non-self-adjoint harmonic oscillators: physical consequences and mathematical pitfalls. *Phys Rev A* 2013;88:032120.
48. Kadison RV, Ringrose JR. *Fundamentals of the Theory of Operator Algebras*. Volume I. New York: Academic Press; 1983.
49. Antoine J-P, Karwowski W. Interpolation theory and refinement of nested Hilbert spaces. *J Math Phys* 1981;22:2489–2496.
50. Kato T. *Perturbation Theory for Linear Operators*. Berlin: Springer-Verlag; 1976.
51. Gitman DM, Tyutin IV, Voronov BL. *Self-Adjoint Extensions in Quantum Mechanics*. Dordrecht, Heidelberg: Birkhäuser, Springer-Verlag; 2012.
52. Barut AO, Rączka R. *Theory of Group Representations and Applications*. Warszawa: PWN — Polish Scientific Publishers; 1977.
53. Nelson E. Analytic vectors. *Ann Math* 1959;70:572–615.
54. Schwartz J. Some non-selfadjoint operators. *Comm Pure Appl Math* 1960;13:609–639.
55. Antoine J-P, Trapani C. Operators on partial inner product spaces: towards a spectral analysis. 2014. Article published online: *Mediterr J Math* 2014; doi 10.1007/s00009-014-0499-6.
56. Bellomonte G, Di Bella S, Trapani C. Operators in rigged Hilbert spaces: some spectral properties. *J Math Anal Appl* 2014;411:931–946.
57. Antoine J-P, Trapani C. The partial inner product space method: a quick overview. *Adv Math Phys* 2010;2010:457635; Erratum, Ibid. 2010;2011:272703.
58. Birkhoff G. *Lattice Theory*. 3rd ed. Providence (RI): American Mathematical Society Colloquium Publications; 1966.
59. Köthe G. *Topological Vector Spaces*. Volumes I, II. Berlin: Springer-Verlag; 1969, 1979.
60. Bellomonte G, Trapani C. Rigged Hilbert spaces and contractive families of Hilbert spaces. *Monatsh Math* 2011;164:1271–285.

61. Mityagin BS, Shvarts AS. Functors in the category of Banach spaces. *Russ Math Surv* 1964;19:65–127.
62. Semadeni Z, Zidenberg H. Inductive and inverse limits in the category of Banach spaces. *Bull Acad Polon Sci Sér Sci Math Astr Phys* 1965;13:579–583.
63. Dieudonné J. Sur les espaces de Köthe. *J Anal Math (Jerusalem)* 1951;I:81–115.
64. Goes S, Welland R. Some remarks on Köthe spaces. *Math Ann* 1968;175:127–131.
65. Antoine J-P, Lambert D, Trapani C. Partial inner product spaces: some categorical aspects. *Adv Math Phys* 2011;2011:957592.
66. Antoine J-P, Trapani C. Some classes of operators on partial inner product spaces. In: Arendt W, et al., editors. *Spectral Theory, Mathematical System Theory, Evolution Equations, Differential and Difference Equations*, Volume 221, *Operator Theory: Advances and Applications* (Proceedings of IWOTA 2010). Basel: Birkhäuser; 2012. p 25–46.

INDEX

Non-Selfadjoint Operators in Quantum Physics: Mathematical Aspects, First Edition.
Edited by Fabio Bagarello, Jean-Pierre Gazeau, Franciszek Hugon Szafraniec and Miloslav Znojil.